CP Violation

Why did the matter in our Universe not annihilate itself with antimatter immediately after its creation? The discovery of **CP** violation may answer this fundamental question. From the basics to the front-line of research, this timely account presents background information and theoretical tools necessary for understanding this phenomenon. Early chapters explore charge conjugation, parity and time reversal symmetries before introducing the Kobayashi–Maskawa ansatz for **CP** violation and examining the theoretical understanding of **CP** violating K meson decays. Following chapters reveal how the discovery of B mesons provides a new laboratory in which to study **CP** violation and predict **CP** asymmetries in B and rare K meson decays. Later chapters continue the search for a new fundamental theory and address the problem of baryogenesis in the big bang universe. The importance of close links with experiment is stressed throughout. Each chapter concludes with problems. Detailed references are included.

This book is suitable for graduate students and researchers in particle physics, atomic and nuclear physics and the history and philosophy of science.

IKAROS BIGI was born in Munich, Germany. Following undergraduate and postgraduate studies at the Universities of Munich, Oxford and Stanford, he has taught and researched at the Max-Planck Institute for Physics, CERN, RWTH Aachen, UCLA, the University of Oregon, SLAC and the University of Notre Dame. He is a former scholarship student of the Maximilianeum Foundation and Scholarship Foundation of the German People and has been appointed both a Heisenberg Fellow and a Max-Kade Fellow.

SANDA ICHIRO was born in Tokyo, and at the age of 14 accompanied his father who was transferred to the United States on business. After a bachelor's degree in physics from the University of Illinois and a PhD from Princeton University, he taught and researched at Columbia University, Fermilab and Rockefeller University. In 1992, after 34 years in the US, he went back to Japan as a professor of physics at Nagoya University. He is now chairman of the physics department. He is a winner of the 10th Inoue Prize (1993) and the 43rd Nishina Memorial Prize (1997). Both prizes have been awarded for his work on **CP** violation, and on B physics.

Since they began their collaboration in 1981, the authors have written 12 papers together and have spent a great deal of energy promoting the crucial importance of comprehensive experimental studies of beauty hadrons. In their first joint paper they explained the special role played by certain decays into flavour-nonspecific final states; among them was the channel $B \to \psi K_s$ upon which considerable and dedicated experimental efforts are focused at CERN, Fermilab, DESY, SLAC and KEK, and pointed out the peculiarities of the experimental situation which required asymmetric B factories, now being built at SLAC and KEK.

CAMBRIDGE MONOGRAPHS ON
PARTICLE PHYSICS,
NUCLEAR PHYSICS AND COSMOLOGY
9

General Editors: T. Ericson, P. V. Landshoff

CP VIOLATION

I. I. BIGI
University of Notre Dame

A. I. SANDA
Nagoya University

CAMBRIDGE
UNIVERSITY PRESS

PUBLISHED BY THE PRESS SYNDICATE OF THE UNIVERSITY OF CAMBRIDGE
The Pitt Building, Trumpington Street, Cambridge, United Kingdom

CAMBRIDGE UNIVERSITY PRESS
The Edinburgh Building, Cambridge CB2 2RU, UK   http://www.cup.cam.ac.uk
40 West 20th Street, New York, NY 10011–4211, USA   http://www.cup.org
10 Stamford Road, Oakleigh, Melbourne 3166, Australia
Ruiz de Alarcón 13, 28014 Madrid, Spain

First published 2000

Printed in the United Kingdom at the University Press, Cambridge

Typeface 11/13pt Times   *System* LATEX   [UPH]

A catalogue record for this book is available from the British Library

Library of Congress Cataloguing in Publication data

Bigi, I. I.
CP violation / I. I. Bigi, A. I. Sanda.
p.   cm. – (Cambridge monographs on particle physics, nuclear physics, and cosmology ; 9)
Includes bibliographical references and index.
ISBN 0 521 44349 0
1. CP violation (Nuclear physics) I. Sanda, A. I. (A. Ichiro)
II. Title. III. Series.
QC793.3.V5B54   1999
539.7.′25–dc21   98–38616   CIP

ISBN 0 521 44349 0 hardback

Dedications

Meiner Mutter Friederike Bigi
and
To my wife Hiroko M. M. Sanda

Contents

ix

Foreword

Some discoveries in the sciences profoundly change how we view nature. The discovery of parity violation in the weak interactions in 1956 certainly falls into this illustrious category. Yet it just started the shift to a new perspective; it was the discovery of **CP** violation in 1964 by Christenson, Cronin, Fitch and Turlay at Brookhaven National Lab – completely unexpected to almost all despite the experience of eight years earlier – that established the new paradigm that even in the microscopic regime symmetries should not be assumed to hold a priori, but have to be subjected to determined experimental scrutiny.

It would seem that after the initial period of discoveries little progress has been achieved, since despite dedicated efforts **CP** violation has not been observed outside the decays of K_L mesons, nor can we claim to have come to a real understanding of this fundamental phenomenon.

We have, however, ample reason to expect imminent dramatic changes. Firstly, direct **CP** violation has been observed in K_L decays. Secondly, our phenomenological and theoretical descriptions have been refined to the point that we can predict with confidence that the known forces of nature will generate huge **CP** asymmetries, which could even be close to 100%, in the decays of so-called beauty mesons. Dedicated experiments are being set up to start taking data that would reveal such effects before the turn of the millennium. What they observe – or do not observe – will shape our knowledge of nature's fundamental forces.

We consider it thus an opportune time to take stock, to represent **CP** invariance and its limitations in its full multi-layered complexity. In our presentation we pursue three goals:

- We want to provide a detailed frame of reference for properly evaluating the role of **CP** violation in fundamental physics and to prepare us for digesting the upcoming observations and discoveries.

- We will show that an in-depth treatment of **CP** violation draws on most concepts and tools of particle physics. It thus serves as an unorthodox introduction to quantum field theories (and beyond).

- We want to communicate to the reader that the quest for understanding **CP** violation is more than just an important scholarly task. It represents a most exciting intellectual adventure of which we do not know the outcome. For this very purpose we provide historical perspectives from the last half century.

Accordingly our intended readership is manifold: we want

- to give (theoretical) guidance to the workers in the field,

- to provide an introduction for people who would like to become researchers in this field or at least educated observers,

- to present material which could serve as a supplementary text for courses on quantum field theory, and

- to allow people interested in the history and development of fundamental science to glean maybe some new insights.

We are not pretending our book makes easy reading. We hope, however, that the committed reader will find gratifying the way we start from the basics, give numerous homework problems as an integral part of the learning process and enrich – we think – the narrative with historical remarks. We actually believe that more than one reading will be necessary for a full understanding. To facilitate such an approach we designate sections which can be left out in a first reading by placing their title between the symbols ♠.

As theorists we cannot do full justice to experimental endeavours. Yet we try to communicate our conviction that physics is so wonderfully exciting exactly because it is an empirical science where theory and experiment play an interactive role.

We have benefitted greatly from interacting with many of our colleagues. In particular we would like to acknowledge Drs N. Uraltsev and Z-Z. Xing for their advice and collaboration, Drs A. Garcia and U. Sarid for their suggestions concerning the text, and Ms M. Ota for her help in preparing the manuscript. We also express our gratitude to Bernie and Theresa Vonderschmidt for their hospitality during the period in which part of the book was written.

PART 1
BASICS OF CP VIOLATION

1

Prologue

All animals are equal.
But some animals are
more equal than others!
G. Orwell, *Animal Farm*

The sciences in general and physics in particular are full of fascinating phenomena; this is why they have attracted intense human interest early on and have kept it ever since. Yet even so we feel that the question to which degree nature is invariant under time reversal and **CP** transformations is so fundamental that it richly deserves its own comprehensive monograph. Two lines of reasoning – different, though not unrelated to each other – lead us to this conclusion. The first relies on multi-layered considerations, the second is based on a property inferred for the whole universe.

- The first line of reasoning centres on the important role symmetries have always played in physics; it has been recognized only this century, though, how central and crucial this role actually is, and this insight forms one of the lasting legacies of modern physics to human perception of nature and thus to human culture. The connection between continuous symmetries – like translational and rotational invariance – and conserved quantities – momentum and angular momentum for these examples – has been formulated through Noether's theorems. The pioneering work of Wigner and others revealed how atomic and nuclear spectra that appeared at first sight quite complicated could be understood through an analysis of underlying symmetry groups, even when they hold only in an approximate sense. This line of reasoning was successfully applied to nuclear and elementary particle physics through the introduction of isospin symmetry $SU(2)$, which was later generalized to $SU(3)$ symmetry in particle physics.

A completely new chapter has been opened through the introduction of *local* gauge theories. In particular Yang–Mills theories were introduced as a formal extrapolation of abelian gauge theories like QED.

It was realized only considerably later that the gauge principle plays an essential role in constructing fully relativistic quantum theories that are both non-trivial and renormalizable. Furthermore, it was understood that symmetry breaking can be realized in two different modes, namely *manifestly* and *spontaneously*[1].

Similarly, *discrete* symmetries have formed an important part of our understanding of the physical world around us – like in crystallography and chemistry. The weight of such considerations is emphasized further with the imposition of permutation symmetries in quantum theory as expressed through Bose–Einstein and Fermi–Dirac statistics. Embedding discrete symmetries into local gauge theories has led to intriguing consequences; among them is the emergence of anomalies, which will be discussed in later chapters.

The primary subject of this book will be discrete symmetries which are of general and fundamental relevance for physics:

- Parity **P** – reflecting the space coordinate $\vec{x}$ into $-\vec{x}$;

- Charge conjugation **C** – transforming a particle into its antiparticle;

- The combined transformation of **CP**;

- Time reversal **T** – changing the time coordinate t into $-t$; it thus amounts to reversal of motion.

We have learnt that nature is largely, but not completely, invariant under these transformations. Although these insights were at first less than eagerly accepted by our community, they form an essential element of what is called the Standard Model (SM) of high energy physics. Yet the story is far from over. For the SM contains 20-odd parameters; those actually exhibit a rather intriguing pattern that cannot be accidental. They must be shaped by some unknown New Physics, and we consider it very likely that a comprehensive analysis of how these discrete symmetries are implemented in nature will reveal the intervention of New Physics.

Furthermore we believe that time reversal, **T**, and the combined transformation of **CP** occupy a very unique place in the pantheon of symmetries. The fact that their violation has been observed in nature has consequences the importance of which cannot be overestimated. Once it was realized that **P** and **C** are violated – and actually violated maximally – it was noted with considerable relief that **CP** was apparently still conserved. For it had been suggested that microscopic **T** invariance follows from

---

[1] The latter case is usually referred to as spontaneous symmetry breaking.

Mach's principle; because of **CPT** invariance that holds so naturally in quantum field theories, **CP** violation could not occur without **T** violation.

There is a subtle point about time reversal that is to be understood. **T** can be viewed as reversal of motion. The notion that the laws of nature might be invariant under such a transformation seems absurd at first sight. When watching a filmed sequence of events we can usually tell with *great confidence* whether the film is being played forward or backward. For example, a house of cards will collapse into a disordered pile rather than rise out from such a pile by itself. However this disparity can be understood by realizing that while both sequences are in principle equally possible – as demanded by microscopic **T** invariance – the second one is so unlikely to occur for a macroscopic system as to make it practically impossible. That is why the expression 'with great confidence' was used above. We will address this point in more detail later on.

It came as a great shock that microscopic **T** invariance is violated in nature, that 'nature makes a difference between past and future' even on the most fundamental level. We might feel that such a statement is sensationalist rather than scientific; yet there is indeed something very special about a violation of the invariance under **T** or **CP**. We offer the following observations in support of our view. Elucidating them will be one of the central themes of our book. They might carry little meaning for the reader at this point; yet we expect this to have changed after she or he has finished the book.

    – **CP** violation is more fundamental than **C** violation in the following sense: **C** violation as it was discovered can be described by saying that only left handed neutrinos and right handed anti-neutrinos interact. This, however, does not allow a genuine distinction of matter and anti-matter[2]: for their difference is expressed in terms of 'left' and 'right handed' which represents a convention – as long as **CP** is conserved! However once **CP** is violated – even if ever so slightly – then matter and antimatter can be distinguished in an absolute, convention-independent way. The practical realization of this general observation goes as follows: while the K_L meson can decay into a positron or an electron together with a pion of the opposite charge and a neutrino or anti-neutrino, it exhibits a slight preference for the mode $K_L \to e^+ \nu_e \pi^-$. The positron is then called an antilepton; matter and antimatter are thus distinguished by *nature* rather than by convention.

---

[2] In our world the electron is defined as matter and the positron is defined as anti-matter.

- Time reversal is described by an antiunitary rather than a unitary operator, which introduces many intriguing subtleties. Among them is Kramer's degeneracy: from $\mathbf{T}^2 = \pm 1$ we deduce that two distinct classes of states can exist; those with $\mathbf{T}^2 = -1$ are interpreted as fermions, in contrast to bosons for which $\mathbf{T}^2 = 1$ holds.

- **CP** violation represents the most delicately broken symmetry observed so far in nature and provides us with a powerful phenomenological probe. Consider the historical precedent: the observation of $K_L \to \pi\pi$ in 1964 led to the prediction (in 1972) that a *third* family of quarks and leptons had to exist – before the existence of the final member of the *second* family, the charm quark, had been accepted. It took until 1995 before the top quark, the last member of the third family, had been discovered – with a mass about 400 times as much as the K_L meson!

- While **P** violation has not been understood in a profound way, it can unequivocably be embedded into the gauge sector through chiral couplings to the gauge bosons. Among other things, we can give a natural and meaningful definition of a 'maximal' violation of **P** or **C** invariance, namely that all interacting neutrinos [anti-neutrinos] are left-handed [right-handed]. Furthermore, the 'see-saw' mechanism provides us with an intriguing dynamical scenario invoking the restauration of **P** invariance at high energies to explain the smallness of neutrino masses. The situation is completely different for **CP** and **T** violation. In general it can enter through gauge or Yukawa interactions. In the Kobayashi–Maskawa (KM) ansatz it is implemented through complex Yukawa couplings; thus it is connected to the least understood part of the SM. The best that can be said is that the SM with three quark families allows for **CP** violation; yet the latter appears as a mere 'add-on'. We are not even able to give real meaning to the notion of 'maximal' **CP** violation.

- The second argument focuses on the observation that while the universe is almost empty, it is not completely empty and actually in a decidedly biased way so! To use a more traditional scientific language: for every trillion or so photons there is just a single baryon in the universe – apparently without any sight of an antibaryon that cannot be explained as the product of a primary collision between matter particles:

$$\text{today}: \quad N(\text{antibaryons}) \ll N(\text{baryons}) \ll N(\text{photons}) \qquad (1.1)$$

Of course we have no reason to complain about this state of affairs. Life could not have developed, we could not exist if nature had been more even-handed in its matter–antimatter distribution.

As first pointed out in a seminal paper by A. Sakharov, there are three essential elements in any attempt to understand the excess of baryons over antibaryons in the universe as a dynamical quantity that can be *calculated* rather than merely *attributed to the initial conditions*:

1 reactions that change baryon number have to occur;

2 they cannot be constrained by **CP** invariance; and

3 they must proceed outside thermal equilibrium.

CP violation is thus one essential element in any attempt to achieve such an ambitious goal, and as it turns out, it is the one area that we can best subject to further experimental scrutiny.

In summary: understanding **CP** and **T** violation will bring with it both practical benefits and profound insights since it represents an essential and unalienable element in the fabric of nature's grand design, as sketched in Fig. 1.1. A second glance at this sketch shows that we are dealing with a highly interwoven as well as dense fabric, as is true for all high quality fabrics. To understand its structure, to exploit the interrelationships among its elements and to interprete data, we obviously need a guiding principle (or two); the concept of symmetry in all its implementations can serve as such. We feel very strongly that progress towards understanding this tapestry can be made only through a feed-back between further dedicated and comprehensive experimental studies and theoretical analysis.

Synopsis

The goal of this book is to show that **T** and **CP** invariance and their violations are much more than exotic phenomena existing in their own little reservation. As indicated above (and discussed in more detail later on) these subjects are, despite their subtle appearance, intimately connected with nature's fundamental structure. Their proper treatment therefore requires a full understanding and usage of our most advanced theoretical tools, namely quantum mechanics and quantum field theory[3]. At the same time we will insist that close contact with experiment has to be maintained.

To pursue this goal our presentation will proceed as follows: first we will describe in considerable detail how **P, C** and **T** transformations are implemented in classical physics, and in theories with first and second quantization. Then we will briefly recapitulate how the study of strange particles initiated the observation of **P** non-invariance and led to the discovery of **CP** violation, before describing the phenomenology of the neutral kaon system in detail. After addressing other

---

[3] Even superstring theories might be called upon in the future to provide the substratum for the relevant field theory, as briefly mentioned in our discussion of **CPT** invariance and of New Physics scenarios.

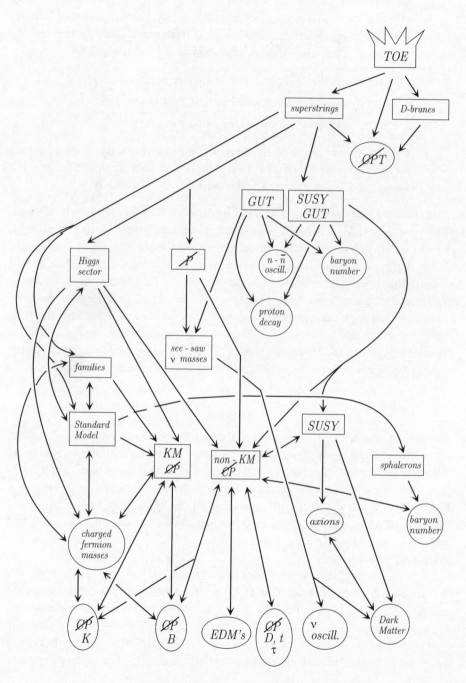

Fig. 1.1.   Nature's grand tapestry.

searches for **T** non-invariance in K decays and through electric dipole moments of neutrons and atoms, we introduce the Kobayashi-Maskawa (KM) ansatz as the minimal implementation of **CP** violation in the SM of high energy physics, and apply it to the description of strange decays and electric dipole moments. We will emphasize how essential it is that a dedicated pursuit of searches for **CP** violation in light quark systems continue in the foreseeable future.

On the other hand, the KM ansatz leads unequivocally to a 'paradigm of large **CP** asymmetries' in B decays. We describe the rich phenomenology anticipated for beauty decays emphasizing five points:

- some predictions enjoy high *parametric* reliability, as in $B_d \to \psi K_S$;

- for others – as in $B_d \to \pi^+\pi^-$ – such reliability can be achieved through measuring related transitions;

- *parametric* reliability can be turned into *numerical* precision;

- there are many promising ways to search for indirect manifestations of 'New Physics', the most obvious one being to analyse $B_s \to \psi\eta$, $\psi\phi$, $D_s\bar{D}_s$;

- completion of such a program requires a long-term commitment.

From today's perspective there are attractive, or at least intriguing, aspects to the KM ansatz:

•*It provides a natural gateway for* **CP** *violation to enter; no new degrees of freedom have to be postulated.* Three complete quark families have been found experimentally. The SM then does *not automatically* conserve **CP**: it has enough structure to support the existence of a physical weak phase. It could still turn out that this phase vanishes; yet within the SM context this would appear to be 'unnatural' – it would have to have a dynamical origin beyond the SM.

•*It can accommodate the observed phenomenology – quite meagre in its positive signals – within the experimental and theoretical uncertainties.* It has even made predictions that are borne out by the data, namely the elusiveness of direct **CP** violation and the tiny size of electric dipole moments. All of this is achieved without any fine-tuning of parameters and with some of the fundamental parameters – V_{td}, V_{ts} and m_t – observed to be of a numerical size that before (when the KM ansatz was conceived and for many years thereafter) would have appeared to be quite unreasonable.

Nevertheless we consider it quite unlikely that the KM ansatz could remain the final word on **CP** violation – far from it! We are willing to stake our reputation[4] on the prediction that dedicated and comprehensive studies of **CP** violation will reveal the presence of New Physics:

---

[4] Of course it is merely a theorists' reputation.

•The KM ansatz constitutes merely a *parametrization* of a profound phenomenon. The KM matrix actually reflects the mismatch in the alignment of the up-type and down-type quark mass matrices; its elements and thus also the origin of **CP** violation are, therefore, related to two of the central mysteries of the SM: why are there quark (and lepton) families and how do their masses get generated? Because of this connection it is not surprising that we cannot claim a true understanding of **CP** violation. On the other hand, without detailed knowledge of the physical elements of the KM matrix, we do not make full use of the information that nature is allowing us to acquire on the dynamics underlying the generation of fermion masses.

•What we already know about these matrix elements – mostly concerning their moduli – strongly suggests that some very specific dynamics was generating them. For the matrix, rather than being merely unitary, exhibits a very peculiar structure, as outlined before, that is quite unlikely to have come about by accident. The matrix thus contains information on New Physics – albeit in a highly coded form.
•Many extensions to the SM have been suggested to cure perceived ills of the SM. **CP** studies can then be employed as highly sensitive probes for manifestations of such New Physics.

This provides us also with an opportunity to address general aspects of the way in which **CP** violation can be realized in nature.

(i) **CP** symmetry can be broken in a 'hard' way, i.e., through dimension-four operators in the Lagrangian, namely gauge couplings to fermions or Yukawa couplings. The KM ansatz is an implementation of the latter variant of such a scenario. For quark mass matrices are derived from the Yukawa couplings through the Higgs mechanism; since in the SM we introduce a single Higgs doublet field, we need the Yukawa couplings to exhibit an irreducible complex phase. This phase is then a free parameter and cannot be calculated.

(ii) **CP** invariance can be broken explicitly in a 'soft' manner, i.e., through operators of dimension below four. We will see that SUSY extensions provide for such scenarios which hold out the promise – or at least the hope – that the basic **CP** violating parameters could be understood dynamically.

(iii) **CP** symmetry is realized in a spontaneous fashion; this is also referred to – sloppily – as spontaneous **CP** violation: while the Lagrangian conserves **CP** (the gauge and Yukawa couplings can be made real), the groundstate does not; the vacuum expectation value of neutral Higgs fields develop complex phases. Again, we entertain the hope that the relevant quantities can be derived from the dynamics – in principle, and some day. Models with an extended Higgs or gauge sector allow us to realize such scenarios.

We sketch various alternatives or extensions of the KM implementation of **CP** violation – among them SUSY scenarios – and describe processes where realistically only the intervention of New Physics can produce observable **CP** asymmetries, namely in

- the decays of charm hadrons and τ leptons,

- production and decay of the top quark,

- ν oscillations.

Finally, we address the most ambitious problem, namely baryogenesis in the big bang universe.

2
Prelude: **C**, **P** and **T** in classical dynamics

*Time reversal – more
than meets the eye*

In this chapter we study **C**, **P** and **T** symmetry in classical mechanics and electrodynamics. First we restate the definitions of these transformations together with some comments:

- Parity transformations **P** change a space coordinate $\vec{x}$ into $-\vec{x}$. This is equivalent to a mirror reflection followed by a rotation. For momentum and angular momentum we have, respectively:

$$\vec{p} \xrightarrow{\mathbf{P}} -\vec{p}, \quad \vec{l} \equiv \vec{x} \times \vec{p} \xrightarrow{\mathbf{P}} \vec{l}. \tag{2.1}$$

  This is an example for a general classification: among vectors, which are defined by their transformations under rotations, we can distinguish between polar vectors that change their sign under parity ($\vec{V} \xrightarrow{\mathbf{P}} -\vec{V}$) and axial vectors that do not ($\vec{A} \xrightarrow{\mathbf{P}} \vec{A}$). Likewise we have scalars S – like $\vec{p}_1 \cdot \vec{p}_2$ – and pseudoscalars P – like $\vec{p} \cdot \vec{l}$ – with $S \xrightarrow{\mathbf{P}} S$ whereas $P \xrightarrow{\mathbf{P}} -P$.

- Time reversal **T** reflects t into $-t$ while leaving $\vec{x}$ unchanged and therefore

$$\vec{p} \xrightarrow{\mathbf{T}} -\vec{p}, \quad \vec{l} \xrightarrow{\mathbf{T}} -\vec{l}. \tag{2.2}$$

  T thus represents *reversal of motion*.

- Charge conjugation **C** transforms a particle into its antiparticle of equal mass, momentum and spin, but opposite quantum numbers like electric charge. The notion of an antiparticle is actually ad-hoc and even foreign in the realm of *non*relativistic dynamics. It arose first in the context of the Dirac equation before it was realized that the existence of such antiparticles is a general necessity in quantum field theory.

2.1 Classical mechanics

The motion of an object with mass m is controlled by Newton's equation

$$\vec{F} = m \frac{d^2 \vec{x}}{dt^2} \tag{2.3}$$

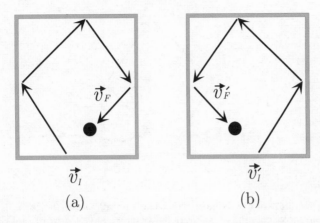

Fig. 2.1.   Take a movie of the motion of a billiard ball shown in (a). We get (b)
by turning the film around and projecting the light from the reverse side. It is
the mirror image of (a).

for given initial position $x(0)$ and velocity $\dot{x}(0)$; $\vec{F}$ denotes the force acting on the
object, and $\vec{x}$ is the coordinate of the object at time t.

2.1.1 Parity

Eq. (2.3) is clearly invariant under parity $\vec{x}(t) \xrightarrow{\mathbf{P}} -\vec{x}(t)$ because the force is
described by a polar vector

$$\vec{F} \xrightarrow{\mathbf{P}} -\vec{F}. \tag{2.4}$$

The impact of parity can be visualized in the following way. Consider the motion
of a billiard ball. The ball is hit with a cue and given an initial velocity $\vec{v}_I$. It moves
around the table bouncing off its side walls and ends up with a final velocity $\vec{v}_F$
as shown in Fig. 2.1(a). Call this sequence the *genuine* motion. If we take a movie
of this motion we can (re)view the genuine motion. If we had made the movie
using an old-fashioned film (rather than a video tape) we could turn the film and
project the light through it *backward*; the screen would then show a *fake* motion
as shown in Fig. 2.1(b), namely a mirror image of the genuine one. However,
just watching the screen, we would *not* be able to tell the difference as long as
parity is conserved. The fake motion constitutes a physically possible sequence:
parity constitutes a symmetry! If we painted the side walls of the billiard table
in different colours and communicated this information to the spectator, the fake
motion could be distinguished from the genuine one. Yet in that case we would
only have exchanged a *convention* for what is left and right!

   At this point the curious reader might wonder how parity *non*invariance could
actually reveal itself in this scenario. Consider a gedanken world permeated
by a time independent and homogeneous electric field and billiard balls that

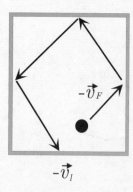

Fig. 2.2.   Motion of the billiard ball when our film is run backwards. Again, this motion looks genuine to us. Over the years we have learned to take the time reversal symmetry of classical mechanics for granted.

are electrically charged. With this electric field pulling charged billiard balls in its direction, their trajectories will be curved rather than straight lines. More importantly, they and their mirror images are quite distinct in their time evolution, even in the points where the balls hit the walls of the table; based on their past experience in such a world the spectatators will without a doubt be able to tell whether the movie shows the real motion or the film has been turned before the projection. This situation is described by saying that parity is broken explicitly by a background field.

You might view this example as contrived since the background field first of all breaks rotational invariance. Let us consider a more refined example *without* a background field. The billiard balls are now riding on an air cushion and spinning around their direction of motion. (The air cushion is introduced simply to separate the spinning from an overall forward motion.) Assume the interaction to be such that when a billiard ball hits a wall it is reflected at a smaller [larger] angle than its incoming angle when it is right handed [left handed], i.e., spinning parallel [anti-parallel] to its motion. Projecting the movie of a *right* handed ball *after* the film tape has been turned would show a *left* handed ball being reflected at a *smaller* angle – in conflict with the past experience of the spectators! The intervention of parity violation would thus allow them to tell whether they were watching a possible or an impossible motion.

2.1.2 Time reversal

If $\dot{x}(t)$ is a solution to the equation of motion, Eq. (2.3), then so is $\dot{x}(-t)$ for this second order differential equation with initial and final conditions switching roles. We should keep in mind, though, that in general $\dot{x}(-t) \neq \pm\dot{x}(t)$; we will give an example in Sec. 2.2.3.

It turns out to be particularly instructive to discuss the example with billiard

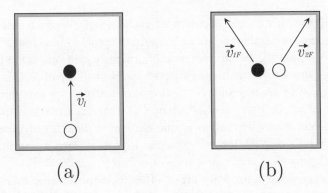

(a)   (b)

Fig. 2.3.   Now imagine two balls on a billiard table. In (a) a white ball is hit
towards the black ball. When they collide, they fly apart (b). Now reverse this
motion and observe it running backwards in time. All of us can tell when the
film is run backwards. This is because we know from experience that adjusting
the motion of two balls so that one of them stops just after the collision is very
very difficult and requires lots of practice.

balls and how their motion would appear in a movie. Consider first the case of a
single billiard ball. In the genuine motion the ball starts out with initial velocity
$\vec{v}_I$, bounces off the side walls of the billiard table and ends up with final velocity
$\vec{v}_F$. However, if we watched the sequence projected onto the screen from a film
running backwards we could not decide if we watched a fake motion or a genuine
one where the ball starts out with $-\vec{v}_F$ and ends up with $-\vec{v}_I$, see Fig. 2.2 or Fig.
2.1(b).

Next we consider the situation in which a stationary ball is hit by another ball,
as shown in Fig. 2.3. Watching the filmed sequence we would – with *considerable
confidence* – single out the genuine motion as the following one: a ball comes
in, hits a stationary ball and both balls move with, in general, different velocities
and under different angles, as shown in Fig. 2.3. For the reverse motion where
two balls come together under different angles and with different velocities and
hit each other causing one of the balls to come to a complete stop is, as we know
from experience, a very improbable one. We have performed this or a similar kind
of experiment before and have learnt that it takes lots of practice and many tries
to realize the reversed motion, since it requires a very careful tuning of the initial
conditions. To summarize our discussion of this example: both the motion and
its reversed version represent sequences *allowed* by the dynamics, i.e., *microscopic*
time reversal invariance holds; however, the reversed version is much *less likely*
to occur, leading to an apparent macroscopic asymmetry! Thus we can identify
the fake motion with considerable confidence, as stated above. If we go one step
further in our experiments, namely hit a collection of densely packed billiard balls
at rest with one other billiard ball and watch the billiard balls getting scattered
in all directions, confidence quickly turns into certainty!

The central message in general terms is the following: the motion of an object is controlled not only by the equation of motion, but also by *initial conditions.* Microscopic time reversal invariance means that the rate for reaction $a \to b$ equals that for $b \to a$ *once the initial configurations,* namely a in one case and b in the other, *have been precisely realized!* However, the probability of realizing a or b as an initial state is in general different, and in fact vastly so for complex processes. The likelihood of the time reversed version of a *complex* reaction to happen is very low indeed. This observation is one element – though not necessarily the only one – in resolving the following puzzle: if physical laws are invariant under time reversal, why can't we build a time machine?

Another experience from daily life is often used to illustrate this point: manoeuvring a car between two other cars standing along the kerbside for parking is typically a considerably harder (and more frustrating) task than leaving the same parking space later. The reason is the following: to park the car you have to fit it into more or less a single cell of final configurations; for leaving you can use any of many possible trajectories to final states outside the parking spot.

2.2 Electrodynamics

Electrodynamics is governed by Maxwell's equations:

$$\vec{\nabla} \cdot \vec{E} = 4\pi\rho$$
$$\vec{\nabla} \times \vec{B} - \frac{1}{c}\frac{\partial \vec{E}}{\partial t} = \frac{4\pi}{c}\vec{J}$$
$$\vec{\nabla} \cdot \vec{B} = 0$$
$$\vec{\nabla} \times \vec{E} + \frac{1}{c}\frac{\partial \vec{B}}{\partial t} = 0. \tag{2.5}$$

2.2.1 Charge conjugation

The equations remain manifestly invariant under sign reversal of charge density, current, electric field and magnetic field:

$$\rho \xrightarrow{C} -\rho, \quad \vec{J} \xrightarrow{C} -\vec{J}$$
$$\vec{E} \xrightarrow{C} -\vec{E}, \quad \vec{B} \xrightarrow{C} -\vec{B}. \tag{2.6}$$

These symmetry transformations define charge conjugation **C** for classical electrodynamics.

2.2.2 Parity

The electric field between two oppositely charged particles changes sign when the positions of these two particles are reversed. Similarly, we expect the current density $\vec{J}$ to change sign. The following transformations thus represent parity

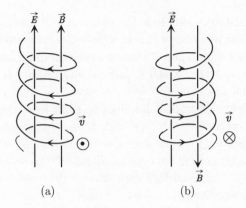

(a)   (b)

Fig. 2.4. (a) Motion of a particle, with $q > 0$, in static electric and magnetic fields, where the initial velocity $\vec{v}$ is out of the plane of the paper as shown. (b) Under time reversal, $\vec{E} \to \vec{E}$, $\vec{B} \to -\vec{B}$, $\vec{v} \to -\vec{v}$. Note that the time reversed motion cannot be obtained by taking a movie of motion (a) and running the film backwards.

reflection **P**:

$$\rho \overset{P}{\to} \rho, \quad \vec{j} \overset{P}{\to} -\vec{j}$$
$$\vec{E} \overset{P}{\to} -\vec{E}, \quad \vec{B} \overset{P}{\to} \vec{B}. \tag{2.7}$$

Note that $\vec{B}$ does not change sign. This is consistent with the fact that $\vec{B} = \vec{\nabla} \times \vec{A}$, where $\vec{A}$ is the vector potential. $\vec{B}$ is said to be an axial vector. Obviously Maxwell's equations are invariant under the combined transformations of Eq. (2.7).

2.2.3 *Time reversal*

Under time reversal, we expect the current and thus the $\vec{B}$ field to reverse direction, whereas charge density and the electric field remain invariant. Maxwell's equations indeed possess a symmetry

$$\rho \overset{T}{\to} \rho, \quad \vec{j} \overset{T}{\to} -\vec{j}$$
$$\vec{E} \overset{T}{\to} \vec{E}, \quad \vec{B} \overset{T}{\to} -\vec{B}, \tag{2.8}$$

consistent with our expectations. A particle carrying charge q and moving with velocity $\vec{v}$ experiences a **T** invariant Lorentz force:

$$\vec{F} = q \left(\vec{E} + \vec{v} \times \vec{B} \right) \overset{T}{\to} \vec{F}. \tag{2.9}$$

As shown in Fig. 2.4 (a), its motion under the influence of homogenous and time-independent electric and magnetic fields *parallel* to each other is described by a *clockwise* (for $q > 0$) screw around the direction of $\vec{E}$ if the initial momentum of the particle was perpendicular to the two fields.

As shown in Fig. 2.4 (b), under **T**, $\vec{B}$ flips its direction, thus becoming *anti*parallel to $\vec{E}$: the particle now describes a *counter-clockwise* screw around $\vec{E}$.

Running the movie of an *upward clockwise* screw motion backwards will show a *downward clockwise* screw, which is intrinsically distinct from the truly time reversed motion which is a *counter-clockwise* screw! So what went wrong?

Without the $\vec{E}$ field, the particle describes a clockwise closed orbit around the $\vec{B}$ field. After time reversal, which flips both $\vec{B}$ and the initial momentum it still describes a clockwise orbit around the new direction of $\vec{B}$, which amounts to a counter-clockwise orbit around the original direction; this sequence is reproduced when the film is played backwards. The problem is the motion of a particle under a static $\vec{E}$ field (Problem 2.6).

2.3 Résumé

The discrete transformations **C**, **P** and **T** can be defined in classical dynamics in a rather straightforward manner (although the concept of charge conjugation is ad-hoc at this level) and simple criteria for them to represent symmetries can be stated. The question of time reversal symmetry requires a clear distinction between, on the one hand, the equation of motion and, on the other, the likelihood for certain initial conditions to be established. *Microscopic* **T** invariance means first of all that if $a \to b$ can happen, so can $b \to a$! Both processes have to occur with equal likelihood only if a and b can equally be realized as *initial* states. The apparent **T** asymmetries encountered in daily life are ascribed to asymmetries in the realizability of the corresponding initial conditions. For further reading we suggest Ref. [1].

Problems

2.1   In elementary courses on electricity and magnetism we learn that the direction of the magnetic field $\vec{B}$ is given by the *right hand rule*: The right hand thumb points to the direction of the current and the other fingers wrapping around the current vector give the direction of the $\vec{B}$ field. Under parity $\vec{J} \to -\vec{J}$. The right hand rule would seem to imply that $\vec{B} \to -\vec{B}$, contradicting Eq. (2.7). Resolve the apparent contradiction.

2.2   Discuss the transformation property of the vector potential $\vec{A}(x, t)$ under **C**, **P** and **T**.

2.3   Discuss the transformation properties of a magnetic monopole under **C**, **P** and **T**.

2.4   Ohm's law

$$\vec{j} = \sigma \vec{E} \tag{2.10}$$

stating that the current density is proportional to the electric field strength, appears to violate time reversal invariance since $\vec{j} \to -\vec{j}$, yet $\vec{E} \to \vec{E}$ under **T**. Resolve the apparent paradox. *Hint*: note that Ohm's law does not represent a microscopic identity; it is based on a time average. Is the underlying equation of motion purely second order in time?

2.5 Since the forces driving our metabolism are electromagnetic in nature, they conserve **T**. This suggests that we can, in principle, get younger. Why has this not been observed?

2.6 Consider the motion of a particle in parallel $\vec{E}$ and $\vec{B}$ fields, as shown in Fig. 2.4(a) where the initial velocity is coming out of the plane of the paper. Convince yourselves that without the $\vec{E}$ field there is no problem, as discussed in the text. Now consider a motion of a particle in the background $\vec{E}$ field. By integrating the equation of motion, we get

$$m\dot{\vec{x}} = q\vec{E}t. \tag{2.11}$$

How can we interpret this equation in connection with the motion obtained by running the film backwards?

3

C, P and T in non-relativistic quantum mechanics

Subtle is the Lord –
but malicious she is not (we hope)!

Einstein

One of the basic concepts of quantum mechanics is the superposition principle for states. If orthogonal vectors $|a\rangle$ and $|b\rangle$ are vectors in a Hilbert space, so is $|\psi\rangle = \alpha|a\rangle + \beta|b\rangle$. If an operator $\mathcal{O}$ represents a symmetry, it requires also $\mathcal{O}|a\rangle$ and $\mathcal{O}|b\rangle$ to be orthogonal vectors in the Hilbert space.

What is a requirement on $\mathcal{O}|\psi\rangle$? If $\mathcal{O}$ represents a symmetry transformation, action of $\mathcal{O}$ does not change the outcome of any measurement. The operator must thus satisfy

$$|\langle\psi|\mathcal{O}^\dagger\mathcal{O}|\psi'\rangle|^2 = |\langle\psi|\psi'\rangle|^2. \tag{3.1}$$

This can be satisfied by[1]

$$\langle\psi|\mathcal{O}^\dagger\mathcal{O}|\psi'\rangle = \langle\psi|\psi'\rangle, \quad \text{or} \quad \langle\psi|\mathcal{O}^\dagger\mathcal{O}|\psi'\rangle = \langle\psi|\psi'\rangle^*. \tag{3.2}$$

The first relation implies that $\mathcal{O}$ is unitary. The second is satisfied if $\mathcal{O}$ is anti-unitary:

$$\mathcal{O}|\psi\rangle = \alpha^*\mathcal{O}|a\rangle + \beta^*\mathcal{O}|b\rangle. \tag{3.3}$$

An anti-unitary operator can be defined by the product

$$\mathcal{O} = UK, \tag{3.4}$$

where U is unitary and K is the complex conjugation operator defined by

$$K[\alpha|a\rangle + \beta|b\rangle] = \alpha^*K|a\rangle + \beta^*K|b\rangle, \tag{3.5}$$

where α and β are complex constants. The hermitian conjugate of K always acts to the left:

$$[\alpha\langle a| + \beta\langle b|]K^\dagger = \alpha^*\langle a|K^\dagger + \beta^*\langle b|K^\dagger. \tag{3.6}$$

The inverse operator is defined by

$$K^2 = 1. \tag{3.7}$$

---

[1] An alert reader may say that an additional arbitrary phase can be introduced on the left hand sides of Eq. (3.2). This is indeed true and is always present. Our argument goes through in spite of this freedom.

We can also define K by its action on the wave function:

$$\langle x|K|\psi\rangle = \psi(x)^*. \tag{3.8}$$

We will see below that parity and charge conjugation are described by unitary operators, whereas time reversal is implemented by an anti-unitary one.

There are actually two questions to be addressed, namely:

- Do the dynamics obey the symmetry?

- Are the admissible initial states invariant?

If both conditions are satisfied, then the possible final states will be symmetric as well.

3.1 Parity

Consider the Schrödinger equation for the state vectors $|\psi;t\rangle$,

$$i\hbar\frac{\partial}{\partial t}|\psi;t\rangle = H|\psi;t\rangle$$

$$H = \frac{|\vec{P}|^2}{2m} + V(\vec{X}), \tag{3.9}$$

where $\vec{X}$ is the position operator, and assume that parity transformations are implemented by a *unitary* operator $\mathbf{P}$ commuting with the Hamilton operator H:

$$[\mathbf{P}, H] = 0. \tag{3.10}$$

Parity is then conserved in Eq. (3.9): if $|\psi;t\rangle$ represents a solution, so does $\mathbf{P}|\psi;t\rangle$! For this conclusion to be correct

$$\mathbf{P}^{-1}i\mathbf{P} = i \tag{3.11}$$

has to hold, i.e., $\mathbf{P}$ has to be linear rather than antilinear.

There are simple rules telling us how observables behave under $\mathbf{P}$. Based on the correspondence principle, we require the expectation value of the position operator $\vec{X}$ to change sign under parity[2],

$$\langle\psi;t|\mathbf{P}^\dagger\vec{X}\mathbf{P}|\psi;t\rangle = -\langle\psi;t|\vec{X}|\psi;t\rangle. \tag{3.12}$$

This is guaranteed to happen if

$$\mathbf{P}^\dagger\vec{X}\mathbf{P} = -\vec{X} \quad \text{or} \quad \{\vec{X}, \mathbf{P}\} = 0, \tag{3.13}$$

where $\{a, b\}$ denotes an anti-commutator, holds as an operator relation. Hence we can deduce

$$\mathbf{P}|\vec{x}\rangle = e^{i\delta}|-\vec{x}\rangle \tag{3.14}$$

---

[2] The unitarity of $\mathbf{P}$ stated above – $\mathbf{P}^\dagger = \mathbf{P}^{-1}$ – is used here and below.

for eigenstates of the position operator with $e^{i\delta}$ being an arbitrary phase. As is easily shown: from $\vec{X}|\vec{x}\rangle = \vec{x}|\vec{x}\rangle$ in conjunction with Eq. (3.13) we obtain

$$\vec{X}\mathbf{P}|\vec{x}\rangle = -\mathbf{P}\vec{X}|\vec{x}\rangle = (-\vec{x})\mathbf{P}|\vec{x}\rangle \; ; \tag{3.15}$$

i.e., $\mathbf{P}|\vec{x}\rangle$ is an eigenvector of $\vec{X}$ with eigenvalue $-\vec{x}$. Conveniently, we adopt the phase convention $e^{i\delta} = 1$. Thus

$$\mathbf{P}^2 = 1, \tag{3.16}$$

meaning that **P** is Hermitian as well as unitary,

$$\mathbf{P}^\dagger = \mathbf{P}^{-1} = \mathbf{P} \tag{3.17}$$

and its eigenvalues are ± 1.

The momentum operator is best introduced as the generator of (infinitesimal) translations $d\vec{x}$ in space:

$$\mathscr{T}(d\vec{x}) \simeq 1 + \frac{i}{\hbar}\vec{P} \cdot d\vec{x} \tag{3.18}$$

Since a translation followed by a parity transformation is equivalent to a parity reflection followed by a translation in the opposite direction, $\mathscr{T}(-d\vec{x})$,

$$\mathbf{P}\mathscr{T}(d\vec{x}) = \mathscr{T}(-d\vec{x})\mathbf{P}, \tag{3.19}$$

that is,

$$\mathbf{P}\left(1 + \frac{i}{\hbar}\vec{P} \cdot d\vec{x}\right) \simeq \left(1 - \frac{i}{\hbar}\vec{P} \cdot d\vec{x}\right)\mathbf{P}, \tag{3.20}$$

we have

$$\mathbf{P}^\dagger\vec{P}\mathbf{P} = -\vec{P} \quad \text{or} \quad \{\vec{P}, \mathbf{P}\} = 0. \tag{3.21}$$

Since **P** anticommutes with both $\vec{X}$ and $\vec{P}$, it leaves the quantization condition

$$[X_i, P_j] = i\hbar\delta_{ij} \tag{3.22}$$

invariant, provided that **P** is linear and thus commutes with i.

Angular momentum $\vec{J}$ is introduced as the generator of rotations. Since rotations and parity commute, so do parity and angular momentum:

$$\mathbf{P}^\dagger\vec{J}\mathbf{P} = \vec{J} \quad \text{or} \quad [\vec{J}, \mathbf{P}] = 0. \tag{3.23}$$

This holds in particular also for the orbital angular momentum operator $\vec{L} = \vec{X} \times \vec{P}$,

$$[\vec{L}, \mathbf{P}] = 0 \tag{3.24}$$

which can be deduced also directly using Eq. (3.13) and Eq. (3.21).

Rewriting Eq. (3.9) in a configuration space representation, we arrive at

$$i\hbar\frac{\partial}{\partial t}\psi(\vec{x}, t) = H_x\psi(\vec{x}, t), \tag{3.25}$$

where

$$\psi(\vec{x}, t) \equiv \langle \vec{x} | \psi; t \rangle \tag{3.26}$$

$$H_x = -\frac{\hbar^2}{2m} \nabla^2 + V(\vec{x}). \tag{3.27}$$

If $\psi(t, \vec{x})$ is a solution of this Schrödinger equation, then $\psi(t, -\vec{x})$ solves

$$i\hbar \frac{\partial}{\partial t} \psi(t, -\vec{x}) = \left[-\frac{\hbar^2}{2m} \nabla^2 + V(-\vec{x}) \right] \psi(t, -\vec{x}). \tag{3.28}$$

Let us now assume that we have a parity even potential

$$V(\vec{X}) = V(-\vec{X}). \tag{3.29}$$

Then $\psi(t, \vec{x})$ and $\psi(t, -\vec{x})$ solve the same equation – as do the combinations $\psi_{\pm}(t, \vec{x}) = \psi(t, \vec{x}) \pm \psi(t, -\vec{x})$; i.e., for a parity even potential we can express all solutions as eigenstates of parity.

So, if $|\psi\rangle$ is a parity eigenstate with eigenvalue $+1$ or -1, then its wavefunction is parity even or odd, respectively:

$$\psi(-\vec{x}) \equiv \langle -\vec{x} | \psi \rangle = \langle \vec{x} | \mathbf{P} | \psi \rangle = \pm \langle \vec{x} | \psi \rangle \equiv \pm \psi(\vec{x}). \tag{3.30}$$

Parity considerations provide many powerful constraints. A few typical ones are listed under Problems 3.1–3.4.

For elementary particles or fields we can define also an *intrinsic* parity. This can be done in a strict and unambiguous manner as long as all forces conserve parity. Yet we can go one step further: we can assign an intrinsic parity as determined by the *strong* forces that produce the particle in question – even if the subsequent *weak* decay does not obey this symmetry. A case in point is pions and kaons. As explained in a later chapter, they are associated with the spontaneous realization of chiral invariance and have to act as pseudoscalar states, i.e., they carry odd intrinsic parity. This is confirmed by direct observation; on the other hand we will see that parity is violated even maximally in weak decays of pions and kaons.

With an S-wave $\pi\pi$ state necessarily carrying even parity – see Problem 3.2 – the decay $K \to \pi\pi$ could *not* occur – if the weak decays conserved parity!

3.2 Charge conjugation

To discuss a non-trivial consequence of **C** invariance we introduce the minimal electromagnetic coupling:

$$H = \frac{|\vec{P}|^2}{2m} - \frac{e}{2mc} [\vec{A} \cdot \vec{P} + \vec{P} \cdot \vec{A}] + \frac{e^2}{2mc^2} \vec{A}^2 + e\phi. \tag{3.31}$$

Here e is the charge of the particle, $\vec{A}$ is the vector potential, and ϕ is the electric potential. Note that the potential $V = H - |\vec{P}|^2/2m$ is velocity or momentum dependent here. The Hamilton operator is invariant under the following

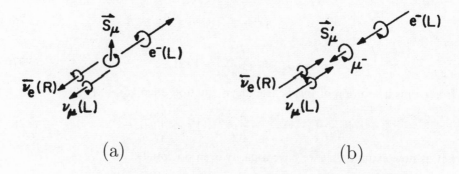

(a)   (b)

Fig. 3.1.   (a) A collinear decay of a vertically polarized muon. (b) The time reversed reaction results in a muon at rest with its polarization vector along the electron momentum. So, in quantum mechanics, we cannot realize the fully time reversed reaction. Reproduced from Ref. [2] by permission of Harwood Academic Publishers.

transformation:

$$e \overset{C}{\to} -e, \quad \vec{A} \overset{C}{\to} -\vec{A}, \quad \phi \overset{C}{\to} -\phi, \tag{3.32}$$

with $\vec{P}$ and $\vec{X}$ remaining unchanged. Thus

$$\mathbf{C}H\mathbf{C}^{-1} = H, \tag{3.33}$$

and **C** must be a linear operator. Note also that the operation defined in Eq. (3.32) is such that operating it twice gets back to the original configuration to within a phase factor: $\mathbf{C}^2 = e^{2i\delta}\mathbf{1}$. As in the case of parity, we can adopt the phase convention $\delta = 0$. Accordingly

$$\mathbf{C}^2 = \mathbf{1}. \tag{3.34}$$

Then the operator **C** has eigenvalues ± 1. We will extend the discussion of charge conjugation later when discussing relativistic quantum theory.

3.3 Time reversal

In our discussion of time reversal in classical mechanics we have emphasized one point in particular. *Microscopic* **T** invariance means that the rates for the two processes $a \to b$ and $b \to a$ are the same if b, which arises as a final state in the first process, is arranged identically – in *all its aspects* – as the initial state for the second process. We pointed out that such an arrangement requires a fine-tuning of the initial conditions that is typically very difficult to achieve. This difficulty in corpuscular mechanics turns into a real impossibility in wave mechanics.

The best example we know of is given by Lee in Ref. [2], see Fig. 3.1. Consider a decay

$$\mu^-(\Uparrow) \to e_L^- + \bar{v}_{e,R} + v_{\mu,L}, \tag{3.35}$$

where the muon is polarized 'up', i.e., perpendicular to the line of flight of the decay products. Now consider a time reversed collision of three particles forming a muon. It leads to a muon polarized in the line-of-flight direction of the initial beams and *not* a muon polarized 'down':

$$e_L^- + \bar{v}_{e,R} + v_{\mu,L} \to \mu^-(\Leftarrow) \neq \mu^-(\Downarrow). \tag{3.36}$$

The fact that we cannot realize the fully time reversed reaction goes well beyond the *practical* impossibility of creating such neutrino beams and colliding them with an electron beam, which we encounter already in classical physics. The underlying reason is a quantum mechanical effect, as can be seen in two ways. (i) Spin 'up' in the, say, x direction is *not* orthogonal to spin 'up' in the z direction. (ii) The truly time reversed version of muon decay has an initial state $e_L^- + \bar{v}_{eR} + v_{\mu,L}$ that consists of incoming *spherical* waves that furthermore have to be coherent; i.e., we had to reverse the momenta and spins of the three leptons in *all possible directions* while maintaining the required phase relationships among their amplitudes! This is obviously impossible rather than 'merely' unfeasible.

Let us now prepare the necessary formalism. The task at hand is to find a unitary or anti-unitary operator that under $t \to \bar{t} = -t$ generates reversal of motion at the operator level, i.e.,

$$
\begin{aligned}
\mathbf{T}^{-1}\vec{X}\mathbf{T} &= \vec{X} &\quad \text{or} \quad & [\vec{X}, \mathbf{T}] = 0 \\
\mathbf{T}^{-1}\vec{P}\mathbf{T} &= -\vec{P} &\quad \text{or} \quad & \{\vec{P}, \mathbf{T}\} = 0 \\
\mathbf{T}^{-1}\vec{J}\mathbf{T} &= -\vec{J} &\quad \text{or} \quad & \{\vec{J}, \mathbf{T}\} = 0,
\end{aligned} \tag{3.37}
$$

and that through its action upon a solution of the Schrödinger equation provides us with another solution, to be interpreted as the time reversed one:

$$|\overline{\psi}; -t\rangle = \mathbf{T}\,|\psi; t\rangle. \tag{3.38}$$

A glance at the most fundamental equation of quantum mechanics

$$[X_i, P_j] = i\hbar\delta_{ij} \tag{3.39}$$

leads us to conclude that something unusual happens. According to Eq. (3.37), the left hand side changes sign. What happens with the right hand side which is a c-number? If $\mathbf{T}$ is antiunitary we have $\mathbf{T}^{-1}i\mathbf{T} = -i$ and the commutation relation is invariant under $\mathbf{T}$ [3]. If we had overlooked that Eq. (3.1) allows an antiunitary operator to implement a symmetry transformation, we would have concluded that time reversal symmetry is inconsistent with quantum mechanics. For a moment consider $\mathbf{T} = K$, the complex conjugation operator. We then have

$$\mathbf{T}^{-1}\vec{X}\mathbf{T} = \vec{X} \tag{3.40}$$

$$\mathbf{T}^{-1}\left(-i\hbar\frac{\partial}{\partial\vec{X}}\right)\mathbf{T} = i\hbar\frac{\partial}{\partial\vec{X}} \quad \text{i.e.,} \quad \mathbf{T}^\dagger\vec{P}\mathbf{T} = -\vec{P}, \tag{3.41}$$

as required by Eq. (3.37). From the defining algebraic relation for angular momentum,

$$[J_i, J_j] = i\hbar\epsilon_{ijk}J_k, \tag{3.42}$$

we read off that

$$\mathbf{T}^{-1}\vec{J}\mathbf{T} = -\vec{J} \tag{3.43}$$

indeed follows with the antiunitary nature of $\mathbf{T}$ again being essential. For the orbital angular momentum operator this property can be derived also directly from the definition $\vec{L} \equiv \vec{X} \times \vec{P}$.

What remains to be seen is the following: how is the solution of the time reversed Schrödinger equation related to the solution of the original one if the Hamilton operator commutes with $\mathbf{T}$? Under time reversal the Schrödinger equation Eq. (3.25) changes into

$$\mathbf{T}H|\psi;t\rangle = \mathbf{T}i\hbar\frac{\partial}{\partial t}|\psi;t\rangle = i\hbar\frac{\partial}{\partial\bar{t}}\mathbf{T}|\psi;t\rangle = H\mathbf{T}|\psi;t\rangle, \tag{3.44}$$

since $\mathbf{T}i\hbar(\partial/\partial t) = -i\hbar(\partial/\partial t)\mathbf{T}$ and $[H, \mathbf{T}] = 0$. With the definition

$$|\overline{\psi};\bar{t} = -t\rangle \equiv \mathbf{T}|\psi;t\rangle, \tag{3.45}$$

this turns into

$$i\hbar\frac{\partial}{\partial\bar{t}}|\overline{\psi};\bar{t}\rangle = H|\overline{\psi};\bar{t}\rangle. \tag{3.46}$$

To see the relationship between the solutions $|\overline{\psi}\rangle$ and $|\psi\rangle$ explicitly we expand the state vectors $|\psi\rangle$ and $|\overline{\psi}\rangle$ in terms of the position eigenstates,

$$|\psi\rangle = \int \mathrm{d}^3x'|x'\rangle\langle x'|\psi\rangle \tag{3.47}$$

$$\mathbf{T}|\psi\rangle = \int \mathrm{d}^3x'\mathbf{T}(|x'\rangle\langle x'|\psi\rangle) = \int \mathrm{d}^3x'|x'\rangle\langle x'|\psi\rangle^*. \tag{3.48}$$

If $\psi(\vec{x}, t) \equiv \langle\vec{x}|\psi;t\rangle$ is a solution of the Schrödinger equation Eq. (3.25), then its complex conjugate $\psi^*(\vec{x}, t) = \langle\vec{x}|\psi;t\rangle^*$ satisfies

$$i\hbar\frac{\partial}{\partial(-t)}\psi^*(\vec{x}, t) = H\psi^*(\vec{x}, t). \tag{3.49}$$

Eq. (3.49) can of course be obtained directly from Eq. (3.25) through complex conjugation, which effectively achieves $t \to -t$ through $i\partial/\partial t \to (i\partial/\partial t)^* =$

$i\partial/\partial(-t)$. What we have shown here is how it follows from the formal properties of the antiunitary operator **T**.

In many cases the ansatz $\mathbf{T} = K$ provides a satisfactory representation of time reversal. However, this simple ansatz is not sufficient when spin degrees of freedom are involved. Consider, for example, a system with spin 1. The spin operator $\vec{S}$ can then be represented as

$$S_x = i\hbar \begin{pmatrix} 0 & 0 & 0 \\ 0 & 0 & -1 \\ 0 & 1 & 0 \end{pmatrix}, \quad S_y = i\hbar \begin{pmatrix} 0 & 0 & 1 \\ 0 & 0 & 0 \\ -1 & 0 & 0 \end{pmatrix},$$

$$S_z = i\hbar \begin{pmatrix} 0 & -1 & 0 \\ 1 & 0 & 0 \\ 0 & 0 & 0 \end{pmatrix}. \tag{3.50}$$

Since in this representation

$$\vec{S}^* = -\vec{S}, \tag{3.51}$$

$\mathbf{T} = K$ transforms $\vec{S}$ in the required manner. Let $\psi(m)$ be the wavefunction of an eigenstate of, for example, S_z:

$$S_z\psi(m) = m\psi(m). \tag{3.52}$$

Taking the complex conjugate, we obtain

$$-S_z\psi^*(m) = m\psi^*(m), \tag{3.53}$$

i.e., $\psi^*(m)$ is then the wavefunction of an eigenstate of S_z with eigenvalue $-m$, and

$$\mathbf{T}\psi(m) = \psi^*(m) = \psi(-m) \tag{3.54}$$

is indeed satisfied for $\mathbf{T} = K$ – if S_z is defined by Eq. (3.50).

Yet the representation of Eq. (3.50) is not very illuminating. In quantum mechanics, we are allowed to have at most one component of $\vec{J}$ commuting with H. To reveal the underlying physics we choose a maximal set of commuting operators, like

$$[H, J^2] = [H, J_z] = [J^2, J_z] = 0. \tag{3.55}$$

It is then more natural to choose a representation in which one of the components of, in our example, the spin operator $\vec{S}$ is diagonal:

$$S_x = \frac{\hbar}{\sqrt{2}} \begin{pmatrix} 0 & 1 & 0 \\ 1 & 0 & 1 \\ 0 & 1 & 0 \end{pmatrix}, \quad S_y = \frac{\hbar}{\sqrt{2}} \begin{pmatrix} 0 & -i & 0 \\ i & 0 & -i \\ 0 & i & 0 \end{pmatrix},$$

$$S_z = \hbar \begin{pmatrix} 1 & 0 & 0 \\ 0 & 0 & 0 \\ 0 & 0 & -1 \end{pmatrix}. \tag{3.56}$$

But now Eq. (3.51) is no longer satisfied, and we must modify the operator **T**:

$$\mathbf{T} = \exp\left[-i\frac{\pi}{\hbar}S_y\right]K, \tag{3.57}$$

where $\exp[-i\frac{\pi}{\hbar}S_y]$ rotates a vector by an angle π around the y axis so that $S_{x,z} \to -S_{x,z}$, but $S_y \to S_y$. The choice of Eq. (3.57) satisfies $\mathbf{T}\vec{S}\mathbf{T}^{-1} = -\vec{S}$ for the representation of Eq. (3.56).

For future reference, we list useful properties of **T**.

1

$$\mathbf{T}^\dagger\mathbf{T} = K^\dagger U^\dagger U K = K^\dagger K. \tag{3.58}$$

2

$$\langle A|\mathbf{T}^\dagger\mathbf{T}|B\rangle = \langle A|K^\dagger K|B\rangle = \langle A|B\rangle^* = \langle B|A\rangle. \tag{3.59}$$

3.4 Kramers' degeneracy

The apparently simple fact that **T** is antiunitary has a very important consequence we might not have anticipated: it tells us that all possible physical states can be divided into two distinct classes. These two classes turn out to be the bosonic and the fermionic degrees of freedom, yet the intriguing aspect is that the reasoning proceeds without reference to spin or the quantization condition; [1, 4, 5].

The argument goes as follows: if we perform the time reversal operation twice, a state can change at most by a phase factor:

$$\mathbf{T}^2 = UKUK = UU^*KK = UU^* = U(U^T)^{-1} = \eta, \tag{3.60}$$

where η is an $n \times n$ diagonal matrix with diagonal elements $(e^{i\eta_1}, \ldots, e^{i\eta_n})$. The last equality in Eq. (3.60) implies

$$U = \eta U^T \quad \text{or} \quad U^T = U\eta. \tag{3.61}$$

Substituting the second relation into the first one, we obtain

$$U = \eta U\eta. \tag{3.62}$$

Then η_i must satisfy

$$1 = e^{i(\eta_i+\eta_j)} \quad \text{for } i, j = 1, \ldots, n. \tag{3.63}$$

Since this equation is valid for all i, j, we set $i = j$ and find that

$$e^{i\eta_j} = \pm 1 \tag{3.64}$$

must hold. Note that there is no such restriction for unitary operators. $\mathbf{P}^2 = e^{i\alpha}$, $\mathbf{C}^2 = e^{i\beta}$, where α and β are arbitrary phases.

There is a clear physical distinction between $\mathbf{T}^2 = +1$ and $\mathbf{T}^2 = -1$. Suppose a system is invariant under time reversal and let $|E\rangle$ and $|E^{(T)}\rangle = \mathbf{T}|E\rangle$ be eigenstates with the same eigenvalue. We now show that if $\mathbf{T}^2 = -1$, $|E\rangle$ and $|E^{(T)}\rangle$ are orthogonal. The proof is very simple:

$$\langle E|E^{(T)}\rangle = \langle E^{(T)}|\mathbf{T}^\dagger\mathbf{T}|E\rangle = \langle E|(\mathbf{T}^\dagger)^2\mathbf{T}|E\rangle = -\langle E|\mathbf{T}|E\rangle, \qquad (3.65)$$

i.e.,

$$\langle E|\mathbf{T}|E\rangle = 0, \qquad (3.66)$$

meaning that the two energy-degenerate vectors $|E\rangle$ and $\mathbf{T}|E\rangle$ do *not* describe the same physical state. They are actually orthogonal to each other. *This is referred to as Kramers' degeneracy.*

This implies that $|E\rangle$ carries an internal degree of freedom that is changed by $\mathbf{T}$, if $\mathbf{T}^2 = -1$. If $\mathbf{T}^2 = +1$ holds, such a conclusion cannot be drawn, since the argument given above then merely yields a tautology. This was all derived *without* reference to spin, half-integer or otherwise. Of course it is completely consistent with results explictly derived for spin degrees of freedom, see Eq. (3.54).

Consider a state with n spin-$\frac{1}{2}$ particles with z component of the polarization s_i. Noting that $(e^{-i\frac{\pi}{2}\sigma_y})^2|\pm\rangle = -|\pm\rangle$, we obtain

$$\mathbf{T}^2|x_1, s_1; \ldots; x_n, s_n\rangle = (-1)^n|x_1, s_1; \ldots; x_n, s_n\rangle, \qquad (3.67)$$

and we conclude that $\mathbf{T}^2 = -1$ applies for a system with an odd number of fermions.

Kramers' degeneracy has two consequences. One is conceptual and we have discussed it above. It can be rephrased in the following way. Physical systems can be eigenstates of $\mathbf{T}^2$ with eigenvalues of either $+1$ or -1. A world where only the eigenvalue $+1$ were realized would be conceivable. However, it turns out that both classes of states are realized in nature. Similarly, it would be conceivable that only integer-spin states occurred in nature, or only systems obeying Bose–Einstein statistics. Yet the general experience is that dynamical structures – unless explicitly forbidden – are implemented, and actually in a very efficient way: odd-integer states are dubbed as fermions[3] and now are also seen as carrying the eigenvalue -1 of $\mathbf{T}^2$.

The second consequence of Kramers' degeneracy is of a more practical nature. Often it merely expresses the degeneracy between odd-integer spin up and down. But it can be more intriguing and emerge as a useful tool for treating problems in solid state physics with few manifest symmetries. Consider, for example, a system of electrons in an external electrostatic field $\vec{E}$. Such a field breaks rotational invariance; therefore neither spin nor orbital angular momentum are conserved in this situation. Yet time reversal invariance is still preserved. No matter how complicated this field is, (at least) a *two-fold degeneracy* is necessarily maintained for an *odd* number of electrons due to their spin-orbit couplings, but not for an *even* one; odd-number and even-number electron systems therefore exhibit

---

[3] In quantum field theory this is understood through the spin-statistics theorem.

very different behaviour in such external fields. This degeneracy is lifted once the electrons interact through their magnetic moments with an *external magnetic* field. For a magnetic field is not invariant under **T**. This observation allows us to use Kramers' degeneracy as a sensitive probe of the electronic states of paramagnetic crystals [6]. As just stated, odd-number systems of electrons exhibit at least a two-fold degeneracy that remains immune to the *electric* fields inside the crystals; it is, however, lifted by magnetic fields. Measuring the energy splittings then allows us to deduce the local magnetic fields inside the crystal.

3.5 Detailed balance

When we are thinking of how to test time reversal symmetry, the first thing we might come up with is to see if a reaction

$$a + b \rightarrow c + d, \tag{3.68}$$

and its reverse

$$c + d \rightarrow a + b, \tag{3.69}$$

occur with equal probability in nature. This symmetry is called *detailed balance*. Microscopically, the time reversal symmetry implies detailed balance, i.e., the equality of S matrix elements for the two processes (see Problems 3.6 and 3.9). The term detailed balance, strictly speaking, is actually used only when there are no spin degrees of freedom; when they are present, one averages over the initial spins and sums over the final ones. For clarity Eq. (3.102) is often referred to as the *reciprocity relation*. We have to keep in mind, however, that time reversal invariance is a sufficient, but not necessary, condition for detailed balance.

- If the **T** violating interaction H_- is so weak that it is relevant in first order only, then the hermiticity of H_- is sufficient to impose detailed balance:

$$|\langle ab|H_-|cd\rangle|^2 = |\langle cd|H_-|ab\rangle|^2. \tag{3.70}$$

- Another symmetry like parity might ensure it in the specific case under study. Consider elastic pion–nucleon scattering $\pi_i + N_i \rightarrow \pi_f + N_f$ where the phase space factors are obviously the same for a given c.m. energy. With two different spin configurations in the initial and final state the transition operator T is described by a 2×2 matrix; thus it can be expanded in terms of the Pauli matrices.

$$T(\vec{p}_{N_f}, p_{\pi_f}; \vec{p}_{N_i}, \vec{p}_{\pi_i}) = g_1(s,t) + g_2(\vec{p}_{N_i} + \vec{p}_{\pi_i}) \cdot \vec{\sigma}$$
$$+ \quad g_3(s,t)(\vec{p}_{N_i} - \vec{p}_{N_f}) \cdot \vec{\sigma} + g_4(s,t)(\vec{p}_{N_i} \times \vec{p}_{N_f}) \cdot \vec{\sigma}, \tag{3.71}$$

where $s = (p_{N_i} + p_{\pi_i})^2$ and $t = (p_{N_i} - p_{N_f})^2$.

With

$$\vec{\sigma} \overset{T}{\leftrightarrow} -\vec{\sigma} \, , \; \vec{p}_{\pi_i} \overset{T}{\leftrightarrow} -\vec{p}_{\pi_f} \, , \; \vec{p}_{N_i} \overset{T}{\leftrightarrow} -\vec{p}_{N_f}, \tag{3.72}$$

time reversal invariance implies

$$g_3(s,t) = 0. \tag{3.73}$$

Yet this term also violates parity. If the scattering interactions conserve parity, detailed balance therefore holds even if time reversal symmetry is violated. For further reading we suggest Refs. [1, 7, 8].

3.6 Electric dipole moments

The most direct tests of **T** invariance arise for single particle transitions. Later we will discuss neutrino oscillations in this context. Here we address a *static* quantity describing a particle or state. For further reading we suggest Ref. [8].

Consider a system, such as an elementary particle, an atom or a molecule in a weak electric field $\vec{E}$. The energy shift of the system due to the external electric field can be expanded in a power series in $\vec{E}$,

$$\Delta\mathscr{E} = d_i E_i + d_{ij} E_i E_j + \cdots. \tag{3.74}$$

The coefficient of the linear term in $\vec{E}$ is called the electric dipole moment (EDM), and that of the quadratic term is called the induced dipole moment.

3.6.1 The neutron EDM

An EDM is a measure of charge polarization within a particle. It is an expectation value of an operator

$$\vec{d} = \sum_i \vec{r}_i e_i. \tag{3.75}$$

For **P** conserved we have:

$$\langle N, s | \vec{d} | N, s \rangle = \langle N, s | \mathbf{P}^\dagger \mathbf{P} \vec{d} \mathbf{P}^\dagger \mathbf{P} | N, s \rangle = -\langle N, s | \vec{d} | N, s \rangle. \tag{3.76}$$

So, it must vanish. A nonvanishing expectation value of this operator then implies that the dynamics acting on the particle violates parity symmetry.

In the early 1950s, Purcell and Ramsey used this argument to test parity conservation in nuclear forces [9]. At the time the existing limit on the neutron EDM, obtained by Rabi, Havens and Rainwater, and by Fermi and Hughes, was $10^{-17}e$ cm. With an experiment at the Oak Ridge reactor, they had improved the limit by a factor of 10^3.

After the discovery of parity violation in 1957 this experiment seemed ideally suited for an independent search for parity violation. Landau, Lee and Schwinger, however, pointed out that a non-vanishing EDM required violation of yet another symmetry: **T**. Since $\vec{d}$ is a vector,

$$\langle N, s|\vec{d}|N, s\rangle = C\langle N, s|\vec{J}|N, s\rangle \tag{3.77}$$

must hold, where $\vec{J}$ is the angular momentum operator. The proportionality follows from the fact that $\vec{J}$ is the only three component object which characterizes a neutron; it is *not* based on treating the neutron as an elementary rather than a bound state. Let us be more precise and specific: if in addition to $\vec{J}$ we could specify a different direction, say $\vec{U}$, for the system under study, to which the EDM could be proportional, then the system would be *degenerate* by its dependance on $\vec{U}$. The issue thus is *non-degeneracy* rather than *elementarity*! Under time reversal invariance, $\vec{d}$ transforms as

$$\mathbf{T}\vec{d}\mathbf{T}^{-1} = \vec{d}, \tag{3.78}$$

compared to

$$\mathbf{T}\vec{J}\mathbf{T}^{-1} = -\vec{J}, \tag{3.79}$$

implying the EDM has to vanish: $C = 0$. So for an EDM to emerge, both parity and time reversal must be violated. This did not discourage Ramsey. He states [9],

> ... just as parity had been an assumed symmetry which should be based on experiment, time reversal symmetry should be similarly based.

The energy shift of a neutron electric dipole in an electric field $\vec{E}(\vec{r})$ is given by

$$\Delta\mathscr{E} = C\vec{J} \cdot \vec{E}(\vec{r}). \tag{3.80}$$

As $\vec{E}(\vec{r})$ is even under **T** and $\vec{J}$ is odd, such a term in a Hamiltonian violates time reversal invariance.

State-of-the-art measurements of the neutron EDM are done with ultra-cold neutrons: neutrons from a reactor are cooled down to an energy of the order of 10^{-7} eV and then stored in a magnetic bottle. This energy is comparable to the kinetic energy gained by a neutron when it drops 1 m in the earth's gravitational field. Ultra-cold neutrons undergo total reflection from the surfaces of most materials. Neutron spin is also preserved during these collisions. When these neutrons are placed in parallel static electric and magnetic fields, the energy shift given by Eq. (3.80) changes the neutron Larmor frequency. This shift in frequency is measured by the magnetic resonance technique [10]. Two groups found

$$\begin{aligned}
d_n &= (-3 \pm 5) \times 10^{-26} \text{ e cm} \quad \text{ILL [11]} \\
d_n &= (2.6 \pm 4 \pm 1.6) \times 10^{-26} \text{ e cm} \quad \text{Gatchina [12]}.
\end{aligned} \tag{3.81}$$

Ramsey ends his talk [9] by saying:

After 43 years of searching for, but not finding, a neutron electric dipole moment, I suppose I should be discouraged and believe that no particle dipole moment will ever be discovered and the search should be abandoned. On the contrary, I am now quite optimistic.... For the most of the past 43 years, the searches have been lonely ones. Now there are promising experiments with atoms, electrons, and protons as well. I sincerely hope that someone will hit the jackpot soon...at age 76 I cannot wait another 43 years unless there is a way to achieve real time reversal for biological clocks.

3.6.2 Atomic EDM

In the previous section, we have pointed out that $\vec{d}$ must be proportional to the total angular momentum $\vec{J}$ of a non-degenerate system. Only then does the non-vanishing of $\vec{d}$ imply **T** violation. Yet for a complex system like an atom or a molecule, there surely must be other vectors characterizing the system. For example, a dumb-bell with a positive charge on one end and a negative one on the other is an electric dipole by definition. In this section, we give an additional constraint on the EDM defined in Eq. (3.74) such that if it is detected for an atom or molecule, it implies that its elementary constituents – electrons or nucleons – must possess an EDM.

At first this seems suspicious. A water molecule has a polarization. Energy shift is linear. How can we say that this electric polarization is not an EDM?

Consider two states $|\pm\rangle$ with opposite parity and energy $\mathscr{E}_{\pm}$, with $\mathscr{E}_{+} < \mathscr{E}_{-}$. In a constant external electromagnetic field along the z direction – $E_z \neq 0 = E_x = E_y$ – these opposite parity states can mix. Denote the mixing matrix element of the electric potential associated with E_z by $\langle +|zE_z|-\rangle = \Delta$. The Hamilton operator can be written in matrix notation as

$$H = \begin{pmatrix} \mathscr{E}_{+} & \Delta \\ \Delta & \mathscr{E}_{-} \end{pmatrix} \tag{3.82}$$

with eigenvalues

$$\mathscr{E}_{1,2} = \frac{1}{2}(\mathscr{E}_{+} + \mathscr{E}_{-}) \pm [\frac{1}{4}(\mathscr{E}_{+} - \mathscr{E}_{-})^2 + \Delta^2]^{\frac{1}{2}}. \tag{3.83}$$

With *sufficiently weak* electric field $\vec{E}$, we can achieve $\langle +|zE_z|-\rangle \equiv \Delta \ll |\mathscr{E}_{-} - \mathscr{E}_{+}|$, leading to

$$\mathscr{E}_1 \simeq \mathscr{E}_{-} + \frac{\Delta^2}{\mathscr{E}_{-} - \mathscr{E}_{+}}, \quad \mathscr{E}_2 \simeq \mathscr{E}_{+} - \frac{\Delta^2}{\mathscr{E}_{-} - \mathscr{E}_{+}} \; ; \tag{3.84}$$

and thus

$$\mathscr{E}_2 - \mathscr{E}_1 = \mathscr{E}_{-} - \mathscr{E}_{+} + \frac{2\Delta^2}{\mathscr{E}_{-} - \mathscr{E}_{+}}, \tag{3.85}$$

i.e., the induced energy shift is *quadratic* in $\vec{E}$ rather than linear. In the terminology of Eq. (3.74), it is an induced dipole moment and it does *not* imply **T** violation.

At this point we might raise the following objection. Consider exact energy degeneracy: $\mathscr{E}_+ = \mathscr{E}_-$. Then, instead of Eq. (3.85), we have

$$\Delta\mathscr{E} = \mathscr{E}_2 - \mathscr{E}_1 = 2\Delta, \tag{3.86}$$

and the eigenstates are $|+\rangle \pm |-\rangle$. For this case

$$d_z = \langle +|z|-\rangle \; ; \tag{3.87}$$

i.e., the energy shift is linear in the electric field. Schiff calls this a permanent EDM [13] as well. Degenerate states can possess a permanent EDM that does *not* imply **T** violation. This happens, for example, for the $2s_{\frac{1}{2}}$, $2p_{\frac{1}{2}}$ states of the hydrogen atom, since those excited states are degenerate. On the other hand we expect all atomic ground states to be non-degenerate.

How do we know that a system is degenerate or not? Now, come back to our example of a water molecule. Denote as $|\uparrow\rangle$ [$|\downarrow\rangle$] a state with polarization aligned along [opposite to] the $+z$ direction. Since electrodynamics conserve parity,

$$|\pm\rangle = |\uparrow\rangle \pm |\downarrow\rangle \tag{3.88}$$

are degenerate parity eigenstates. The ground state of the water molecule is degenerate, and Eq. (3.86) implies that the energy shift is linear. Using Schiff's definition, a water molecule possesses an EDM but it does not signal **T** violation ; actually, all polarized molecules have degenerate ground states with uninteresting EDM.

We prefer to introduce an additional criterion for the existence of an EDM: we require $\Delta\mathscr{E}$ to change sign under a **P** transformation. In our example, $\Delta \to -\Delta$ under parity. But the states transform as $|+\rangle + |-\rangle \leftrightarrow |+\rangle - |-\rangle$ and the label for the states, $1 \leftrightarrow 2$ so that $\mathscr{E}_2 - \mathscr{E}_1$ is invariant under parity. Eq. (3.87) does not satisfy such a requirement for an EDM. If an EDM were proportional to $\vec{J}$, $\Delta\mathscr{E} \propto \vec{J} \cdot \vec{E}$ would satisfy the strict requirement of EDM.

3.6.3 Schiff's theorem

Effects due to an electron EDM get screened in an atom. The result is quite natural, at least qualitatively: the charge carriers inside an atom placed in an external electric field get shifted; this distortion creates a polarization and thus an induced electric field that shields the electron.

Schiff's theorem states that this shielding is *complete* and lists the conditions for its validity [13, 14]: an atomic EDM vanishes under the following conditions:

1 Atoms consist of *non*-relativistic particles, which interact only electro-statically.

2 The EDM distribution of each atomic constituent is identical to its charge distribution.

The electric field at r_i – the coordinate of electron i – is given by

$$\vec{E}_i = -\vec{\nabla}_{r_i} V(r_1, \ldots, r_n), \tag{3.89}$$

and the term in the Hamilton operator associated with the EDM interaction reads

$$H_{\text{edm}} = -d \sum_i \vec{\sigma} \cdot \vec{\nabla}_{r_i} V(r_1, \ldots, r_n) = -id[Q, H_0], \tag{3.90}$$

where $H_0 = \frac{p^2}{2mc} + V$ and $Q = -i \sum_i \vec{\sigma} \cdot \vec{\nabla}_{r_i}$. Then we can write

$$H = H_0 - id[Q, H_0] = e^{-idQ} H_0 e^{idQ} + \mathcal{O}(Q^2). \tag{3.91}$$

If ψ_n is an eigenstate of H_0 with eigenvalue E_n, then obviously so is $e^{-idQ}\psi_n$ for H – with the *same* eigenvalue E_n! Thus there is no shift in the atomic levels of non-relativistic systems due to a possible EDM of the atomic constituents.

In heavy and thus high Z atoms, on the other hand, valence electrons feel strong Coulomb fields when they come close to the nucleus. Their motion is then highly relativistic and one of the conditions of Schiff's theorem breaks down. As the EDM interaction violates parity, it generates strong mixing between opposite parity states at short distance. We might have expected that the complete shielding of the electron EDM expressed in Schiff's theorem gets translated into a partial shielding when relativistic corrections are included. Yet it turns out that for heavy atoms such as Cs, Tl, Rb, and Fr, there actually emerge huge *enhancement* factors that are as large as two orders of magnitude or even more! The best measurement to date is that of Cs, with an enhancement factor of about 100 [15]:

$$d_{\text{Cs}} \simeq 100 \cdot d_e \implies d_e = (-0.3 \pm 0.8) \times 10^{-26} e \text{ cm.} \tag{3.92}$$

We can realistically expect further improvements in sensitivity by orders of magnitude in the foreseeable future, due to advances in experimental techniques and the identification of atoms with even larger enhancement factors relating the EDM of the atom to that of the electron.

3.7 Résumé

Defining charge conjugation, parity and time reversal transformations in quantum mechanics and stating which criteria have to be satisfied for **C**, **P** and **T** to represent symmetries of the Hilbert space requires more technical ado than for classical dynamics. A special feature arises in that time reversal is described by an *anti*linear operator. Of special interest is Kramer's degeneracy, which anticipates the existence of fermions.

Problems

3.1   Show that a necessary condition for

$$\langle \phi | \vec{X} | \psi \rangle \neq 0 \tag{3.93}$$

is that $|\psi\rangle$ and $|\phi\rangle$ are eigenstates of *opposite* parity. This is referred to as *Laporte's rule*.

3.2   Consider an eigenstate of $\vec{L}^2$ and L_z denoted by $|l, m\rangle$: Prove that

$$\mathbf{P}|l, m\rangle = (-1)^l |l, m\rangle \tag{3.94}$$

holds.

3.3   Show that if the dynamics conserve parity – $[H, \mathbf{P}] = 0$ – and if $|\psi\rangle$ denotes a *non-degenerate* eigenstate of H, then it is also a parity eigenstate.

3.4   Demonstrate that for $[H, \mathbf{P}] = 0$ no transition can occur between states of opposite parity.

3.5   For an *orbital* angular momentum eigenstate, establish the following identity:

$$\mathbf{T}\,|l, m\rangle = (-1)^m |l, -m\rangle \tag{3.95}$$

from the properties of spherical harmonics.

3.6   Prove that

$$\langle \psi_1 | H | \psi_2 \rangle = \langle \psi_2 | \mathbf{T}^\dagger H \mathbf{T} | \psi_1 \rangle \tag{3.96}$$

if $\mathbf{T}$ commutes with H.

3.7   Show that the charge conjugation operator $\mathbf{C}$ must be linear.

3.8   Suppose

$$\sum_n |n; \text{out}\rangle\langle n; \text{out}| = \sum_n |n; \text{in}\rangle\langle n; \text{in}| = 1, \tag{3.97}$$

– i.e., both 'in' and 'out' states form a complete set of states – and furthermore

$$\sum_n \mathcal{O}|n; \text{out}\rangle\langle n; \text{out}|\mathcal{O}^\dagger = 1, \tag{3.98}$$

where $\mathcal{O}$ stands for $\mathbf{C}$, $\mathbf{P}$ or $\mathbf{T}$. Finally, define

$$\Gamma_{12} = 2\pi \sum_F \delta(M_P - M_F)\langle P^0 | H_{\text{weak}} | n; \text{out}\rangle\langle n; \text{out} | H_{\text{weak}} | \bar{P}^0 \rangle. \tag{3.99}$$

Show that $\mathbf{CP}$, $\mathbf{T}$ and $\mathbf{CPT}$ symmetries imply $\Gamma_{12} = \Gamma_{21}$, $\Gamma_{12} = \Gamma_{12}^*$ and $\Gamma_{12} = \Gamma_{21}^*$, respectively.

3.9 The time reversal operator transforms

$$\mathbf{T}|\vec{p}_a, \vec{s}_a; \vec{p}_b, \vec{s}_b; \text{in}\rangle = |-\vec{p}_a, -\vec{s}_a; -\vec{p}_b, -\vec{s}_b; \text{out}\rangle, \qquad (3.100)$$

and the scattering $a + b \rightarrow c + d$ is controled by the S matrix

$$S(a + b \rightarrow c + d) = \langle \vec{p}_c, \vec{s}_c; \vec{p}_d, \vec{s}_d; \text{out}|\vec{p}_a, \vec{s}_a; \vec{p}_b, \vec{s}_b; \text{in}\rangle. \qquad (3.101)$$

Show that $\mathbf{T}$ invariance implies:

$$S(a + b \rightarrow c + d) = S(c + d \rightarrow a + b), \qquad (3.102)$$

where

$$S(c + d \rightarrow a + b) = \langle -\vec{p}_a, -\vec{s}_a; -\vec{p}_b, -\vec{s}_b; \text{out}|-\vec{p}_c, -\vec{s}_c; -\vec{p}_d, -\vec{s}_d; \text{in}\rangle.$$
$$(3.103)$$

3.10 Consider an electron at rest and and write an operator $R(2\pi)$ which rotates the spin of the electron by 2π. Discuss the relationship between $R(2\pi)$ and $\mathbf{T}^2$.

4

C, P and T in relativistic quantum theories

In describing symmetries in relativistic quantum theories we usually start out by showing how relativistic extensions of the Schrödinger equation like the Klein–Gordon and Dirac equations transform under the symmetry; subsequently we discuss quantum field theory. We do not feel that following the historical development is always the most illuminating way to introduce new concepts. Instead we will begin with quantum field theories, i.e., second quantized theory, and subsequently come back to the first quantized version.

To introduce symmetries in quantum field theory, we assume that there are operators which transform states in a way that leave all physical observables unchanged.

In particular, we formulate the theory such that

1 The groundstate or 'vacuum' remains invariant:

$$\mathbf{P} \, |0\rangle = |0\rangle, \quad \mathbf{C} \, |0\rangle = |0\rangle, \quad \mathbf{T} \, |0\rangle = |0\rangle. \tag{4.1}$$

2 Likewise for the action:

$$S = \int \mathrm{d}^4 x \, \mathscr{L}(t, \vec{x}) \to S \tag{4.2}$$

or equivalently,

$$[\mathbf{P}, H] = [\mathbf{C}, H] = [\mathbf{T}, H] = 0. \tag{4.3}$$

3 The quantization conditions must remain invariant.

As can be inferred from the title of the book, the symmetries we are discussing are not exact. We must eventually consider how to break them. When symmetries are broken, the above conditions are relaxed. Nature has chosen two ways to do this. We say that a symmetry is broken

- *explicitly* if condition 2 is no longer valid;

- *spontaneously* if condition 1 is no longer valid. It is actually more appropriate to call it a spontaneous realization of the symmetry, since the dynamics still obey the latter. We will use both terms interchangeably.

The gauge symmetry of the SM is broken spontaneously; QCD with massless quarks possesses an $SU(3)_L \times SU(3)_R$, chiral symmetry. This symmetry is broken spontaneously to $SU(3)_{L+R}$, giving rise to massless Goldstone bosons that

36

double as pion and kaon pseudoscalar mesons; those acquire masses when chiral symmetry is broken explicitly through quark masses.

How **C**, **P** and **T** are broken is an interesting question to which we do not know the answer. Let us hope that one of the readers will eventually settle this question. With this comment, we shall postpone the discussion of symmetry breaking to later chapters.

4.1 Notation

Before we go on, let us define some notation [16]. Throughout the remainder of this book we will adopt the unit system that is most convenient for high energy physics, where the speed of light provides the natural scale and quantum effects are not necessarily small; i.e., we set

$$c = 1 = \hbar \tag{4.4}$$

We write four-vectors with upper and lower indices, where the metric tensor $g_{\mu\nu}$, defined by

$$g_{\mu\nu} = \text{diag}[1, -1, -1, -1] \tag{4.5}$$

is used to raise or lower indices:

$$x^\mu = (t, \vec{x}), \qquad x_\mu = g_{\mu\nu} x^\nu = (t, -\vec{x}), \tag{4.6}$$

with the usual convention of summing over repeated indices. Likewise

$$
\begin{aligned}
p^\mu &= (E, \vec{p}), \quad p_\mu = (E, -\vec{p}) \\
\partial^\mu &= \frac{\partial}{\partial x_\mu}, \quad \partial_\mu = \frac{\partial}{\partial x^\mu},
\end{aligned}
\tag{4.7}
$$

and for the identification of the four-momentum operator with the derivative operator in configuration space:

$$p^\mu = i\left(\frac{\partial}{\partial t}, -\frac{\partial}{\partial x^1}, -\frac{\partial}{\partial x^2}, -\frac{\partial}{\partial x^3}\right) = i\partial^\mu. \tag{4.8}$$

For the Dirac matrices we shall choose the representation

$$\gamma^0 = \begin{pmatrix} 1 & 0 \\ 0 & -1 \end{pmatrix}, \quad \gamma^i = \begin{pmatrix} 0 & \sigma^i \\ -\sigma^i & 0 \end{pmatrix}, \tag{4.9}$$

where

$$\mathbf{1} = \begin{pmatrix} 1 & 0 \\ 0 & 1 \end{pmatrix}, \sigma^1 = \begin{pmatrix} 0 & 1 \\ 1 & 0 \end{pmatrix}, \sigma^2 = \begin{pmatrix} 0 & -i \\ i & 0 \end{pmatrix}, \sigma^3 = \begin{pmatrix} 1 & 0 \\ 0 & -1 \end{pmatrix} \tag{4.10}$$

are the 2×2 unit matrix and Pauli matrices, respectively. Also, $\gamma^5 = i\gamma^0\gamma^1\gamma^2\gamma^3 = \begin{pmatrix} 0 & 1 \\ 1 & 0 \end{pmatrix}$.

Finally, all operators which are products of field operators are understood to be normal ordered.

4.2 Spin-1 fields

Maxwell's equations can be expressed in a compact and manifestly Lorentz covariant fashion:

$$\partial_\mu F^{\mu\nu} = eJ^\nu, \tag{4.11}$$

with $F^{\mu\nu} \equiv \partial^\mu A^\nu - \partial^\nu A^\mu$ being the field strength tensor and J^ν the current. The equation of motion, Eq. (4.11), can be obtained from the Lagrangian:

$$\mathcal{L} = -\frac{1}{4} F_{\mu\nu} F^{\mu\nu} - eJ^\mu A_\mu. \tag{4.12}$$

Electrodynamics are quantized by first introducing a gauge condition; for example, the Coulomb gauge condition reads:

$$\vec{\nabla} \cdot \vec{A} = 0. \tag{4.13}$$

The conjugate momentum associated with $\vec{A}$ is given by

$$\vec{\Pi}(t,\vec{x}) = -\frac{\partial}{\partial t}\vec{A}(t,\vec{x}) - \vec{\nabla}A_0 = \vec{E}, \tag{4.14}$$

and we postulate the quantization condition

$$[A_i(t,\vec{x}),\Pi_j(t,\vec{y})] = -i\left(\delta_{ij}\delta^{(3)}(\vec{x}-\vec{y}) + \frac{\partial^2}{\partial x^i \partial x^j}\frac{1}{4\pi|\vec{x}-\vec{y}|}\right), \tag{4.15}$$

$$[A_i(t,\vec{x}),A_j(t,\vec{y})] = 0 = [\Pi_i(t,\vec{x}),\Pi_j(t,\vec{y})], \quad i,j = 1,2,3. \tag{4.16}$$

The last term on the right hand side of Eq. (4.15) is required for the quantization condition to be consistent with Gauss's Law: $\vec{\nabla} \cdot \vec{E} = 0$.

To show that quantum electrodynamics is invariant under **P**, **C** and **T**, we have to investigate the symmetry of the Lagrangian as well as the quantization conditions.

From our experience with classical fields, we can postulate the transformation properties for the quantized fields:

$$\begin{aligned}
&\mathbf{P}\,A_\mu(t,\vec{x})\mathbf{P}^\dagger = A^\mu(t,-\vec{x}), && \mathbf{P}\,J^\mu(t,\vec{x})\mathbf{P}^\dagger = J_\mu(t,-\vec{x}),\\
&\mathbf{C}\,A_\mu(t,\vec{x})\mathbf{C}^\dagger = -A_\mu(t,\vec{x}), && \mathbf{C}\,J^\mu(t,\vec{x})\mathbf{C}^\dagger = -J^\mu(t,\vec{x}),\\
&\mathbf{T}\,A_\mu(t,\vec{x})\mathbf{T}^{-1} = A^\mu(-t,\vec{x}), && \mathbf{T}\,J^\mu(t,\vec{x})\mathbf{T}^{-1} = J_\mu(-t,\vec{x}).
\end{aligned} \tag{4.17}$$

Now let us see if the action remains invariant under these transformations. First observe that

$$\mathbf{P}\partial_\mu\mathbf{P}^\dagger = \partial^\mu, \tag{4.18}$$

where the raising of the Lorentz indices should be noted. Thus

$$\begin{aligned}
\mathbf{P}\mathcal{L}(t,\vec{x})\mathbf{P}^\dagger &= \mathcal{L}(t,-\vec{x}) & (4.19)\\
\mathbf{C}\mathcal{L}(t,\vec{x})\mathbf{C}^\dagger &= \mathcal{L}(t,\vec{x}) & (4.20)\\
\mathbf{T}\mathcal{L}(t,\vec{x})\mathbf{T}^{-1} &= \mathcal{L}(-t,\vec{x}). & (4.21)
\end{aligned}$$

Since the sign of integration variables can be changed, as we integrate over all space–time, the action Eq. (4.2) is invariant under these symmetries.

P, **C** and **T** leave the quantization conditions unchanged. Let us illustrate this by showing that **T** leaves Eq. (4.15) invariant:

$$\mathbf{T}[A_i(t,\vec{x}), \Pi_j(t,\vec{y})]\mathbf{T}^{-1} = [-A_i(-t,\vec{x}), \Pi_j(-t,\vec{y})]$$

$$= i\left(\delta_{ij}\delta^{(3)}(\vec{x}-\vec{y}) + \frac{\partial^2}{\partial x^i \partial x^j}\frac{1}{4\pi|\vec{x}-\vec{y}|}\right). \qquad (4.22)$$

Note that the right hand side of Eq. (4.22) acquires a minus sign since **T** is antiunitary. On the left hand side we have used the fact that

$$\mathbf{T}\vec{\Pi}(t,\vec{x})\mathbf{T}^{-1} = \frac{\partial}{\partial t}\vec{A}(-t,\vec{x}) - \vec{\nabla}A_0(-t,\vec{x})$$

$$= \vec{\Pi}(-t,\vec{x}). \qquad (4.23)$$

The invariance of the quantization condition under **C** and **P** is left as an exercise (see Problem 4.1).

A photon field operator can be expanded in terms of creation and annihilation operators:

$$A^\mu(t,\vec{x}) = \int \frac{d^3\vec{p}}{\sqrt{(2\pi)^3 2E_p}} \sum_{s=\pm} [\mathbf{a}(\vec{p},s)\epsilon_s^\mu e^{-ip\cdot x} + \mathbf{a}^\dagger(\vec{p},s)\epsilon_s^{\mu*} e^{ip\cdot x}], \qquad (4.24)$$

where $\epsilon_s^\mu(\vec{p})$ and $\mathbf{a}_s^\dagger$ are a polarization vector, and a creation operator for a photon with momentum $\vec{p}$ and polarization s, respectively.

For an on-shell photon with four-momentum $p^\mu = (p,0,0,p)$, the polarization vector is given by [18]:

$$\epsilon_\pm^\mu = \frac{1}{\sqrt{2}}(0,1,\pm i,0), \qquad (4.25)$$

where we have chosen $\epsilon_\pm^i$ to be an eigenstate of S_z as given in Eq. (3.50). The photon polarization must transform as spin:

$$\mathbf{T}\vec{\epsilon}_\pm\mathbf{T}^{-1} = (\vec{\epsilon}_\pm)^* = \vec{\epsilon}_\mp$$

$$\mathbf{P}\vec{\epsilon}_\pm\mathbf{P}^{-1} = \vec{\epsilon}_\pm$$

$$\mathbf{C}\vec{\epsilon}_\pm\mathbf{C}^{-1} = \vec{\epsilon}_\pm. \qquad (4.26)$$

Now transforming $A^\mu(t,\vec{x})$ under **C**, **P** and **T** and changing variables, we can deduce

$$\mathbf{P}\mathbf{a}(\vec{p},\pm)\mathbf{P}^\dagger = -\mathbf{a}(-\vec{p},\pm),$$

$$\mathbf{C}\mathbf{a}(\vec{p},\pm)\mathbf{C}^\dagger = -\mathbf{a}(\vec{p},\pm),$$

$$\mathbf{T}\mathbf{a}(\vec{p},\pm)\mathbf{T}^{-1} = -\mathbf{a}(-\vec{p},\mp). \qquad (4.27)$$

Thus we can say that a one-photon state carries odd *intrinsic* parity and

C parity:

$$\mathbf{P}|\gamma;\vec{p}\rangle = -|\gamma;-\vec{p}\rangle \qquad (4.28)$$

$$\mathbf{C}|\gamma;\vec{p}\rangle = -|\gamma;\vec{p}\rangle. \qquad (4.29)$$

Other spin-1 fields can carry internal quantum numbers – as is the case for non-abelian gauge bosons. For those we cannot define intrinsic **C** parity. However, they can still carry intrinsic parity; the latter is odd [even] for vector [axial vector] fields.

4.3 Spin-0 fields

A charged spinless particle can be described by a complex field $\phi(t,\vec{x})$. Consider a Lagrangian for such a spin-0 field in an electromagnetic field,

$$\mathcal{L} = \frac{1}{2}(\partial_\mu + ieA_\mu)\phi^\dagger(t,\vec{x})(\partial^\mu - ieA^\mu)\phi(t,\vec{x}) - \frac{1}{2}m^2\phi^\dagger(t,\vec{x})\phi(t,\vec{x}) \qquad (4.30)$$

with quantization conditions

$$[\phi(t,\vec{x}), \phi(t,\vec{y})] = [\phi^\dagger(t,\vec{x}), \phi^\dagger(t,\vec{y})] = 0, \qquad (4.31)$$

$$[\phi(t,\vec{x}), \partial_t\phi^\dagger(t,\vec{y})] = [\phi^\dagger(t,\vec{x}), \partial_t\phi(t,\vec{y})] = i\delta^{(3)}(\vec{x} - \vec{y}). \qquad (4.32)$$

As in the previous example, the field can be expanded in terms of creation and annihilation operators:

$$\phi(t,\vec{x}) = \int \frac{d^3\vec{p}}{\sqrt{(2\pi)^3 2E_p}}[\mathbf{b}(\vec{p})e^{-ipx} + \mathbf{d}^\dagger(\vec{p})e^{ipx}], \qquad (4.33)$$

where $\mathbf{b}(\vec{p})$, and $\mathbf{d}(\vec{p})$ are annihilation operators for particle and anti-particle, respectively.

We shall find that the quantized theory exhibits **C**, **P** and **T** symmetry.

The transformation properties of $\phi(t,\vec{x})$ can be guessed by examining the current:

$$J^\mu(t,\vec{x}) = i[\phi^\dagger(t,\vec{x})\partial^\mu\phi(t,\vec{x}) - \phi(t,\vec{x})\partial^\mu\phi^\dagger(t,\vec{x})]. \qquad (4.34)$$

4.3.1 Parity

Let us postulate

$$\phi^P(t,\vec{x}) \equiv \mathbf{P}\phi(t,\vec{x})\mathbf{P}^\dagger = \phi(t,-\vec{x}). \qquad (4.35)$$

With this definition, using Eq. (4.17) we can check that the action remains invariant. The same goes for the quantization conditions Eq. (4.31) and Eq. (4.32).

The transformation property, Eq. (4.35), leads to

$$\mathbf{P}\mathbf{b}(\vec{p})\mathbf{P}^\dagger = \mathbf{b}(-\vec{p}), \quad \mathbf{P}\mathbf{d}(\vec{p})\mathbf{P}^\dagger = \mathbf{d}(-\vec{p}). \tag{4.36}$$

So, an n-particle state transforms as:

$$\mathbf{P}|\vec{p}_1,\ldots,\vec{p}_n\rangle = |-\vec{p}_1,\ldots,-\vec{p}_n\rangle. \tag{4.37}$$

4.3.2 Charge conjugation

The charge conjugation transformation for the current is given by Eq. (4.17), and this can be realized by

$$\phi^C(t,\vec{x}) \equiv \mathbf{C}\phi(t,\vec{x})\mathbf{C}^\dagger = \phi^\dagger(t,\vec{x}). \tag{4.38}$$

By taking the Hermitian conjugate of Eq. (4.31) and Eq. (4.32), we can easily show that $\phi^C(t,\vec{x})$ satisfies the quantization condition. Also, by taking the Hermitian conjugate of $\mathscr{L}(t,\vec{x})$ given in Eq. (4.30), we see that the Lagrangian written in terms of $\phi^C(t,\vec{x})$ remains invariant except for the sign of the charge.

Under $\mathbf{C}$, the creation and annihilation operators transform into each other:

$$\mathbf{C}\mathbf{b}(\vec{p})\mathbf{C}^\dagger = \mathbf{d}(\vec{p}). \tag{4.39}$$

Thus charge conjugation $\mathbf{C}$ changes a particle into its antiparticle.

4.3.3 Time reversal

Let us show that

$$\phi^T(t,\vec{x}) \equiv \mathbf{T}\phi(t,\vec{x})\mathbf{T}^{-1} = \phi(-t,\vec{x}), \quad \mathbf{T}\phi^\dagger(t,\vec{x})\mathbf{T}^{-1} = \phi^\dagger(-t,\vec{x}), \tag{4.40}$$

gives the correct transformation property of the current, Eq. (4.17). The time component of the current transforms as

$$\begin{aligned}
\mathbf{T}J^0(t,\vec{x})\mathbf{T}^{-1} &= -i[\phi^\dagger(-t,\vec{x})\partial^0\phi(-t,\vec{x}) - \phi(-t,\vec{x})\partial^0\phi^\dagger(-t,\vec{x})] \\
&= J^0(-t,\vec{x}).
\end{aligned} \tag{4.41}$$

The space component can be analyzed in a similar manner, and indeed Eq. (4.17) is satisfied. The transformation properties of the creation and annihilation operators can be deduced from Eq. (4.40),

$$\mathbf{T}\mathbf{b}(\vec{p})\mathbf{T}^{-1} = \mathbf{b}(-\vec{p}), \quad \mathbf{T}\mathbf{d}(\vec{p})\mathbf{T}^{-1} = \mathbf{d}(-\vec{p}). \tag{4.42}$$

At this point the reader should notice a curious difference between how a wave function and a field operator transform under $\mathbf{T}$. In the previous chapter, we pointed out that a wave function transforms as

$$\Psi(t,\vec{x},\vec{p}) \xrightarrow{\mathbf{T}} \Psi^*(-t,\vec{x},-\vec{p}) \tag{4.43}$$

in contrast to Eq. (4.40). This will be addressed later in Sec. 4.7.

The quantization conditions Eq. (4.31) and Eq. (4.32) remain invariant under time reversal, whereas the Lagrangian transforms as

$$
\begin{aligned}
\mathbf{T}\mathscr{L}(t,\vec{x})\mathbf{T}^{-1} &= \frac{1}{2}(\partial_\mu - ieA^\mu(-t,\vec{x}))\phi^\dagger(-t,\vec{x})(\partial^\mu + ieA_\mu(-t,\vec{x}))\phi(-t,\vec{x}) \\
&\quad -\frac{1}{2}m^2\phi^\dagger(-t,\vec{x})\phi(-t,\vec{x}) \\
&= \frac{1}{2}(\bar{\partial}^\mu + ieA^\mu(-t,\vec{x}))\phi^\dagger(-t,\vec{x})(\bar{\partial}_\mu - ieA_\mu(-t,\vec{x}))\phi(-t,\vec{x}) \\
&\quad -\frac{1}{2}m^2\phi^\dagger(-t,\vec{x})\phi(-t,\vec{x}) \\
&= \mathscr{L}(-t,\vec{x}), \tag{4.44}
\end{aligned}
$$

where $\bar{\partial}$ denotes ∂ with the sign of time reversed. Thus the action remains invariant.

4.4 Spin-$\frac{1}{2}$ fields

The Lagrangian for a free spin-1/2 field reads as follows:

$$\mathscr{L} = \overline{\psi}(t,\vec{x})\left(i\gamma_\mu\partial^\mu - m\right)\psi(t,\vec{x}), \tag{4.45}$$

and the current is given by

$$J^\mu(t,\vec{x}) = \overline{\psi}(t,\vec{x})\gamma^\mu\psi(t,\vec{x}). \tag{4.46}$$

We postulate the canonical anti-commutation relations for the fields are given by

$$
\begin{aligned}
\{\psi_\alpha(t,\vec{x}), \psi_\beta^\dagger(t,\vec{y})\} &= \delta^3(\vec{x}-\vec{y})\delta_{\alpha\beta} \\
\{\psi_\alpha(t,\vec{x}), \psi_\beta(t,\vec{y})\} &= \{\psi_\alpha^\dagger(t,\vec{x}), \psi_\beta^\dagger(t,\vec{y})\} = 0. \tag{4.47}
\end{aligned}
$$

The spinor field can conveniently be expressed through its Fourier components:

$$\psi_\alpha(t,\vec{x}) = \int \frac{d^3\vec{p}}{(2\pi)^{\frac{3}{2}}} \sqrt{\frac{m}{E_p}} \sum_{s=\pm} [\mathbf{b}(\vec{p},s)u_\alpha(\vec{p},s)e^{-ip\cdot x} + \mathbf{d}^\dagger(\vec{p},s)v_\alpha(\vec{p},s)e^{ip\cdot x}], \tag{4.48}$$

with $\mathbf{b}[\mathbf{b}^\dagger]$ and $\mathbf{d}[\mathbf{d}^\dagger]$ denoting annihilation [creation] operators for particles and antiparticles, respectively.

The Dirac spinors $u(\vec{p},s)$ and $v(\vec{p},s)$ represent solutions to the Dirac equation in momentum space:

$$(\not{p}-m)u(\vec{p},s) = 0, \tag{4.49}$$

$$(\not{p}+m)v(\vec{p},s) = 0. \tag{4.50}$$

They are given by

$$u(\vec{p},+) = \sqrt{\frac{E+m}{2m}}\begin{pmatrix} 1 \\ 0 \\ \frac{p_z}{E+m} \\ \frac{p_x+ip_y}{E+m} \end{pmatrix} ; u(\vec{p},-) = \sqrt{\frac{E+m}{2m}}\begin{pmatrix} 0 \\ 1 \\ \frac{p_x-ip_y}{E+m} \\ \frac{-p_z}{E+m} \end{pmatrix}$$

$$v(\vec{p},+) = \sqrt{\frac{E+m}{2m}}\begin{pmatrix} \frac{p_z}{E+m} \\ \frac{p_x+ip_y}{E+m} \\ 1 \\ 0 \end{pmatrix} ; v(\vec{p},-) = \sqrt{\frac{E+m}{2m}}\begin{pmatrix} \frac{p_x-ip_y}{E+m} \\ \frac{-p_z}{E+m} \\ 0 \\ 1 \end{pmatrix},$$

$$(4.51)$$

with the spin quantized along the z axis.

The following relations can easily be read off from Eq. (4.51):

$$\begin{aligned} u(-\vec{p},s) &= \gamma_0 u(\vec{p},s) \\ v(-\vec{p},s) &= -\gamma_0 v(\vec{p},s) \end{aligned}$$

$$(4.52)$$

$$\begin{aligned} i\gamma^2 u(\vec{p},s)^* &= sv(\vec{p},-s) \\ i\gamma^2 v(\vec{p},s)^* &= -su(\vec{p},-s) \end{aligned}$$

$$(4.53)$$

$$\begin{aligned} \gamma_1\gamma_3 u(\vec{p},s)^* &= -su(-\vec{p},-s) \\ \gamma_1\gamma_3 v(\vec{p},s)^* &= -sv(-\vec{p},-s). \end{aligned}$$

$$(4.54)$$

Eq. (4.51) to Eq. (4.54), which were derived for free Dirac spinors, can be employed also in discussing *asymptotic* fields.

4.4.1 Parity

We want to find a transformation law for $\psi(t,\vec{x})$ which guarantees the transformation property of the current given in Eq. (4.17). Noting that

$$\gamma_0\gamma^\mu\gamma_0 = \gamma_\mu,$$

$$(4.55)$$

we make the ansatz:

$$\psi^{\mathbf{P}}(t,\vec{x}) \equiv \mathbf{P}\psi(t,\vec{x})\mathbf{P}^\dagger = \gamma_0\psi(t,-\vec{x}).$$

$$(4.56)$$

Then we have

$$\overline{\psi^{\mathbf{P}}} \equiv \mathbf{P}\overline{\psi}(t,\vec{x})\mathbf{P}^\dagger = \overline{\psi}(t,-\vec{x})\gamma_0,$$

$$(4.57)$$

and the current transforms as

$$\mathbf{P}J^\mu(t,\vec{x})\mathbf{P}^\dagger = \overline{\psi}(t,-\vec{x})\gamma_0\gamma^\mu\gamma_0\psi(t,-\vec{x}) = J_\mu(t,-\vec{x}),$$

$$(4.58)$$

as desired. Since

$$\mathbf{P}\mathscr{L}(t,\vec{x})\mathbf{P}^\dagger = \mathscr{L}(t,-\vec{x}),$$

$$(4.59)$$

the action remains invariant. Likewise for the anti-commutation relations:

$$
\begin{aligned}
\{\psi_\alpha^{\mathbf{P}}(t,\vec{x}),(\psi^{\mathbf{P}})_\beta^\dagger(t,\vec{y})\} &= (\gamma_0)_{\alpha\rho}(\gamma_0)_{\sigma\beta}\{\psi_\rho(t,-\vec{x}),\psi_\sigma^\dagger(t,-\vec{y})\} \\
&= (\gamma_0)_{\alpha\rho}(\gamma_0)_{\rho\beta}\delta^{(3)}(\vec{x}-\vec{y}) \\
&= \delta^{(3)}(\vec{x}-\vec{y})\delta_{\alpha\beta}.
\end{aligned}
\tag{4.60}
$$

Likewise,

$$
\{\psi_\alpha^{\mathbf{P}}(t,\vec{x}),\psi_\beta^{\mathbf{P}}(t,\vec{y})\} = 0 = \{(\psi^{\mathbf{P}})_\alpha^\dagger(t,\vec{x}),(\psi^{\mathbf{P}})_\beta^\dagger(t,\vec{y})\}.
\tag{4.61}
$$

The transformation properties of the fields can also – and maybe more transparently – be expressed in terms of the creation and annihilation operators. Applying the parity operator to the Dirac field, we have

$$
\begin{aligned}
\mathbf{P}\,\psi(t,\vec{x})\mathbf{P}^\dagger &= \int \frac{\mathrm{d}^3\vec{p}}{(2\pi)^{\frac{3}{2}}}\sqrt{\frac{m}{E_p}}\sum_{s=\pm}\Big[\mathbf{P}\,b(\vec{p},s)\mathbf{P}^\dagger u(\vec{p},s)e^{-ip\cdot x} \\
&\quad + \mathbf{P}\,d^\dagger(\vec{p},s)\mathbf{P}^\dagger v(\vec{p},s)e^{ip\cdot x}\Big] \\
&= \int \frac{\mathrm{d}^3\vec{p}}{(2\pi)^{\frac{3}{2}}}\sqrt{\frac{m}{E_p}}\sum_{s=\pm}\Big[\mathbf{P}\,b(-\vec{p},s)\mathbf{P}^\dagger u(-\vec{p},s)e^{-i(E_p t+\vec{p}\cdot\vec{x})} \\
&\quad + \mathbf{P}\,d^\dagger(-\vec{p},s)\mathbf{P}^\dagger v(-\vec{p},s)e^{i(E_p t+\vec{p}\cdot\vec{x})}\Big] \\
&= \int \frac{\mathrm{d}^3\vec{p}}{(2\pi)^{\frac{3}{2}}}\sqrt{\frac{m}{E_p}}\sum_{s=\pm}\Big[\mathbf{P}\,b(-\vec{p},s)\mathbf{P}^\dagger\gamma_0 u(\vec{p},s)e^{-i(E_p t+\vec{p}\cdot\vec{x})} \\
&\quad - \mathbf{P}\,d^\dagger(-\vec{p},s)\mathbf{P}^\dagger\gamma_0 v(\vec{p},s)e^{i(E_p t+\vec{p}\cdot\vec{x})}\Big].
\end{aligned}
\tag{4.62}
$$

To satisfy Eq. (4.56) with

$$
\begin{aligned}
\gamma_0\psi(t,-\vec{x}) &= \gamma_0\int \frac{\mathrm{d}^3\vec{p}}{(2\pi)^{\frac{3}{2}}}\sqrt{\frac{m}{E_p}}\sum_{s=\pm}[b(\vec{p},s)u(\vec{p},s)e^{-i(E_p t+\vec{p}\cdot\vec{x})} \\
&\quad + d^\dagger(\vec{p},s)v(\vec{p},s)e^{i(E_p t+\vec{p}\cdot\vec{x})}]\,,
\end{aligned}
\tag{4.63}
$$

we have to require

$$
\mathbf{P}\,b(\vec{p},s)\mathbf{P}^\dagger = b(-\vec{p},s),\quad \mathbf{P}\,d(\vec{p},s)\mathbf{P}^\dagger = -d(-\vec{p},s).
\tag{4.64}
$$

The second equality in Eq. (4.64) is particularly intriguing: we have found that fermions and antifermions carry *opposite* intrinsic parity.

4.4.2 Charge conjugation

In analogy with the spin-0 case, where $C\phi C^\dagger = \phi^\dagger$, we seek a transformation

$$
\psi^C(t,\vec{x}) = C\psi(t,\vec{x})C^\dagger = C\overline{\psi}^{\,\mathrm{tr}}(t,\vec{x}),
\tag{4.65}
$$

where the superscript tr denotes transposition, whereas T refers to time reversal. Noting that

$$
\overline{\psi^C}(t,\vec{x}) = \psi^{\mathrm{tr}}\gamma^0 C^\dagger\gamma^0,
\tag{4.66}
$$

we find

$$
\begin{aligned}
\mathbf{C}J^{\mu}(t,\vec{x})\mathbf{C}^{\dagger} &= \overline{\psi^{C}}(t,\vec{x})\gamma^{\mu}\psi^{C}(t,\vec{x}) \\
&= \psi_{\alpha}(t,\vec{x})[\gamma^{0}C^{\dagger}\gamma^{0}\gamma^{\mu}C]_{\alpha\beta}\overline{\psi}_{\beta}(t,\vec{x}) \\
&= -\overline{\psi}_{\beta}(t,\vec{x})[\gamma^{0}C^{\dagger}\gamma^{0}\gamma^{\mu}C]_{\alpha\beta}\psi_{\alpha}(t,\vec{x}) \\
&= -J^{\mu}(t,\vec{x}),
\end{aligned}
\tag{4.67}
$$

provided that

$$
\gamma^{0}C^{\dagger}\gamma^{0}\gamma_{\mu}C = \gamma_{\mu}^{\text{tr}}.
\tag{4.68}
$$

So, if

$$
C = i\gamma^{2}\gamma^{0},
\tag{4.69}
$$

Eq. (4.68) is satisfied, and we recover the transformation property of the current under **C** given in Eq. (4.17).

With our choice of phase, $C^{2} = -1$ and

$$
C = -C^{\dagger} = -C^{-1} = -C^{\text{tr}}.
\tag{4.70}
$$

The free Lagrangian Eq. (4.45) is invariant under **C**.

$$
\begin{aligned}
\mathbf{C}\mathscr{L}\mathbf{C}^{\dagger} &= -\psi(t,\vec{x})_{\alpha}[C^{\dagger}\left(i\gamma_{\mu}\partial^{\mu} - m\right)C]_{\alpha\beta}\overline{\psi}_{\beta}(t,\vec{x}) \\
&= \overline{\psi}(t,\vec{x})_{\beta}[C^{\dagger}\left(-i\gamma_{\mu}\overleftarrow{\partial}^{\mu} - m\right)C]_{\alpha\beta}\psi_{\alpha}(t,\vec{x}) \\
&= \mathscr{L}.
\end{aligned}
\tag{4.71}
$$

Through an integration by parts in evaluating the action, we see that it is invariant under **C**.

We find again that the anticommutators stay the same:

$$
\begin{aligned}
\{\psi_{\alpha}^{C}(t,\vec{x}),(\psi^{C})_{\beta}^{\dagger}(t,\vec{y})\} &= (C\gamma_{0})_{\alpha\rho}(\gamma_{0}C^{\dagger})_{\sigma\beta}\{\psi_{\rho}^{*}(t,\vec{x}),\psi_{\sigma}^{\text{tr}}(t,\vec{y})\} \\
&= (C\gamma_{0})_{\alpha\rho}(\gamma_{0}C^{\dagger})_{\rho\beta}\delta^{(3)}(\vec{x}-\vec{y}) \\
&= \delta^{(3)}(\vec{x}-\vec{y})\delta_{\alpha\beta},
\end{aligned}
\tag{4.72}
$$

and likewise

$$
\{\psi_{\alpha}^{C}(t,\vec{x}),\psi_{\beta}^{C}(t,\vec{y})\} = 0 = \{(\psi^{C})_{\alpha}^{\dagger}(t,\vec{x}),(\psi^{C})_{\beta}^{\dagger}(t,\vec{y})\}.
\tag{4.73}
$$

The transformation properties of the creation and annihilation operators are deduced by considering the expression in momentum space:

$$
\begin{aligned}
\mathbf{C}\psi(t,\vec{x})\mathbf{C}^\dagger &= \sum_s \int \frac{\mathrm{d}^3 p}{(2\pi)^{\frac{3}{2}}} \sqrt{\frac{m}{E_p}} [\mathbf{C}\mathbf{b}(\vec{p},s)\mathbf{C}^\dagger u(\vec{p},s)e^{-ip\cdot x} \\
&\quad + \mathbf{C}\mathbf{d}^\dagger(\vec{p},s)\mathbf{C}^\dagger v(\vec{p},s)e^{ip\cdot x}] \\
&= \sum_s \int \frac{\mathrm{d}^3 p}{(2\pi)^{\frac{3}{2}}} \sqrt{\frac{m}{E_p}} [\mathbf{b}^\dagger(\vec{p},s)i\gamma^2 u^*(\vec{p},s)e^{ip\cdot x} \\
&\quad + \mathbf{d}(\vec{p},s)i\gamma^2 v^*(\vec{p},s)e^{-ip\cdot x}] \\
&= \sum_s \int \frac{\mathrm{d}^3 p}{(2\pi)^{\frac{3}{2}}} \sqrt{\frac{m}{E_p}} [\mathbf{b}^\dagger(\vec{p},s)(s)v(\vec{p},-s)e^{ip\cdot x} \\
&\quad + \mathbf{d}(\vec{p},s)(-s)u(\vec{p},-s)e^{-ip\cdot x}].
\end{aligned}
\tag{4.74}
$$

This leads to the transformation property:

$$
\mathbf{C}\mathbf{b}(\vec{p},s)\mathbf{C}^\dagger = s\mathbf{d}(\vec{p},-s).
\tag{4.75}
$$

4.4.3 Time reversal

Let us consider the time reversal operator

$$
\psi^{\mathbf{T}}(t,\vec{x}) = \mathbf{T}\psi(t,\vec{x})\mathbf{T}^{-1} = U\psi(-t,\vec{x}).
\tag{4.76}
$$

The electromagnetic current transforms as

$$
\begin{aligned}
\mathbf{T}J^\mu(t,\vec{x})\mathbf{T}^{-1} &= \mathbf{T}\overline{\psi}(t,\vec{x})\mathbf{T}^{-1}\gamma^{\mu*}\mathbf{T}\psi(t,\vec{x})\mathbf{T}^{-1} \\
&= \overline{\psi}(-t,\vec{x})U^\dagger \gamma^{\mu*} U\psi(-t,\vec{x}).
\end{aligned}
\tag{4.77}
$$

So, we need to find a matrix $\mathbf{T} = UK$ such that

$$
U\gamma^{\mu*} U^\dagger = \gamma_\mu.
\tag{4.78}
$$

A simple computation shows that

$$
U = \gamma^1 \gamma^3
\tag{4.79}
$$

satisfies Eq. (4.78).

Since

$$
\begin{aligned}
\mathbf{T}\mathscr{L}\mathbf{T}^{-1} &= \overline{\psi}(-t,\vec{x})U^\dagger \left(-i\gamma_\mu^* \partial^\mu - m\right) U\psi(-t,\vec{x}) \\
&= \overline{\psi}(-t,\vec{x})\left(-i\gamma^\mu \partial^\mu - m\right)\psi(-t,\vec{x}) \\
&= \mathscr{L}(-t,\vec{x}),
\end{aligned}
\tag{4.80}
$$

the action remains invariant under **T**. Likewise for the quantization conditions;

Table 4.1. Summary of how charged fields transform under **C**, **P** and **T**.

F	$PFP^\dagger$	$CFC^\dagger$	TFT^{-1}	$CPFCP^\dagger$	$CPTFCPT^{-1}$
$S^+(t,\vec{x})$	$S^+(t,-\vec{x})$	$S^-(t,\vec{x})$	$S^+(-t,\vec{x})$	$S^-(t,-\vec{x})$	$S^-(-t,-\vec{x})$
$P^+(t,\vec{x})$	$-P^+(t,-\vec{x})$	$P^-(t,\vec{x})$	$P^+(-t,\vec{x})$	$-P^-(t,-\vec{x})$	$-P^-(-t,-\vec{x})$
$V_\mu^+(t,\vec{x})$	$V^{+\mu}(t,-\vec{x})$	$-V_\mu^-(t,\vec{x})$	$V^{+\mu}(-t,\vec{x})$	$-V^{-\mu}(t,-\vec{x})$	$-V_\mu^-(-t,-\vec{x})$
$A_\mu^+(t,\vec{x})$	$-A^{+\mu}(t,-\vec{x})$	$A_\mu^-(t,\vec{x})$	$A^{+\mu}(-t,\vec{x})$	$-A^{-\mu}(t,-\vec{x})$	$-A_\mu^-(-t,-\vec{x})$

Eq. (4.47) can be easily verified. We obtain

$$\{\psi_\alpha^T(t,\vec{x}),(\psi^T)_\beta^\dagger(t,\vec{y})\} = (\gamma^1\gamma^3)_{\alpha\sigma}(\gamma^3\gamma^1)_{\kappa\beta}\{\psi_\sigma(-t,\vec{x}),\psi_\kappa^\dagger(-t,\vec{y})\}$$
$$= (\gamma^1\gamma^3)_{\alpha\sigma}(\gamma^3\gamma^1)_{\sigma\beta}\delta^{(3)}(\vec{x}-\vec{y}) = \delta^{(3)}(\vec{x}-\vec{y})\delta_{\alpha\beta} \quad (4.81)$$

$$\{\psi_\alpha^T(t,\vec{x}),\psi_\beta^T(t,\vec{y})\} = 0 = \{(\psi^T)_\alpha^\dagger(t,\vec{x}),(\psi^T)_\beta^\dagger(t,\vec{y})\}. \quad (4.82)$$

In momentum space, we expect

$$\begin{aligned}
\mathbf{T}\psi(t,\vec{x})\mathbf{T}^{-1} &= \sum_s \int \frac{d^3p}{(2\pi)^{\frac{3}{2}}}\sqrt{\frac{m}{E_p}}[\mathbf{Tb}(\vec{p},s)\mathbf{T}^{-1}u^*(\vec{p},s)e^{ipx} \\
&\quad + \mathbf{Td}^\dagger(\vec{p},s)\mathbf{T}^{-1}v^*(\vec{p},s)e^{-ipx}] \\
&= \sum_s \gamma^1\gamma^3 s \int \frac{d^3p}{(2\pi)^{\frac{3}{2}}}\sqrt{\frac{m}{E_p}}[\mathbf{Tb}(\vec{p},s)\mathbf{T}^{-1}u(-\vec{p},-s)e^{ipx} \\
&\quad + \mathbf{Td}^\dagger(\vec{p},s)\mathbf{T}^{-1}v(-\vec{p},-s)e^{-ipx}]. \quad (4.83)
\end{aligned}$$

Using a relation given in Eq. (4.76),

$$\begin{aligned}
\psi^T(t',\vec{x}) &= \sum_s \int \frac{d^3p}{(2\pi)^{\frac{3}{2}}}\sqrt{\frac{m}{E_p}}\gamma^1\gamma^3[\mathbf{b}(\vec{p},s)u(\vec{p},s)e^{-i(-E_pt-\vec{p}\cdot\vec{x})} \\
&\quad + \mathbf{d}(\vec{p},s)^\dagger v(\vec{p},s)e^{i(-E_pt-\vec{p}\cdot\vec{x})}] \\
&= \gamma^1\gamma^3 \sum_s \int \frac{d^3p}{(2\pi)^{\frac{3}{2}}}\sqrt{\frac{m}{E_p}}[\mathbf{b}(-\vec{p},-s)u(-\vec{p},-s)e^{-i(-E_pt+\vec{p}\cdot\vec{x})} \\
&\quad + \mathbf{d}^\dagger(-\vec{p},-s)v(-\vec{p},-s)e^{i(-E_pt+\vec{p}\cdot\vec{x})}]. \quad (4.84)
\end{aligned}$$

Comparing these two equations, we obtain

$$\mathbf{Tb}(\vec{p},s)\mathbf{T}^{-1} = s\mathbf{b}(-\vec{p},-s), \qquad \mathbf{Td}(\vec{p},s)\mathbf{T}^{-1} = s\mathbf{d}(-\vec{p},-s). \quad (4.85)$$

4.5 **CP** and **CPT** transformations

The transformation properties of scalar, pseudoscalar, vector and axialvector fields – S, P, V_μ and A_μ, respectively – are summarized in Table 4.1. Next we

Table 4.2.   Summary of how fermion bilinears transform under **C**, **P** and **T**.

Γ	$\Gamma^{\mathbf{P}}$	$\Gamma^{\mathbf{C}}$	$\Gamma^{\mathbf{T}}$	$\Gamma^{\mathbf{CP}}$	$\Gamma^{\mathbf{CPT}}$
	$\gamma_0\Gamma\gamma_0$	$(\gamma^2\gamma^0\Gamma\gamma^2\gamma^0)^{tr}$	$-\gamma^1\gamma^3\Gamma^*\gamma^1\gamma^3$	$-(\gamma^2\Gamma\gamma^2)^{tr}$	$(\gamma_5\gamma_0\Gamma^\dagger\gamma_0\gamma_5)^{tr}$
1	1	1	1	1	1
γ_5	$-\gamma_5$	γ_5	γ_5	$-\gamma_5$	$-\gamma_5$
γ_μ	γ^μ	$-\gamma_\mu$	γ^μ	$-\gamma^\mu$	$-\gamma_\mu$
$\gamma_\mu\gamma_5$	$-\gamma^\mu\gamma_5$	$\gamma_\mu\gamma_5$	$\gamma^\mu\gamma_5$	$-\gamma^\mu\gamma_5$	$-\gamma_\mu\gamma_5$
$\sigma_{\mu\nu}$	$\sigma^{\mu\nu}$	$-\sigma_{\mu\nu}$	$-\sigma^{\mu\nu}$	$-\sigma^{\mu\nu}$	$\sigma_{\mu\nu}$

consider how fermion bilinears transform under these discrete symmetries. First define

$$\overline{\psi}_a^{\mathbf{P}}(t,\check{x})\Gamma_i\psi_b^{\mathbf{P}}(t,\check{x}) = \overline{\psi}_a(t,-\check{x})\Gamma_i^{\mathbf{P}}\psi_b(t,-\check{x});$$
$$\overline{\psi}_a^{\mathbf{C}}(t,\check{x})\Gamma_i\psi_b^{\mathbf{C}}(t,\check{x}) = \overline{\psi}_b(t,\check{x})\Gamma_i^{\mathbf{C}}\psi_a(t,\check{x});$$
$$\overline{\psi}_a^{\mathbf{T}}(t,\check{x})\Gamma_i\psi_b^{\mathbf{T}}(t,\check{x}) = \overline{\psi}_a(-t,\check{x})\Gamma_i^{\mathbf{T}}\psi_b(-t,\check{x});$$
$$\overline{\psi}_a^{\mathbf{CP}}(t,\check{x})\Gamma_i\psi_b^{\mathbf{CP}}(t,\check{x}) = \overline{\psi}_b(t,-\check{x})\Gamma_i^{\mathbf{CP}}\psi_a(t,-\check{x});$$
$$\overline{\psi}_a^{\mathbf{CPT}}(t,\check{x})\Gamma_i\psi_b^{\mathbf{CPT}}(t,\check{x}) = \overline{\psi}_b(-t,-\check{x})\Gamma_i^{\mathbf{CPT}}\psi_a(-t,-\check{x}), \tag{4.86}$$

where Γ stands for all possible combinations of γ matrices. Table 4.2 summarizes these transformation properties. Note that we can identify

$$\overline{\psi}\psi = S, \qquad \overline{\psi}\gamma_\mu\psi = V_\mu,$$
$$\overline{\psi}\gamma_5\psi = P, \qquad \overline{\psi}\gamma_\mu\gamma_5\psi = A_\mu. \tag{4.87}$$

The transformation properties of a Lagrangian written in terms of bosons is thus precisely the same as that for a Lagrangian involving fermion bilinears. We shall, therefore, confine our discussion to the transformation properties of a Lagrangian expressed through boson fields. The notation is simpler that way and the essence of the argument becomes more transparent.

Let us consider the toy interaction Lagrangian

$$\mathscr{L}_T = aV_\mu^+(t,\check{x})V^{\mu,-}(t,\check{x}) + bA_\mu^+(t,\check{x})A^{\mu,-}(t,\check{x})$$
$$+cV_\mu^+(t,\check{x})A^{\mu,-}(t,\check{x}) + c^*A_\mu^+(t,\check{x})V^{\mu,-}(t,\check{x}), \tag{4.88}$$

which under **CP** transforms as follows:

$$\mathbf{CP}\mathscr{L}_T\mathbf{CP}^\dagger = aV^{\mu,-}(t,-\check{x})V_\mu^+(t,-\check{x}) + bA^{\mu,-}(t,-\check{x})A_\mu^+(t,-\check{x})$$
$$+cV^{\mu,-}(t,-\check{x})A_\mu^+(t,-\check{x}) + c^*A^{\mu,-}(t,-\check{x})V_\mu^+(t,-\check{x}), \tag{4.89}$$

i.e., **CP** *is conserved if the coupling parameter c is real!* Similar statements can be made for Lagrangians written in terms of scalar and pseudoscalar boson fields.

(See Problem 4.12). We can generalize this and consider a Hamiltonian:

$$\mathscr{L} = \sum_i a_i \mathcal{O}_i + \text{h.c.} \tag{4.90}$$

where $\mathbf{CP}\mathcal{O}_i\mathbf{CP}^\dagger = \mathcal{O}_i^\dagger$ and **CP** is conserved for a_i real.

Reconciling the demands of quantum mechanics with those of special relativity within a *local* description requires the existence of antiparticles. A considerably stronger statement can actually be made concerning the relationship between particles and antiparticles: the combined transformation **CPT** can always be defined – as an antiunitary operator – in such a way for a local quantum field theory that it represents a symmetry [17], i.e.,

$$\mathbf{CPT}\,\mathscr{L}(t,\vec{x})(\mathbf{CPT})^{-1} = \mathscr{L}(-t,-\vec{x}). \tag{4.91}$$

This important theorem can be proven rigorously in axiomatic field theory based on the assumptions of

- Lorentz invariance,

- the existence of a unique vacuum state, and

- weak local commutativity obeying the 'right' statistics.

The Lagrangian of Eq. (4.88) transforms as

$$\begin{aligned}
\mathbf{CPT}&\mathscr{L}_T\mathbf{CPT}^{-1}\\
&= aV^{\mu-}(-t,-\vec{x})V_\mu^+(-t,-\vec{x}) + bA^{\mu-}(-t,-\vec{x})A_\mu^+(-t,-\vec{x})\\
&\quad +c^*V^{\mu-}(-t,-\vec{x})A_\mu^+(-t,-\vec{x}) + cA^{\mu-}(-t,-\vec{x})V_\mu^+(-t,-\vec{x})
\end{aligned} \tag{4.92}$$

i.e., **CPT** is indeed conserved, no matter what the coupling parameters a, b and c are.

The argument is easily repeated for fermions by noting again that each bosonic field can be written in terms of fermionic bilinears which transform in exactly in the same way under **C**, **P** and **T**; likewise for more realistic Lagrangians.

4.6 Some consequences of the CPT theorem

Although the proof of this theorem, at least for the basic cases, appears rather simple, its consequences are far-reaching. The most celebrated ones are:

- The equality of masses and total widths or lifetimes for particles P and antiparticles $\overline{P}$:

$$M(P) = M(\overline{P}), \quad \Gamma(P) = \Gamma(\overline{P}), \tag{4.93}$$

which are easily proved:

$$
\begin{aligned}
M(P) &= \langle P|H|P\rangle \\
&= \langle P|(\mathbf{CPT})^\dagger \mathbf{CPT}\, H(\mathbf{CPT})^{-1}\mathbf{CPT}\,|P\rangle^* \\
&= \langle \overline{P}|\mathbf{CPT}H\mathbf{CPT}^{-1}|\overline{P}\rangle^* \\
&= \langle \overline{P}|H|\overline{P}\rangle^* = M(\overline{P}).
\end{aligned}
\tag{4.94}
$$

− Under time reversal, an incoming spherical wave transforms to an outgoing spherical wave. So, under time reversal multi-particle states transform like

$$
\mathbf{T}|p_1, p_2, \ldots ; \text{out}\rangle = |-p_1, -p_2, \ldots ; \text{in}\rangle.
\tag{4.95}
$$

$$
\begin{aligned}
\Gamma(P) &= 2\pi \sum_f \delta(M_P - E_f)|\langle f; \text{out}|H_{\text{decay}}|P\rangle|^2 \\
&= 2\pi \sum_f \delta(M_P - E_f)|\langle f; \text{out}|(\mathbf{CPT})^\dagger \mathbf{CPT} \\
&\quad\ H_{\text{decay}}(\mathbf{CPT})^{-1}\mathbf{CPT}\,|P\rangle^*|^2 \\
&= 2\pi \sum_f \delta(M_P - E_f)|\langle \overline{f}; \text{in}|H_{\text{decay}}|\overline{P}\rangle|^2 \\
&= 2\pi \sum_f \delta(M_P - E_f)|\langle \overline{f}; \text{out}|H_{\text{decay}}|\overline{P}\rangle|^2 \\
&= \Gamma(\overline{P}).
\end{aligned}
\tag{4.96}
$$

We have used the fact that both 'in' and 'out' states form complete sets of states:

$$
\sum_f |f; \text{in}\rangle\langle f; \text{in}| = \sum_f |f; \text{out}\rangle\langle f; \text{out}| = 1.
\tag{4.97}
$$

Proofs for these relations are quite elementary. Nevertheless their contents are far from trivial, namely that these equalities are guaranteed already by **CPT** symmetry irrespective of whether **CP** is conserved or not.

Noting that the completeness condition (see Problem 3.8) was used in the proof given above, we can deduce the following: all decay channels can be divided into classes F_i containing final states $f_\alpha^{(i)}$ that are mutually distinct under the strong interactions; i.e., for $f_\alpha^{(i)} \in F_i$ and $f_\rho^{(j)} \in F_j$ with $i \neq j$ neither $f_\alpha^{(i)} \to f_\rho^{(j)}$ nor $f_\rho^{(j)} \to f_\alpha^{(i)}$ can be driven by the strong interactions, irrespective of α and ρ. **CPT** invariance then implies

$$
\sum_{f_\alpha^{(i)} \in F_i} \Gamma(P \to f_\alpha^{(i)}) = \sum_{\overline{f}_\alpha^{(i)} \in \overline{F}_i} \Gamma(\overline{P} \to \overline{f}_\alpha^{(i)}),
\tag{4.98}
$$

where $\overline{f}_\alpha^{(i)}$ and $\overline{F}_i$ are **CP** conjugate to $f_\alpha^{(i)}$ and F_i, respectively.

– As an example, let us neglect *weak* and *electromagnetic* final state interactions. Then, 'in' and 'out' states cannot be distinguished, and we have a weaker form of **CPT** invariance. Under this assumption **CPT** implies not only that the *total* sum of all partial decay rates be the same for particles and antiparticles (Problem 4.10); it actually tells us that the sums over certain *sub*sets of all decay rates have to coincide for particles and antiparticles. For example, **CPT** symmetry already enforces

$$\Gamma(\mu^- \to e^- \bar{v}_e v_\mu) = \Gamma(\mu^+ \to e^+ v_e \bar{v}_\mu) \qquad (4.99)$$

beyond

$$\Gamma_{tot}(\mu^-) = \Gamma_{tot}(\mu^+). \qquad (4.100)$$

Likewise,

$$\Gamma(K^- \to e^- \bar{v} \pi^0) = \Gamma(K^+ \to e^+ v \pi^0), \qquad (4.101)$$

$$\Gamma(K^- \to \pi^- \pi^0) = \Gamma(K^+ \to \pi^+ \pi^0), \qquad (4.102)$$

$$\Gamma(K^0 \to \pi^+ \pi^- + \pi^0 \pi^0) = \Gamma(\overline{K}^0 \to \pi^+ \pi^- + \pi^0 \pi^0). \qquad (4.103)$$

On the other hand, we *cannot* ignore *strong* final state interaction and the equality of $\Gamma(K^0 \to \pi^+ \pi^-)$ and $\Gamma(\overline{K}^0 \to \pi^+ \pi^-)$ is *not* guaranteed by **CPT** invariance.

• Magnetic dipole moments are equal in magnitude, but opposite in sign for particles and antiparticles (Problem 4.10)

$$\mu_{mag}(P) = -\mu_{mag}(\overline{P}). \qquad (4.104)$$

4.7 ♠ Back to first quantization ♠

With the discrete transformations implemented in a quantum field theory, we can translate them easily into the language of quantum mechanics, i.e., a theory with first quantization only, as expressed through the Dirac or Klein–Gordon equations etc. for wave functions. The groundstate is assumed to be invariant under **C**, **P** and **T**:

$$\mathbf{C}|0\rangle = |0\rangle, \quad \mathbf{P}|0\rangle = |0\rangle, \quad \mathbf{T}|0\rangle = |0\rangle. \qquad (4.105)$$

The transformation properties for an *n*-particle state are determined by the transformation properties of the appropriate creation operators; i.e., for a single electron [positron] state with momentum p and spin s:

$$|e^-(\vec{p}, s)\rangle = \mathbf{b}^\dagger(\vec{p}, s)|0\rangle, \qquad [|e^+(\vec{p}, s)\rangle = \mathbf{d}^\dagger(\vec{p}, s)|0\rangle].$$

The wavefunction Ψ describing such a one-particle state produced by the field operator ψ is then given by

$$\Psi(t, \vec{x}) \equiv \langle 0|\psi(t, \vec{x})|\vec{p}, s\rangle, \qquad (4.106)$$

which is easily generalized to the case of an n-particle state

$$|\vec{p}_1, s_1; \ldots; \vec{p}_n, s_n\rangle.$$

A transformation is described by an operator $\mathcal{O}$ acting in the following way:

$$\Psi(t, \vec{x}) = \langle 0|\psi(t, \vec{x})|\vec{p}_1, s_1; \ldots; \vec{p}_n, s_n\rangle$$
$$\xrightarrow{\mathcal{O}} \Psi^{\mathcal{O}}(t, \vec{x}) = \langle 0|\mathcal{O}\psi(t, \vec{x})\mathcal{O}^{-1}|\vec{p}_1, s_1; \ldots; \vec{p}_n, s_n\rangle. \tag{4.107}$$

In quantum mechanics we have found

$$\Psi(t, \vec{x}) \xrightarrow{\mathbf{T}} \Psi^*(-t, \vec{x}). \tag{4.108}$$

For time reversal symmetry, using its antiunitary property, we obtain

$$\begin{aligned}
\Psi(t, \vec{x}) &= \langle 0|\mathbf{T}^\dagger \mathbf{T}\psi(t, \vec{x})\mathbf{T}^{-1}\mathbf{T}|\vec{p}_1, s_1; \ldots; \vec{p}_n, s_n\rangle^* \\
&\longrightarrow \langle 0|\psi(\vec{x}, -t)| - \vec{p}_1, -s_1; \ldots; -\vec{p}_n, -s_n\rangle^* \\
&= \Psi^*(-t, \vec{x}),
\end{aligned} \tag{4.109}$$

which is consistent with Eq. (4.108)!

A similar relationhip between the operator equation Eq. (4.56) and Eq. (3.30) can easily be obtained.

4.8 ♠ Phases and phase conventions for C and P ♠

From the discussion of Kramer's degeneracy we have found that $\mathbf{T}^2$ must be ± 1, so we are not free to change the phase of $\mathbf{T}$. But no such constraint exist for **C** and **P**.

We have been somewhat cavalier about the phases for **C** and **P** up to now. For example, Eq. (4.56), and Eq. (4.65) can be replaced by

$$\begin{aligned}
\mathbf{P}\psi(t, \vec{x})\mathbf{P}^\dagger &= \eta^P \gamma^0 \psi(t, -\vec{x}) \\
\mathbf{C}\psi(t, \vec{x})\mathbf{C}^\dagger &= \eta^C C \overline{\psi}^{tr}(t, \vec{x}),
\end{aligned} \tag{4.110}$$

where η_P and η_C are arbitrary phases; similarly for fields with spin 0 or 1. It can easily be seen that these phases do not change any of the conclusions reached above in an essential way.

There is a reason for not worrying too much about the phases other than innate sloppiness. For there is no unique definition of, say, parity: it can be redefined by combining it with any conserved internal quantum number like baryon number B, lepton number L or electric charge Q:

$$\mathbf{P}' = \mathbf{P}\, e^{i(bB + lL + qQ)}, \tag{4.111}$$

with b, l and q being arbitrary real numbers (as long as interactions do not change B, L, or Q). (See Problem 4.4.) **P** and **P**$'$ can lay equivalent claims to be *the* parity operator.

For the neutral pion with $B = L = Q = 0$ the two definitions coincide; its parity is thus well defined ab initio (and turns out to be negative). The situation is intrinsically different for the proton p, neutron n, electron e or the charged pion: $\mathbf{P}$ and $\mathbf{P}'$ are not equivalent for those states. This apparent vice can however be turned into a virtue. By judiciously adjusting the parameters b, q and l we can assign an intrinsic parity of $+1$ to p, n and e. With this definition, the intrinsic parity of the charged pion can be determined from [7]

$$\pi^- + d \rightarrow n + n, \tag{4.112}$$

where π^- is captured in an S wave state. Since π^- is spin-less and deutrium has $J = 1$, the parity of the initial state is equal to the intrinsic parity of π^-. The final two-neutron state forms a $^3P_1$ configuration which carries odd parity; thus

$$\mathbf{P}|\pi^-\rangle = -|\pi^-\rangle. \tag{4.113}$$

The parity of π^0 can be detemined by analyzing correlations between the polarizations of the two gamma rays in $\pi^0 \rightarrow \gamma\gamma$ decay [7]. Experiments give

$$\mathbf{P}|\pi^0\rangle = -|\pi^0\rangle. \tag{4.114}$$

This is gratifying since it is consistent with combining π^+, π^0 and π^- into an iso-triplet.

For charge conjugation transformation, we recall that $\mathbf{C}|\gamma\rangle = -|\gamma\rangle$. Since the decay

$$\pi^0 \rightarrow \gamma\gamma \tag{4.115}$$

is the major decay mode of π^0, we have

$$\mathbf{C}|\pi^0\rangle = +|\pi^0\rangle \tag{4.116}$$

and the decay $\pi^0 \rightarrow 3\gamma$ is forbidden by charge conjugation symmetry.

4.9 ♠ Internal symmetries ♠

Continuous internal symmetries like isospin and $SU(3)_{\mathrm{Fl}}$ introduce other charge-like quantum numbers that change under charge conjugation. That is described in terms of a multiplet of complex scalar fields, ϕ_i, where 'i' labels elements of the group representation under study: the charge conjugation operation can be defined as

$$\mathbf{C}\phi_i(t, \vec{x})\mathbf{C}^\dagger = \eta_i^C \phi_i^\dagger(t, \vec{x}) \tag{4.117}$$

i.e., we can choose a different phase η_i^C for each field. This freedom is important, for example, when we consider the following $SU(3)_{Fl}$ octet fields:

$$\pi^{\pm} = \frac{1}{\sqrt{2}}(\phi_1 \mp i\phi_2), \quad K^{\pm} = \frac{1}{\sqrt{2}}(\phi_4 \mp i\phi_5),$$

$$K^0 = \frac{1}{\sqrt{2}}(\phi_6 - i\phi_7), \quad \overline{K} = \frac{1}{\sqrt{2}}(\phi_6 + i\phi_7),$$

$$\pi^0 = \phi_3, \qquad\qquad \eta = \phi_8. \tag{4.118}$$

Note that (K^+, K^0) is an $SU(2)$ doublet. Under an $SU(2)$ rotation around an axis $\hat{n}$ by an angle θ, a doublet representation transforms as follows:

$$\begin{pmatrix} \phi_+ \\ \phi_- \end{pmatrix} \rightarrow e^{i\sum_i \theta \sigma^i n_i} \begin{pmatrix} \phi_+ \\ \phi_- \end{pmatrix}$$

$$i\sigma^2 \begin{pmatrix} \phi_+^* \\ \phi_-^* \end{pmatrix} \rightarrow e^{i\sum_i \sigma^i \theta_i} i\sigma^2 \begin{pmatrix} \phi_+^* \\ \phi_-^* \end{pmatrix}. \tag{4.119}$$

It turns out to be more convenient to introduce an extra minus sign into the definition of the charge conjugated doublet:

$$\begin{pmatrix} K^+ \\ K^0 \end{pmatrix}^C = \begin{pmatrix} -\overline{K}^0 \\ K^- \end{pmatrix}, \tag{4.120}$$

i.e., making use of the freedom defined in Eq. (4.117), we can adopt a different phase convention for the charge conjugate of (K^0, K^+) since they form a distinct iso-doublet.

$$\mathbf{C}|K^+\rangle = +|K^-\rangle$$
$$\mathbf{C}|K^0\rangle = -|\overline{K}^0\rangle. \tag{4.121}$$

For our later discussion the concept of G-parity will turn out to be very powerful. Consider the operator $exp[-i\pi T_i]$, where T_i is a generator of isospin rotations around the axis i. Under this rotation,

$$e^{-i\pi T_2} \begin{pmatrix} \phi_1 \\ \phi_2 \\ \phi_3 \end{pmatrix} = \begin{pmatrix} -\phi_1 \\ \phi_2 \\ -\phi_3 \end{pmatrix}. \tag{4.122}$$

This has the effect of

$$e^{-i\pi T_2}|\pi^{\pm}\rangle = -|\pi^{\mp}\rangle, \quad e^{-i\pi T_2}|\pi^0\rangle = -|\pi^0\rangle. \tag{4.123}$$

So, defining a G-parity as $G = \mathbf{C}exp[-i\pi T_2]$, we find

$$G|\pi^{\pm}\rangle = -|\pi^{\pm}\rangle, \quad G|\pi^0\rangle = -|\pi^0\rangle. \tag{4.124}$$

It follows immediately that a state with n pions is an eigenstate of G-parity:

$$G|\pi_1 \dots \pi_n\rangle = (-1)^n |\pi_1 \dots \pi_n\rangle. \tag{4.125}$$

With the strong interaction obeying isospin and charge invariance, G-parity is also conserved.

4.10 The role of final state interactions

4.10.1 T *invariance and Watson's theorem*

A strangeness changing transition has to be *initiated* by weak forces which can be treated perturbatively. On the other hand, the final state is shaped largely by strong dynamics which is mostly beyond the reach of a perturbative description. Yet, even so, we can make some reliable theoretical statements based on symmetry considerations.

On the one hand, the G-parity introduced above tells us that a state of an *even* number of pions cannot evolve *strongly* into a state with an *odd* number. The two-step process where a K meson decays *weakly* into two pions which subsequently interact *strongly* to evolve into a three pion final state therefore cannot happen:

$$K \xrightarrow{H_{\text{weak}}} 2\pi \xrightarrow{H_{\text{str}}} 3\pi \ . \tag{4.126}$$

On the other hand, the two pions emerging from the weak decay $K \to \pi\pi$ are not asymptotic states yet; due to the strong interactions they will undergo rescattering before they lose sight of each other. Deriving the properties of these strong final state interactions from first principles is beyond our present computational capabilities. However, we can relate some of their properties reliably to other observables.

First we assume the weak interactions to be invariant under time reversal:

$$\mathbf{T} H_W \mathbf{T}^{-1} = H_W.$$

The amplitude for $K^0 \to 2\pi$ is then denoted by

$$\langle (2\pi)_I^{\text{out}}|H_W|K^0 \rangle = |A_I|e^{i\phi_I}, \tag{4.127}$$

where I denotes the isospin of the 2π state. We will show now that the phase ϕ_I generated by the strong interactions actually coincides with the S wave $\pi\pi$ phase shift δ_I taken at energy M_K [19]. That is, the amplitude is real except for the fact that the two pions interact before becoming asymptotic states.

With $\mathbf{T}$ being an *antiunitary* operator and using Eq. (3.58) we can write down:

$$\begin{aligned}
\langle (\pi\pi)_I ; \text{out}|H_W|K \rangle &= \langle (\pi\pi)_I ; \text{out}|\mathbf{T}^\dagger \mathbf{T} \, H_W \mathbf{T}^{-1} \mathbf{T} \, |K \rangle^* \\
&= \langle (\pi\pi)_I ; \text{in}|H_W|K \rangle^* ,
\end{aligned} \tag{4.128}$$

since for a single particle state – K in this case – there is no distinction between an in and an out state. Next we insert a complete set of out states:

$$\langle (\pi\pi)_I ; \text{out}|H_W|K \rangle = \sum_n (\langle (\pi\pi)_I ; \text{in}|n; \text{out} \rangle \langle n; \text{out}|H_W|K \rangle)^*. \tag{4.129}$$

$\langle n; \text{out}|(\pi\pi)_I ; \text{in} \rangle$ is an S matrix element which contains the energy momentum conserving delta function.

We can now analyse the possible final states. Kinematically, the only allowed hadronic states in the sum over n are 2π and 3π combinations. G-parity, which

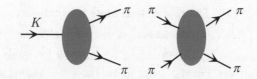

Fig. 4.1.   Only 2π states contribute to the sum of intermediate states in the amplitude for $K \to \pi\pi$. $K \to 3\pi$ is forbidden by G-parity and $K \to 4\pi$ is forbidden by energy conservation.

is conserved by the strong interactions, enforces

$$\langle(\pi\pi)_I ; \text{in}|(3\pi); \text{out}\rangle = 0. \tag{4.130}$$

Therefore only the 2π out state can contribute in the sum:

$$\langle(\pi\pi)_I ; \text{in}|n; \text{out}\rangle = 0 \quad \text{for} \quad n \neq (2\pi)_I. \tag{4.131}$$

This is usually referred to as the condition of *elastic* unitarity, and is shown in Fig. 4.1. With the S matrix for $(\pi\pi)_I \to (\pi\pi)_I$ given by

$$S_{\text{elastic}} \equiv \langle(\pi\pi)_I ; \text{out}|(\pi\pi)_I ; \text{in}\rangle = e^{2i\delta_I}, \tag{4.132}$$

where δ_I denotes the $\pi\pi$ phase shift, we find from Eq. (4.129)

$$\langle(\pi\pi)_I ; \text{out}|H_W|K\rangle = e^{2i\delta_I}\langle(\pi\pi)_I ; \text{out}|H_W|K\rangle^*, \tag{4.133}$$

or equivalently

$$\langle(\pi\pi)_I ; \text{out}|H_W|K^0\rangle = |\langle(\pi\pi)_I ; \text{out}|H_W|K^0\rangle|e^{i\delta_I} \equiv A_I e^{i\delta_I}. \tag{4.134}$$

These arguments can be retraced for $\overline{K}^0 \to \pi\pi$, yielding

$$\langle(\pi\pi)_I ; \text{out}|H_W|\overline{K}^0\rangle = |\langle(\pi\pi)_I ; \text{out}|H_W|\overline{K}^0\rangle|e^{i\delta_I} \equiv \overline{A}_I e^{i\delta_I}. \tag{4.135}$$

As long as H_W conserves **T**, the two amplitudes A_I and $\overline{A}_I$ remain real after having the strong phase shift factored out. Once, however, **T** is violated in H_W, additional phases appear in A_I and $\overline{A}_I$, as stated before and discussed later on.

Strong final state interactions effect also the decays of heavy flavour hadrons, yet we *cannot* apply Watson's theorem there blindly. In particular there is no reason why *elastic* unitarity should apply in two-body or even quasi-two-body *beauty* decays: strong final state interactions are actually quite likely to generate additional hadrons in the final state. The decays of *charm* hadrons represent a borderline case: while the final state interactions can change the identity of the emerging particles and can produce additional hadrons, their impact is somewhat moderated since the available phase space is less than abundant. As discussed in more detail in the chapter on charm decays, introducing the concept of *absorption* might provide a useful approximation here.

4.10.2 T-*odd correlations and the impact of final state interactions*

In decay processes it is impossible to realize the time reversed sequence, as discussed in Sec. 3.3; tests of detailed balance thus cannot be performed. Instead we search for T-odd correlations. Consider, for example, $\Lambda \to N + \pi$ and $\overline{\Lambda} \to \overline{N} + \pi$ decays. The decay amplitude is given by

$$
\begin{aligned}
\langle P\pi^-|\mathscr{H}|\Lambda\rangle &= \bar{u}(p_N, s_N)(A_S + A_P\gamma_5)u(p_\Lambda, s_\Lambda) \\
\langle \overline{P}\pi^+|\mathscr{H}|\overline{\Lambda}\rangle &= \bar{v}(p_{\overline{N}}, s_{\overline{N}})(\overline{A}_S + \overline{A}_P\gamma_5)v(p_{\overline{\Lambda}}, s_{\overline{\Lambda}}).
\end{aligned} \tag{4.136}
$$

A_S, A_P, $\overline{A}_S$ and $\overline{A}_P$ are scalar functions reflecting the weight of the S- and P-wave decay amplitudes. These amplitudes possess strong interaction phases due to final state strong interaction. Assuming **CPT** symmetry, we can write amplitudes in terms of isospin amplitudes,

$$
A_S = \mathscr{A}^S_{1/2}e^{i\delta^S_{1/2}} + \mathscr{A}^S_{3/2}e^{i\delta^S_{3/2}} \tag{4.137}
$$

$$
\overline{A}_S = -(\mathscr{A}^S_{1/2})^* e^{i\delta^S_{1/2}} - (\mathscr{A}^S_{3/2})^* e^{i\delta^S_{3/2}} \tag{4.138}
$$

$$
A_P = \mathscr{A}^P_{1/2}e^{i\delta^P_{1/2}} + \mathscr{A}^P_{3/2}e^{i\delta^P_{3/2}} \tag{4.139}
$$

$$
\overline{A}_P = (\mathscr{A}^P_{1/2})^* e^{i\delta^P_{1/2}} + (\mathscr{A}^P_{3/2})^* e^{i\delta^P_{3/2}} \tag{4.140}
$$

Here δ^L_I denotes the strong phase of the final state with isospin I and orbitral angular momentum L. Watson's theorem allows us to identify δ^L_I with the πp phase shift. Note that the phase shifts are equal for particle and antiparticle reactions as strong interaction conserves **CP**. It is easy to check that **CP** symmetry implies (see Problem 4.13)

$$
\mathscr{A}_S = \mathscr{A}^*_S \quad \text{and} \quad \mathscr{A}_P = \mathscr{A}^*_P \tag{4.141}
$$

In computing $|\langle N|\phi_\pi|\Lambda\rangle|^2$, we can introduce the spin projection operator $(1 + \not{s}\gamma_5)/2$ for both Λ and N and sum over spins of Λ and N. Among other terms, this gives a term

$$
\text{Re}\,[e^{i(\delta_S - \delta_P)}\mathscr{A}_S\mathscr{A}^*_P i\epsilon_{\alpha\beta\gamma\delta}p^\alpha_\Lambda p^\beta_N s^\gamma_\Lambda s^\delta_N] \tag{4.142}
$$

in the differential decay distribution for $\Lambda \to N\pi$ decay. In the rest frame of Λ, the above reduces to

$$
-\,\text{Im}\,[e^{i(\delta_S - \delta_P)}(\mathscr{A}_S\mathscr{A}^*_P)]M_\Lambda \vec{p}_N \cdot (\vec{s}_\Lambda \times \vec{s}_N). \tag{4.143}
$$

Of course, a similar term arises in the antiparticle decay. A triple correlation that is *odd* under **T** can arise in two fundamentally different ways (and any combination thereof):

- The intervention of **T** violating forces can create complex phases in $\mathscr{A}_P$ relative to $\mathscr{A}_S$.

- On the other hand, even **T** conserving dynamics (like the strong and electromagnetic forces) can induce $\text{Im}[e^{i(\delta_S - \delta_P)}] \neq 0$.

Measurements show indeed that $\langle \vec{P}_N \cdot (\sigma_N \times \sigma_\Lambda)\rangle \neq 0$, and they yield

$$\arg\frac{\mathscr{A}_P}{\mathscr{A}_S} + \delta_P - \delta_S = -6.5 \pm 3.5°, \tag{4.144}$$

which is quite consistent with what we independently know about πN phase shifts, $\delta_P - \delta_S$.

The reader is likely to raise the following question: how come the observation of a **P**-odd correlation establishes **P** violation, whereas that of a **T**-odd correlation does not carry the same unequivocal message about **T** invariance?

The answer is built on several elements of a conceptual as well as technical nature:

- As seen from Eq. (4.95), the substitution $(\vec{p}, \vec{s}) \longrightarrow (-\vec{p}, -\vec{s})$ does not fully implement time reversal for decay and scattering experiments – an incoming spherical wave is transformed to an outgoing spherical wave.

- If the interactions under study are so weak that only first order effects have to be considered, then a **T**-odd correlation does actually signal **T** violation.

- The situation changes, however, qualitatively, if higher order contributions are – or have to be – included. Those effects are derived through iteration factors $\exp[-i \int dt\, H]$.

  These evolution operators do *not* commute with an *anti*linear operator like **T**, even if H does.

To disentangle a **CP** violating correlation from a trivial correlation due to final state interaction, we can compare **CP** conjugate observables. If **CP** is conserved, we get

$$\text{Im}\,[e^{i(\delta_S - \delta_P)}(A_S A_P^*)] = -\text{Im}\,[e^{i(\delta_S - \delta_P)}(\overline{A}_S \overline{A}_P^*)], \tag{4.145}$$

i.e., equal **T**-odd correlations for particle and antiparticle decays.

4.10.3 Final state interaction and partial widths

Final state interactions becomes indispensible in making **CP** violation observable in certain partial widths. Consider a weak decay channel $P \to f$ receiving contributions from two coherent processes. The transition amplitudes then reads as follows:

$$A(P \to f) = e^{i\phi_1^{\text{weak}}} e^{i\delta_1^{\text{FSI}}} |\mathscr{A}_1| + e^{i\phi_2^{\text{weak}}} e^{i\delta_2^{\text{FSI}}} |\mathscr{A}_2|, \tag{4.146}$$

where $\phi_{1,2}^{\text{weak}}$ are phases due to the weak decay dynamics and $\delta_{1,2}^{\text{FSI}}$ are due to strong (or electromagnetic) final state interactions. For **CP** conjugate amplitude, we have

$$A(\overline{P} \to \overline{f}) = e^{-i\phi_1^{\text{weak}}} e^{i\delta_1^{\text{FSI}}} |\mathscr{A}_1| + e^{-i\phi_2^{\text{weak}}} e^{i\delta_2^{\text{FSI}}} |\mathscr{A}_2|, \tag{4.147}$$

and therefore, for the partial widths,

$$\frac{\Gamma(P \to f) - \Gamma(\overline{P} \to \overline{f})}{\Gamma(P \to f) + \Gamma(\overline{P} \to \overline{f})} =$$
$$-\frac{2\sin(\Delta\phi_W)\sin(\Delta\delta^{FSI})|\mathscr{A}_2/\mathscr{A}_1|}{1 + |\mathscr{A}_2/\mathscr{A}_1|^2 + 2|\mathscr{A}_2/\mathscr{A}_1|\cos(\Delta\phi_W)\cos(\Delta\delta^{FSI})} \qquad (4.148)$$

where $\Delta\phi_W = \phi_1^{weak} - \phi_2^{weak}$ and $\Delta\delta^{FSI} = \delta_1^{FSI} - \delta_2^{FSI}$.

So, there are two requirements for a decay to reveal **CP** violating effects:

1　**CP** violation enters through weak dynamics: $\Delta\phi_W = \phi_1^{weak} - \phi_2^{weak}$.

2　Final state interactions induce a non-trivial phase shift: $\Delta\delta^{FSI} = \delta_1^{FSI} - \delta_2^{FSI}$.

The asymmetry gets larger if the two interfering amplitudes are of comparable size: $\mathscr{A}_2/\mathscr{A}_1 \sim \mathcal{O}(1)$. So, the search is on to look for decays which satisfy these conditions.

4.11 Résumé and outlook

Quantum field theories provide us with all the necessary tools for implementing discrete transformations in a natural way. Charge conjugation is no longer ad-hoc; the existence of antiparticles is required by the demands of Lorentz invariance. The combined transformation **CPT** constitutes an almost unavoidable symmetry of any quantum field theory. **C** and **P** invariance, on the other hand, can easily be broken. **CP** symmetry can be violated, but requires attention to subtle details.

Due to **CPT** constraints, **CP** violation manifests itself through complex phases in the underlying dynamics. These phases enter the amplitudes of a reaction and its **CP** conjugate version with *opposite* signs and are customarily referred to as *weak* phases. In our preceding discussion of Watson's theorem we have learnt that a *second* class of phases arises through final state interaction, i.e., wherever we treat an interaction beyond the lowest order. This applies in particular to the strong force, but sometimes also to electromagnetic forces. They generate phase shifts that are the same for a given process and its **CP** conjugate.

This second class of phases can induce a **T**-odd correlation even with vanishing weak phases, i.e., if the underlying dynamics is **T** invariant. This effect can be disentangled by comparing such correlation in **CP** conjugate transitions. On the other hand, in partial widths asymmetries, these phase shifts act as a necessary evil – necessary since, in their absence, an existing **CP** violation will remain unobservable, and evil since their values cannot be computed.

Problems

4.1　Show that the quantization condition Eq. (4.15) is invariant under **C** and **P**.

4.2 Going back to the definition $\mathbf{T} = UK$, show that the two equations given in Eq. (4.40) are consistent with each other.

4.3 Show that $[\psi(t,\check{x})^{\mathbf{C}}]^{\mathbf{C}} = \psi(t,\check{x})$ and $[\psi(t,\check{x})^{\mathbf{T}}]^{\mathbf{T}} = \psi(t,\check{x})$ for a spinor field ψ.

4.4 Assume that a theory is invariant under parity and also under a global gauge transformation. Then show that the parity operation is defined only up to a phase. For example

$$\mathbf{P}\phi(t,\check{x})\mathbf{P}^{\dagger} = e^{i\alpha}\phi(t,-\check{x}).$$

4.5 Consider scalar QED where the field equation for a (complex) scalar field reads

$$[(i\partial_{\mu} - eA_{\mu})(i\partial^{\mu} - eA^{\mu}) - m^2]\phi(t,\check{x}) = 0. \tag{4.149}$$

Show that if ϕ is a solution to Eq. (4.149), then so is $\phi^{\dagger}$ – with the same mass, but opposite charge $-e$.

4.6 Apply **T** to Eq. (4.149) and derive

$$\left[-\partial^2 + ie\left(\bar{\partial}_{\mu}A^{\mu}(\bar{t},\check{x}) + A_{\mu}(\bar{t},\check{x})\bar{\partial}^{\mu}\right) + e^2A_{\mu}(\bar{t},\check{x})A^{\mu}(\bar{t},\check{x}) - m^2\right]$$
$$\times \mathbf{T}\phi(t,\check{x})\mathbf{T}^{-1} = 0, \tag{4.150}$$

where $\bar{\partial} = (\partial_{\bar{t}}, \vec{\partial})$. Thus we can conclude the following: if $\phi(t,\check{x})$ – as a solution to the Klein–Gordon equation – describes the motion of a scalar particle with charge e and mass m in an electromagnetic field, then $\mathbf{T}\phi(-t,\check{x})\mathbf{T}^{-1}$ traces the original motion of the particle backwards in time.

4.7 The behaviour of a spin-1/2 field coupled to an electromagnetic field is described by

$$\left[i\gamma_{\mu}\left(\partial^{\mu} + ieA^{\mu}\right) - m\right]\psi(t,\check{x}) = 0; \tag{4.151}$$

show that

$$\left[i\gamma_{\mu}\left(\partial^{\mu} + ie\mathbf{P}A^{\mu}\mathbf{P}^{\dagger}\right) - m\right]\mathbf{P}\psi(t,\check{x})\mathbf{P}^{\dagger} = 0; \tag{4.152}$$

where $\mathbf{P}\psi(t,\check{x})\mathbf{P}^{\dagger} = \gamma_0\psi(t,-\check{x})$.

4.8 Take the complex conjugate of Eq. (4.151). Show that $\psi^*(x)$ satisfies the following equation:

$$[i(\not{\partial} - ie\,\not{A} - m]\gamma_2\psi^*(x) = 0. \tag{4.153}$$

Discuss the physical interpretation of $\gamma^2\psi^*(x)$.

4.9 Applying the anti-unitary operator **T** to the Dirac equation, show that

$$\left[i\gamma_0 \left(\partial_{-t} + ieA_0(-t, \vec{x}) \right) - i\vec{\gamma} \cdot \left(\vec{\partial} + ie\vec{A}(-t, \vec{x}) \right) - m \right] \mathbf{T}\psi(t, \vec{x})\mathbf{T}^{-1} = 0.$$

$$(4.154)$$

Thus we have

$$\mathbf{T}\psi(t, \vec{x})\mathbf{T}^{-1} = \psi(-t, \vec{x}). \tag{4.155}$$

4.10 Investigate how the magnetic moment $\vec{\mu}_{\text{mag}}(P)$, defined by

$$\vec{\mu}_{\text{mag}}(P) \cdot \vec{B} = \langle P, \vec{s} | \vec{\sigma} | P, \vec{s} \rangle \cdot \vec{B},$$

transforms under **CPT**.

4.11 By using the completeness of the 'in' and 'out' states, show that **CPT** symmetry implies Eq. (4.102)

4.12 Consider the P, S, V_μ and A_μ fields defined in Eq. (4.87). Discuss how they transform under **C**, **P** and **T**. Consider a toy Hamiltonian

$$H = (aS + bP)(aS + bP)^\dagger + (cV_\mu + dA_\mu) \cdot (cV^\mu + dA^\mu)^\dagger. \tag{4.156}$$

Under what condition is H invariant under **C**, **P**, **T**, **CP** and **CPT**?

4.13 Consider a Hamiltonian for hyperon nonleptonic decay $\Lambda \to \pi^- p$:

$$H_W = \overline{\psi}_P(A_S + A_P\gamma_5)\psi_\Lambda + \phi_\pi^+ \overline{\psi}_\Lambda(A_S^* - A_P^*\gamma_5)\psi_P \phi_\pi^- \tag{4.157}$$

Now the transformation property of $\phi_\pi(t, \vec{x})$ is:

$$\begin{aligned}
\mathbf{P}\phi_\pi^\pm(t, \vec{x})\mathbf{P}^\dagger &= -\phi_\pi^\pm(t, -\vec{x}) \\
\mathbf{C}\phi_\pi^\pm(t, \vec{x})\mathbf{C}^\dagger &= +\phi_\pi^\mp(t, \vec{x}) \\
\mathbf{T}\phi_\pi^\pm(t, \vec{x})\mathbf{T}^{-1} &= \phi_\pi^\pm(-t, \vec{x})
\end{aligned} \tag{4.158}$$

Check that the action obtained from H_W given in Eq. (4.157) is invariant under **CPT**. Eq. (4.157) is invariant under **CPT** except for an overall sign. How does this sign affect the decay rate? Show that (a) **P** invariance implies $A_S = \overline{A}_S = 0$; (b) **CP** invariance implies $\overline{A}_S = -A_S$, $\overline{A}_P = A_P$; (c) **T** invariance implies $\text{Im}\,A_S = \text{Im}\,A_P = 0$. Of course the phases we have chosen in Eq. (4.158) are quite arbitrary. Choose $\mathbf{C}\phi_\pi^\pm(t, \vec{x})\mathbf{C}^\dagger = -\phi_\pi^\mp(t, \vec{x})$ and discuss how above conclusions are modified.

5

The arrival of strange particles

The discovery of hadrons with the internal quantum number 'strangeness' marks the beginning of a most exciting epoch in particle physics that even now, fifty years later, has not yet found its conclusion. As we will describe in detail, the discoveries made in this area of research have proved essential in formulating the SM of high energy physics. This is *not* to say, however, that it has been a particularly glorious chapter for theorists always making successful predictions. On the contrary, we have to admit that by and large experiments have driven the development, and that major discoveries came unexpectedly or even against expectations expressed by theorists. Theorists on the whole have fared better in making *post*dictions than *pre*dictions[1]. Some of the relevant questions remain unanswered today and/or have led to even more profound puzzles.

These discoveries were not spread out evenly over time; some periods were clearly more fertile than others. In this chapter we will give a brief overview of important developments. We do not aim at giving a complete history of this period. Instead we try to give a flavour of the excitement accompanying these discoveries, thus illustrating the thrill and the unexpected turns that can occur in the basic sciences in general and in high energy physics in particular [20].

5.1 The discovery of strange particles

Tracking chambers, being based on ionization, cannot record the passage of a neutral particle. Yet when it decays into two charged particles, the tracks of the decay products can be recorded, and a so-called V pattern arises. Such a pattern was observed by Rochester and Butler in October 1946 when they exposed cloud chambers (the state-of-the-art tracking chambers of that period) to cosmic rays. They concluded that they had observed the decay of a primary particle with a mass of (435 ± 100) MeV into two secondary particles with a mass around 100 MeV; or in today's language:

$$K^0 \to \pi^+ \pi^-. \tag{5.1}$$

The discovery of charged pions – in a different experiment – was reported later, in May 1947, and Rochester and Butler reported their observation in December

---

[1] Such an achievement, which is not so uncommon in the basic sciences, should not be belittled, though!

1947 after having found one event with a kink in its charged track; the latter event today would be referred to as

$$K^+ \to \pi^+\pi^0. \tag{5.2}$$

It should be noted that the neutral pion was officially not discovered until 1950.

These objects were produced and studied in experiments at the new accelerators at Brookhaven National Laboratory (hereafter referred to as BNL) starting in 1953, and at the Berkeley Laboratory from 1955 on. Another neutral particle was seen to decay into two charged particles, this time with a mass somewhat larger than the proton mass. Today we refer to it as the Λ hyperon, and thus

$$\Lambda \to p\pi^-. \tag{5.3}$$

This led immediately to a puzzle: it was observed that the production rate for these new particles greatly exceeded their decay rate. To be more specific: hyperons are produced copiously in πp collisions; i.e.,

$$\pi p \to \Lambda + X \tag{5.4}$$

is driven by the *strong* interactions, whereas the decay of Eq. (5.3) exhibits a lifetime of about 10^{-10} s, which represents an eternity relative to strong transition times of order 10^{-23} s. This apparent paradox, which gave rise to the name 'strange' particles, was resolved by Pais in 1952 through the concept of *associated production* [21]. A new quantum number – not surprisingly called 'strangeness' – is introduced, which is conserved by the strong, though *not* the weak, interaction. Particles carrying this quantum number are then produced pairwise by the strong interactions from a non-strange initial state, like in

$$pp \to K\overline{K} + X$$
$$\pi p \to \Lambda K + X. \tag{5.5}$$

The decays $K \to \pi\pi$ or $\Lambda \to p\pi$ changing this quantum number have to proceed *weakly*. There are actually two ways in which strangeness can be introduced: we could assign a 'strange parity' +1 to all non-strange particles like nucleons and pions and −1 to all strange particles like Λ and K. Hadronic collisions would yield an even number of strange particles, whereas $\Lambda \to p\pi$ and $K \to \pi\pi$ have to proceed weakly. Yet the discovery of the so-called cascade particles through their weak decay

$$\Xi^- \to \Lambda\pi^- \tag{5.6}$$

closed this option of strangeness being a *multiplicative* quantum number. With the weak forces flipping the sign of the strange parity, we had to assign strange parity +1 to Ξ^- due to Eq. (5.6). Yet then the transition

$$\Xi^- \to n\pi^- \tag{5.7}$$

would not only be allowed, it would proceed *strongly* and thus dominate weak transitions. However, Eq. (5.7) has never been observed! Thus – as pointed out independently by Gell-Mann [22] and Nakano and Nishijima [23] in 1953 – we have to introduce strangeness as an *additive* quantum number: we assign strangeness -1 to Λ and K^-, $+1$ to K^+, -2 to Ξ^-, and 0 to all non-strange hadrons. This satisfies the observed pattern of production and decay.

5.2 The $\theta - \tau$ puzzle

The second period is characterized by the $\theta - \tau$ puzzle. Two decay reactions had been found for charged strange mesons, namely

$$\theta^+ \rightarrow \pi^+\pi^0,$$
$$\tau^+ \rightarrow \pi^+\pi^+\pi^-. \tag{5.8}$$

With θ^+ being spinless, the π^+ and π^0 have to form an S-wave; the 2π final state thus necessarily carries positive parity. The angular distributions of the three pions from τ^+ decay reveal the final state to carry zero total angular momentum as well, but with negative parity! It was assumed that parity, like angular momentum, was conserved by the relevant forces. The parity of the initial state then coincides with that of the final state. With θ and τ exhibiting different parity, they had to be distinct objects and thus indeed deserved different names. The problem arose when ever more precise measurements failed to find any significant difference in either the mass or the lifetime of the θ and τ mesons. This constituted the $\theta - \tau$ puzzle: how could nature – short of being malicious – assign the same mass to two distinct particles? Or even more baffling: how could nature contrive to generate the same lifetime to two distinct particles, the major decay channels of which possess a totally different phase space?

This Gordian knot could be cut in one stroke if parity were not absolutely conserved. For then θ and τ could represent merely two decay modes of the same particle. Parity conservation had been tested extensively in strong and electromagnetic transitions. Yet the breakthrough came when Lee and Yang pointed out in 1956 [24] that this symmetry had *not* been probed yet in weak transitions; they proceeded to suggest a list of relevant tests. Soon after this theoretical analysis, Wu and collaborators [25] found in nuclear β decays that parity and charge conjugation invariance were indeed broken by the weak forces. This epochal discovery was almost instantly confirmed by other groups in different processes, and this allowed the identification of both θ^+ and τ^+ with the K^+ meson.

5.3 The existence of two different neutral kaons

At about the same time another important lesson was learnt, again in response to a challenge. As mentioned in the beginning of this chapter, neutral strange mesons had been found to exist:

$$K^{\text{neut}} \rightarrow \pi^+\pi^-. \tag{5.9}$$

Since these mesons carry non-zero strangeness they cannot be – unlike the π^0 – their own antiparticle. Therefore two neutral kaons had to exist, K^0 and $\overline{K}^0$, differing by two units of strangeness. The obvious question then arises: how can you establish their separate existence? This challenge was successfully taken up by Gell-Mann and Pais [26]. Their analysis – through careful quantum mechanical reasoning – has yielded some of the more glorious pages in the theoretical development.

It appeared natural to assume **CP** invariance for the analysis[2] – partly for simplicity and partly because illustrious theorists had made strong-worded pronouncements why nature had better obey that symmetry. We will come back to that aspect in the next chapter.

Manifest **CP** symmetry can be realized in one of two scenarios: the asymptotic states, i.e., the states with definite mass and lifetimes, either are

- **CP** *eigenstates* or

- pairs of *mass degenerate* particles and antiparticles with equal mass.

This will be illustrated by the following explicit discussion. Let us, for now, turn off weak forces. K^0 and $\overline{K}^0$ can then neither decay nor transform into each other. Denote a time dependent $K^0 - \overline{K}^0$ state by

$$\Psi(t) = a(t)|K^0\rangle + b(t)|\overline{K}^0\rangle \equiv \begin{pmatrix} a(t) \\ b(t) \end{pmatrix}. \tag{5.10}$$

We consider the free Schrödinger equation for the $\Psi(t)$:

$$i\hbar \frac{\partial}{\partial t}\Psi = H\Psi, \tag{5.11}$$

where H denotes the Hamilton operator. For the single particle system under discussion, H is a generalized mass matrix which is diagonal and real in the absence of weak forces:

$$H = \begin{pmatrix} M_K & 0 \\ 0 & M_K \end{pmatrix}, \tag{5.12}$$

where **CPT** symmetry enforces the equality of the two diagonal elements; K^0 and $\overline{K}^0$ are thus two *degenerate* mass eigenstates.

The situation changes *qualitatively* once weak forces are included that can change strangeness: decay processes with $|\Delta S| = 1$ take place. Consider, for example, decays:

$$K^0 \rightarrow \pi\pi, \quad \overline{K}^0 \rightarrow \pi\pi. \tag{5.13}$$

The dynamics then mixes K^0 and $\overline{K}^0$ through the chain $K^0 \rightarrow \pi\pi \rightarrow \overline{K}^0$. We denote the mixing term of the Hamiltonian as Δ. For now, let us forget the

---

[2] Gell-Mann and Pais actually assumed **C** conservation at first.

fact that we have to enlarge our Hilbert space to include (multi)pion states, and discuss the mixing effect – we shall come back to this point in Chapter 6. Then

$$H = \begin{pmatrix} M_K & \Delta \\ \Delta & M_K \end{pmatrix}. \tag{5.14}$$

Since Δ is second order in the weak interactions, it is truly infinitesimal compared to M_K. Nevertheless it dictates the eigenstate to be

$$|K_1\rangle \equiv \frac{1}{\sqrt{2}} \left(|K^0\rangle + |\overline{K}^0\rangle \right),$$

$$|K_2\rangle \equiv \frac{1}{\sqrt{2}} \left(|K^0\rangle - |\overline{K}^0\rangle \right). \tag{5.15}$$

With

$$\mathbf{CP}|K^0\rangle = |\overline{K}^0\rangle, \tag{5.16}$$

we have $\mathbf{CP}|K_{\frac{1}{2}}\rangle = \pm|K_{\frac{1}{2}}\rangle$. It can be shown that (Problem 5.1)

$$\mathbf{CP}|\pi\pi\rangle = +|\pi\pi\rangle. \tag{5.17}$$

Therefore the 2π final state is fed by K_1 decays only,

$$K_1 \to 2\pi,$$

$$K_2 \not\to 2\pi. \tag{5.18}$$

The leading nonleptonic channel for K_2 is then

$$K_2 \to 3\pi. \tag{5.19}$$

The phase space for Eq. (5.19) is very restricted – $3 \cdot M_\pi \simeq 420$ MeV vs $M(K_2) \simeq 500$ MeV. Thus we expect the lifetime for the $\mathbf{CP}$ odd state K_2 to be much longer than for the $\mathbf{CP}$ even K_1. The long lived meson thus predicted was discovered by Lederman and his collaborators in 1956 [27]. Since K_1 and K_2 possess quite different lifetimes, it is customary to refer to them as K_S and K_L, respectively, with S [L] referring to short-lived [long-lived]. Likewise we use $M(K_1) = M(K_S) = M_S$ and $M(K_2) = M(K_L) = M_L$. State-of-the-art measurements yield for their lifetimes [28]:

$$\tau_S \equiv \tau(K_S) = (0.8926 \pm 0.0012) \times 10^{-10} \text{ s}$$

$$\tau_L \equiv \tau(K_L) = (5.15 \pm 0.04) \times 10^{-8} \text{ s}. \tag{5.20}$$

It should be kept in mind, though, that this huge difference in lifetimes, namely $\tau(K_L)/\tau(K_S) \sim 600$, reflects a dynamical accident. If pions were massless, $\tau_L \sim \tau_S$ and we might not know about $\mathbf{CP}$ violation even today!

5.4 CP-invariant particle–antiparticle oscillations

There are many more intriguing facets than the existence of two (vastly) different lifetimes for neutral kaons. Their behaviour can be described in terms of two equivalent bases, namely the strong interaction or strangeness eigenstates K^0 and $\overline{K}^0$ or the mass eigenstates K_L and K_S. The latter have a simple exponential evolution in (proper) time t:

$$\begin{aligned} |K_L(t)\rangle &= e^{-i(M_L - \frac{i}{2}\Gamma_L)t}|K_L(0)\rangle \\ |K_S(t)\rangle &= e^{-i(M_S - \frac{i}{2}\Gamma_S)t}|K_S(0)\rangle. \end{aligned} \tag{5.21}$$

Thus

$$\begin{aligned} I(K_S \to K_S;t) &= I_0 \cdot |\langle K_S|K_S(t)\rangle|^2 = I_0 e^{-\Gamma_S t} \\ I(K_S \to K_L;t) &= I_0 \cdot |\langle K_L|K_S(t)\rangle|^2 = 0, \end{aligned} \tag{5.22}$$

i.e., an initially pure K_S beam of intensity I_0 will subsequently contain only K_S mesons plus their decay products; likewise for initial K_L beams.

Yet for an initially pure K^0 beam the pattern is much more complex. The time evolution for a $|K^0(t)\rangle$ is obtained by inverting Eq. (5.15), i.e.,

$$\begin{aligned} |K^0(t)\rangle &= \frac{1}{\sqrt{2}}\left(|K_S(t)\rangle + |K_L(t)\rangle\right) \\ &= \frac{1}{\sqrt{2}}\left(e^{-i(M_S - \frac{i}{2}\Gamma_S)t}|K_S(0)\rangle + e^{-i(M_L - \frac{i}{2}\Gamma_L)t}|K_L(0)\rangle\right) \\ &= f_+(t)|K^0\rangle + f_-(t)|\overline{K}^0\rangle, \end{aligned} \tag{5.23}$$

and likewise for the antiparticle

$$\begin{aligned} |\overline{K}^0(t)\rangle &= \frac{1}{\sqrt{2}}\left(|K_S(t)\rangle - |K_L(t)\rangle\right) \\ &= \frac{1}{\sqrt{2}}\left(e^{-i(M_S - \frac{i}{2}\Gamma_S)t}|K_S(0)\rangle - e^{-i(M_L - \frac{i}{2}\Gamma_L)t}|K_L(0)\rangle\right) \\ &= f_-(t)|K^0\rangle + f_+(t)|\overline{K}^0\rangle, \end{aligned} \tag{5.24}$$

where

$$f_\pm(t) = \frac{1}{2}\left[e^{-i(M_S - \frac{i}{2}\Gamma_S)t} \pm e^{-i(M_L - \frac{i}{2}\Gamma_L)t}\right]. \tag{5.25}$$

The probability of finding a K_S in an initially pure K^0 beam, $I(K^0 \to K_S;t)$, or $\overline{K}^0$ beam, $I(\overline{K}^0 \to K_S;t)$, still follows an exponential decay law:

$$I(K^0 \to K_S;t) = I_0 \cdot |\langle K_S|K^0(t)\rangle|^2 = \frac{1}{2}I_0 e^{-\Gamma_S t} = I(\overline{K}^0 \to K_S;t), \tag{5.26}$$

with the factor $1/2$ reflecting the presence of the K_L component in the beam:

$$I(K^0 \to K_L;t) = \frac{1}{2}I_0 e^{-\Gamma_L t} = I(\overline{K}^0 \to K_L;t). \tag{5.27}$$

A non-trivial pattern arises when we ask for the probability of finding a K^0 in an initially pure K^0 beam:

$$\begin{aligned} I(K^0 \to K^0;t) &= I_0 \cdot |\langle K^0|K^0(t)\rangle|^2 = I_0 \cdot |f_+(t)|^2 \\ &= \frac{1}{4}I_0\left[e^{-\Gamma_L t} + e^{-\Gamma_S t} + 2e^{-\frac{1}{2}(\Gamma_S + \Gamma_L)t}\cos\Delta M_k t\right], \end{aligned} \tag{5.28}$$

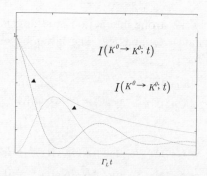

Fig. 5.1.   The probability of finding K^0 in an initial K^0 beam as a function of time, and the probability of finding $\overline{K}^0$ in the same beam.

where

$$\Delta M_K = M_L - M_S. \tag{5.29}$$

The last term in the square bracket of Eq. (5.28) reflects the fact that it cannot be decided as a matter of principle whether the observed K^0 came from the K_S or K_L component. Therefore $I(K^0 \to K^0; t)$ is *not* given by the average of $I(K^0 \to K_S; t)$ and $I(K^0 \to K_L; t)$ – an interference term is also present. The law of probabilities is not violated of course, since we have

$$I(K^0 \to \overline{K}^0; t) = \frac{1}{4} I_0 \left[e^{-\Gamma_L t} + e^{-\Gamma_S t} - 2 e^{-\frac{1}{2}(\Gamma_S + \Gamma_L)t} \cos \Delta M_k t \right] \tag{5.30}$$

and therefore

$$I(K^0 \to \overline{K}^0; t) + I(K^0 \to K^0; t) = \frac{1}{2} I_0 \left[e^{-\Gamma_L t} + e^{-\Gamma_S t} \right]. \tag{5.31}$$

These functions are shown in Fig. 5.1. From Eq. (5.30) we read off that $I(K^0 \to \overline{K}^0; t = 0) = 0$, as it has to be since the initial beam consists of K^0 mesons only. Yet at later times $t > 0$ we have

$$I(K^0 \to \overline{K}^0; t > 0) > 0 \; ; \tag{5.32}$$

i.e., a state that is orthogonal to the initial state gets spontaneously 'generated' at later times!

From the preceding discussion it is clear how the K_S and K_L are identified, namely through the decay modes $K_S \to 2\pi$ and $K_L \to 3\pi$ exhibiting the lifetimes τ_S and τ_L, respectively. The states of definite strangeness, namely K^0 and $\overline{K}^0$, can be identified through the flavour-specific semileptonic channels

$$K^0 \to l^+ \nu \pi^-, \quad \overline{K}^0 \to l^- \overline{\nu} \pi^+. \tag{5.33}$$

It is an experimental fact [28] that semileptonic decays of $K^0 [\overline{K}^0]$ always contain $l^+[l^-]$:

$$x = \frac{\langle l^-\bar{v}\pi^+|\mathscr{H}|K^0\rangle}{\langle l^+v\pi^-|\mathscr{H}|K^0\rangle} \qquad (5.34)$$

$$\text{Re}\,x = 0.006 \pm 0.018$$
$$\text{Im}\,x = -0.003 \pm 0.026. \qquad (5.35)$$

For the hyperon decay,

$$\frac{\Gamma(\Sigma^+ \to ne^+v)}{\Gamma(\Sigma^+ \to \text{all})} < 5 \times 10^{-6}. \qquad (5.36)$$

Within the quark model it is automatic since semileptonic strange decays are driven by

$$s \to ul^-\bar{v}\,. \qquad (5.37)$$

Changes in strangeness and the quark charge in this decay are then correlated:

$$\Delta S = 0 - (-1) = +1$$
$$\Delta Q = +\tfrac{2}{3} - (-\tfrac{1}{3}) = +1\,; \qquad (5.38)$$

i.e., it obeys the $\Delta S = \Delta Q = +1$ rule.

Thus we find

$$\Gamma(K^0(t) \to l^-\bar{v}\pi^+) = |f_-(t)|^2 \Gamma(\overline{K}^0 \to l^-\bar{v}\pi^+)$$
$$\Gamma(\overline{K}^0(t) \to l^-\bar{v}\pi^+) = |f_+(t)|^2 \Gamma(\overline{K}^0 \to l^-\bar{v}\pi^+). \qquad (5.39)$$

Analysing the decay rate evolution of these modes in time, we can extract ΔM_K in addition to Γ_S and Γ_L. The present value is

$$\Delta M_K = M_L - M_S = (3.4782 \pm 0.0176) \times 10^{-12} \text{ MeV}$$
$$\Delta \Gamma_K = \Gamma_S - \Gamma_L = (7.2823 \pm .0098) \times 10^{-12} \text{ MeV}. \qquad (5.40)$$

It should be seen as utterly amazing that such a tiny mass difference ΔM_K can be measured, in particular if we quote it relative to the average of the K_S and K_L mass

$$\frac{\Delta M_K}{M_K} \simeq 7 \times 10^{-15}. \qquad (5.41)$$

What this shows is the power of quantum mechanical interference effects tracked over 'macroscopic' distances. To understand it, first note that

$$\cos(\Delta M_k t) = \cos\left(\frac{\Delta M_K}{\Gamma_S} \frac{t}{\tau_S}\right). \qquad (5.42)$$

If the beam energy is chosen high enough then the neutral kaon – after its production and before its decay – travels a distance that can be resolved and measured. Knowing its momentum we can determine t, its proper time of decay. Its scale is set by $\tau_S = \Gamma_S^{-1}$ and the scale for ΔM_K is then given by Γ_S. The width Γ_S is obviously of second order in the weak interactions. ΔM_K represents a $\Delta S = 2$ transition; as explained later in detail, it is generated by iterating two $\Delta S = 1$ reactions and thus is also of second order in the weak interactions. Therefore ΔM_K and Γ_S can be expected to be of roughly comparable order of magnitude. Since $\Gamma_S \simeq 7.3 \cdot 10^{-12}$ MeV such experiments are sensitive to mass differences $\sim \mathcal{O} \left(10^{-12} \text{ MeV}\right)$. Indeed we find

$$\frac{\Delta M_K}{\Gamma_S} \simeq 0.49. \tag{5.43}$$

Finally, we should note that in the 1960s tracking technologies were not sufficiently refined for resolving reliably the differences in the production and decay vertices for particles with $c\tau_S \simeq 2.7$ cm. Another technique was used then, to be discussed next.

5.5 Regeneration – which is heavier: K_L or K_S?

So far we have implicitly assumed that the neutral kaons travel through a trivial medium, namely empty space, which does not affect the particle-antiparticle balance. However, if they travel through nuclear matter, the situation will change dramatically since K^0 and $\overline{K}^0$ interact quite differently with a nucleus. A K^0 with the quark content $\bar{s}d$ can, at low energies, interact only (quasi-)elastically with nucleons N:

$$K^0 + N \to K^0 + N^{(*)}, \tag{5.44}$$

where the notation $N^{(*)}$ allows for the nucleon to be excited into a non-strange resonance N^*. A $\overline{K}^0 = (s\bar{d})$ on the other hand can – in addition to its elastic reaction $\overline{K}^0 + N \to \overline{K}^0 + N^{(*)}$ – also excite a nucleon into a Λ or Σ hyperon or a Y^* resonance:

$$\overline{K}^0[s\bar{d}] + N[uud/udd] \to \pi^0[d\bar{d}] + \Lambda/\Sigma/Y^*[suu/sud] . \tag{5.45}$$

K^0 and $\overline{K}^0$ will therefore propagate quite differently through nuclear matter. To obtain an intuitive picture, let us assume that $\overline{K}^0$ is totally absorbed by a nucleus whereas K^0 is not. The nucleus then acts like a Stern–Gerlach filter letting K^0 pass but not $\overline{K}^0$. Pais and Piccioni [29] presented a beautiful analysis of this basic quantum mechanical problem briefly sketched here.

Consider a π^- beam hitting a target and producing a mixture of K^0 and $\overline{K}^0$ or, equivalently, of K_S and K_L. Traveling through a vacuum, the mass eigenstates will decay through $K_S \to 2\pi$ and $K_L \to 3\pi$. Since $\tau_S \ll \tau_L$, the K_S component will quickly decay away leaving a practically pure K_L beam behind. If the latter hits nuclear matter, only its K^0 component will not be absorbed; the emerging state is thus the linear combination $|K_S\rangle + |K_L\rangle$ – i.e., the previously extinct K_S component has been *regenerated* through rescattering with nuclear matter! This is depicted in Fig. 5.2.

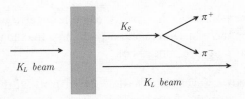

Fig. 5.2. Regeneration of K_S from a K_L beam going through matter. K_L is a coherent linear combination of the K^0 and $\overline{K}^0$ states. When it travels through matter, it interacts with nuclei through strong interaction. The strong interaction eigenstates K^0 and $\overline{K}^0$ scatter differently and the coherence is lost. The resulting beam is a mixture of K_L and K_S – K_S is regenerated.

To be more specific, we can write down the time evolution for a K^0 or a $\overline{K}^0$ traveling through a medium by modifying Eq. (5.23) and Eq. (5.24) as follows:

$$|K^0(t_{\mathrm{med}})\rangle = f_+(t_{\mathrm{med}})g|K^0\rangle + f_-(t_{\mathrm{med}})\overline{g}|\overline{K}^0\rangle$$
$$|\overline{K}^0(t_{\mathrm{med}})\rangle = f_-(t_{\mathrm{med}})g|K^0\rangle + f_+(t_{\mathrm{med}})\overline{g}|\overline{K}^0\rangle. \qquad (5.46)$$

The quantities g and $\overline{g}$ are the forward scattering amplitudes for K^0 and $\overline{K}^0$ in the medium, respectively. We then find

$$\Gamma(K^0(t_{\mathrm{med}})) \sim e^{-\Gamma_L t_{\mathrm{med}}}|g_-|^2 + e^{-\Gamma_S t_{\mathrm{med}}}|g_+|^2$$
$$+ \ 2e^{-\frac{1}{2}(\Gamma_L+\Gamma_S)t_{\mathrm{med}}}\mathrm{Re}\,(e^{-i\Delta M_K t_{\mathrm{med}}}g_- g_+^*), \qquad (5.47)$$

where $g_- = \frac{1}{2}(g-\overline{g})$ and $g_+ = \frac{1}{2}(g+\overline{g})$. By introducing plates of varying thickness, measurements of $-\Delta M_K \tau + arg(g_+) - arg(g_-)$ can be made. Finally, the phases of g_+ and g_- are determined from known phase shifts of $K^0 N$ and $\overline{K}^0 N$ scattering. This allows us to determine not only $|\Delta M_K|$, but even the sign of ΔM_K. It turns out that

$$M_L > M_S, \qquad (5.48)$$

leading to the nice mnemonic that in K_L and K_S, L means both *larger* (mass) and *longer* (lifetime) whereas S denotes *smaller* (mass) and *shorter* (lifetime).

5.6 The quiet before the storm

We might forgive theorists of that time for feeling quite smug about the situation. Many essential insights had been gained from analysing the dynamics of strange particles, or were at least prompted by it:

- A new quantum number – strangeness – conserved by the strong and electromagnetic forces had been discovered. In modern parlance we would say that the first hadrons of the second family had been found.

- It had been realized that parity as well as charge conjugation were violated by the weak forces and actually in a maximal way (as far as the charged currents are concerned).

- It appeared however that we did not have to fall back all the way to **CPT** symmetry: the violations observed for **C** and **P** were such that they compensated each other, i.e., **CP** invariance apparently was satisfied.

- General arguments had been advanced as to why nature had to be **CP** symmetric.

- The predictions concerning the existence of two kinds of neutral kaons with vastly different lifetimes had been confirmed.

- The mass difference for those two states had been determined with spectacular sensitivity through a judicious application of ingenious quantum mechanical reasoning and beautiful experimentation.

Yet this harmonious picture was about to receive a shattering blow![3]

5.7 The discovery of CP violation

At a conference held in 1989 celebrating the XXVth anniversary of CP violation, Pais stated in his retrospective lecture [20]: 'It was a very good year, 1964, both in theory and in experimental high energy physics.' For the following discoveries and breakthroughs were made that year:

1 The Higgs mechanism for the *spontaneous* realization of a symmetry was first developed.

2 The quark model and the first elements of current algebra were put forward.

3 The charm quark was first introduced to establish quark–lepton symmetry.

4 $SU(6)$ symmetry was proposed.

5 The first storage ring for e^+e^- collisions was built in Frascati.

6 The Ω^- baryon was found at Brookhaven National Laboratory.

7 CP violation was discovered at the same lab.

---

[3] It should be noted that not everybody subscribed to this orthodox view. Okun, in his 1963 text book [30], explicitly listed the search for $K_L \to \pi\pi$ as a priority for the future then. In response to his suggestion, an experiment was performed; however, it did not succeed in accumulating statistics sufficient to see a signal.

Finding the Ω^- was seen as a breakthrough confirmation of the 'eight-fold way' of $SU(3)$ symmetry. Being composed of three strange quarks, it also exhibits rather directly the need for colour as a new internal degree of freedom. The theoretical concepts listed under items 1–3 turn out to be very relevant for our attempts to deal with **CP** invariance and its limitations.

As mentioned in the preceding chapter, **CP** violation had not been expected. Its observation thus came as a shock to the community, as illustrated in Pais' personal account [20]:

> I had been aware of that result since one morning in early June when I was having breakfast in the old, rather dilapidated, Brookhaven Cafeteria which some of us remember with slight nostalgia. That morning Jim and Val had walked up to me and had asked could they talk to me. Naturally, I said. They proceeded to tell me that they had found K_2-decays into $\pi^+\pi^-$. How could that be I asked; it violates CP-invariance. They knew that, they said, but there it was. Why was the effect not due to regeneration of short-lived K_S, I wanted to know. Because, they said, that effect was far too small in the helium bag where the 2π had been found. I asked many more questions, why were they not seeing $2\pi\gamma$ decays with a soft photon, or $\pi\mu\nu$ decays with a soft neutrino and perhaps some confusion about mass. They had thought long and hard about these and other alternatives (the actual experiment had been concluded the previous July) and had ruled them out one by one. After they left I had another coffee. I was shaken by the news. I knew quite well that a small amount of CP-violation would not drastically alter the earlier discussions, based on CP-invariance, of the neutral K-complex. Also, the experience of seeing a symmetry fall by the way side was not new to anyone who had lived through the 1956–7 period. At that time, however, there had at once been the consolation that P- and C-violation could be embraced by a new pretty concept, CP-invariance. What shook all concerned was that with CP gone there was nothing elegant to replace it with. The very smallness of the 2π rate, CP-invariance as a near miss, made the news even harder to digest.

While this discovery was completely unexpected, it did not occur in a vacuum, but marked the culmination of a long series of experiments, the highlights of which were the following [31]

- At the BNL Cosmotron, the BNL-Columbia group discovered what is today known as K_L [27].

- Piccioni and his collaborators [32] demonstrated the regeneration phenomenon at the Berkeley Betatron.

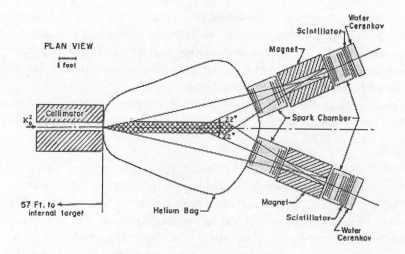

Fig. 5.3.   A schematic drawing of an apparatus used by Cronin, Fitch and collaborators. K^0 and $\overline{K}^0$ are created on the internal target. They travel through the collimator so that only the ones traveling parallel to the axis of the apparatus are accepted. Also, most of the K_S component has decayed away and we have almost a pure K_L beam. Two detectors which consist of magnets and spark chambers measure three-momenta of the two body decay product $K_L \to \pi^+\pi^-$. The invariant mass of the $\pi^+\pi^-$ pair as well as the direction of K_L have been checked.

- Adair and his collaborators [33] claimed to have seen an excess of K_S regeneration in the forward direction.

Cronin, Fitch, Christenson, and Turlay [34] set out to investigate this anomalous K_S regeneration. Their apparatus and its location with respect to the AGS proton beam is shown in Fig. 5.3[4].

K^0 and $\overline{K}^0$ mesons are produced in a target 57 feet in front of the spectrometer. Over that distance the K_S component has practically completely faded away and a pure K_L beam remains. The latter passes through a lead collimator into a bag filled with helium gas. Knowing the parameters controlling regeneration we compute the K_S regeneration rate due to K_L scattering off helium nuclei: it is found to be entirely negligible.

By inserting a regenerator they could study the excess K_S production in the forward direction reported by Adair *et al.* [33]. With much better statistics, they were able to rule out any anomalous regeneration of K_S mesons. As a secondary objective they set out to put a better upper limit on the **CP**-violating $K_L \to 2\pi$ channel. They had claimed in their proposal that it '... can set a limit of about

---

[4] Reminiscing in March 1997 about how the experiment was operated, Val Fitch revealed a certain nonchalance concerning safety: 'Sometimes we had to duck in and out of the beam line. But we were of course protected by our film badges.'

one in a thousand for the partial decay of $K_2 \to 2\pi$ in one hour of operation'
[35]. Instead they saw, to their astonishment, the decay $K_L \to 2\pi$.

Since the experimenters were acutely aware of the revolutionary nature of their
discovery if it were true (and of their embarrassment if it were not) they explored
even unlikely alternative interpretations of their signal – like exotic regeneration
of K_S or a misidentification of the final state, as referred to in Pais' conversation
with Fitch and Cronin. Having done that they reported the result

$$\frac{\Gamma(K_L \to \pi^+ \pi^-)}{\Gamma(K_S \to \pi^+ \pi^-)} = [(2.0 \pm 0.4) \times 10^{-3}]^2, \qquad (5.49)$$

which is completely consistent with what is known today.

It appears that the *experimental findings* per se were readily accepted. As Pais
put it: 'The perpetrators were widely regarded as real pros.' On the other hand
there was considerable reluctance in the theoretical community to part with
the notion of absolute **CP** and **T** invariance. Alternative *interpretations* were
suggested.

- Observing $K_L \to \pi\pi$ to occur *by itself* does *not* establish **CP** violation.
  What is needed for that conclusion to take hold is to have two mass
  non-degenerate states K_L and K_S[5] with $K_S \to \pi\pi$ and $K_L \to 2\pi$, 3π –
  within the framework of *linear* quantum mechanics with its superposition
  principle! We could attempt to reconcile the observations with **CP** symmetry by going beyond the usual quantum mechanics and introduce
  non-linear terms into the Schrödinger equation. This would violate the
  superposition principle – yet that had not been tested to that sensitivity
  before. This escape route was actually tried [36] – which actually shows
  once more how unprepared the theoretical community was for accepting
  CP violation.

- It was suggested that a new particle U emerged unobserved from the
  decay:

$$K_L \to K_S + U \to (\pi\pi)_{K_S} + U. \qquad (5.50)$$

  Observation of an interference between $K_L \to \pi\pi$ and $K_S \to \pi\pi$ with
  K_S being coherently regenerated from the K_L beam also ruled out this
  scenario [36]. It is curious to note – from a historical perspective – that
  here it has turned out to be correct to abandon an apparently sacrosanct
  symmetry principle rather than introduce an exotic new particle. In an
  earlier situation when faced with the paradox of *continuous* electron
  spectra in β decay one – actually Pauli – had been vindicated in retaining
  the symmetry leading to momentum conservation even at the price of

---

[5] It does not matter at this point how tiny ΔM_K is.

postulating an exotic particle, the neutrino, with apparently little prospect of ever establishing its existence.

- It was pointed out that the observed **CP** asymmetry might be an *environmental* effect due to the preponderance of matter over antimatter in our corner of the universe. More specifically, the existence of a new long-range force of cosmological origin was postulated.

While not all conceivable alternatives were ruled out experimentally in a completely rigorous fashion, the community came quickly and decisively around to accepting **CP** violation as a fact.

The observations can be summarized by saying that the quantum mechanical K_L state contains a small admixture of a **CP** even component in addition to its dominant **CP** odd part:

$$|K_L\rangle = \frac{1}{\sqrt{1 + |\bar{\epsilon}|^2}} \left(|K_2\rangle + \bar{\epsilon}|K_1\rangle \right). \tag{5.51}$$

CPT invariance then tells us that the K_S state in turn contains a **CP** odd component controlled by the same (complex) impurity parameter $\bar{\epsilon}$:

$$|K_S\rangle = \frac{1}{\sqrt{1 + |\bar{\epsilon}|^2}} \left(|K_1\rangle - \bar{\epsilon}|K_2\rangle \right). \tag{5.52}$$

It should be noted that because $\mathrm{Br}(K_L \rightarrow \pi\pi)$ is so tiny, the argument used to explain the large lifetime difference still applies.

Problems

5.1 Show that the final state in $K \rightarrow \pi^+\pi^-, \pi^0\pi^0$ is **CP** even. Next consider $K \rightarrow 3\pi^0, \pi^+\pi^-\pi^0$; what can you now say about the **CP** parity of the final state?

5.2 Imagine a world in which the pion is massless. Do you think **CP** violation would have been discovered?

5.3 Imagine instead a world with $M_\pi = 200$ MeV. How would this affect the $\tau - \theta$ puzzle and the discovery of **CP** violation?

5.4 By diagonalizing the Hamiltonian shown in Eq. (5.14), show that $M_{K_1} - M_{K_2} = 2\Delta$. Is the sign of Δ observable? Can you distinguish the states K_1 and K_2 physically? If so, how? In discussing Eq. (3.74) we had argued that for $E_+ = E_-$, the sign of Δ there had no meaning. Compare the two situations.

6

Quantum mechanics of neutral particles

In the preceding chapter we have described the discovery of **CP** violation from a historical point of view; we depended largely on intuition rather than on the formalism. The latter will be discussed in detail now.

6.1 The effective Hamiltonian

For the sake of simplicity, we have been intentionally sloppy in describing $K - \overline{K}$ mixing in Sec. 5.3. We simply stated the time dependences, Eq. (5.21), of $|K_L(t)\rangle$ and $|K_S(t)\rangle$ states assuming **CP** was conserved.

Here we shall present the quantum mechanical formalism for particle–antiparticle oscillations, and show how **CP** violation can manifest itself there. The formalism of particle–antiparticle oscillations presented in this section is very general. It describes a situation where particles and antiparticles – unlike the case with π^0 mesons – are distinguished by an internal quantum number like strangeness, beauty, charm, etc., or lepton number, baryon number, etc.; weak or superweak interactions not conserving these quantum numbers can drive a particle–antiparticle transition where the internal quantum number changes by two units. We will not contemplate violation of electric charge conservation and therefore study only electrically neutral states P^0 and $\overline{P}^0$. Yet P^0 may be K^0, D^0, B^0, a neutron or a neutrino[1]. For the original discussion, see Ref. [37, 38]. P^0 and $\overline{P}^0$ are distinguished merely by an internal quantum number F with the property that $\Delta F = 0$ for H_{strong} and H_{QED} while $\Delta F \neq 0$ for H_{weak}. Let us suppose that H_{weak} couples P^0 and $\overline{P}^0$ to a common state I: Then $H_{\text{weak}} \otimes H_{\text{weak}}$ will induce a transition $P^0 \leftrightarrow \overline{P}^0$ as a one-step process or as an iteration of two $\Delta F = 1$ reactions through an intermediate state I:

$$P^0 \xrightarrow{\Delta F=1} I \xrightarrow{\Delta F=1} \overline{P}^0 \quad \text{or} \quad \overline{P}^0 \xrightarrow{\Delta F=-1} I \xrightarrow{\Delta F=-1} P^0. \tag{6.1}$$

I can represent a real on-shell state common to P^0 and $\overline{P}^0$ decays or it can denote a virtual off-shell state. All this can be taken into account by considering the Hamiltonian for a state carrying the quantum number F:

$$H = H_{\Delta F=0} + H_{\Delta F=1} + H_{\Delta F=2}. \tag{6.2}$$

---

[1] The formalism described here applies directly to neutron–antineutron and $\nu - \overline{\nu}$ oscillations; as shown later it can easily be generalized to the more 'conventional' case of $\nu_\alpha - \nu_\beta$ oscillations.

$H_{\Delta F=0}$ contains the strong and electromagnetic forces conserving F^2; $H_{\Delta F=1}$ denotes the weak forces changing F by one unit; $H_{\Delta F=2}$ allows for the presence of some so-called superweak forces changing F by two units, thus producing $P^0 \to \overline{P}^0$ as a one-step process.

The time evolution of the $P^0 \leftrightarrow \overline{P}^0$ system, including its decays, is given by a vector in Hilbert space

$$|\tilde{\Psi}(t)\rangle = a(t)|P^0\rangle + b(t)|\overline{P}^0\rangle + c(t)|\pi\pi\rangle + d(t)|3\pi\rangle + e(t)|\pi l\ \bar{\nu}_l\rangle + \cdots, \qquad (6.3)$$

where l stands for an electron or muon, and the Schrödinger equation reads

$$i\hbar \frac{\partial}{\partial t}\tilde{\Psi} = H\tilde{\Psi}, \qquad (6.4)$$

where H is an *infinite*-dimensional Hermitian matrix in the Hilbert space. The full time dependence of $\tilde{\Psi}(t)$ cannot be obtained rigorously since that would require subjugating strong dynamics completely to theoretical control, which at present is beyond our capabilities.

The situation simplifies dramatically and becomes tractable if we reduce our demands. Let us consider the following scenario [37, 38]:

- The initial state is a linear combination of P^0 and $\overline{P}^0$ alone: $|\Psi(0)\rangle = a(0)|P^0\rangle + b(0)|\overline{P}^0\rangle$.

- We are interested only in $a(t)$ and $b(t)$ and not in any other coefficients.

- We restrict ourselves to times that are much larger than a typical strong interaction scale. This is called the Weisskopf–Wigner approximation [39].

Then we can write

$$i\hbar \frac{\partial}{\partial t}\Psi(t) = \mathcal{H}\Psi(t) \qquad (6.5)$$

where $\Psi(t)$ is restricted to the subspace of P^0 and $\overline{P}^0$:

$$\Psi(t) = \begin{pmatrix} a(t) \\ b(t) \end{pmatrix}. \qquad (6.6)$$

The matrix $\mathcal{H}$ is given by

$$\mathcal{H} = \mathbf{M} - \frac{i}{2}\mathbf{\Gamma} = \begin{pmatrix} M_{11} - \frac{i}{2}\Gamma_{11} & M_{12} - \frac{i}{2}\Gamma_{12} \\ M_{21} - \frac{i}{2}\Gamma_{21} & M_{22} - \frac{i}{2}\Gamma_{22} \end{pmatrix}, \qquad (6.7)$$

---

[2] There are also weak corrections with $\Delta F = 0$, but they can safely be ignored here.

where

$$M_{11} = M_P + \sum_n \mathscr{P}\left[\frac{|\langle n; \text{out}|H_{\Delta F=1}|P^0\rangle|^2}{M_P - M_n}\right]$$

$$M_{22} = M_{\overline{P}} + \sum_n \mathscr{P}\left[\frac{|\langle n; \text{out}|H_{\Delta F=1}|\overline{P}^0\rangle|^2}{M_{\overline{P}} - M_n}\right]$$

$$M_{12} = M_{21}^* = \langle P^0|H_{SW}|\overline{P}^0\rangle$$

$$\qquad + \sum_n \mathscr{P}\left[\frac{\langle P^0|H_{\Delta F=1}|n; \text{out}\rangle\langle n; \text{out}|H_{\Delta F=1}|\overline{P}^0\rangle}{M_P - M_n}\right]$$

$$\Gamma_{11} = 2\pi \sum_n \delta(M_P - M_n)|\langle n; \text{out}|H_{\Delta F=1}|P^0\rangle|^2$$

$$\Gamma_{22} = 2\pi \sum_n \delta(M_P - M_n)|\langle n; \text{out}|H_{\Delta F=1}|\overline{P}^0\rangle|^2$$

$$\Gamma_{12} = \Gamma_{21}^* = 2\pi \sum_n \delta(M_P - M_n)\langle P^0|H_{\Delta F=1}|n; \text{out}\rangle\langle n; \text{out}|H_{\Delta F=1}|\overline{P}^0\rangle.$$

$$(6.8)$$

$\mathscr{P}$ stands for the *principal part* prescription[3]. A detailed derivation can be found in [40], and Ref. [41] gives an alternative approach.

6.2 Constraints from CPT, CP and T

Let us assume that H commutes with **C**, **P** and **T**. We shall define these operators so that

$$\mathbf{C}|P^0\rangle = -|\overline{P}^0\rangle, \quad \mathbf{P}|P^0\rangle = -|P^0\rangle, \quad \mathbf{T}|P^0\rangle = |P^0\rangle, \qquad (6.9)$$

with

$$\mathbf{CP}|n; \text{out(in)}\rangle = |\overline{n}; \text{out(in)}\rangle$$
$$\mathbf{T}|n; \text{in(out)}\rangle = |n; \text{out(in)}\rangle. \qquad (6.10)$$

Assuming as usual that the 'in' and 'out' states form equivalent complete bases, we can easily show (see Problem 6.1) that $M_{11} = M_{22}$ and $\Gamma_{11} = \Gamma_{22}$ from **CP** and **CPT** symmetry. Also, $M_{12} = M_{21}$ and $\Gamma_{12} = \Gamma_{21}$ from **CP** and **T** symmetry. So, we have (see Problems 3.8 and 6.1)

$$\mathbf{CPT} \text{ or } \mathbf{CP} \text{ invariance} \implies M_{11} = M_{22},\ \Gamma_{11} = \Gamma_{22},$$
$$\mathbf{CP} \text{ or } \mathbf{T} \text{ invariance} \implies \text{Im}\,M_{12} = 0 = \text{Im}\,\Gamma_{12}. \qquad (6.11)$$

---

[3] A comment on the normalization of states is in order. By $\langle \vec{k}_1 \ldots \vec{k}_n|H_{\Delta F=1}|P^0\rangle$, we mean

$$\frac{(2\pi)^3\delta^3(\vec{p} - \ldots \vec{k}_n)\langle \vec{k}_1 \ldots \vec{k}_n|H(0)|P\rangle}{\sqrt{2MV2E_1V\ldots 2E_nV}},$$

where $\langle \vec{k}_1 \ldots \vec{k}_n|H(0)|P\rangle$ is a Lorentz scalar, and $H_{\Delta F=1} = \int d^3x\, H(x)$.

6.3 Spherical coordinates

The Schrödinger equation is best solved by diagonalizing the matrix $\mathscr{H}$; the solutions to the then decoupled equations represent the two mass eigenstates. Since $\mathbf{M}$ and $\mathbf{\Gamma}$ are Hermitian they can be diagonalized by a unitary transformation. It is however not immediately obvious that the same transformation can diagonalize both $\mathbf{M}$ and $\mathbf{\Gamma}$. The formulism for tackling this problem was developed in [42] and [43].

Any 2×2 matrix can be expanded in terms of the three Pauli matrices σ_i and the unit matrix using complex coefficients:

$$\mathbf{M} - \frac{i}{2}\mathbf{\Gamma} = E_1\sigma_1 + E_2\sigma_2 + E_3\sigma_3 - iD\mathbf{1}, \tag{6.12}$$

where the Pauli matricies are as defined in Eq. (4.10). Comparing both sides of Eq. (6.12) we arrive at

$$
\begin{aligned}
E_1 &= \operatorname{Re} M_{12} - \frac{i}{2}\operatorname{Re}\Gamma_{12} \\
E_2 &= -\operatorname{Im} M_{12} + \frac{i}{2}\operatorname{Im}\Gamma_{12} \\
E_3 &= \frac{1}{2}(M_{11} - M_{22}) - \frac{i}{4}(\Gamma_{11} - \Gamma_{22}) \\
D &= \frac{i}{2}(M_{11} + M_{22}) + \frac{1}{4}(\Gamma_{11} + \Gamma_{22}).
\end{aligned}
\tag{6.13}
$$

We can define *complex* numbers E, θ, and ϕ such that

$$E = \sqrt{E_1^2 + E_2^2 + E_3^2}, \tag{6.14}$$

$$E_1 = E\sin\theta\cos\phi, \quad E_2 = E\sin\theta\sin\phi, \quad E_3 = E\cos\theta. \tag{6.15}$$

Cosine and sine of a complex number z are defined in the usual way:

$$\cos z = \frac{1}{2}\left(e^{iz} + e^{-iz}\right), \quad \sin z = \frac{1}{2i}\left(e^{iz} - e^{-iz}\right). \tag{6.16}$$

The condition for $P^0 - \overline{P}^0$ oscillations to occur is then expressed by

$$E \neq 0, \quad \sin\theta \neq 0. \tag{6.17}$$

The constraints on $\mathbf{M}$ and $\mathbf{\Gamma}$ imposed by the discrete symmetries are expressed as follows:

$$
\begin{aligned}
\mathbf{CPT} \text{ or } \mathbf{CP} \text{ invariance} &\implies \cos\theta = 0 \\
\mathbf{CP} \text{ or } \mathbf{T} \text{ invariance} &\implies \phi = 0.
\end{aligned}
\tag{6.18}
$$

It is important to note that these arrows apply in one direction only; for example we can adopt a phase convention in which $\cos\theta = 0$ or $\phi = 0$ even when $\mathbf{CP}$ is violated.

This representation in (complex) spherical coordinates allows us to write down the mass eigenstates in a rather compact fashion:

$$
\begin{aligned}
|P_1\rangle &= p_1|P^0\rangle + q_1|\overline{P}^0\rangle, \\
|P_2\rangle &= p_2|P^0\rangle - q_2|\overline{P}^0\rangle,
\end{aligned}
\tag{6.19}
$$

where

$$
\begin{aligned}
p_1 &= N_1 \cos\frac{\theta}{2}, & q_1 &= N_1 e^{i\phi} \sin\frac{\theta}{2}, \\
p_2 &= N_2 \sin\frac{\theta}{2}, & q_2 &= N_2 e^{i\phi} \cos\frac{\theta}{2},
\end{aligned}
\tag{6.20}
$$

with the normalization factors (θ and ϕ are complex in general!):

$$
\begin{aligned}
N_1 &= \frac{1}{\sqrt{|\cos\frac{\theta}{2}|^2 + |e^{i\phi}\sin\frac{\theta}{2}|^2}}, \\
N_2 &= \frac{1}{\sqrt{|\sin\frac{\theta}{2}|^2 + |e^{i\phi}\cos\frac{\theta}{2}|^2}}.
\end{aligned}
\tag{6.21}
$$

With **CPT** invariance imposing $\cos\theta = 0$, considerable simplifications arise: $p_1 = p_2$ and $q_1 = q_2$, and we can drop the subscripts. The states

$$
\begin{aligned}
|P_1\rangle &= p|P^0\rangle + q|\overline{P}^0\rangle, \\
|P_2\rangle &= p|P^0\rangle - q|\overline{P}^0\rangle,
\end{aligned}
\tag{6.22}
$$

are mass eigenstates with eigenvalues

$$
\begin{aligned}
M_1 - \frac{i}{2}\Gamma_1 &= -iD + E = M_{11} - \frac{i}{2}\Gamma_{11} + \frac{q}{p}\left(M_{12} - \frac{i}{2}\Gamma_{12}\right), \\
M_2 - \frac{i}{2}\Gamma_2 &= -iD - E = M_{11} - \frac{i}{2}\Gamma_{11} - \frac{q}{p}\left(M_{12} - \frac{i}{2}\Gamma_{12}\right).
\end{aligned}
\tag{6.23}
$$

with

$$
\left(\frac{q}{p}\right)^2 = \frac{M_{12}^* - \frac{i}{2}\Gamma_{12}^*}{M_{12} - \frac{i}{2}\Gamma_{12}}
\tag{6.24}
$$

Obviously there exist *two* solutions:

$$
\frac{q}{p} = \pm\sqrt{\frac{M_{12}^* - \frac{i}{2}\Gamma_{12}^*}{M_{12} - \frac{i}{2}\Gamma_{12}}}.
\tag{6.25}
$$

Choosing the negative rather than the positive sign in Eq. (6.25) is equivalent to interchanging the labels $1 \leftrightarrow 2$ of the mass eigenstates, see Eq. (6.22), Eq. (6.23).

6.4 ♠ On phase conventions ♠

The binary ambiguity just mentioned is a special case of a more general one. For antiparticles are defined up to a phase only; adopting a different phase convention – e.g., going from $|\overline{P}^0\rangle$ to $e^{i\xi}|\overline{P}^0\rangle$ – will modify the off-diagonal elements of $\mathbf{M}$ and $\mathbf{\Gamma}$:

$$M_{12} - \frac{i}{2}\Gamma_{12} \longrightarrow e^{i\xi}\left(M_{12} - \frac{i}{2}\Gamma_{12}\right) \tag{6.26}$$

and thus

$$\frac{q}{p} \longrightarrow e^{-i\xi}\frac{q}{p} \tag{6.27}$$

yet leave their product invariant:

$$\frac{q}{p}\left(M_{12} - \frac{i}{2}\Gamma_{12}\right) \longrightarrow \frac{q}{p}\left(M_{12} - \frac{i}{2}\Gamma_{12}\right). \tag{6.28}$$

This is as it should be since the differences in mass and width, see Eq. (6.23),

$$M_2 - M_1 = -2\text{Re}\left(\frac{q}{p}(M_{12} - \frac{i}{2}\Gamma_{12})\right)$$

$$\Gamma_2 - \Gamma_1 = 4\text{Im}\left(\frac{q}{p}(M_{12} - \frac{i}{2}\Gamma_{12})\right), \tag{6.29}$$

being observables, have to be insensitive to the *arbitrary* phase of $\overline{P}^0$!

The following comments will turn out to be important for the subsequent discussion on **CP** asymmetries:

- The mass eigenstates P_1 and P_2 will in general not be orthogonal to each other without **CP** invariance:

$$\langle P_1|P_2\rangle = |p|^2 - |q|^2 \neq 0 \tag{6.30}$$

  for $|q/p| \neq 1$ or $\text{Im}\phi \neq 0$.

- At this point, the physical meaning of labels '1' and '2' are not clear. Once we know the sign of $M_1 - M_2$, it becomes an empirical question whether

$$\Gamma_2 > \Gamma_1 \quad \text{or} \quad \Gamma_2 < \Gamma_1 \tag{6.31}$$

  holds, i.e., whether the heavier state is shorter or longer lived.

- In the limit of **CP** invariance the two mass eigenstates are **CP** eigenstates as well, and we can raise another meaningful question: is the heavier state **CP** even or odd? With **CP** symmetry implying $\arg(M_{12}/\Gamma_{12}) = 0$, q/p becomes a pure phase: $|q/p| = 1$. It is then convenient to adopt a phase *convention* s.t. M_{12} is real; $q/p = \pm 1$ and $\text{CP}|P^0\rangle = \pm|\overline{P}^0\rangle$ are then the remaining choices.

– With $\frac{q}{p} = +1$ we have

$$|P_1\rangle = \frac{1}{\sqrt{2}}\left(|P^0\rangle + |\overline{P}^0\rangle\right)$$

$$|P_2\rangle = \frac{1}{\sqrt{2}}\left(|P^0\rangle - |\overline{P}^0\rangle\right). \tag{6.32}$$

For $\mathbf{CP}|P^0\rangle = |\overline{P}^0\rangle$, P_1 and P_2 are $\mathbf{CP}$ even and odd, respectively; therefore

$$M_- - M_+ = M_2 - M_1 = -2\mathrm{Re}\left(\frac{q}{p}\left(M_{12} - \frac{i}{2}\Gamma_{12}\right)\right) = -2M_{12} \tag{6.33}$$

For $\mathbf{CP}|P^0\rangle = -|\overline{P}^0\rangle$, P_1 and P_2 switch roles; i.e., P_1 and P_2 are now $\mathbf{CP}$ odd and even. Thus

$$M_- - M_+ = M_1 - M_2 = 2M_{12}. \tag{6.34}$$

– Alternatively we can set $\frac{q}{p} = -1$:

$$|P_1\rangle = \frac{1}{\sqrt{2}}\left(|P^0\rangle - |\overline{P}^0\rangle\right)$$

$$|P_2\rangle = \frac{1}{\sqrt{2}}\left(|P^0\rangle + |\overline{P}^0\rangle\right) \tag{6.35}$$

while maintaining $\mathbf{CP}|P^0\rangle = |\overline{P}^0\rangle$; P_1 and P_2 are then $\mathbf{CP}$ odd and even, respectively. Accordingly

$$M_- - M_+ = M_1 - M_2 = 2\mathrm{Re}\left(\frac{q}{p}\left(M_{12} - \frac{i}{2}\Gamma_{12}\right)\right) = -2M_{12}. \tag{6.36}$$

– We leave the fourth possibility as an exercise.

– Eq. (6.33) and Eq. (6.36) on one side and Eq. (6.34) on the other apparently do not coincide; yet below we will see that the theoretical prediction for M_{12} changes sign depending on the choice of $\mathbf{CP}|P^0\rangle = \pm|\overline{P}^0\rangle$. These expressions therefore all agree, of course.

If the lifetime difference is too small to be observed or $\mathbf{CP}$ is too badly broken to be of use in classifying states, then we are limited to stating that one mass eigenstate is heavier than the other.

● Later we will discuss how to evaluate M_{12} within a given theory for the $P - \overline{P}$ complex. The examples above have illustrated that some care has to be applied in interpreting such a result:

– Expressing the mass eigenstates explicitly in terms of flavour eigenstates involves some conventions. Once we adopt a certain definition, we have to stick with it; yet our choice cannot influence observables.

6.5 ΔM and $\Delta \Gamma$

The *relative* phase between M_{12} and Γ_{12} represents an observable quantity describing indirect **CP** violation. The following notation will turn out to be convenient:

$$M_{12} = \overline{M}_{12} e^{i\xi}, \quad \Gamma_{12} = \overline{\Gamma}_{12} e^{i\xi} e^{i\zeta}, \quad \frac{\Gamma_{12}}{M_{12}} = \frac{\overline{\Gamma}_{12}}{\overline{M}_{12}} e^{i\zeta} = r e^{i\zeta}. \quad (6.37)$$

The signs of $\overline{M}_{12}$ and $\overline{\Gamma}_{12}$ are fixed such that both ξ and $\xi + \zeta$ are restricted to lie between $-\frac{\pi}{2}$ and $\frac{\pi}{2}$; i.e., the real quantities $\overline{M}_{12}$ and $\overline{\Gamma}_{12}$ are a priori allowed to be negative as well as positive. A relative minus sign between M_{12} and Γ_{12} is of course physically significant, while the absolute sign is not. Yet we will see that the absolute sign provides us with a useful bookkeeping device.

Throughout this book, we will set

$$\Delta M = M_2 - M_1, \quad \text{and} \quad \Delta \Gamma = \Gamma_1 - \Gamma_2 \quad (6.38)$$

since then both ΔM and $\Delta \Gamma$ are positive for kaons. Let us sketch three complementary scenarios:

- For the K meson system, we have (See Problem 6.9)

$$0 < \Delta M_K = M_L - M_S = -2\overline{M}_{12}^K$$
$$0 < \Delta \Gamma_K = \Gamma_S - \Gamma_L = 2\overline{\Gamma}_{12}^K. \quad (6.39)$$

The data, Eq. (5.43), and the fact that $\Delta \Gamma \simeq \Gamma_S$ due to $\Gamma_S \gg \Gamma_L$ lead to $r \sim 2$, where r is defined in Eq. (6.37). In this approximation,

$$\frac{q}{p} \simeq \sqrt{\frac{M_{12}^*}{M_{12}}} \left(1 - \frac{(1+i)\zeta_K}{2} \right) \quad (6.40)$$

and

$$\left| \frac{q}{p} \right| \simeq 1 - \frac{\zeta_K}{2}. \quad (6.41)$$

- We can have $\Delta \Gamma \ll \Delta M$ together with $\Delta \Gamma \ll \Gamma$; this is the situation predicted for the B_d complex, as explained later. We then have

$$\Delta M_B = M_2 - M_1 = -2\,\overline{M}_{12}^B$$
$$\Delta \Gamma = \Gamma_1 - \Gamma_2 = 2\,\overline{\Gamma}_{12}^B \cos \zeta_B \quad (6.42)$$

$$\frac{q}{p} \simeq \sqrt{\frac{M_{12}^*}{M_{12}}} (1 - \frac{r}{2} \sin \zeta_B), \quad \left| \frac{q}{p} \right| \simeq 1 - \frac{r}{2} \sin \zeta. \quad (6.43)$$

We find $|q/p| \simeq 1$ similar to the K^0 case, yet for a different reason, namely $r \ll 1$ rather than $\zeta \ll 1$. With $\Delta\Gamma/\Gamma$ too small to be observable and/or ζ sizeable, the mass eigenstates can be characterized by their masses only without any other empirical label.

- A slight variation arises if $\Delta\Gamma \ll \Gamma$ holds, yet $\Delta\Gamma/\Gamma$ can still be measured. While the expressions listed above apply, their consequences are different. $\Delta\Gamma$ is an observable now in *practice* rather than merely in principle. This situation might be realized for B_s mesons, to be discussed later.

The authors of [44] define

$$\tilde{\Delta}M \equiv M_H - M_L, \tag{6.44}$$

where the subscript H [L] stands for heavy [light]; $\tilde{\Delta}M$ is thus positive by definition. They also define

$$\begin{aligned}|P_L\rangle &= p|P\rangle + q|\overline{P}\rangle \\ |P_H\rangle &= p|P\rangle - q|\overline{P}\rangle\end{aligned} \tag{6.45}$$

with

$$\frac{q}{p} = \pm\sqrt{\frac{M_{12}^* - \frac{i}{2}\Gamma_{12}^*}{M_{12} - \frac{i}{2}\Gamma_{12}}}. \tag{6.46}$$

If **CP** symmetry (approximately) holds we can sensibly ask if the **CP** even state is heavier or lighter than the **CP** odd state. If the calculation yields a negative value for the difference $M_2 - M_1$ – yet we want to keep the convention $\mathbf{CP}|P^0\rangle = |\overline{P}^0\rangle$ – we must choose the minus sign in Eq. (6.46) to remain consistent with Eq. (6.44). To say it differently: if we require $\tilde{\Delta}M > 0$ we have to compute the sign of $M_2 - M_1$ to decide on the sign for q/p in Eq. (6.46). Alternatively we can keep the plus sign if we adopt $\mathbf{CP}|P^0\rangle = -|\overline{P}^0\rangle$. In either case we have to calculate the sign of $M_2 - M_1$ *before* we can fix the convention for q/p or the phase of $|\overline{P}^0\rangle$. We feel it is more natural to define a convention independent of prior theoretical calculations.

6.6 Time evolution

A procedure similar to the one used in deriving Eq. (5.23) and Eq. (5.24) leads to expressions for the time evolution of states starting out as P^0 or $\overline{P}^0$:

$$\begin{aligned}|P^0(t)\rangle &= f_+(t)|P^0\rangle + \frac{q}{p}f_-(t)|\overline{P}^0\rangle \\ |\overline{P}^0(t)\rangle &= f_+(t)|\overline{P}^0\rangle + \frac{p}{q}f_-(t)|P^0\rangle,\end{aligned} \tag{6.47}$$

with

$$f_\pm(t) = \frac{1}{2}e^{-iM_1 t}e^{-\frac{1}{2}\Gamma_1 t}\left[1 \pm e^{-i\Delta M t}e^{\frac{1}{2}\Delta\Gamma t}\right]. \tag{6.48}$$

Denoting by $A(f)$ and $\overline{A}(f)$ the amplitude for the decay of P^0 and $\overline{P}^0$, respectively, into a final state f, and by $\overline{\rho}(f)$ and $\rho(f)$ their ratios, i.e.,

$$A(f) = \langle f|H_{\Delta F=1}|P^0\rangle, \quad \overline{A}(f) = \langle f|H_{\Delta F=1}|\overline{P}^0\rangle,$$

$$\overline{\rho}(f) = \frac{\overline{A}(f)}{A(f)} = \frac{1}{\rho(f)}, \tag{6.49}$$

we write down

$$\Gamma(P^0(t) \to f) \quad \propto \quad e^{-\Gamma_1 t}|A(f)|^2\left[K_+(t) + K_-(t)\left|\frac{q}{p}\right|^2|\overline{\rho}(f)|^2\right.$$

$$+ \quad 2\mathrm{Re}\left[L^*(t)\left(\frac{q}{p}\right)\overline{\rho}(f)\right]\right] \tag{6.50}$$

$$\Gamma(\overline{P}^0(t) \to f) \quad \propto \quad e^{-\Gamma_1 t}|\overline{A}(f)|^2\left[K_+(t) + K_-(t)\left|\frac{p}{q}\right|^2|\rho(f)|^2\right.$$

$$+ \quad 2\mathrm{Re}\left[L^*(t)\left(\frac{p}{q}\right)\rho(f)\right]\right], \tag{6.51}$$

where

$$|f_\pm(t)|^2 = \frac{1}{4}e^{-\Gamma_1 t}K_\pm(t),$$

$$f_-(t)f_+^*(t) = \frac{1}{4}e^{-\Gamma_1 t}L^*(t),$$

$$K_\pm(t) = 1 + e^{\Delta\Gamma t} \pm 2e^{\frac{1}{2}\Delta\Gamma t}\cos\Delta M t,$$

$$L^*(t) = 1 - e^{\Delta\Gamma t} + 2ie^{\frac{1}{2}\Delta\Gamma t}\sin\Delta M t. \tag{6.52}$$

6.7 CP violation

The decay rate evolution is in general rather complex where the time dependence is described in terms of factors $\cos\Delta M t$, $\sin\Delta M t$, $e^{-\Gamma t}$, $e^{-\Gamma t}\cos\Delta M t$ and $e^{-\Gamma t}\sin\Delta M t$.

Staring at the most general expression is not always very illuminating. The physics can become clearer through examining simpler special cases; we will discuss three complementary categories.

(A) No (or no appreciable) oscillations occur, i.e.,

$$\Delta M = \Delta\Gamma = 0. \tag{6.53}$$

Then we have

$$K_+(t) \equiv 4, \quad K_-(t) \equiv L(t) \equiv 0. \tag{6.54}$$

This situation is depicted in Fig. 6.1(A). The time evolutions are then purely exponential in t; a CP asymmetry can still arise, if

$$|A(f)| \neq |\overline{A}(\overline{f})| \tag{6.55}$$

were to hold.

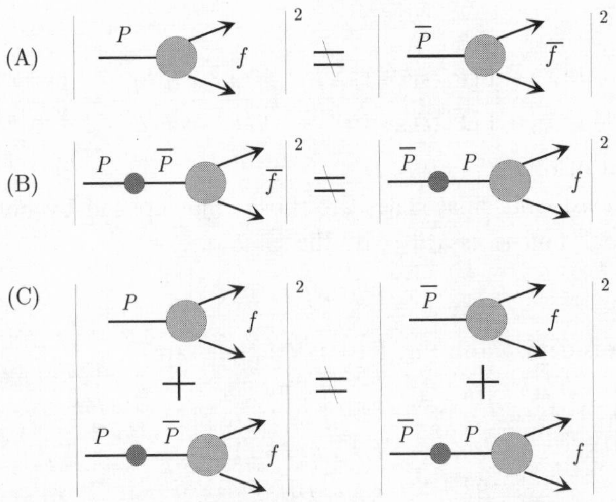

Fig. 6.1.   Three ways in which **CP** violation can show its face. (A) Without $P-\overline{P}$ mixing, **CP** asymmetry may arise if there is some **CP** violating mechanism in the $P \rightarrow f$ and $\overline{P} \rightarrow \overline{f}$ decay amplitudes. This is called direct **CP** violation. (B) In the presence of $P-\overline{P}$ oscillations, flavour specific final states can reveal **CP** violation in M_{12}. (C) If we choose a **CP** eigenstate f, both $P \rightarrow f$ and $\overline{P} \rightarrow f$ decays occur. Then there are two possible decay chains for $P \rightarrow f$ and $\overline{P} \rightarrow f$ decays. These two decay chains interfere and this may result in a **CP** asymmetry.

(B) *Flavour-specific* decays are those that can come from either P^0 or $\overline{P}^0$, but not both:

$$P^0 \rightarrow f \nleftarrow \overline{P}^0 \quad \text{or} \quad P^0 \nrightarrow f \leftarrow \overline{P}^0. \qquad (6.56)$$

This situation is depicted in Fig. 6.1(B). Prominent flavour-specific channels for neutral mesons like K^{neut}, D^{neut} or B^{neut} are provided by semileptonic decays. This is condensed into the following notation:

$$P^0 \rightarrow l^+ + X \nleftarrow \overline{P}^0, \quad P^0 \nrightarrow l^- + X \leftarrow \overline{P}^0, \qquad (6.57)$$

i.e.,

$$|A(l^+X)| = |\overline{A}(l^-X)| \equiv A_{SL}, \quad A(l^-X) = \overline{A}(l^+X) = 0, \qquad (6.58)$$

with **CPT** invariance enforcing $|A(l^+X)| = |\overline{A}(l^-X)|$.

Eq. (6.50) and Eq. (6.51) then yield

$$\Gamma(P^0(t) \rightarrow l^+ + X) \quad \propto \quad e^{-\Gamma_1 t}K_+(t)|A_{SL}|^2$$

$$\Gamma(P^0(t) \rightarrow l^- + X) \quad \propto \quad e^{-\Gamma_1 t}K_-(t)\left|\frac{q}{p}\right|^2 |A_{SL}|^2$$

$$\Gamma(\overline{P}^0(t) \rightarrow l^- + X) \quad \propto \quad e^{-\Gamma_1 t}K_+(t)|A_{SL}|^2$$

$$\Gamma(\overline{P}^0(t) \rightarrow l^+ + X) \quad \propto \quad e^{-\Gamma_1 t}K_-(t)\left|\frac{p}{q}\right|^2 |A_{SL}|^2. \qquad (6.59)$$

Note that

$$\frac{\Gamma(P^0(t) \to l^- X) - \Gamma(\overline{P}^0(t) \to l^+ X)}{\Gamma(P^0(t) \to l^- X) + \Gamma(\overline{P}^0(t) \to l^+ X)} = \frac{|q/p|^2 - |p/q|^2}{|q/p|^2 + |p/q|^2} = \frac{1 - |p/q|^4}{1 + |p/q|^4} \qquad (6.60)$$

is independent of time.

(C) *Flavour-nonspecific* final states are those that are fed by both P^0 and $\overline{P}^0$ decays, although not necessarily with the same rate:

$$P^0 \to f \leftarrow \overline{P}^0. \qquad (6.61)$$

This situation is depicted in Fig. 6.1(C). Examples are

$$
\begin{aligned}
K^0 &\to \pi\pi \leftarrow \overline{K}^0 \\
D^0 &\to K\overline{K}, \pi\pi, K\pi, \overline{K}\pi \leftarrow \overline{D}^0 \\
B^0 &\to \psi K_S, D\overline{D}, \pi\pi \leftarrow \overline{B}^0.
\end{aligned} \qquad (6.62)
$$

CP eigenstates – $\mathbf{CP}|f_\pm\rangle = \pm|f_\pm\rangle$ – fall into this category, but such final states are not necessarily **CP** eigenstates: for example, doubly Cabibbo suppressed transitions allow the channels $K^-\pi^+$ and $K^+\pi^-$ to be fed from D^0 as well as $\overline{D}^0$ decays; this subject will be addressed in more detail later on. Yet when the final state is a **CP** eigenstate the relevant expressions simplify considerably – especially if

$$|A(f)| = |\overline{A}(f)| \qquad (6.63)$$

holds.

The important point to note here is that there are sources of **CP** violation other than Imϕ. Consider a case $r \ll 1$ which is relevant for B decays. Then $|q/p| = 1$. Let us further suppose that Eq. (6.63) holds and $|\overline{\rho}(f)| = 1$. The master equations Eq. (6.50) and Eq. (6.51) yield:

$$\Gamma(P^0(t) \to f) \propto 2e^{-\Gamma_1 t}|A(f)|^2$$
$$\times \left(1 + e^{\Delta\Gamma t} + \text{Re}\left(\frac{q}{p}\overline{\rho}(f)\right)[1 - e^{\Delta\Gamma t}] - 2\text{Im}\left(\frac{q}{p}\overline{\rho}(f)\right)e^{\frac{1}{2}\Delta\Gamma t}\sin\Delta Mt\right)$$
$$(6.64)$$

$$\Gamma(\overline{P}^0(t) \to f) \propto 2e^{-\Gamma_1 t}|A(f)|^2$$
$$\times \left(1 + e^{\Delta\Gamma t} + \text{Re}\left(\frac{q}{p}\overline{\rho}(f)\right)[1 - e^{\Delta\Gamma t}] + 2\text{Im}\left(\frac{q}{p}\overline{\rho}(f)\right)e^{\frac{1}{2}\Delta\Gamma t}\sin\Delta Mt\right)$$
$$(6.65)$$

and

$$\frac{\Gamma(P^0(t) \to f) - \Gamma(\overline{P}^0(t) \to f)}{\Gamma(P^0(t) \to f) + \Gamma(\overline{P}^0(t) \to f)} = \frac{-2\sin(\phi_{\Delta F=2} + \phi_{\Delta F=1})e^{\frac{1}{2}\Delta\Gamma t}\sin\Delta Mt}{1 + e^{\Delta\Gamma t} + \cos(\phi_{\Delta F=2} + \phi_{\Delta F=1})[1 - e^{\Delta\Gamma t}]}$$
$$(6.66)$$

where we have written $q/p = e^{i\phi_{\Delta F=2}}$ and $\bar{\rho} = e^{i\phi_{\Delta F=1}}$, and used the fact that $\frac{q}{p}\bar{\rho}(f) = (\frac{p}{q}\rho(f))^*$. Thus we see that even if $|q/p| = 1$ and $|A(f)| = |\bar{A}(f)|$ hold, a CP asymmetry arises if two conditions are satisfied:

- $P^0 - \overline{P}^0$ oscillations occur generating $\Delta M \neq 0$;

- $\arg(q/p) + \arg(\bar{\rho}(f)) \neq 0$.

6.8 On the sign of the CP asymmetry

At first sight we might think that outside the kaon complex the observable asymmetry $\sin\Delta Mt \ \mathrm{Im}\frac{q}{p}\bar{\rho}(f)$ contains an ambiguity concerning its sign. For the term $\sin\Delta Mt = \sin(M_2 - M_1)t$ flips its sign under exchanging $1 \leftrightarrow 2$, and the subscripts 1 and 2 are mere labels as long as we *cannot* distinguish P_1 and P_2 through their lifetimes. However, further reflection shows that the exchange $1 \leftrightarrow 2$ flips the sign of ΔM as well as that of $\frac{q}{p}\bar{\rho}(f)$. The observable $\sin\Delta Mt \ \mathrm{Im}\frac{q}{p}\bar{\rho}(f)$ thus remains *unaffected* by such a switch in labels. This can be seen in the following manner:

- Changing $q/p \rightarrow -q/p$ maintains the defining property $(q/p)^2 = (M_{12}^* - \frac{i}{2}\Gamma_{12}^*)/(M_{12} - \frac{i}{2}\Gamma_{12})$, see Eq. (6.24).

- Yet the two mass eigenstates labeled by subscripts 1 and 2 exchange places, see Eq. (6.22).

- The mass difference $\Delta M \equiv M_2 - M_1 = -2\mathrm{Re}[\frac{q}{p}(M_{12} - \frac{i}{2}\Gamma_{12})]$ then flips its sign – yet $\sin\Delta Mt \ \mathrm{Im}\frac{q}{p}\bar{\rho}(f)$ remains invariant!

- Alternatively, $|\overline{P}^0\rangle \rightarrow -|\overline{P}^0\rangle$ leads to the switch $1 \leftrightarrow 2$, see Eq. (6.19). Thus $\Delta M \rightarrow -\Delta M$, $\bar{\rho}(f) \rightarrow -\bar{\rho}(f)$ again without affecting $\sin\Delta Mt \ \mathrm{Im}\frac{q}{p}\bar{\rho}(f)$!

Hence we infer that *within a given theory of $\Delta F = 1, 2$ dynamics we can calculate the product $\sin\Delta Mt \ \mathrm{Im}\frac{q}{p}\bar{\rho}(f)$ including its sign.* Later this will be demonstrated explicitly. Here we want to make only two additional statements:

- Consider a CP eigenstate $f_\pm$ as final state: $\mathbf{CP}|f_\pm\rangle = \pm|f_\pm\rangle$. We then have

$$\langle f_\pm|H|\overline{P}^0\rangle = \langle f_\pm|(\mathbf{CP})^\dagger\mathbf{CP} \ H(\mathbf{CP})^\dagger\mathbf{CP}|\overline{P}^0\rangle$$
$$= \pm\langle f_\pm|H^{CP}|P^0\rangle \tag{6.67}$$

and therefore

$$\bar{\rho}(f_\pm) = \pm\frac{\langle f_\pm|H^{CP}|P^0\rangle}{\langle f_\pm|H^{CP}|P^0\rangle} ; \tag{6.68}$$

i.e., the sign of the asymmetry depends on the **CP** quantum number of the final state.

- Consider a decay $B \rightarrow f_\pm$. It is important to distinguish this decay from $B \rightarrow f_\pm \pi^0$, which is often experimentally non-trivial. Because $\mathbf{CP}|\pi^0\rangle = -|\pi^0\rangle$, certain kinematical configurations will lead to $\overline{\rho}(f_\pm \pi^0) = -\overline{\rho}(f_\pm)$. This will have an effect of reducing the asymmetry.

6.9 What happens if you don't observe the decay time?

For the K meson system it is relatively easy to observe the time dependence of the instantaneous decay rate. Yet the tracks of short lived particles such as *B* mesons, which live for only about 1.5 ps, are much more difficult to measure since $c\tau_B \sim 0.5$ mm. As explained later, so-called microvertex detectors have been developed that allow us to track such decays; yet typical detectors effectively integrate over all times of decay. Therefore we give here time integrated versions of our predictions. We can express time integrated versions of Eq. (6.50)–Eq. (6.51) and Eq. (6.64)–Eq. (6.65) in terms of

$$\langle K_\pm \rangle \equiv \int_0^\infty dt e^{-\Gamma_1 t} K_\pm(t)$$

$$= \frac{2}{\Gamma}\left[\frac{1}{1-y^2} \pm \frac{1}{1+x^2}\right],$$

$$\langle L^* \rangle \equiv \int_0^\infty dt e^{-\Gamma_1 t} L^*(t)$$

$$= \frac{2}{\Gamma}\left[-\frac{y}{1-y^2} + i\frac{x}{1+x^2}\right], \tag{6.69}$$

where

$$x = \frac{\Delta M}{\Gamma}, \qquad y = \frac{\Delta \Gamma}{2\Gamma}, \quad \text{and} \quad \Gamma = \frac{1}{2}(\Gamma_1 + \Gamma_2) \tag{6.70}$$

denote the oscillation rate calibrated by the decay rate. We record here the time integrated version of Eq. (6.59), Eq. (6.64) and Eq. (6.65), the latter two for $y = 0$:

$$\int_0^\infty \Gamma(P^0(t) \rightarrow l^+ + X)dt \quad \propto \quad \frac{2+x^2-y^2}{(1-y^2)(1+x^2)}|A_{SL}|^2,$$

$$\int_0^\infty \Gamma(P^0(t) \rightarrow l^- + X)dt \quad \propto \quad \frac{x^2+y^2}{(1-y^2)(1+x^2)}\left|\frac{q}{p}\right|^2 |A_{SL}|^2,$$

$$\int_0^\infty \Gamma(\overline{P}^0(t) \rightarrow l^- + X)dt \quad \propto \quad \frac{2+x^2-y^2}{(1-y^2)(1+x^2)}|A_{SL}|^2,$$

$$\int_0^\infty \Gamma(\overline{P}^0(t) \rightarrow l^+ + X)dt \quad \propto \quad \frac{x^2+y^2}{(1-y^2)(1+x^2)}\left|\frac{p}{q}\right|^2 |A_{SL}|^2. \tag{6.71}$$

If $|\bar{\rho}| = |\frac{q}{p}| = 1$ and $y = 0$,

$$\int_0^\infty \Gamma(P^0(t) \to f) dt \quad \propto \quad 4|A(f)|^2 \cdot \left(1 - \frac{x}{1+x^2} \mathrm{Im}\frac{q}{p}\bar{\rho}(f)\right),$$

$$\int_0^\infty \Gamma(\bar{P}^0(t) \to f) dt \quad \propto \quad 4|A(f)|^2 \cdot \left(1 + \frac{x}{1+x^2} \mathrm{Im}\frac{q}{p}\bar{\rho}(f)\right). \qquad (6.72)$$

The observable **CP** asymmetry thus depends primarily on the ratio x rather than on ΔM. This is as expected: for with the **CP** asymmetry being due to the interference between the decay and the oscillation amplitude, it is the *relative* weight of the latter that counts. On the other hand, it remains highly desirable to resolve the decay vertices and thus track the time evolution of these decays. Later we will explain why this is even absolutely essential in certain cases like $\Upsilon(4S) \to B^0\bar{B}^0$.

6.10 Regeneration

We consider how a beam made up of P states evolves in time when passing through nuclear matter along the z axis. So, z and t are related by $z = \gamma\beta t$, $\gamma = 1/\sqrt{1-\beta^2}$; β denotes the velocity of P. We start with defining $\Psi(t)$ similarly to Eq. (5.10) with K replaced by P. The behaviour of P^0 and $\bar{P}^0$ passing through a medium can be described in terms of indices of refraction n and $\bar{n}$; those are defined by [43]

$$\frac{d|\Psi(t)\rangle}{dz} = i\,k \begin{pmatrix} n-1 & 0 \\ 0 & \bar{n}-1 \end{pmatrix} |\Psi(t)\rangle, \qquad (6.73)$$

where k is the momentum of the particle. The evolution equations then read as follows:

$$\frac{d|\Psi(t)\rangle}{dt} = -\left[(\tfrac{1}{2}\Gamma + i\,\mathbf{M}) - ik\beta\gamma \begin{pmatrix} n-1 & 0 \\ 0 & \bar{n}-1 \end{pmatrix}\right] |\Psi(t)\rangle. \qquad (6.74)$$

Refraction effectively adds a term $-k\beta\gamma(n-1)$ to M_{11} and $-k\beta\gamma(\bar{n}-1)$ to M_{22}, while keeping the off-diagonal matrix element unchanged:

$$\frac{d|\Psi(t)\rangle}{dt} = -\left[(\tfrac{1}{2}\Gamma + i\,\tilde{\mathbf{M}})\right] |\Psi(t)\rangle \qquad (6.75)$$

$$\tilde{\mathbf{M}} = \begin{pmatrix} M_{11} + -k\beta\gamma(n-1) & M_{12} \\ M_{12}^* & M_{22} + -k\beta\gamma(\bar{n}-1) \end{pmatrix}. \qquad (6.76)$$

We can then employ the same procedure as developed in the preceding chapter to obtain decoupled differential equations. The eigenvectors and eigenvalues of $\tilde{\mathbf{M}}$ will differ from those of $\mathbf{M}$. Yet there is one important *qualitative* distinction: since

$$\tilde{M}_{11} \neq \tilde{M}_{22} \qquad (6.77)$$

we are dealing with a situation where **CPT** invariance is *effectively* broken: the fact that P^0 and $\overline{P}^0$ propagate through matter rather than antimatter mimics **CPT** violation. This observation recalls a theme we have sounded before in analysing time reversal invariance (and to which we will return): *an observed difference in two conjugate transition rates can be caused by an asymmetry in the prevailing boundary conditions rather than a difference in the fundamental dynamics.*

Using spherical coordinates as introduced in Section 6.3, we have

$$\tilde{E}_1 = E_1, \qquad\qquad \tilde{E}_2 = E_2,$$
$$\tilde{E}_3 = -\frac{1}{2}k\beta\gamma(n-\bar{n}), \qquad \tilde{E} = \sqrt{E_1^2 + E_2^2 + \tilde{E}_3^2}, \qquad (6.78)$$

and therefore

$$\cos\tilde{\theta} = \frac{\tilde{E}_3}{\tilde{E}} \neq 0, \qquad (6.79)$$

representing the effective **CPT** asymmetry alluded to above. We then find for a state that started out as pure $|P_1\rangle$ at time $t = 0$, at later times

$$|P_1(t)\rangle = (\tilde{f}_+(t) + \sin\tilde{\theta}\tilde{f}_-(t))|P_1\rangle + \cos\tilde{\theta}\tilde{f}_-(t)|P_2\rangle, \qquad (6.80)$$

where the tilde over $\tilde{f}_\pm(t)$ reminds us that these functions are determined by the eigenvalues of the mass matrix $\tilde{\mathbf{M}}$ given in Eq. (6.76). Likewise when the initial state is pure $|P_2\rangle$:

$$|P_2(t)\rangle = (\tilde{f}_+(t) - \sin\tilde{\theta}\tilde{f}_-(t))|P_2\rangle + \cos\tilde{\theta}\tilde{f}_-(t)|P_1\rangle. \qquad (6.81)$$

This means that due to $\cos\tilde{\theta} \neq 0$ a P_2 [P_1] is regenerated from an initially pure P_1 [P_2] beam when traversing matter.

This is also of practical interest for actual tests of **CPT** invariance. Real experiments are undertaken not in a perfect vacuum, but in an environment that is dominated by matter; in particular the detector is *not* **CPT** invariant! This places some inherent limitations on **CPT** tests.

6.11 The Bell–Steinberger inequality

With P_1 and P_2 denoting the two mass eigenstates of the $P^0 - \overline{P}^0$ complex we have

$$\langle P_1|\tfrac{1}{2}\boldsymbol{\Gamma} + i\mathbf{M}|P_2\rangle = (\tfrac{1}{2}\Gamma_2 + iM_2)\langle P_1|P_2\rangle,$$
$$\langle P_2|\tfrac{1}{2}\boldsymbol{\Gamma} + i\mathbf{M}|P_1\rangle = (\tfrac{1}{2}\Gamma_1 + iM_1)\langle P_2|P_1\rangle. \qquad (6.82)$$

Taking the complex conjugate of the first equation and adding it to the second one yields

$$\langle P_2|\boldsymbol{\Gamma}|P_1\rangle = \left[\frac{1}{2}(\Gamma_1 + \Gamma_2) + i(M_1 - M_2)\right]\langle P_2|P_1\rangle. \qquad (6.83)$$

Recalling the definition Eq. (6.8), we have

$$\langle P_2|\Gamma|P_1\rangle = 2\pi \sum_f \delta(M_P - M_f)\langle f|H|P_2\rangle^* \langle f|H|P_1\rangle, \qquad (6.84)$$

and using the Schwartz inequality,

$$|\langle P_2|\Gamma|P_1\rangle|^2 \leq \sum_f \Gamma_1^f \Gamma_2^f, \qquad (6.85)$$

we finally arrive at [45]

$$|\langle P_2|P_1\rangle| \leq \sqrt{\frac{\sum_f 4\Gamma_1^f \Gamma_2^f}{(\Gamma_1 + \Gamma_2)^2 + 4(M_1 - M_2)^2}}. \qquad (6.86)$$

Eq. (6.86) is called the Bell–Steinberger relation or inequality, which was first derived for neutral kaons. In the preceding section we have learnt that the two mass eigenstates are no longer orthogonal to each other if **CP** (or **CPT**) invariance is violated. The inequality in Eq. (6.86) tells us that the amount of that non-orthogonality is constrained by a sum over exclusive P_1 and P_2 widths. We see that this inequality is numerically quite relevant for kaons.

$$|\langle K_L|K_S\rangle| \leq \sqrt{\frac{\sum_f \Gamma_L^f \Gamma_S^f}{(\Gamma_L + \Gamma_S)^2 + 4(\Delta M_K)^2}} \simeq \sqrt{\frac{\sum_f 2\Gamma_L^f \Gamma_S^f}{\Gamma_S^2}}, \qquad (6.87)$$

where again we have used $\Gamma_L \ll \Gamma_S \simeq 2\Delta M_K$. Since $\Gamma_L \Gamma_S = \sum_{f_L, f_S} \Gamma_L^{f_L} \Gamma_S^{f_S} \geq \sum_f \Gamma_L^f \Gamma_S^f$, we obtain, using the experimental result Eq. (5.20), a very conservative bound

$$|\langle K_L|K_S\rangle| \leq \sqrt{\frac{2\Gamma_L}{\Gamma_S}} \simeq 0.06. \qquad (6.88)$$

Since **CP** invariance requires $\langle K_L|K_S\rangle = 0$, the following message is contained in Eq. (6.88): unitarity as expressed through the Bell–Steinberger relation together with the experimental findings $\Gamma_L \ll \Gamma_S \sim 2\Delta M_K$ already imposes a near-orthogonality of K_L and K_S, irrespective of **CP** violation!

6.12 Résumé on $P^0 - \overline{P}^0$ oscillations

Now we are in a position to summarize our discussion on $P^0 - \overline{P}^0$ oscillations. There are four classes of **CP** asymmetries:

1

$$|A(f)| \neq |\overline{A}(\overline{f})|, \qquad (6.89)$$

or for f being **CP** self-conjugate: $|\overline{\rho}_f| \neq 1$. This condition is obviously independent of the phase convention for $\overline{P}^0$. Such an asymmetry unambiguously reflects $\Delta F = 1$ dynamics and can occur also in the decays of charged mesons, baryons and leptons. It is referred to as *direct* **CP** *violation*.

2

$$\left|\frac{q}{p}\right| \neq 1 .$$ (6.90)

This can be detected experimentally by measuring the lepton asymmetry, see Eq. (6.60). This is unambiguously driven by the dynamics in the $\Delta F = 2$ sector. It thus represents **CP** violation in $P^0 - \overline{P}^0$ oscillations and is often referred to as *superweak* **CP** *violation*, as explained later.

3

$$\text{Im}\left[\frac{q}{p}\overline{\rho}_f\right] = \left|\frac{q}{p}\overline{\rho}_f\right| \sin(\arg(q/p) + \arg\overline{\rho}(f)) \neq 0 .$$ (6.91)

It reflects the combined effect of $\Delta F = 2$ and $\Delta F = 1$ dynamics and can be referred to as *CP violation involving* $P^0 - \overline{P}^0$ *oscillations*. As long as such an asymmetry has been studied for a *single* final-state f only (or in a single pair of **CP** conjugate states f and $\overline{f}$), it is meaningless to differentiate between the effects of $\Delta F = 2$ and $\Delta F = 1$ forces, i.e. between *superweak* and *direct* **CP** violation. For a change in the phase convention will shift the weight between $\arg(q/p)$ and $\arg\overline{\rho}(f)$. Changing

$$\mathbf{CP}|P^0\rangle \equiv |\overline{P}^0\rangle$$ (6.92)

to the equivalent definition

$$\mathbf{CP}|P^0\rangle \equiv e^{i\xi}|\overline{P}^0\rangle , \quad \xi \text{ real},$$ (6.93)

will obviously have no effect on $|\overline{\rho}_f|$ or $|q/p|$. Yet q/p and $\overline{\rho}_f$ are affected by it: on one hand we have

$$\overline{\rho}(f) = \frac{\overline{A}(f)}{A(f)} = \frac{\langle f|H_{\Delta F=1}|\overline{P}^0\rangle}{\langle f|H_{\Delta F=1}|P^0\rangle} \to e^{2i\xi}\overline{\rho}(f),$$ (6.94)

whereas on the other hand we find

$$(M_{12}, \Gamma_{12}) \to e^{2i\xi}(M_{12}, \Gamma_{12}),$$ (6.95)

leading to

$$\frac{q}{p} = \sqrt{\frac{M_{12}^* - \frac{i}{2}\Gamma_{12}^*}{M_{12} - \frac{i}{2}\Gamma_{12}}} \to e^{-i2\xi}\sqrt{\frac{M_{12}^* - \frac{i}{2}\Gamma_{12}^*}{M_{12} - \frac{i}{2}\Gamma_{12}}} = e^{-i2\xi}\frac{q}{p}.$$ (6.96)

Yet their product remains invariant:

$$\frac{q}{p}\overline{\rho}(f) \to \frac{q}{p}\overline{\rho}(f),$$ (6.97)

i.e., the sum $\arg(q/p) + \arg\overline{\rho}(f)$ – in contrast to its individual terms – is *not* sensitive to changes in the phase convention for the antiparticle and thus qualifies as an observable – as is clear also from the explicit calculation leading to Eq. (6.66).

4 Once we have found a **CP** asymmetry in two different final states f_1 and f_2 (or in two different conjugate pairs), the issue of superweak vs direct **CP** violation can be addressed in a meaningful way. For if we observe

$$\sin(\arg(q/p) + \arg\overline{\rho}(f_1)) \neq \sin(\arg(q/p) + \arg\overline{\rho}(f_2)), \qquad (6.98)$$

then we know unambiguously that direct **CP** violation is present.

These observations suggest the following more meaningful characterization: *if we can choose the phase for the antiparticle in such a way that all* **CP** *asymmetries for this particle–antiparticle complex can be assigned to* $\phi_{\Delta F=2}$, *then we are dealing with a* superweak *scenario.*

Problems

6.1   Derive $M_{12} = M_{21}$ from **CP** and **T** symmetry.

6.2   Considering the normalization of the state, show that

$$\Gamma_{11} = \frac{1}{2M} \int \prod_{i=1}^{n} \frac{d^3 k_i}{(2\pi)^3 2E_i} (2\pi)^4 \delta^4(P - k_1 - \cdots - k_n)$$

$$|\langle \vec{k}_1 \ldots k_n | \mathcal{H}(0) | P \rangle|^2 \qquad (6.99)$$

as expected for the width of P decay into n particles.

6.3   In quantum mechanics we have learnt that two eigenstates which correspond to different eigenvalues are orthogonal. Explain Eq. (6.30).

6.4   With $|P_{1,2}\rangle \sim |P^0\rangle \pm e^{i\phi} |\overline{P}^0\rangle$, derive Eq. (6.80) and Eq. (6.81).

6.5   Explain how the experimental fact $3M_\pi \simeq M_K$ makes the Bell–Steinberger bound very powerful.

6.6   For $\Delta\Gamma \ll \Delta M$, show that

$$f_\pm(t) = \begin{pmatrix} \cos\frac{\Delta M t}{2} \\ i \sin\frac{\Delta M t}{2} \end{pmatrix} e^{-\frac{1}{2}\Gamma_1 t} e^{-i\frac{1}{2}(M_1 + M_2)t}. \qquad (6.100)$$

$$\Gamma(P^0(t) \to l^+ + X) \quad \propto \quad e^{-\Gamma_1 t} \cos^2 \frac{\Delta M t}{2} |A_{SL}|^2$$

$$\Gamma(P^0(t) \to l^- + X) \quad \propto \quad e^{-\Gamma_1 t} \sin^2 \frac{\Delta M t}{2} \left|\frac{q}{p}\right|^2 |A_{SL}|^2$$

$$\Gamma(\overline{P}^0(t) \to l^- + X) \quad \propto \quad e^{-\Gamma_1 t} \cos^2 \frac{\Delta M t}{2} |A_{SL}|^2$$

$$\Gamma(\overline{P}^0(t) \to l^+ + X) \quad \propto \quad e^{-\Gamma_1 t} \sin^2 \frac{\Delta M t}{2} \left|\frac{p}{q}\right|^2 |A_{SL}|^2. \qquad (6.101)$$

6.7   Show that for $\Delta\Gamma \ll \Delta M$, Eq. (6.51), Eq. (6.50), and Eq. (6.66) simplify to

$$\Gamma(P^0 \to f) \ \sim \ e^{-\Gamma t}|A(f)|^2 \left[1 - \mathrm{Im} \left(\frac{q}{p}\rho(f) \right) \sin \Delta M t \right]$$

$$\Gamma(\overline{P}^0 \to f)| \ \sim \ e^{-\Gamma t}|A(f)|^2 \left[1 + \mathrm{Im} \left(\frac{q}{p}\rho(f) \right) \sin \Delta M t \right] \quad (6.102)$$

$$\frac{\Gamma(P^0(t) \to f) - \Gamma(\overline{P}^0(t) \to f)}{\Gamma(P^0(t) \to f) + \Gamma(\overline{P}^0(t) \to f)} = -\sin(\phi_{\Delta F=2} + \phi_{\Delta F=1})\sin\Delta M t. \quad (6.103)$$

6.8   There is another special case of considerable interest for neutral kaons: waiting long enough for the K_S component to decay away and thus preparing a K_L beam, we track its transition rate into a $\pi\pi$ final state. Show that

$$\left(\frac{\Gamma(K^0 \to \pi\pi)}{\Gamma(\overline{K}^0 \to \pi\pi)} \right) \propto e^{-\Gamma_S t} + e^{-\Gamma_L t}|\eta|^2 \pm 2e^{-\frac{1}{2}(\Gamma_L+\Gamma_S)t}\mathrm{Re}\left[e^{-i\Delta M t}\eta\right], \quad (6.104)$$

where the third term describes the $K_L - K_S$ interference.

6.9   Diagonalize the Hamiltonian

$$\begin{pmatrix} M & \Delta e^{i\phi} \\ \Delta e^{-i\phi} & M \end{pmatrix} \quad (6.105)$$

where Δ is a positive or negative real number and ϕ is the phase of the off-diagonal element, which is defined to be between 0 and π. Show that the mass eigenstate is $(1, \pm e^{-i\phi})$ with eigenvalue $M \pm \Delta$.

7

CP phenomenology of strange decays

As long as **CPT** invariance holds, both **CP** and **T** violation are represented by complex relative phases in the effective coupling constants. For a **CP** or **T** asymmetry to become observable we thus need (at least) two amplitudes containing different weak couplings, yet contributing *coherently*.

This can be achieved in numerous ways, which can be grouped into several categories. First we will list these categories together with relevant data; then we will describe the phenomenology relating these observables.

7.1 The landscape

7.1.1 Existence

CP violation manifested itself first through the *existence* of a certain decay mode, $K_L \to \pi\pi$. Defining

$$\eta_{+-} \equiv \frac{\langle \pi^+\pi^-|H_W|K_L\rangle}{\langle \pi^+\pi^-|H_W|K_S\rangle}, \quad \eta_{00} \equiv \frac{\langle \pi^0\pi^0|H_W|K_L\rangle}{\langle \pi^0\pi^0|H_W|K_S\rangle}, \tag{7.1}$$

we see that **CP** invariance implies $\eta_{+-} = 0 = \eta_{00}$. Present data yield:

$$|\eta_{+-}| = (2.275 \pm 0.019) \times 10^{-3} \quad [46]$$
$$|\eta_{00}| = (2.285 \pm 0.019) \times 10^{-3}. \quad [47] \tag{7.2}$$

7.1.2 Difference in integrated rates

CP *asymmetries* were subsequently found in semileptonic K_L decays [48]:

$$\delta_l \equiv \frac{\Gamma(K_L \to l^+\nu_l\pi^-) - \Gamma(K_L \to l^-\bar{\nu}_l\pi^+)}{\Gamma(K_L \to l^+\nu_l\pi^-) + \Gamma(K_L \to l^-\bar{\nu}_l\pi^+)} = (3.27 \pm 0.12) \times 10^{-3}, \tag{7.3}$$

where we have take the average $\delta_l \equiv \langle \delta_e + \delta_\mu \rangle$.

7.1.3 Direct test of **T** invariance

Very recently the so-called Kabir test [50] has revealed a time asymmetry[1] for the first time [51]:

$$A_T = \frac{\text{rate}(\overline{K}^0 \to K^0) - \text{rate}(K^0 \to \overline{K}^0)}{\text{rate}(\overline{K}^0 \to K^0) + \text{rate}(K^0 \to \overline{K}^0)} = (6.6 \pm 1.3 \pm 1.0) \times 10^{-3}. \tag{7.4}$$

---

[1] We have assumed that **CPT** is conserved in these semileptonic decays (see Chapter 14).

This measurement contains a few special features. When analysing K_L decays 'patience pays off'; i.e., because of $\Gamma(K_S) \gg \Gamma(K_L)$ we can wait till practically all K_S mesons have decayed away. The task we face in comparing $K^0 \to \overline{K}^0$ with $\overline{K}^0 \to K^0$ is quite different: the decay products reveal the flavour identity of the *final* state kaon:

$$K^0 \Rightarrow \overline{K}^0 \to l^- \nu \pi^+ \quad \text{vs} \quad \overline{K}^0 \Rightarrow K^0 \to l^+ \nu \pi^-. \tag{7.5}$$

We then need independent information on the flavour identity of the *initial* kaon. This is achieved through correlations imposed by associated production. The CPLEAR collaboration studied low energy proton–antiproton annihilation:

$$p\bar{p} \to K^+ \overline{K}^0 \pi^- \quad \text{vs} \quad p\bar{p} \to K^- K^0 \pi^+. \tag{7.6}$$

It is the sign of the charged kaon that reveals the flavour identity of the neutral kaon produced in association with it. In the future we will also study $K^0 \overline{K}^0$ production at DAΦNE [52] in

$$e^+ e^- \to \phi(1020) \to K^0 \overline{K}^0. \tag{7.7}$$

7.1.4 Decay rate evolution

The decay rate evolution for $K^0 \to \pi^+ \pi^-$ as a function of (proper) time of decay differs from that for $\overline{K}^0 \to \pi^+ \pi^-$:

$$A_{+-}(t) = \frac{\Gamma(K^0(t) \to \pi^+ \pi^-) - \Gamma(\overline{K}^0(t) \to \pi^+ \pi^-)}{\Gamma(K^0(t) \to \pi^+ \pi^-) + \Gamma(\overline{K}^0(t) \to \pi^+ \pi^-)}. \tag{7.8}$$

The recent experimental result from CPLEAR [53] is shown in Fig. 7.1. We feel the desire to show you this figure because it is quite analogous to what we search for in B^0 decays.

The underlying physics can be illuminated by the following observation: within the approximation discussed in Sec. (6.1), **CP** violation is established by finding the decay rate evolution for the decay of a neutral meson P^0 into a **CP** *eigenstate* $f_\pm$ to be different from any *single pure* exponential; i.e.,

$$\frac{\mathrm{d}}{\mathrm{d}t} e^{\Gamma t} \mathrm{rate}(P^0 \to f_\pm; t) \neq 0 \quad \forall \text{ real } \Gamma \quad \Longrightarrow \quad \textbf{CP} \text{ violation.} \tag{7.9}$$

The proof is completely elementary. Assume **CP** to be conserved. Then the mass eigenstates of P^0 are **CP** eigenstates as well: $\textbf{CP}|P_{1 \atop 2}^0\rangle = \pm|P_{1 \atop 2}^0\rangle$; furthermore P_1^0 can decay into f_+, but P_2^0 cannot. Therefore

$$\mathrm{rate}(P^0 \to f_+; t) = \frac{N_1}{N_1 + N_2} \cdot \mathrm{rate}(P^0 \to f_+; t) = e^{-\Gamma_1 t} \cdot \mathrm{const.}, \tag{7.10}$$

where $N_1[N_2]$ denotes the original number of $P_1^0[P_2^0]$ mesons in the beam, Q.E.D. Such a deviation from a pure exponential time evolution in $K^0 \to \pi\pi$ has been

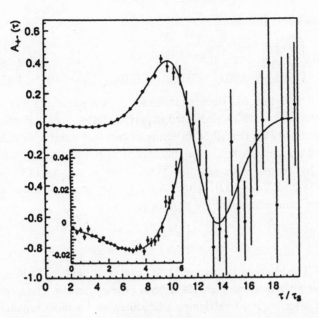

Fig. 7.1. Time dependence of $A_{+-}(t)$ in units of K_S lifetime. Later we shall insist that the time dependent asymmetry between $B \to \Psi K_S$ and $\overline{B} \to \Psi K_S$ is the gold-plated mode for **CP** violation searches in B decays. This figure represents the equivalent asymmetry for K meson decays. The problems are that $\tau_B \sim 1.5$ ps and the data can be taken only for $\tau/\tau_B \leq 4$, and that the branching ratio $Br(B \to \Psi K_S) \sim 10^{-4}$. The only thing going for us in the B system is that the asymmetry may be as large as 50% as opposed to 0.2% for the K meson system. This figure was reproduced from [53] Physics Letters by permission of Elsevier Science.

observed, as shown in Fig. 7.2; the departure from the exponential at large t which demonstrates **CP** violation is caused by the $K_L \to \pi\pi$ amplitude.

7.1.5 Direct **CP** violation

Direct **CP** violation can manifest itself through a difference in the normalized decay amplitudes for $K_L \to \pi^+\pi^-$ vs $\pi^0\pi^0$: $|\eta_{+-}| \neq |\eta_{00}|$. The conventional notation

$$\eta_{+-} \equiv \epsilon + \epsilon', \quad \eta_{00} \equiv \epsilon - 2\epsilon' \qquad (7.11)$$

makes this explicit: ϵ' describes the *channel dependent* **CP** violation whereas ϵ characterizes the decaying object K_L *produced by* $K^0 - \overline{K}^0$ *oscillations*. By measuring only the rates we obtain

$$\mathrm{Re}\frac{\epsilon'}{\epsilon} = \frac{1}{6}\frac{|\eta_{+-}|^2 - |\eta_{00}|^2}{|\eta_{+-}|^2}. \qquad (7.12)$$

The experimental verdict was ambiguous, but suddenly became very clear as the book was going into press:

$$\mathrm{Re}\frac{\epsilon'}{\epsilon} = \begin{cases} (23 \pm 6.5) \times 10^{-4} & \text{NA31[54]} \\ (7.4 \pm 5.9) \times 10^{-4} & \text{E731[55]} \\ (28.0 \pm 0.30_{\mathrm{stat.}} \pm 0.26_{\mathrm{syst.}} \pm 0.10_{\mathrm{MC\ stat.}}) \times 10^{-4} & \text{E832[56]} \end{cases}$$

i.e., direct **CP** violation has been established! A new paradigm has thus emerged after more than 30 years of dedicated experimentation: the **CP** phenomenology of K_L decays has to be described in terms of *two* real numbers rather than one!

A negative conclusion holds for differences in partial decay rates of charged kaons (and hyperons) – for example $\Gamma(K^+ \to \pi^+\pi^-\pi^+)$ vs $\Gamma(K^- \to \pi^-\pi^+\pi^-)$, see Eq. (7.91) and Eq. (7.92) below. As yet, only a much lower experimental sensitivity has been achieved here than in $K_L \to \pi\pi$.

7.1.6 **T** *odd correlations*

Examples of **T** odd correlations are Dalitz plot asymmetries in $K \to 3\pi$ and $P_\perp$, the transverse polarization of muons from $K \to \pi\mu\nu$ decays. The latter represents, as explained later, a most intriguing phenomenon. Its most sensitive study was performed a long time ago at Brookhaven National Laboratory [57]:

$$\langle \vec{s}(\mu) \cdot (\vec{p}(\mu) \times \vec{p}(\pi)) \rangle \equiv P_\perp(K^+ \to \pi^0\mu^+\nu) = (-4.2 \pm 6.2) \times 10^{-3}. \tag{7.13}$$

7.2 **Semileptonic decays**

Rather than following the historical order, we will first discuss semileptonic K_L decays since the dynamical situation is less complex there than in nonleptonic transitions.

Within the quark model semileptonic kaon decays obey, as discussed in Sec. 5.4, the $\Delta S = \Delta Q$ rule, which makes semileptonic modes flavour specific; i.e., transitions yielding 'wrong sign' leptons have to proceed via $K^0 - \overline{K}^0$ oscillations. This is completely consistent with the observation so far; any future deviation would reveal the intervention of New Physics.

Let us define the following general amplitudes [58]:

$$\begin{aligned} \langle l^+\nu\pi^- |\mathcal{H}_W|K^0\rangle &= F_l(1-y_l) \\ \langle l^+\nu\pi^- |\mathcal{H}_W|\overline{K}^0\rangle &= x_l F_l(1-y_l) \\ \langle l^-\bar{\nu}\pi^+ |\mathcal{H}_W|K^0\rangle &= \bar{x}_l^* F_l^*(1+y_l^*) \\ \langle l^-\bar{\nu}\pi^+ |\mathcal{H}_W|\overline{K}^0\rangle &= F_l^*(1+y_l^*). \end{aligned} \tag{7.14}$$

Discrete symmetries place the following restrictions on these parameters:

$\Delta S = \Delta Q$ rule:	$x_l = \bar{x}_l = 0$
CP invariance:	$x_l = \bar{x}_l^*; \quad F_l = F_l^*; \quad y_l = -y_l^*$
T invariance:	$\mathrm{Im}\, F = \mathrm{Im}\, y_l = \mathrm{Im}\, x_l = \mathrm{Im}\, \bar{x}_l = 0$
CPT invariance:	$y_l = 0, \ x_l = \bar{x}_l.$

We postpone a discussion of experimental tests of the $\Delta S = \Delta Q$ rule and **CPT** symmetry to Chapter 14. *Till then we assume* **CPT** *symmetry and the* $\Delta S = \Delta Q$ *rule to hold.* We then have

$$\langle l^+ v_l \pi^- | \mathscr{H}_W | K_L \rangle = p \langle l^+ v_l \pi^- | \mathscr{H}_W | K^0 \rangle$$
$$\langle l^- \overline{v}_l \pi^+ | \mathscr{H}_W | K_L \rangle = -q \langle l^- \overline{v}_l \pi^+ | \mathscr{H}_W | \overline{K}^0 \rangle, \qquad (7.15)$$

with **CPT** invariance implying

$$\langle l^+ v_l \pi^- | \mathscr{H}_W | K^0 \rangle = \langle l^- v_l \pi^+ | \mathscr{H}_W | \overline{K}^0 \rangle^* . \qquad (7.16)$$

The asymmetry δ_l defined in Eq. (7.3) is driven by $\Delta S = 2$ dynamics and can be expressed by

$$\delta_l = \frac{|p|^2 - |q|^2}{|p|^2 + |q|^2} \simeq \frac{1}{2} \zeta_K = (3.27 \pm 0.12) \times 10^{-3}, \qquad (7.17)$$

where Eq. (6.41) was used.

As for the Kabir test, the flavour identity of the final state kaon is deduced from its semileptonic decay; its analysis therefore fits in here. Using Eq. (6.47), the probability for $K^0(t)$ to be detected as $\overline{K}^0$, and the probability for $\overline{K}^0(t)$ to be detected as K^0 are given by

$$\text{rate}(K^0 \to \overline{K}^0) = \int_0^\infty \left| \frac{q}{p} f_-(t) \right|^2 dt$$
$$\text{rate}(\overline{K}^0 \to K^0) = \int_0^\infty \left| \frac{p}{q} f_-(t) \right|^2 dt, \qquad (7.18)$$

respectively, so that A_T defined in Eq. (7.4) is given by

$$A_T = \frac{|p/q|^2 - |q/p|^2}{|p/q|^2 + |q/p|^2} = \frac{1 - |q/p|^4}{1 + |q/p|^4} \simeq \zeta_K . \qquad (7.19)$$

Thus we get $A_T = 2\delta_l$, making the results of Eq. (7.3) and Eq. (7.4), quite consistent.

7.3 Non-leptonic K_L decays

7.3.1 Decay amplitudes

The two pion states in K decays can be classified in terms of their isospin I. Due to Bose statistics two S wave pions in $K_{L,S} \to \pi\pi$ can form an $I = 0$ or 2 configuration. As explained in Chapter 4, Watson's theorem tells us that the amplitude for a K^0 decaying into two pions with total isospin I can be expressed as follows:

$$\langle (2\pi)_I | H_W | K^0 \rangle = A_I e^{i\delta_I}, \qquad (7.20)$$

where δ_I denotes the S wave $(\pi\pi)_I$ phase shift at energy M_K and A_I is real if H_W conserves **CP**; i.e., in that case the phase of the decay amplitude is determined completely by the strong final state interactions. Likewise we find for the **CP** conjugate reactions

$$\langle (2\pi)_I | H_W | \overline{K}^0 \rangle = \overline{A}_I e^{i\delta_I}. \tag{7.21}$$

The same phase δ_I appears in both Eq. (7.20) and Eq. (7.21) since the strong forces obey **CP** and **T** invariance.

With **CP** violation arising in the weak dynamics an additional phase will emerge in A_I and $\overline{A}_I$; it will be referred to as *weak* phase.

Throughout this section we will assume **CPT** invariance until explicitly stated otherwise. Thus we have

$$
\begin{aligned}
\langle (2\pi)_I | H_W | K_L \rangle &= e^{i\delta_I} \left[p A_I - q \overline{A}_I \right] \\
\langle (2\pi)_I | H_W | K_S \rangle &= e^{i\delta_I} \left[p A_I + q \overline{A}_I \right].
\end{aligned}
\tag{7.22}
$$

We now want to write the decay amplitudes in terms of isospin amplitudes:

$$
\begin{aligned}
A(K^0 \to \pi^+ \pi^-) &= \frac{1}{\sqrt{3}} (A_2 e^{i\delta_2} + \sqrt{2} A_0 e^{i\delta_0}) \\[2mm]
A(K^0 \to \pi^0 \pi^0) &= \sqrt{\frac{2}{3}} (-\sqrt{2} A_2 e^{i\delta_2} + A_0 e^{i\delta_0}) \\[2mm]
A(K^+ \to \pi^+ \pi^0) &= \sqrt{\frac{3}{2}} A_2 e^{i\delta_2},
\end{aligned}
\tag{7.23}
$$

where the phases are obtained using Watson's theorem.

Expressing the state $\pi^+ \pi^-$ in terms of $(2\pi)_{I=0,2}$, we have

$$
\begin{aligned}
\langle \pi^+ \pi^- | H_W | K_L \rangle &= \sqrt{\frac{2}{3}} \langle (2\pi)_0 | H_W | K_L \rangle + \frac{1}{\sqrt{3}} \langle (2\pi)_2 | H_W | K_L \rangle \\[2mm]
&= \sqrt{\frac{2}{3}} e^{i\delta_0} p A_0 \left[\left(1 - \frac{q}{p} \frac{\overline{A}_0}{A_0} \right) + \frac{1}{\sqrt{2}} e^{i(\delta_2 - \delta_0)} \frac{p A_2 - q \overline{A}_2}{p A_0} \right] \\[2mm]
&= \sqrt{\frac{2}{3}} e^{i\delta_0} p A_0 \left[\Delta_0 + \frac{1}{\sqrt{2}} e^{i(\delta_2 - \delta_0)} \omega \Delta_2 \right],
\end{aligned}
\tag{7.24}
$$

where

$$
\Delta_I = 1 - \frac{q}{p} \frac{\overline{A}_I}{A_I}, \quad \omega = \frac{A_2}{A_0}.
$$

Similarly,

$$
\begin{aligned}
\langle \pi^0 \pi^0 | H_W | K_L \rangle &= -\frac{1}{\sqrt{3}} \langle (2\pi)_0 | H_W | K_L \rangle + \sqrt{\frac{2}{3}} \langle (2\pi)_2 | H_W | K_L \rangle \\[2mm]
&= \sqrt{\frac{2}{3}} e^{i\delta_0} p A_0 \left[\left(1 - \frac{q}{p} \frac{\overline{A}_0}{A_0} \right) - \sqrt{2} e^{i(\delta_2 - \delta_0)} \frac{p A_2 - q \overline{A}_2}{p A_0} \right] \\[2mm]
&= \sqrt{\frac{2}{3}} e^{i\delta_0} p A_0 \left[\Delta_0 - \sqrt{2} e^{i(\delta_2 - \delta_0)} \omega \Delta_2 \right].
\end{aligned}
\tag{7.25}
$$

Since we know already from Eq. (7.2) that $|\eta_{+-}|$ and $|\eta_{00}|$ are tiny quantities, it suffices to treat these amplitude ratios merely to first order in **CP** violation. Therefore we need to retain the **CP** conserving parts in the $K_S \to \pi\pi$ amplitude only. Remembering that $p = q = \frac{1}{\sqrt{2}}$ if we ignore **CP** violation,

$$
\begin{aligned}
\langle \pi^+\pi^- | H_W | K_S \rangle &= \sqrt{\frac{2}{3}} \langle (2\pi)_0 | H_W | K_S \rangle + \frac{1}{\sqrt{3}} \langle (2\pi)_2 | H_W | K_S \rangle \\
&= \frac{2}{\sqrt{3}} e^{i\delta_0} A_0 \left(1 + \frac{1}{\sqrt{2}} \omega e^{i(\delta_2 - \delta_0)} \right) \\
\langle \pi^0\pi^0 | H_W | K_S \rangle &= \sqrt{\frac{2}{3}} \langle (2\pi)_0 | H_W | K_S \rangle - \frac{2}{\sqrt{3}} \langle (2\pi)_2 | H_W | K_S \rangle \\
&= \frac{2}{\sqrt{3}} e^{i\delta_0} A_0 \left(1 - \sqrt{2} \omega e^{i(\delta_2 - \delta_0)} \right).
\end{aligned}
\tag{7.26}
$$

Since $|\omega| \simeq 1/20$, we ignore terms of $\mathcal{O}(\omega^2)$ and find

$$
\begin{aligned}
\eta_{+-} &= \frac{1}{2} \left(\Delta_0 - \frac{1}{\sqrt{2}} \omega e^{i(\delta_2 - \delta_0)} (\Delta_0 - \Delta_2) \right) \\
\eta_{00} &= \frac{1}{2} \left(\Delta_0 + \sqrt{2} \omega e^{i(\delta_2 - \delta_0)} (\Delta_0 - \Delta_2) \right),
\end{aligned}
\tag{7.27}
$$

or

$$
\begin{aligned}
\epsilon &= \frac{1}{2} \Delta_0 = \frac{1}{2} \left(1 - \frac{q}{p} \frac{\overline{A}_0}{A_0} \right) \\
\epsilon' &= -\frac{1}{2\sqrt{2}} \omega e^{i(\delta_2 - \delta_0)} (\Delta_0 - \Delta_2) = \frac{1}{2\sqrt{2}} \omega e^{i(\delta_2 - \delta_0)} \frac{q}{p} \left(\frac{\overline{A}_0}{A_0} - \frac{\overline{A}_2}{A_2} \right).
\end{aligned}
\tag{7.28}
$$

A few comments can illuminate the content of these relations:

- ϵ and ϵ' are written in terms of $\frac{q}{p} \frac{\overline{A}_I}{A_I}$, which has been shown to be independent of phase conventions, see Eq. (6.97). So, these quantities are phase convention independent – as they should be since they are experimentally measurable quantities.

- In the literature you will often come across the Wu–Yang phase convention, where the phase of K_0 is chosen so that A_0 is real. Note that our formalism allows us to discuss **CP** violation without chosing a phase convention. This is because our ϵ and ϵ' are independent of them.

- The quantity ϵ defined by the ratios of **CP** violating and conserving decay amplitudes, see Eq. (7.11), should *not* be confused with the **CP** impurity parameter, $\bar{\epsilon}$, in the K_L state, which is often also denoted by

$$
|K_L\rangle = \frac{1}{\sqrt{2(1 + |\bar{\epsilon}|^2)}} [(1 + \bar{\epsilon})|K^0\rangle - (1 - \bar{\epsilon})|\overline{K}^0\rangle].
\tag{7.29}
$$

Within the Wu–Yang phase convention, and only in this phase convention,

$$\epsilon = \bar{\epsilon}. \tag{7.30}$$

- The expression for ϵ' makes explicit what was said before: ϵ' vanishes unless

$$\frac{\bar{A}_0}{A_0} \neq \frac{\bar{A}_2}{A_2}, \tag{7.31}$$

i.e., ϵ' represents differences in the **CP** asymmetries for different decay channels thus implying that $H_W(\Delta S = 1)$ violates **CP**.

7.3.2 Constraints on A_I and $\bar{A}_I$

In the derivation of Watson's theorem we have used that the strong interactions conserve **T** and **CP**. Now we analyse the constraints that discrete symmetries of the weak Hamiltonian H_W place on A_I and $\bar{A}_I$.

Invoking **CPT** invariance, we deduce

$$
\begin{aligned}
A_I e^{i\delta_I} &= \langle (2\pi)_I ; \mathrm{out} | H_W | K^0 \rangle \\
&= \langle (2\pi)_I ; \mathrm{out} | (\mathbf{CPT})^\dagger \mathbf{CPT}\, H_W (\mathbf{CPT})^{-1} \mathbf{CPT}\, | K^0 \rangle^* \\
&= \sum_n \langle (2\pi)_I ; \mathrm{in} | n ; \mathrm{out} \rangle^* \langle n ; \mathrm{out} | H_W | \bar{K}^0 \rangle^* \\
&= \bar{A}_I^* e^{i\delta_I}.
\end{aligned} \tag{7.32}
$$

In Eq. (7.32), we have used the fact that $\langle (2\pi)_I ; \mathrm{out} | (2\pi)_I ; \mathrm{in} \rangle = S((\pi\pi)_I)$, and in the low energy region, $S((\pi\pi)_I) = e^{2i\delta_I}$.

If, on the other hand, H_W obeys **CP** symmetry – $[H_W, \mathbf{CP}] = 0$ – we conclude:

$$
\begin{aligned}
A_I e^{i\delta_I} &= \langle (2\pi)_I ; \mathrm{out} | H_W | K^0 \rangle \\
&= \langle (2\pi)_I ; \mathrm{out} | (\mathbf{CP})^{-1} \mathbf{CP}\, H_W (\mathbf{CP})^{-1} \mathbf{CP}\, | K^0 \rangle \\
&= \langle (2\pi)_I ; \mathrm{out} | H_W | \bar{K}^0 \rangle = \bar{A}_I e^{i\delta_I},
\end{aligned} \tag{7.33}
$$

using our usual phase convention $\mathbf{CP}\, | K^0 \rangle = | \bar{K}^0 \rangle$. Finally, **T** invariance implies

$$
\begin{aligned}
A_I e^{i\delta_I} &= \langle (2\pi)_I ; \mathrm{out} | \mathbf{T}^\dagger \mathbf{T}\, H_W \mathbf{T}^{-1} \mathbf{T}\, | K^0 \rangle^* \\
&= \langle (2\pi)_I ; \mathrm{in} | (2\pi)_I ; \mathrm{out} \rangle^* \langle (2\pi)_I ; \mathrm{out} | H_W | K^0 \rangle^* \\
&= A_I^* e^{i\delta_I},
\end{aligned} \tag{7.34}
$$

i.e., A_I has to be real; likewise for $\bar{A}_I$.

In summary[2]:

$$[\mathbf{CPT}, H_W] = 0 \quad \Longrightarrow \quad A_I = \overline{A}_I^*$$

$$[\mathbf{CP}, H_W] = 0 \quad \Longrightarrow \quad A_I = \overline{A}_I$$

$$[\mathbf{T}, H_W] = 0 \quad \Longrightarrow \quad A_I = A_I^* . \tag{7.35}$$

7.4 Relating ϵ to $\mathbf{M} - \frac{i}{2}\mathbf{\Gamma}$

With $\frac{q}{p} = e^{i\phi}$, ϕ complex, and **CPT** invariance implying $\overline{A}_I = A_I^*$, as just shown, we have

$$\Delta_I \equiv 1 - \frac{q}{p}\frac{\overline{A}_I}{A_I} = 1 - \exp[i(\phi - 2\arg A_I)] \simeq -i\phi + 2i\,\arg A_I, \tag{7.36}$$

since **CP** violation is small. Then we find, see Chapter 6 for notation and details:

$$\phi \simeq \frac{E_2}{E_1} = \frac{-\mathrm{Im}M_{12} + \frac{i}{2}\mathrm{Im}\Gamma_{12}}{\mathrm{Re}M_{12} - \frac{i}{2}\mathrm{Re}\Gamma_{12}}. \tag{7.37}$$

Denoting

$$\xi_I = \arg A_I, \tag{7.38}$$

we obtain

$$\epsilon \simeq i\left[\frac{1}{2}\frac{\mathrm{Im}M_{12} - \frac{i}{2}\mathrm{Im}\Gamma_{12}}{\mathrm{Re}M_{12} - \frac{i}{2}\mathrm{Re}\Gamma_{12}} + \xi_0\right]$$

$$\epsilon' \simeq \frac{-i}{\sqrt{2}}e^{i(\delta_2 - \delta_0)}\omega(\xi_0 - \xi_2). \tag{7.39}$$

The expression for ϵ can be simplified further, for the difference in the two eigenvalues of $\mathbf{M} - \frac{i}{2}\mathbf{\Gamma}$ can be well approximated by

$$M_S - M_L - \frac{i}{2}(\Gamma_S - \Gamma_L) = 2E \simeq 2E_1, \tag{7.40}$$

since $E_1 = E\cos\phi \simeq E$ for small ϕ. Finally we arrive at

$$\epsilon \simeq -i\left(\frac{\mathrm{Im}M_{12} - \frac{i}{2}\mathrm{Im}\Gamma_{12}}{\Delta M_K + \frac{i}{2}\Delta\Gamma_K} - \xi_0\right). \tag{7.41}$$

We have to keep in mind that Eq. (7.41) was derived using approximations that are quite specific for the $K^0 - \overline{K}^0$ system and do *not* hold for a general $P^0 - \overline{P}^0$ complex.

---

[2] We assume here that strong dynamics obey all these symmetries.

7.4.1 The phase of ϵ

Again using approximations that apply specifically to K^{neut} decays we can determine the phase of ϵ. For that purpose we consider the states which can contribute to the sum in Γ_{12} defined in Eq. (6.8). These intermediate states are

$$(2\pi)_0, \ (2\pi)_2, \ \pi^+\pi^-\pi^0, \ 3\pi^0, \ \pi l \nu, \ldots \tag{7.42}$$

As we will show in the next section, to within 10% accuracy we can ignore all but the $(2\pi)_0$ state

$$\Gamma_{12} \simeq 2\pi \rho_{2\pi} A_0^* \overline{A}_0 \,, \tag{7.43}$$

where $\rho_{2\pi}$ is the phase space factor.

With $\overline{A}_0 = A_0^* = e^{-i\xi_0}|A_0|$ imposed by **CPT** invariance, see Eq. (7.35), we then have

$$\Gamma_{12} \simeq 2\pi \rho_{2\pi} e^{-2i\xi_0} |A_0|^2 \times \text{ phase space} \,. \tag{7.44}$$

Retaining **CP** violation to first order only, we can set

$$\langle 2\pi | H_W | K^0 \rangle = \frac{1}{\sqrt{2}} \langle 2\pi | H_W | K_S \rangle \tag{7.45}$$

and obtain

$$\Gamma_{12} \simeq \frac{1}{2} e^{-2i\xi_0} \Gamma_S \simeq \frac{1}{2} e^{-2i\xi_0} \Delta\Gamma_K \,. \tag{7.46}$$

Inserting this expression into Eq. (7.41) leads to

$$\epsilon \simeq \frac{1}{\sqrt{1 + (\frac{\Delta\Gamma}{2\Delta M})^2}} e^{i\phi_{SW}} \left(-\frac{\text{Im}M_{12}}{\Delta M_K} + \xi_0 \right) \,, \tag{7.47}$$

where

$$\phi_{SW} = \tan^{-1} \frac{2\Delta M_K}{\Delta\Gamma_K} \,. \tag{7.48}$$

Using the experimental numbers given in Eq. (5.40) we arrive at

$$\phi_{SW} = \tan^{-1} \frac{2\Delta M_K}{\Delta\Gamma_K} = (43.59 \pm 0.15)^\circ. \tag{7.49}$$

This phase can be measured in a cleverly conceived experiment first performed by Bohm *et al.* at CERN [59]. A proton beam hits a (nuclear) target producing a neutral kaon that subsequently decays into two pions:

$$p + A \rightarrow K^{\text{neut}} + X \rightarrow \pi\pi + X. \tag{7.50}$$

The important point to note is that we do not identify K^{neut} as a K^0 or $\overline{K}^0$ through an observation of X. As seen in Eq. (6.104), the decay rate for an initial

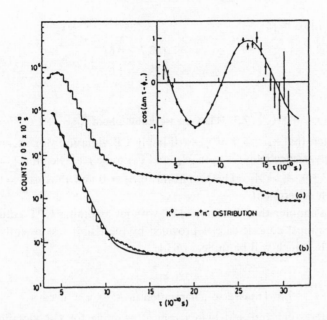

Fig. 7.2. The Bohm method was used by Geweniger et al. [60]. They measured the $K \to \pi\pi$ rate as a function of time in a reaction $p + A \to K+$ anything. The inset shows the interference term where the exponential factors have been taken out. Note that ϕ_η can be extracted totally independently of the normalization. This figure was reproduced from *Physics Letters* by permission of Elsevier Science.

K^0 or $\overline{K}^0$ state is expressed by

$$N(K^0(t) \to \pi\pi) = N_0 \left(e^{-\Gamma_S} + e^{-\Gamma_L}|\eta|^2 + 2\cos(\Delta M_K t - \phi_\eta)e^{-\frac{1}{2}(\Gamma_L+\Gamma_S)}|\eta| \right)$$

$$N(\overline{K}^0(t) \to \pi\pi) = \overline{N}_0 \left(e^{-\Gamma_S} + e^{-\Gamma_L}|\eta|^2 - 2\cos(\Delta M_K t - \phi_\eta)e^{-\frac{1}{2}(\Gamma_L+\Gamma_S)}|\eta| \right),$$

$$(7.51)$$

where ϕ_η is the phase of η and N_0 and $\overline{N}_0$ denote the initial number of K^0 and $\overline{K}^0$, respectively; we have ignored here terms of order $|\eta|^2$. The number of observed decays is given by

$$N(p + A \to [\pi\pi]_{K^{\text{neut}}} + X) \sim$$

$$e^{-\Gamma_S} + \frac{2(N^0 - \overline{N}^0)}{N^0 + \overline{N}^0}\cos(\Delta M_K t - \phi_\eta)e^{-\frac{1}{2}(\Gamma_S+\Gamma_L)t}|\eta| + e^{-\Gamma_L t}|\eta|^2. \quad (7.52)$$

From the *time dependence* of these rates we can extract ϕ_η, the phase of η, *irrespective* of N_0 and $\overline{N}_0$, the numbers of K^0 and $\overline{K}^0$ (as long as $N_0 \neq \overline{N}_0$), respectively, produced in the proton collisions. We should note here that the *sign* of $\Delta M t - \phi_\eta$ cannot be measured, and the measurement of the phase is up to $mod(\pi)$ in this early experiment. Recent experimental data yield for the phases of

η_{+-} and η_{00} [61, 62]:

$$\phi_{+-} = (43.7 \pm 0.6)°$$
$$\phi_{00} = (43.5 \pm 1.0)°. \tag{7.53}$$

7.5 What do we know about T?

We have seen that $\eta_{+-} \neq 0$ or $\eta_{00} \neq 0$ imply **CP** violation: for according to Eq. (6.11) **CP** invariance requires both M_{12} and Γ_{12} to be real, yielding $\frac{q}{p} = 1$; likewise from Eq. (7.35), $A_I = \bar{A}_I$. This would mean $\Delta_I = 0$ and thus $\eta_{+-} = 0 = \eta_{00}$ – in conflict with experiment.

Can we also infer that **T** is violated – without assuming **CPT** symmetry? Yes, we can. The most concise case is provided by the Kabir test recently performed by CPLEAR, which will be in Sec. 14.3.1.

7.6 Transverse polarization in $K \to \pi\mu\nu$ decays

Another interesting and and highly topical example for **CP** violation affecting final state distributions is provided by $K \to \pi\mu\nu$ modes where we measure the polarization of the muon. In the kaon rest frame there are three independent vectors, namely the pion and muon momenta – $\vec{p}_\pi$ and $\vec{p}_\mu$, respectively – and the muon spin vector $\vec{\sigma}_\mu$. We can thus define four non-trivial correlations, namely the scalar $\vec{p}_\pi \cdot \vec{p}_\mu$ and the three pseudoscalars $\vec{\sigma}_\mu \cdot \vec{p}_\mu$, $\vec{\sigma}_\mu \cdot \vec{p}_\pi$ and $\vec{\sigma}_\mu \cdot (\vec{p}_\mu \times \vec{p}_\pi)$. The correlations $\vec{\sigma}_\mu \cdot \vec{p}_\mu$, and $\vec{\sigma}_\mu \cdot \vec{p}_\pi$ violate parity.

The term $\vec{\sigma}_\mu \cdot (\vec{p}_\mu \times \vec{p}_\pi)$, which counts the net polarization transverse to the decay plane, constitutes also a **T**-odd correlation, i.e., it changes sign under time reversal. Observation of a **P**-odd correlation *unequivocally* establishes parity violation, yet with **T**-odd correlations the situation is not so straightforward, as explained in Sec. 3.6. The physical origin of this difference is that a space-inverted sequence of a decay can readily be implemented, yet a time-inverted one in practice cannot.

7.6.1 The final state interactions mimicking the effect of **T** violation

In Sec. 4.10.2, we discussed how the final state interaction can generate transverse polarization even when the dynamics conserves **T**. The effect of final state interaction has been computed and the results are:

$$P_\perp \equiv \langle \frac{\vec{s}_\mu \cdot (\vec{p}_\mu \times \vec{p}_\pi)}{|\vec{p}_\pi \times \vec{p}_\mu|} \rangle \simeq 10^{-3} \text{ for } K^0 \to \pi^-\mu^+\nu \text{ [63]}$$

$$\simeq 10^{-6} \text{ for } K^+ \to \pi^0\mu^+\nu, \text{ [64]} \tag{7.54}$$

where we have assumed that the underlying dynamics obey time reversal invariance; i.e., the stated non-vanishing values for $P_\perp$ are generated purely by final state interaction. They are of quite different nature in the two transitions: in

$K^0 \to \pi^-\mu^+\nu$ it is mainly Coulomb exchange between the two charged particles in the final state and – not surprisingly – the effect is of order α/π; in $K^+ \to \pi^0\mu^+\nu$, on the other hand, the final state interactions are driven by weak forces, which allows for a tiny effect only; for all practical purposes an observation of $P_\perp \neq 0$ in $K^+ \to \pi^0\mu^+\nu$ is an unambiguous sign for **T** violation![3]

7.6.2 Real **T** violation

The formalism for describing the transverse muon polarization in $K \to \pi\mu\bar{\nu}$ decay has been studied long ago [64, 65, 66] within the context of $V - A$ theory. With the KM ansatz yielding, as discussed later, only an unobservably small value for $P_\perp$, we are interested in looking for effects beyond the SM. Therefore we start from the general expression for the transition amplitude

$$A(K^- \to \pi^0\mu^-\bar{\nu}) = \bar{\mu}(p_\mu; \vec{s}_\mu)[F_S + iF_P\gamma_5 + F_V \not{P}_K + F_A \not{P}_K\gamma_5]\nu(p_\nu) \qquad (7.55)$$

with $\not{P}_{K,\pi,\mu,\nu}$ denoting K, π, μ, ν momenta, respectively, and $\vec{s}_\mu$ the muon spin. We then obtain for the differential decay width [67]

$$d\Gamma = d\Gamma_0 + (s \cdot p_\nu)\, d\Gamma_\nu + (s \cdot p_K)\, d\Gamma_K + \epsilon_{\alpha\beta\gamma\delta}p_K^\alpha s^\beta p_\mu^\gamma p_\nu^\delta\, d\Gamma_\perp, \qquad (7.56)$$

where

$$d\Gamma_0 \;\propto\; \frac{1}{2}(p_\mu \cdot p_\nu)[|F_S|^2 + |F_P|^2 - M_K^2(|F_V|^2 + |F_A|^2)]$$
$$+(p_\mu \cdot p_K)(p_\nu \cdot p_K)(|F_V|^2 + |F_A|^2)$$
$$+m_\mu(p_\nu \cdot p_K)[\mathrm{Re}\,(F_S F_V^*) - \mathrm{Im}\,(F_P F_A^*)]$$

$$d\Gamma_\nu \;\propto\; -m_\mu[\mathrm{Im}\,(F_S F_P^*) + M_K^2\,\mathrm{Re}\,(F_V F_A^*)]$$
$$+(p_\mu \cdot p_K)[\mathrm{Im}\,(F_P F_V^*) - \mathrm{Re}\,(F_S F_A^*)]$$

$$d\Gamma_K \;\propto\; 2m_\mu(p_\nu \cdot p_K)\mathrm{Re}\,(F_V F_A^*)$$
$$+(p_\nu \cdot p_\mu)[\mathrm{Re}\,(F_S F_A^*) - \mathrm{Im}\,(F_P F_V^*)]$$

$$d\Gamma_\perp \;\propto\; \mathrm{Re}\,(F_P F_A^*) + \mathrm{Im}\,(F_S F_V^*). \qquad (7.57)$$

These quantities are most conveniently evaluated in the kaon rest frame. The muon polarization s_μ there is related to its vector $\vec{\sigma}$ in the muon rest frame as follows:

$$s_0 = \frac{\vec{\sigma} \cdot \vec{p}_\mu}{m_\mu}, \quad \vec{s}_\parallel = \frac{E_\mu}{m_\mu}\vec{n}_\mu(\vec{\sigma} \cdot \vec{n}_\mu) \quad \vec{s}_\perp = \vec{\sigma} - (\vec{\sigma} \cdot \vec{n}_\mu)\vec{n}_\mu \qquad (7.58)$$

---

[3] **CP** invariance unequivocally predicts $P_\perp(K^+ \to \pi^0\mu^+\nu) = -P_\perp(K^- \to \pi^0\mu^-\bar{\nu})$ and $P_\perp(K^0 \to \pi^-\mu^+\nu) = -P_\perp(\overline{K}^0 \to \pi^+\mu^-\bar{\nu})$. Yet these relations do not provide practical tests since precise measurements of the polarisation of negatively charged muons are not feasible.

where $\vec{n}_\mu$ is a unit vector in the direction of $\vec{p}_\mu$. In terms of these vectors,

$$(s \cdot p_K) = \frac{\vec{\sigma} \cdot \vec{p}_\mu}{m_\mu} M_K$$

$$(s \cdot p_\nu) = \vec{\sigma} \cdot \vec{p}_\mu \left(\frac{M_K - E_\pi}{m_\mu} + \frac{\vec{p}_\mu \cdot \vec{p}_\pi}{m_\mu (E + m_\mu)} \right) + \vec{\sigma} \cdot \vec{p}_\pi$$

$$\epsilon_{\alpha\beta\gamma\delta} p_K^\alpha s^\beta p_\mu^\gamma p_\nu^\delta = M_K \vec{\sigma} \cdot (\vec{p}_\pi \times \vec{p}_\mu). \tag{7.59}$$

It is now trivial to derive the muon polarization in a general direction.

Now we specialize to the $V - A$ theory, where the transition amplitude is given by.

$$A(K^- \to \pi^0 \mu^- \bar{\nu}) = \frac{G_F}{\sqrt{2}} \langle \pi; p_\pi | \bar{u} \gamma_\alpha (1 - \gamma_5) s | K; p_K \rangle \bar{\mu}(p_\mu; \vec{s}_\mu) \gamma^\alpha (1 - \gamma_5) \nu(p_\nu). \tag{7.60}$$

Since no axial vector can be constructed from the two available vectors p_π and p_K, only the vector part contributes through two form factors:

$$\langle \pi; p_\pi | \bar{u} \gamma_\alpha (1 - \gamma_5) s | K; p_K \rangle = f_+(q^2)(p_K + p_\pi)_\alpha + f_-(q^2)(p_K - p_\pi)_\alpha. \tag{7.61}$$

So, inserting Eq. (7.61) into Eq. (7.60) we obtain

$$A(K \to \pi \mu(\vec{s}) \nu) \simeq \frac{G_F}{\sqrt{2}} [f_+(q^2) \bar{\mu}(p_\mu; \vec{s})(1 + \gamma_5)(2\slashed{k}) \nu(p_\nu)$$

$$+ (f_-(q^2) - f_+(q^2)) m_\mu \bar{\mu}(p_\mu; \vec{s})(1 - \gamma_5) \nu(p_\nu)]; \tag{7.62}$$

i.e., in the general notation of Eq. (7.55)

$$F_S = f_+(\xi - 1)m_\mu, \qquad F_P = if_+(\xi - 1)m_\mu, \qquad F_V = 2f_+, \qquad F_A = -2f_+ \tag{7.63}$$

where we have defined

$$\xi = \frac{f_-}{f_+} \tag{7.64}$$

and therefore [66]

$$d\Gamma \sim a_0 - a_1(\vec{\sigma} \cdot \vec{p}_\mu) + a_2(\vec{\sigma} \cdot p_\nu) - m_\mu M_K \operatorname{Im} \xi \vec{\sigma} \cdot (\vec{p}_\pi \times \vec{p}_\mu) \tag{7.65}$$

with

$$a_0 = M_K(E^* - E_\pi)(b^2 m_\mu^2 - M_K^2) + 2M_K^2 E_\nu E_\mu + M_K m_\mu E_\nu (\operatorname{Re} b + \operatorname{Im} b)$$

$$a_1 = 2M_K^2 [E_\nu + (E^* - E_\pi)\operatorname{Re} b]$$

$$a_2 = m_\mu [M_K^2 + |b|^2 m_\mu^2 + 2E_\mu M_K \operatorname{Re} b]$$

$$b = \frac{1}{2}(\xi - 1)$$

$$E^* = \frac{M_K^2 + M_\pi^2 - m_\mu}{2M_K}$$

$$P_\perp \simeq \operatorname{Im} \xi \frac{m_\mu}{M_K} \langle \frac{|\vec{p}_\mu|}{E_\mu + \vec{p}_\mu \cdot \vec{p}_\nu / |\vec{p}_\nu| - m_\mu^2 / M_K} \rangle. \tag{7.66}$$

The average for the transverse muon polarization can thus be expressed through Im ξ, with

$$P_\perp \sim (0.2 \sim 0.3) \text{Im } \xi. \tag{7.67}$$

Typically, the best available value reads [68]:

$$\text{Im } \xi = -0.01 \pm 0.019 \;\; \triangleq \;\; P_\perp \simeq (-1.85 \pm 3.6) \cdot 10^{-3}. \tag{7.68}$$

We note that there has been no new result for nearly 15 years! As mentioned before and shown explicitly later, observation of a signal here – at the level of $P_\perp \sim \mathcal{O}(10^{-3})$ down to $\mathcal{O}(10^{-5})$ – would be an unequivocal signal for the presence of New Physics. There is thus a strong motivation for a new round of experimentation.

A new KEK experiment [69], KEK E246, is expected to reach the sensitivity of $P_\perp = 7 \times 10^{-4}$. An experiment has been proposed at BNL that could even approach 10^{-4}.

7.7 ♠ $K \to 3\pi$ ♠

7.7.1 $K_S \to 3\pi^0$

The transition $K_S \to 3\pi^0$ (like $K_L \to \pi\pi$) requires **CP** violation: with Bose statistics enforcing the three neutral pions to form a symmetric configuration the final state has to be **CP** odd since $\mathbf{CP}|\pi^0\rangle = -|\pi^0\rangle$. A nonzero value for $\eta_{000} \equiv A(K_S \to 3\pi^0)/A(K_L \to 3\pi^0)$ thus constitutes an unambiguous measure for **CP** violation that can be related to $\Gamma(K_L \to 2\pi)$ using **CPT** symmetry:

$$\eta_{000} = \epsilon + i[(\text{Im}A(K_S \to 3\pi^0)/\text{Re}A(K_S \to 3\pi^0)]. \tag{7.69}$$

The CPLEAR collaboration has obtained [70]:

$$\begin{aligned}
\text{Im } \eta_{000} &= 0.15 \pm 0.20 \pm 0.03 \\
\text{Re } \eta_{000} &= 0.18 \pm 0.14 \pm 0.06
\end{aligned}$$

$$\tag{7.70}$$

as well as

$$\frac{\Gamma(K_S \to 3\pi^0)}{\Gamma(K_L \to 3\pi^0)} \le 0.053. \tag{7.71}$$

7.7.2 $K_S \to \pi^+\pi^-\pi^0$

The situation is more complex with $K_S \to \pi^+\pi^-\pi^0$ since the mere existence of this channel does not establish **CP** violation: the **CP** parity of the final state depends, as illustrated in Fig. 7.3, on the orbital angular momentum l of the charged pion pair:

$$\mathbf{CP}|(\pi^+\pi^-)_l\pi^0\rangle = -(-1)^l|(\pi^+\pi^-)_l\pi^0\rangle \; ; \tag{7.72}$$

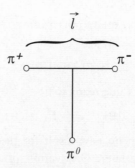

Fig. 7.3.   Final state configuration for $K^0 \rightarrow \pi^+\pi^-\pi^0$. $\vec{l}$ is the angular momentum vector for the $\pi^+\pi^-$ state. Note that $\mathbf{CP}|\pi^+\pi^-\rangle = +|\pi^+\pi^-\rangle$ (see Problem 5.1), $\mathbf{CP}|\pi^0\rangle = -|\pi^0\rangle$. So, $\mathbf{CP}|(\pi^+\pi^-)_l\pi^0\rangle = -(-1)^l|\pi^+\pi^-\pi^0\rangle$.

Table 7.1.   **CP**, l, and isospin properties of the final state $\pi^+\pi^-\pi^0$.

Isospin of $\pi^+\pi^-$	0	1	2
l of $\pi^+\pi^-$	even	odd	even
Isospin of 3π	1	0, 2	1, 3
CP of 3π	odd	even	odd

for l odd – a configuration allowed for $\pi^+\pi^-$, albeit suppressed by the centrifugal barrier – the mode $K_S \rightarrow \pi^+\pi^-\pi^0$ conserves $\mathbf{CP}$[4].

The angular momentum of the charged pion pair is correlated with the isospin of the three pion state, as shown in Table 7.1. The final state is most conveniently analysed in terms of Dalitz plot variables. In terms of pion energies in the rest frame of K:

$$X = \frac{s_{\pi^+} - s_{\pi^-}}{M_\pi^2}$$

$$Y = \frac{s_{\pi^0} - s_0}{M_\pi^2}, \tag{7.73}$$

where

$$s_i = (p_K - p_i)^2, \quad i = 1, 2, 3, \quad s_0 = \frac{1}{3}\sum_i s_i, \tag{7.74}$$

which are odd and even under **CP**, respectively. Defining

$$G(t, X, Y) = \frac{d\Gamma(K^0(t) \rightarrow \pi^+\pi^-\pi^0)}{dXdY}$$

$$\overline{G}(X, Y) = \frac{d\Gamma(\overline{K}^0(t) \rightarrow \pi^+\pi^-\pi^0)}{dXdY}, \tag{7.75}$$

---

[4] Likewise, $K_L \rightarrow \pi^+\pi^-\pi^0$ violates **CP** for l odd; it is also kinematically suppressed relative to the $l = 0$ mode.

we can express their difference as follows:

$$\Delta_{+-0}(t, X, Y) = \frac{G(t, X, Y) - \overline{G}(t, X, Y)}{G(t, X, Y) + \overline{G}(t, X, Y)}$$

$$= \frac{Ae^{-\Gamma_S t} + Be^{-\Gamma_L t} + Ce^{-(\Gamma_S + \Gamma_L)t/2}\cos\Delta Mt + De^{-(\Gamma_S + \Gamma_L)t/2}\sin\Delta Mt}{|A_{S+-0}(X, Y)|^2 e^{-\Gamma_S t} + |A_{L+-0}(X, Y)|^2 e^{-\Gamma_L t}}$$

(7.76)

with

$$A(X, Y) = |A_{S+-0}(X, Y)|^2 \left(1 - \left|\frac{q_2}{p_2}\right|^2\right)$$

$$B(X, Y) = |A_{L+-0}(X, Y)|^2 \left(1 - \left|\frac{q_1}{p_1}\right|^2\right)$$

$$C(X, Y) = -2\mathrm{Re}\,(A_{S+-0}(X, Y)A_{L+-0}(X, Y)^*)$$

$$D(X, Y) = 2\mathrm{Im}\,(A_{S+-0}(X, Y)A_{L+-0}(X, Y)^*).$$

(7.77)

Here we have used the most general decomposition of the kaon state vectors that does not assume **CPT** symmetry:

$$|K^0\rangle = \frac{1}{p_1 q_2 + q_1 p_2}[q_1|K_L\rangle + q_2|K_S\rangle]$$

$$|\overline{K}^0\rangle = \frac{1}{p_1 q_2 + q_1 p_2}[-p_1|K_L\rangle + p_2|K_S\rangle] ;$$

(7.78)

$A_{S,+-0}(X, Y)$ and $A_{L,+-0}(X, Y)$ denote the decay amplitudes for $K_S \to \pi^+\pi^-\pi^0$ and $K_L \to \pi^+\pi^-\pi^0$, respectively.

We want to emphasize three points about Eq. (7.76):

- Nonvanishing $A(X, Y)$ and $B(X, Y)$ implies *indirect* **CP** violation:

$$\left|\frac{q_1}{p_1}\right|^2 \neq 1 \neq \left|\frac{q_2}{p_2}\right|^2.$$

(7.79)

-

$$\left|\frac{q_1}{p_1}\right|^2 \neq \left|\frac{q_2}{p_2}\right|^2$$

(7.80)

  signals **CPT** violation as well. Information on $|q_1/p_1|$ is obtained by studying $\Delta_{+-0}(t, X, Y)$ for $t \gg \Gamma_S^{-1}$. Recalling that $|q_2/p_2|$ is extracted from $K_L \to \pi\pi$ or $K_L \to l^\pm \nu \pi^\mp$ it is amusing to note that both $|q_1/p_1|$ and $|q_2/p_2|$ can be determined from K_L transitions.

- A nonvanishing $C(X, Y)$ or $D(X, Y)$ does not necessarily imply **CP** violation, since, as pointed out above, there is a **CP** conserving component to $K_S \to \pi^+\pi^-\pi^0$.

The last point needs to be explained in more detail. With the final state $\pi^+\pi^-\pi^0$ being **CP** even and odd the transition amplitude is even and odd in X, respectively, see Eq. (7.73) [71]:

$$
\begin{aligned}
A_{S,+-0}(X,Y) &= A_S^+(X,Y) + A_S^-(X,Y) \\
A_{L,+-0}(X,Y) &= A_L^+(X,Y) + A_L^-(X,Y),
\end{aligned} \tag{7.81}
$$

where the superscripts $\pm$ denote the **CP** eigenvalue of $\pi^+\pi^-\pi^0$:

$$
A_{S[L]}^\pm(X,Y) \equiv \frac{1}{2}\left(A_{S[L],+-0}(X,Y) \pm A_{S[L],+-0}(-X,Y)\right); \tag{7.82}
$$

$A_S^-(X,Y)$ and $A_L^+(X,Y)$ are **CP** violating and the other two are **CP** conserving. The product of amplitudes appearing in $C(X,Y)$ and $D(X,Y)$, see Eq. (7.77), then has four components:

$$
\begin{aligned}
A_{S+-0}(X,Y)A_{L+-0}(X,Y)^* &= A_S^+(X,Y)A_L^-(X,Y)^* \quad &\text{cons. cons.} \\
&+ A_S^+(X,Y)A_L^+(X,Y)^* \quad &\text{cons. viol.} \\
&+ A_S^-(X,Y)A_L^-(X,Y)^* \quad &\text{viol. cons.} \\
&+ A_S^-(X,Y)A_L^+(X,Y)^* \quad &\text{viol. viol.}
\end{aligned} \tag{7.83}
$$

With the first and the last term being odd under the interchange $X \leftrightarrow -X$, they can be eliminated by including events at X and $-X$:

$$
\sum_{X,-X} A_{S+-0}(X,Y)A_{L+-0}(X,Y)^* = A_S^+(|X|,Y)A_L^+(|X|,Y)^* \\
+ A_S^-(|X|,Y)A_L^-(|X|,Y)^*. \tag{7.84}
$$

The **CP** violating $|A_S^-(X,Y)|$ and $|A_L^+(X,Y)|$ are small relative to the **CP** conserving $|A_L^-(X,Y)|$; likewise for $|A_S^+(X,Y)|$, though for a different reason: with the $\pi^+ - \pi^-$ pair being in an angular momentum $l=1$ configuration and the π^0 also in an $l=1$ state in the 3π center of mass frame, see Table 7.1, this amplitude is reduced by a centrifugal barrier. Therefore we will retain only terms at most linear in $A_{S,+-0}(X,Y)$ in the following. Eq. (7.76) then reads as follows:

$$
\Delta_{+-0}(t,|X|,Y) \equiv \frac{\sum_{X,-X} G(t,X,Y) - \sum_{X,-X} \overline{G}(t,X,Y)}{\sum_{X,-X} G(t,X,Y) + \sum_{X,-X} \overline{G}(t,X,Y)} \tag{7.85}
$$

$$
= \left(1 - \left|\frac{q_1}{p_1}\right|^2\right) + 2e^{-(\Gamma_S+\Gamma_L)t/2}
$$

$$
\times \left\{ -\cos\Delta Mt\, \mathrm{Re}\left[\frac{A_S^-(|X|,Y)}{A_L^-(|X|,Y)} + R\frac{A_L^+(|X|,Y)^*}{A_L^-(|X|,Y)^*}\right]\right.
$$

$$
\left. + \sin\Delta Mt\, \mathrm{Im}\left[\frac{A_S^-(|X|,Y)}{A_L^-(|X|,Y)} + R\frac{A_L^+(|X|,Y)^*}{A_L^-(|X|,Y)^*}\right]\right\}, \tag{7.86}
$$

where

$$R(|X|, Y) = \frac{A_S^+(|X|, Y)}{A_L^-(|X|, Y)}. \tag{7.87}$$

Note that symmetrization in X must be done before we form the asymmetry.

The second term in the square brackets, being the product of two small ratios, can then be dropped, and we arrive at:

$$\Delta_{+-0}(t, |X|, Y) = \left(1 - \left|\frac{q_1}{p_1}\right|^2\right) + 2e^{-(\Gamma_S + \Gamma_L)t/2}$$
$$\times [-\cos \Delta M t \operatorname{Re} \eta_{+-0} + \sin \Delta M t \operatorname{Im} \eta_{+-0}], \tag{7.88}$$

where

$$\eta_{+-0} = \frac{A_S^-(|X|, Y)}{A_L^-(|X|, Y)} \tag{7.89}$$

represents a **CP** violating parameter.

The difference $\Delta_{+-0}(t, |X|, Y)$ between the transition rates $G(t, |X|, Y)$ and $\bar{G}(t, |X|, Y)$ symmetrized in the Dalitz plot variable X thus becomes a **CP** asymmetry and allows the determination of the amplitude ratio η_{+-0}. A main element in this derivation was the observation that $\frac{|A_S^+(|X|,Y)|}{|A_L^-(|X|,Y)|}$ is small due to the centrifugal barrier of the three pion final state and $\frac{|A_L^+(|X|,Y)|}{|A_L^-(|X|,Y)|}$ is tiny due to the approximate validity of **CP** symmetry.

An analysis similar to the derivation of η_{+-} leads to

$$\eta_{+-0} = \frac{1}{2}\left(1 + \frac{q_1}{p_1}\frac{\bar{A}(\pi^+\pi^-\pi^0)^-}{A(\pi^+\pi^-\pi^0)^-}\right), \tag{7.90}$$

where $A(\pi^+\pi^-\pi^0)^-$ and $\bar{A}(\pi^+\pi^-\pi^0)^-$ are $K \to \pi^+\pi^-\pi^0$ and $\bar{K} \to \pi^+\pi^-\pi^0$ amplitudes with the three pions forming a **CP** $= -1$ final state, respectively.

We recover $\eta_{+-0} = 0$ in the limit of **CP** invariance, since in that case $q_1 A_I(\pi^+\pi^-\pi^0) = -p_1 \bar{A}_I(\pi^+\pi^-\pi^0)$ for $I = 1, 3$.

The experimental values are [71]:

$$\operatorname{Re} \eta_{+-0} = \left(-2 \pm 7 \, {}^{+4}_{-1}\right) \times 10^{-3} \tag{7.91}$$

$$\operatorname{Im} \eta_{+-0} = \left(-2 \pm 9 \, {}^{+2}_{-1}\right) \times 10^{-3}. \tag{7.92}$$

7.7.3 $K^\pm \to \pi^\pm\pi^+\pi^-$

In $K^+ \to \pi^+\pi^+\pi^-$ vs. $K^- \to \pi^-\pi^-\pi^+$ we can search for direct **CP** violation through an analysis of the Dalitz plot population. We can use the following parametrization [72]:

$$|\mathcal{M}|^2 \propto 1 + g\frac{s_3 - s_0}{am_{\pi^+}^2} + h\left(\frac{s_3 - s_0}{am_{\pi^+}^2}\right)^2 + j\frac{s_2 - s_1}{am_{\pi^+}^2} + ak\left(\frac{s_2 - s_1}{am_{\pi^+}^2}\right)^2 + \cdots. \tag{7.93}$$

The index 3 is used for the 'odd pion out', i.e., the π^- [π^+] in K^+ [K^-] decays. The following constraints are imposed by **CP** invariance on the expansion coefficients:

$$j(K^+) = 0 = j(K^-)\,,\ g(K^+) = g(K^-)\,,$$
$$h(K^+) = h(K^-)\,,\ k(K^+) = k(K^-). \tag{7.94}$$

The data have become somewhat 'long in the tooth' [73]:

$$\frac{g(K^+) - g(K^-)}{g(K^+) + g(K^-)} = -0.70 \pm 0.53,$$

and some information, though inconclusive, is also available on h and k. Considering that we are dealing here with direct **CP** violation in nonleptonic decays, which is also probed through searches for ϵ'/ϵ it seems unlikely that an effect will be observed here or a meaningful bound established. Theory will then be spared another embarrassment since it is quite unclear how a reliable description can be developed here after a less than sterling experience in the simpler case of $K_L \to \pi^+\pi^-$ vs $K_L \to \pi^0\pi^0$.

7.8 Hyperon decays

In Sec. 4.10.2 we have discussed how the final state interaction generates **T** odd correlations in hyperon decays. In this section, we revisit this problem and discuss how we can separate the final state interaction effects and **T** violating effects. Here we restrict our discussion to $\Lambda \to N\pi$ decay.

Everything applies, however, to $\Xi \to \Lambda\pi$ decay as well as other nonleptonic hyperon decays.

The decay amplitude for $\Lambda \to N\pi$ is given by

$$A(\Lambda \to P\pi) = \bar{u}_p(\vec{p}_P, \vec{s}_P)[A'_S + A'_P\gamma_5]u_\Lambda(\vec{p}_\Lambda, \vec{s}_\Lambda). \tag{7.95}$$

A'_S and A'_P stand for S and P wave amplitudes, which are parity violating and conserving, respectively.

To derive an expression for the polarization of the final state proton, let us rewrite Eq. (7.95) as

$$\psi_p = [A_S + A_P\vec{\sigma} \cdot \hat{p}_p]\psi_\Lambda, \tag{7.96}$$

where ψ_p, and ψ_Λ are proton and Λ spinors in their rest frame, respectively; $\hat{p}$ is a unit vector along the direction of $\vec{p}$. Finally $A_S = \sqrt{\frac{E+M}{2M}}A'_S$, and $A_P = \sqrt{\frac{E-M}{2M}}A'_P$. The density matrix

$$\rho_P = \psi_p\psi_p^\dagger = \rho_0(\mathbf{1} + \langle\vec{s}_P\rangle \cdot \vec{\sigma}) \tag{7.97}$$

gives the polarization of the proton in its rest frame. Writing $\psi_p = \mathcal{M}\psi_\Lambda$, $\mathcal{M} \equiv A_S + A_P\vec{\sigma} \cdot \hat{p}_p$, we have

$$\rho_P = \mathcal{M}\rho_\Lambda\mathcal{M}^\dagger \tag{7.98}$$

where $\rho_\Lambda = \mathbf{1} + \langle\vec{s}_\Lambda\rangle \cdot \vec{\sigma}$ gives the polarizationof Λ in its rest frame.

The polarization of the proton is given by [74, 2],

$$\langle \vec{s}_P \rangle = \frac{(\alpha - \langle \vec{s}_\Lambda \rangle \cdot \vec{p}_\Lambda)\vec{p}_\Lambda + \beta(\langle \vec{s}_P \rangle \times \vec{p}_\Lambda) + \gamma(\vec{p}_\Lambda \times (\langle \vec{s}_P \rangle \times \vec{p}_\Lambda))}{1 + \alpha(\langle \vec{s}_P \rangle \cdot \vec{p}_\Lambda)}. \tag{7.99}$$

Using the notation defined in Eq. (4.136) we have

$$\alpha = \frac{2\mathrm{Re}\,(A_S^* A_P e^{i(\delta_S - \delta_P)})}{|A_S|^2 + |A_P|^2}, \quad \beta = \frac{2\mathrm{Im}\,(A_S^* A_P e^{i(\delta_S - \delta_P)})}{|A_S|^2 + |A_P|^2}, \quad \gamma = \frac{|A_S|^2 - |A_P|^2}{|A_S|^2 + |A_P|^2}. \tag{7.100}$$

Non-vanishing α or β, which implies non-vanishing A_S, means **P** is violated. But non-vanishing β does not immediately imply **T** violation. It may be due to the final state interaction effect. So, we have to work harder. Let's examine how weak and final state interaction phases enter into these correlations. In terms of amplitudes with definite isospin and angular momentum, we have:

$$\mathscr{A}^S = |\mathscr{A}_{1/2}^S| e^{i(\delta_{1/2}^S + \phi_{1/2}^S)} + |\mathscr{A}_{3/2}^S| e^{i(\delta_{3/2}^S + \phi_{3/2}^S)} \tag{7.101}$$

$$\overline{\mathscr{A}}_S = -|\mathscr{A}_{1/2}^S| e^{i(\delta_{1/2}^S - \phi_{1/2}^S)} - |\mathscr{A}_{3/2}^S| e^{i(\delta_{3/2}^S - \phi_{3/2}^S)} \tag{7.102}$$

$$\mathscr{A}^P = |\mathscr{A}_{1/2}^P| e^{i(\delta_{1/2}^P + \phi_{1/2}^P)} + |\mathscr{A}_{3/2}^P| e^{i(\delta_{3/2}^P + \phi_{3/2}^P)} \tag{7.103}$$

$$\overline{\mathscr{A}}_P = |\mathscr{A}_{1/2}^P| e^{i(\delta_{1/2}^P - \phi_{1/2}^P)} + |\mathscr{A}_{3/2}^P| e^{i(\delta_{3/2}^P - \phi_{3/2}^P)}. \tag{7.104}$$

Here δ and ϕ denote the strong and weak phases, respectively. Note that the weak phases change their sign under **CP**. It is easily seen that, if **CP** is conserved,

$$\alpha = -\overline{\alpha}, \qquad \beta = -\overline{\beta}. \tag{7.105}$$

So,

$$A = \frac{\alpha + \overline{\alpha}}{\alpha - \overline{\alpha}}, \quad B = \frac{\beta + \overline{\beta}}{\beta - \overline{\beta}} \tag{7.106}$$

thus have to vanish – unless **CP** invariance is violated! For recent discussion including theoretical predictions, see [103].

Problems

7.1   Let us take the definition of the **CP** transformation to be:

$$\mathbf{CP}|K^0\rangle = i|K^0\rangle. \tag{7.107}$$

Assuming **CP** symmetry, show that

$$\overline{\epsilon} = \frac{1+i}{1-i}. \tag{7.108}$$

So, $\overline{\epsilon}$ as defined in Eq. (7.29) depends on phase convention and is not a physical observable. Note also that $\mathrm{Im}\,M_{12}$ and $\mathrm{Im}\,\Gamma_{12}$ are also nonvanishing. Are they observables? How does $E_1^2 + E_2^2$ get modified? Are ΔM and $\Delta\Gamma$ observables? How does ϵ get modified?

7.2   In Eq. (7.47), we have derived ϵ in a way which is independent of phase conventions. In the literature, we often come across the Wu–Yang phase convention in which A_0 is real. If A_0 had a phase $e^{i\xi_0}$ in some phase convention, show that we must make

$$|K^0\rangle \to e^{-i\xi_0}|K^0\rangle \qquad (7.109)$$

to make A_0 real. How does this change M_{12}? Show that Eq. (7.47) remains invariant under this change of phase convention.

7.3   Compute $|A(K \to \pi\mu(\hat{s})v)|^2$. The spin dependence can be computed by inserting the spin projection operator $\Lambda_\pm = \frac{1}{2}(1 \pm \gamma_5 \,\hat{s})$ and summing over the spin. Show that:

$$|A(K \to \pi\mu(\hat{s})v)|^2 = \frac{G_F^2}{2m_\mu m_v} \Big[|f_+(q^2)|^2$$
$$\times \mathrm{Tr}((\not{p}_\mu + m_\mu)\gamma_+(\not{p}_K + \not{p}_\pi)\not{p}_v(\not{p}_K + \not{p}_\pi)\gamma_- \Lambda_-)$$
$$+ |f_-(q^2)|^2 \mathrm{Tr}((\not{p}_\mu + m_\mu)\gamma_- \not{p}_v \Lambda_-)$$
$$+ 2\mathrm{Re}\left(f_+(q^2)f_-^*(q^2)\mathrm{Tr}((\not{p}_\mu + m_\mu)\gamma_+(\not{p}_K + \not{p}_\pi)\not{p}_v \Lambda_-) \right) \Big].$$

$$(7.110)$$

Show that the third term contains $\epsilon_{\alpha\beta\gamma\delta}p_K^\alpha s^\beta p_\mu^\gamma p_\pi^\delta$, which is proportional to $\hat{s} \cdot \vec{p}_K \times \vec{p}_\pi$, in the rest frame of the K meson.

7.4   Show that $\mathbf{CP}|\pi^+\pi^-\pi^0\rangle = (-1)^I |\pi^+\pi^-\pi^0\rangle$ holds for $K_S \to \pi^+\pi^-\pi^0$ decay, where I is the isospin of the 3π state.

7.5   Writing $\frac{q}{p} = 1 + i\phi$, show that

$$\eta_{+-0} = -\frac{i}{2}\phi + \frac{1}{2}\frac{\sum_I \omega_I e^{i(\delta_I - \delta_1)}\mathrm{Im}\frac{\overline{A}_I(\pi^+\pi^-\pi^0)}{A_I(\pi^+\pi^-\pi^0)}}{\sum_I \omega_I e^{i(\delta_I - \delta_1)}}. \qquad (7.111)$$

Here ω_I is the Clebsch-Gordan coefficient corresponding to the $(3\pi)_I$ state. If only the final state with $(\pi^+\pi^-)_{I=0}$ contributes:

$$\eta_{+-0} = -\frac{i}{2}\phi + \frac{1}{2}\mathrm{Im}\left(\frac{\overline{A}_1(\pi^+\pi^-\pi^0)}{A_1(\pi^+\pi^-\pi^0)} \right). \qquad (7.112)$$

PART 2

THEORY OF CP VIOLATION

8

The KM implementation of **CP** violation

The KM phase is like the Scarlet Pimpernel:
Sometimes here, sometimes there,
Sometimes everywhere!

8.1 A bit of history

As described before, the discovery of **CP** violation had not been anticipated – it actually ran counter to rather firm expectations. Nevertheless, the experimental findings as well as their interpretation were soon accepted. The relevant phenomenology for strange decays was speedily developed, as described in the two preceeding sections, and the community took on the challenge of designing dynamical models for **CP** and **T** violation.

A general *classification* of dynamical models was defined using a general decomposition of the Hamiltonian into **CP** even and odd parts:

$$H = H^+ + H^-, \quad (\mathbf{CP})^\dagger H^\pm \mathbf{CP} = \pm H^\pm. \tag{8.1}$$

- *Millistrong*: **CP** violation is implemented as a small correction of order 10^{-3} relative to the strong interactions:

$$H^- = H_{MS}^- \sim 10^{-3} \cdot H_S. \tag{8.2}$$

H_{MS}^- conserves strangeness and parity like H_S does. $K_L \to \pi\pi$ is understood as a three-step process:

$$K_L \xrightarrow{H_W} I_- \xrightarrow{H_{MS}^-} I_+ \xrightarrow{H_S} \pi\pi, \tag{8.3}$$

where the states $I_\pm$ are **CP** eigenstates with strangeness zero.

- *Milliweak*: Here we assume

$$H_{MW}^- \sim |\epsilon| \cdot H_W \sim 10^{-3} \cdot H_W \tag{8.4}$$

119

and H_{MW}^- can change parity and strangeness. It then follows that H_{MW}^- constitutes a $\Delta S = 1$ interaction like H_W, and $\Delta S = 2$ transitions are produced by iteration:

$$\langle K^0|H_{MW}^-|\overline{K}^0\rangle = 0 \tag{8.5}$$

$$\arg\frac{M_{12}}{\Gamma_{12}} \sim \frac{|H_W + H_{MW}^-|^2}{H_W^2} \sim \operatorname{Re}\epsilon. \tag{8.6}$$

- *Superweak*: We postulate the existence of a new $\Delta S = 2$ Hamiltonian that is highly suppressed:

$$H_{SW}^- \sim 10^{-9} \cdot H_W. \tag{8.7}$$

Instead of Eq. (8.5) we have

$$\langle K^0|H_{SW}^-|\overline{K}^0\rangle \neq 0 \tag{8.8}$$

to lowest order.

(i) Various *millistrong* models were proposed and analysed soon after the Fitch–Cronin experiment. Their phenomenology is rather accessible experimentally, for we expect **CP** and **T** asymmetries to be roughly of order 10^{-3} in every process unless forbidden by some other reasons. This class of models was quickly ruled out! (ii) The *superweak* scenario was first introduced by Wolfenstein [75] in August 1964. Historically and intellectually it had considerable impact, as we are going to describe. With the demise of the millistrong models it was the only ansatz left standing at that time. (iii) The KM ansatz that came later is of the *milliweak* type.

Apparently it was not realized for quite some time that **CP** violation cannot arise from the charged current interactions with u, d and s quarks only[1]. The community can be forgiven for not worrying about a numerically tiny effect – $\operatorname{Br}(K_L \to \pi^+\pi^-) \simeq 2\times10^{-3}$ – when calculations of $\Gamma(K_L \to \mu^+\mu^-, \gamma\gamma)$ etc. yielded infinities in the absence of a renormalizable theory for the weak interactions.

Yet once the renormalizable Glashow–Salam–Weinberg theory had been formulated and scored a spectacular success in predicting neutral currents, the evaluation of the situation had to change significantly: it is quite remarkable that nobody noticed for some time that even the addition of charm quarks, rendering the theory renormalizable, did not allow for the implementation of **CP** violation, i.e., that New Physics was needed! We do not understand how such a blind spot could arise apart from offering the following observation. Most physicists (Gell-Mann included [76]) thought for a long time that quarks were merely mathematical objects. Furthermore, even if we accepted quarks to be real objects, only three of them had been found. The existence of charm quarks to explain

---

[1] This no-go argument will be explained below.

the absence of strangeness changing neutral currents – as suggested by Glashow, Illiopolous and Maiani (GIM) [77] – was not widely accepted; such a sentiment was expressed through the quote 'Nature must be smarter than Shelley'[2]. This general scepticism (or even agnosticism) may have been the reason for most to be content with a superweak model as explanation for **CP** violation. The situation at the physics department of Nagoya University was quite different, both on the theoretical and experimental side. Sakata, with his co-workers and students, believed in quarks as real objects. In cosmic ray experiments Niu [78] had found what he called X particles, now known as charm particles. In Nagoya there was thus no mental barrier in thinking about the four quark system. Once it was realized that phases of Yukawa couplings could be rotated away if there were only two families of quarks, it was natural to postulate more quarks [79].

8.2 The Standard Model

It is interesting to reflect upon the status of **CP** violation in particle physics now and compare it to that soon after its discovery. The cartoon of Fig. 8.1 was shown by Cabibbo [80] at the 1966 Berkeley Conference. At that time, physicists thought that the theory of the weak interactions was in good shape since they 'understood'[3] how to compute radiative correction to μ decay. In comparison, **CP** violation was like an atomic bomb blowing up in the background. We have made tremendous progress in understanding weak interactions since then. Similarly, we have a much better grasp on **CP** violation. The *relative* understanding of **CP** violation and of weak dynamics, however, is as it was 30 years ago!

The SM of the strong and electroweak forces is based on the gauge group $SU(3)_C \times SU(2)_L \times U(1)$; the electroweak part is referred to as the Glashow–Salam–Weinberg model[4]. We shall not attempt to discuss the basic principles which lead to the SM; instead we concentrate on those features relevant to our subsequent discussion. A more comprehensive and detailed description can be found in textbooks [81, 82].

8.2.1 QCD

QCD describes the strong forces as gauge interactions between quarks and gluons; its Lagrangian is given by

$$\mathscr{L}_{QCD} = -\frac{1}{4} G^a_{\mu\nu} G^{\mu\nu,a} + \overline{Q} i \not{D} Q - \overline{Q}_R \mathscr{M} Q_L + \text{h.c.} \tag{8.9}$$

where $G^a_{\mu\nu}$ denotes the gluon field strength tensor

$$G^a_{\mu\nu} = \partial_\mu A^a_\nu - \partial_\nu A^a_\mu + g_S f^{abc} A^b_\mu A^c_\nu \tag{8.10}$$

---

[2] The creator of this aphorism prefers to remain anonymous.
[3] The definition of 'understood' is clear from the cartoon.
[4] Calling it a model is actually a misnomer since it represents a self-consistent theory.

Fig. 8.1. This cartoon was presented by Cabibbo at the Berkeley conference in 1966. Today, we think that, indeed, the SM of electroweak interaction describes all of electromagnetic and weak interactions. But, do we understand **CP** violation? We gained a tremendous amount of 'understanding' over the past 30 years. This is real progress. But the relative situation between understanding weak interaction and understanding **CP** violation has not changed. The cartoon is relevant even today. This figure was reproduced from *Proceedings of the XIIth International Conference in High Energy Physics* by permission from N. N. Cabibbo.

and D_μ the covariant derivative

$$D_\mu = \partial_\mu - ig_S t^a A_\mu^a \tag{8.11}$$

expressed with the gluon fields A_μ^a and $SU(3)$ generators t^a, $a = 1, ..., 8$; Q represents the column vector containing quark fields in all their colour and flavour reincarnations, with the couplings being flavour-diagonal. Note that once we define a set of Hermitian fields $t^a A_\mu^a$, g_S must be defined as real.

Thus it would seem that **CP** is naturally conserved by QCD, which obviously would be an attractive feature. Alas – upon further reflection it was realized that the situation is less clear-cut: there exists a gauge-invariant and renormalizable operator that was ignored in Eq. (8.9), namely $G_{\mu\nu}^a \tilde{G}^{\mu\nu,a}$, where $\tilde{G}^{\mu\nu,a}$ denotes the dual field strength tensor:

$$\tilde{G}_{\mu\nu}^a \equiv \frac{1}{2}\epsilon_{\mu\nu\alpha\beta} G^{\alpha\beta,a}. \tag{8.12}$$

The operator $G_{\mu\nu}^a \tilde{G}^{\mu\nu,a}$ violates **P**, **T** and **CP**. Even if its coefficient in $\mathscr{L}$ is set to zero at the tree-level, radiative corrections will resurrect it with a coefficient containing an ultraviolet divergence. On the other hand we deduce from the upper bound placed on the neutron electric dipole moment that this coefficient has to be 'unnaturally' tiny. This is called the 'strong **CP** problem'; we will address it in detail later on.

8.2.2 The Glashow–Salam–Weinberg model

Let g, τ_i and W^i_μ, $i = 1, 2, 3$, be the gauge coupling, generators and gauge fields of the $SU(2)$ group, respectively; the corresponding quantities for the $U(1)$ group are g', hypercharge Y and B_μ, respectively. The pure gauge Lagrangian is given by

$$\mathscr{L}_{SU(2)\times U(1),\text{gauge}} = -\frac{1}{4}\left(\sum_{i=1}^{3} W^i_{\mu\nu} W^{\mu\nu,i} + B_{\mu\nu} B^{\mu\nu}\right), \qquad (8.13)$$

where

$$\begin{aligned}
W^i_{\mu\nu} &= \partial_\mu W^i_\nu - \partial_\nu W^i_\mu + g\epsilon^{ijk} W^j_\mu W^k_\nu \\
B_{\mu\nu} &= \partial_\mu B_\nu - \partial_\nu B_\mu
\end{aligned} \qquad (8.14)$$

are the $SU(2)$ and $U(1)$ field strength tensors, respectively. Fermions come in families of left-handed doublets and right-handed singlets

$$Q_L = \begin{pmatrix} U \\ D \end{pmatrix}_L, \quad E_L = \begin{pmatrix} \nu_l \\ l^- \end{pmatrix}_L; \quad U_R, D_R, l_R^-, \qquad (8.15)$$

where $U = (u, c, ...)$, $D = (d, s, ...)$ and $l = (e^-, \mu^-, ...)$ denote the flavour eigenstates of up-type, down-type quarks and charged leptons, respectively, with the number of families unspecified at this point; the chiral fermion fields are defined by $\psi_{\binom{L}{R}} = \frac{1}{2}(1 \mp \gamma_5)\psi$.

The fermionic gauge interactions are expressed through charged currents

$$J^+_\mu = J^1_\mu + iJ^2_\mu = \overline{U}_L\gamma_\mu D_L + \bar{l}_L\gamma_\mu\nu_L, \qquad (8.16)$$

neutral weak currents

$$J^3_\mu = \frac{1}{2}(\overline{U}_L\gamma_\mu U_L - \overline{D}_L\gamma_\mu D_L + \bar{\nu}_L\gamma_\mu l_L - \bar{l}_L\gamma_\mu l_L) \qquad (8.17)$$

and the electromagnetic current

$$J^{\text{em}}_\mu = \frac{2}{3}\overline{U}_L\gamma_\mu U_L - \frac{1}{3}\overline{D}_L\gamma_\mu D_L - \bar{l}_L\gamma_\mu l_L \qquad (8.18)$$

in

$$\begin{aligned}
\mathscr{L}_f &= \mathscr{L}_{CC} + \mathscr{L}_{NC} \\
\mathscr{L}_{CC} &= \frac{g}{\sqrt{2}}\left(J^+_\mu W^{-,\mu} + J^-_\mu W^{+,\mu}\right) \\
\mathscr{L}_{NC} &= eJ^{\text{em}}_\mu A^\mu + \frac{g}{\cos\theta_W}\left(J^3_\mu - \sin^2\theta_W J^{\text{em}}_\mu\right) Z^\mu
\end{aligned} \qquad (8.19)$$

where

$$\frac{g'}{g} = \tan\theta_W, \quad \frac{e}{g} = \sin\theta_W, \qquad (8.20)$$

with θ_W denoting the weak or Weinberg angle and

$$Z_\mu = \cos\theta_W A_\mu^3 - \sin\theta_W B_\mu, \quad A_\mu = \sin\theta_W A_\mu^3 + \cos\theta_W B_\mu. \tag{8.21}$$

As discussed in the case of QCD, the gauge couplings must be defined as real. It would then seem that again no **CP** violation can emerge in the gauge forces. However, this would be a premature conclusion. For there is another dynamical sector in the theory, namely the one involving Higgs fields, which adds another layer of complexity.

The SM contains a single $SU(2)$ doublet of Higgs fields[5]. The Yukawa interactions of Higgs fields with quarks are described by:

$$\begin{aligned} L_{\text{Yukawa}} = \quad & \sum_{i,j}(G_U)_{ij}(\overline{U}_{i,L}, \overline{D}_{i,L}) \begin{pmatrix} \phi^0 \\ -\phi^- \end{pmatrix} U_{j,R} \\ & + \sum_{i,j}(G_D)_{ij}(\overline{U}_{i,L}, \overline{D}_{i,L}) \begin{pmatrix} \phi^+ \\ \phi^0 \end{pmatrix} D_{j,R} \quad +\text{h.c.} \end{aligned} \tag{8.22}$$

The indices i and j run over 1 to n, the number of families. Once the neutral Higgs field acquires a vacuum expectation value (=VEV), $\langle\phi^0\rangle = v$, fermion masses arise.

The mass matrices for the up-type and down-type quarks are then proportional to the corresponding Yukawa couplings with the scale set by v:

$$\mathcal{M}_U = vG_U \text{ and } \mathcal{M}_D = vG_D. \tag{8.23}$$

Since the Yukawa couplings are quite arbitrary, so are the mass matrices, and in general they will contain complex elements. In Sec. 4.6 we have seen that some coupling constants must be complex in order for **CP** to be violated. Complex Yukawa couplings raise the possibility of **CP** violation emerging from this sector – somehow. This question will be analysed next.

8.3 The KM ansatz

8.3.1 *The mass matrices*

Apart from a possible $G \cdot \tilde{G}$ term mentioned before, **CP** violation can enter SM dynamics only through the mass matrices $\mathcal{M}_U$ and $\mathcal{M}_D$. Their physical interpretation becomes more transparent when we write the Lagrangian in terms of the mass eigenstates of quarks. We diagonalize the mass matrices with the help of two unitary matrices each – $T_{U,L}$, $T_{U,R}$ and $T_{D,L}$, $T_{D,R}$ – acting in family space:

$$T_{U,L}\mathcal{M}_U T_{U,R}^\dagger = \mathcal{M}_U^{\text{diag}}, \quad T_{D,L}\mathcal{M}_D T_{D,R}^\dagger = \mathcal{M}_D^{\text{diag}} \tag{8.24}$$

---

[5] Adding more doublets (which can easily be arranged) has phenomenological consequences like the existence of charged Higgs fields; such scenarios will be addressed later.

thus transforming the left- and right-handed quark fields, respectively, into their mass eigenstates:

$$U^m_{L[R]} = T_{U,L[R]} U_{L[R]}, \quad D^m_{L[R]} = T_{D,L[R]} D_{L[R]}. \tag{8.25}$$

T_L and T_R can be found by diagonalizing the Hermitian matrices $\mathcal{M}\mathcal{M}^\dagger$ and $\mathcal{M}^\dagger \mathcal{M}$.

$$\mathcal{M}^2_{\text{diag}} = T_R \mathcal{M}^\dagger \mathcal{M} T^\dagger_R = T_L \mathcal{M} \mathcal{M}^\dagger T^\dagger_L. \tag{8.26}$$

The expressions for the *neutral* currents – electroweak as well as strong – remain mainfestly the same whether they are expressed in terms of *flavour* or *mass* eigenstates (see Problem 8.1) and *no* flavour-changing neutral currents arise on the tree-level; this is referred to as the GIM mechanism [77].

For the *charged* currents the situation is quite different; we find

$$\overline{U}_L \gamma_\mu D_L = \overline{U}^m_L T_{U,L} \gamma_\mu T^\dagger_{D,L} D^m_L = \overline{U}^m_L \gamma_\mu \mathbf{V} D^m_L, \tag{8.27}$$

where

$$\mathbf{V} = T_{U,L} T^\dagger_{D,L} \tag{8.28}$$

is called the CKM matrix. Note that $\mathbf{V}$ is unitary, which reflects weak universality, to be explained below.

8.3.2 Parameters of consequence

The mere fact that the quark mass matrices and thus also $\mathbf{V}$ contain complex phases does not necessarily mean that they generate observable consequences like **CP** asymmetries; for we are allowed to redefine the phases of the quark fields. In their paper Kobayashi and Maskawa [79] analysed the question under which conditions *not all* of these phases can be rotated away.

We start by determining the number of independent physical parameters contained in $\mathbf{V}$.

- A general $n \times n$ complex matrix contains $2n^2$ real parameters.

- Unitarity implies

$$\sum_j \mathbf{V}_{ij} \mathbf{V}^*_{jk} = \delta_{ik}, \tag{8.29}$$

  yielding n constraints for $i = k$ and $2 \cdot \frac{1}{2} \cdot n \cdot (n-1) = n^2 - n$ for $i \neq k$. A unitary $n \times n$ matrix thus contains n^2 independent real parameters.

- The phases of the quark fields can be rotated freely:

$$U^m_i \to e^{i\phi^U_i} U^m_i, \quad D^m_j \to e^{i\phi^D_j} D^m_j, \tag{8.30}$$

leading to

$$
\mathbf{V} \rightarrow \begin{pmatrix} e^{-i\phi_1^U} & \cdots & 0 \\ \vdots & \ddots & \vdots \\ 0 & \cdots & e^{-i\phi_n^U} \end{pmatrix} \mathbf{V} \begin{pmatrix} e^{i\phi_1^D} & \cdots & 0 \\ \vdots & \ddots & \vdots \\ 0 & \cdots & e^{i\phi_n^D} \end{pmatrix}, \tag{8.31}
$$

where $\mathbf{V}$ is multiplied by two diagonal matrices whose elements are pure phases. Since the overall phase is irrelevant, $2n - 1$ *relative* phases can be removed from $\mathbf{V}$ in this way. Accordingly, $\mathbf{V}$ contains $(n-1)^2$ independent physical parameters.

- A general orthogonal $n \times n$ matrix is constructed from $\frac{1}{2}n(n-1)$ quantities describing the independent rotation angles:

$$
N_{\text{angles}} = \frac{1}{2}n(n-1). \tag{8.32}
$$

- The number of independent phases in $\mathbf{V}$ is:

$$
N_{\text{phases}} = (n-1)^2 - \frac{1}{2}n(n-1) = \frac{1}{2}(n-1)(n-2). \tag{8.33}
$$

- For *two* families we have $N_{\text{phase}} = 0$, $N_{\text{angles}} = 1$, i.e., just the Cabibbo angle. There is then *no* **CP** violation through $\mathbf{V}$.

- For *three* families we have

$$
N_{\text{angles}} = 3, \quad N_{\text{phases}} = 1, \tag{8.34}
$$

i.e., in addition to the three Euler mixing angles we encounter one irreducible phase that represents genuine **CP** violation.

- Beyond $n = 3$ there is a rapid proliferation of physical parameters: six angles and three phases arise for $n = 4$.

- If all up-type quarks were mass-degenerate, obviously we could not distinguish them by their masses. Any linear combination of mass eigenstates would still be a mass eigenstate, and we could make the transformation $\overline{U}_L^m \rightarrow \overline{U}_L^m \mathbf{V}^\dagger$ to remove the effect of $\mathbf{V}$. Similarly, if two up-type quarks were degenerate, we can make $\overline{U}_L^m \rightarrow \overline{U}_L^m A^\dagger$, where A is a block-diagonal matrix which mixes the two degenerate quarks. Clearly, the parameters in A can be adjusted to remove the complex phase in $\mathbf{V}$ (see Problem 8.2).

There is an infinity of ways to express the elements of $\mathbf{V}$ in terms of three rotation angles and one phase. Also the phase can be made to appear in many different elements; thus the reference to the Scarlet Pimpernel in the motto over this chapter.

One representation has been sanctioned by the particle data group [28]:

$$V = \begin{pmatrix} c_{12}c_{13} & s_{12}c_{13} & s_{13}e^{-i\delta_{13}} \\ -s_{12}c_{23} - c_{12}s_{23}s_{13}e^{i\delta_{13}} & c_{12}c_{23} - s_{12}s_{23}s_{13}e^{i\delta_{13}} & s_{23}c_{13} \\ s_{12}s_{23} - c_{12}c_{23}s_{13}e^{i\delta_{13}} & -c_{12}s_{23} - s_{12}c_{23}s_{13}e^{i\delta_{13}} & c_{23}c_{13} \end{pmatrix} \qquad (8.35)$$

where $c_{ij} = \cos\theta_{ij}$ and $s_{ij} = \sin\theta_{ij}$ for the Euler angles θ_{ij} with i and j being family labels. Different parametrizations lead to the same physics, of course. Without some specific ideas about the mechanism for flavour generation, none is intrinsically superior for theoretical reasons; on the other hand, some can be more convenient on phenomenological grounds. Later we will present one prime example, the Wolfenstein representation.

8.3.3 Describing weak phases through unitarity triangles

For the case of only three families, a geometric representation can greatly facilitate an intuitive understanding. The unitarity of the CKM matrix leads to two types of relations:

$$\sum_{i=1}^{3} |V_{ij}|^2 = 1; \qquad j = 1, \dots, 3 \qquad (8.36)$$

$$\sum_{i=1}^{3} V_{ji}V_{ki}^* = 0 = \sum_{i=1}^{3} V_{ij}V_{ik}^*; \qquad j, k = 1, \dots, 3, \quad j \neq k. \qquad (8.37)$$

The relations of Eq. (8.36) are of fundamental importance: they are often referred to as expressing *weak universality*, meaning that the overall charged current couplings of each quark – say an up-type quark – to all the down-type quarks is of universal strength. This concept first put forward in [80] deserves continuing experimental scrutiny. Within SM it is a consequence of the universality of non-abelian gauge couplings (of $SU(2)_L$). Yet Eq. (8.36) tells us nothing about the weak phases essential for **CP** violation. That information is contained in Eq. (8.37). Since the CKM matrix elements are complex, these relations imply that they form triangles in a complex plane. Two nice features of this representation can be stated right away:

- Changing the phase *convention* for one of the quark fields will (at most) rotate the whole triangle in the complex plane.

- The six triangles representing Eq. (8.37) are of very different shapes (as we will discuss in considerable detail later on). *Yet they all possess the same area*, as can be seen quite easily. Consider the second equation of Eq. (8.37) with $j = d$ and $k = b$ and multiply it by the phase factor $\mathbf{V}_{ub}^* \mathbf{V}_{ud}$:

$$|\mathbf{V}_{ub}\mathbf{V}_{ud}|^2 + \mathbf{V}_{ub}^*\mathbf{V}_{ud}\mathbf{V}_{cd}^*\mathbf{V}_{cb} + \mathbf{V}_{ub}^*\mathbf{V}_{ud}\mathbf{V}_{td}^*\mathbf{V}_{tb} = 0,$$

and thus

$$\text{Im}\mathbf{V}_{ub}^*\mathbf{V}_{ud}\mathbf{V}_{cd}^*\mathbf{V}_{cb} = -\text{Im}\mathbf{V}_{ub}^*\mathbf{V}_{ud}\mathbf{V}_{td}^*\mathbf{V}_{tb}. \tag{8.38}$$

Multiplying by $\mathbf{V}_{cb}^*\mathbf{V}_{cd}$ instead shows that

$$\text{Im}\mathbf{V}_{cb}^*\mathbf{V}_{cd}\mathbf{V}_{td}^*\mathbf{V}_{tb} = -\text{Im}\mathbf{V}_{cb}^*\mathbf{V}_{cd}\mathbf{V}_{ud}^*\mathbf{V}_{ub}.$$

Repeating this procedure for the other relations of Eq. (8.37), we arrive at the following findings:

$$|\text{Im}\mathbf{V}_{km}^*\mathbf{V}_{lm}\mathbf{V}_{kn}\mathbf{V}_{ln}^*| = |\text{Im}\mathbf{V}_{mk}^*\mathbf{V}_{ml}\mathbf{V}_{nk}\mathbf{V}_{nl}^*| = J, \tag{8.39}$$

irrespective of the indices k, l, m, n! The quantity J is obviously invariant under changes in the phase conventions $|q\rangle \to e^{i\phi_q}|q\rangle$ for any of the quark fields (See Problems 8.3 and 8.5).

$$\text{area (every triangle)} = \frac{1}{2}J. \tag{8.40}$$

This is the geometric translation of the algebraic result obtained before, that with three families there exists only a single irreducible phase [83].

Our discussion above shows that **CP** violation cannot be implemented unless

$$m_u \neq m_c \neq m_t, \ m_d \neq m_s \neq m_b \tag{8.41}$$

$$\theta_{12}, \theta_{13}, \theta_{23} \neq 0, \frac{\pi}{2}. \tag{8.42}$$

In addition

$$\delta \neq 0, \frac{\pi}{2} \tag{8.43}$$

has to hold. It should be noted that all these conditions, Eq. (8.41), Eq. (8.42) and Eq. (8.43) can be summarized in a compact manner. Define:

$$iC \equiv [\mathcal{M}_U\mathcal{M}_U^\dagger, \mathcal{M}_D\mathcal{M}_D^\dagger]. \tag{8.44}$$

Then we can put

$$\begin{aligned}\det C = &-2J(m_t^2 - m_c^2)(m_c^2 - m_u^2)(m_u^2 - m_t^2)\\ &\times(m_b^2 - m_s^2)(m_s^2 - m_d^2)(m_d^2 - m_b^2) \neq 0.\end{aligned} \tag{8.45}$$

Non-vanishing of $\det C$ is a necessary condition for **CP** violation [83].

8.4 The pundits' judgement

In summarizing the preceding discussion let us add some new twists:

- The KM analysis is a prime example for deducing the existence of New Physics *indirectly*: the transition $K_L \to \pi\pi$ involves known particles only, yet it cannot proceed within a two-family SM. New degrees of freedom had to be postulated, namely the beauty and top quarks; it took ten and twenty years, respectively, to find them.

It is tempting to speculate what would have happened if history had proceeded somewhat differently, if the charm quark and τ leptons had been discovered a bit sooner, *before* the KM paper had been written. Furthermore, let us consider $m_\pi \sim 0$, in which case it is hard to distinguish K_L from K_S and discover **CP** violation.

With the third family having made its first appearance, someone would have realized that now **CP** violation can be embedded into the SM. In that hypothetical scenario the direct observation of New Physics, specifically of the τ lepton, would have lead to the prediction of a *qualitatively* new transition between Known Physics states: $K_L \to \pi\pi$.

- In the SM quark masses are generated from the Yukawa couplings with a *single* Higgs doublet. This means that the Yukawa couplings are complex, i.e. that there is a *hard* breaking of **CP** symmetry, namely in dimension-four operators in the non-gauge sector of the Lagrangian. Yet the KM construction is more general than that: algebraically it works equally well if the 3×3 quark mass matrices arise from a more complex Higgs sector with multiple Yukawa couplings and vacuum expectation values; in such scenarios the breaking of **CP** invariance can occur also in a spontaneous fashion. We will return to these considerations when discussing alternative models for **CP** violation.

- Let us suppose for a moment that the KM model correctly describes future data. Does it mean that we have understood **CP** violation? The answer is *No!* In that case we have merely found an economical and phenomenologically successful way of embedding **CP** violation into the SM. **V** is derived from the quark mass matrices. To understand it we have to figure out the origin of quark masses.

Problems

8.1 Show that the neutral current interaction $\overline{U}_L \gamma_\mu U_L$ remains invariant when we express it in terms of mass eigenstates. By looking at the form of the weak neutral current given in Eq. (8.17), show that no flavour changing interaction can be generated.

8.2 Show that if any of the quark masses become degenerate, the CKM matrix can be made real. Show also that if any one of the CKM matrix elements vanishes, then the phases can be adjusted so that the CKM matrix is real.

8.3 Show that J is rephasing invariant.

8.4   Take a 2×2 unitary matrix and show explicitly that the phases can be rotated away by redefining the quark phases. Try the same procedure for a 3×3 unitary matrix and show that not all the phases can be rotated away.

8.5   Show that all unitarity triangles have the same area.

8.6   Consider a representation of **V** with a 2×2 sub-matrix that is completely real. Show that such a matrix does *not* contain **CP** violation.

9

The theory of $K_L \to \pi\pi$ decays

In this chapter, we present theoretical expectations for various **CP** violating observables in $K \to \pi\pi$ decays, where we must face up to the challenge of explaining the $\Delta I = 1/2$ rule [28]:

$$\frac{\Gamma(K_S \to \pi^+\pi^-)}{\Gamma(K^+ \to \pi^+\pi^0)} \sim 450, \qquad (9.1)$$

as mentioned in Sec. 7.3. Bose statistics constrain the 2π system to carry isospin 0 or 2 with the charged combination $\pi^+\pi^0$ forming a pure I=2 state. $K^+ \to \pi^+\pi^0$ is thus a pure $\Delta I = 3/2$ transition, whereas for $K_S \to \pi\pi$, a $\Delta I = 1/2$ amplitude can contribute. Eq. (9.1) thus reveals a huge enhancement of $\Delta I = 1/2$ transitions over $\Delta I = 3/2$ ones. Gell-Mann and Pais [84] coined the name $\Delta I = 1/2$ rule for this effect over 40 years ago.

A theoretical description of decay rates involves two main steps:

- evaluating the underlying Lagrangian at ordinary hadronic scales;

- calculating the matrix elements of the emerging operators.

In the following two sections we mainly state the results on operator renormalization and on the size of the relevant matrix elements giving few explanations. A more detailed treatment is found in [85].

9.1 The low-energy $\Delta S = 1$ non-leptonic Lagrangian

The electroweak Lagrangian given in Eq. (8.19) is defined at an energy scale M_W. At lower scales the W bosons are no longer real dynamical entities, yet their virtual exchange generates an effective coupling between charged currents:

$$\mathscr{H}_{\text{weak}}^{CC} = \frac{G_F}{\sqrt{2}} J_\mu^\dagger J^\mu, \ J^\mu = \begin{pmatrix} \overline{u} \\ \overline{c} \\ \overline{t} \end{pmatrix}^{tr} \gamma^\mu (1-\gamma_5) \mathbf{V} \begin{pmatrix} d \\ s \\ b \end{pmatrix}. \qquad (9.2)$$

QCD corrections renormalize this interaction at scales below M_W. Evolving the Hamiltonian $\mathscr{H}$ to a lower scale μ means that physical degrees of freedom whose characteristic frequencies exceed μ are integrated out; i.e., they are eliminated as dynamical fields from the effective *operators* in $\mathscr{H}$; they can contribute only through quantum corrections which affect the *c-number coefficients* of those operators. 'Soft' degrees of freedom with characteristic frequencies below μ provide the fields from which the operators are built. The scale μ is introduced for separating short-distance effects to be included in the coefficient functions from long-distance dynamics in the matrix elements:

$$\text{long distance} > \mu^{-1} > \text{short distance.} \tag{9.3}$$

In addition to being natural, this separation offers practical advantages as well: because QCD is asymptotically free, a perturbative treatment of it should yield a good approximation for short-distance dynamics. Accordingly we can calculate the QCD corrections to an operator in the short-distance domain and thus evaluate the dependence of its coefficient on μ for $\mu \gg \Lambda_{QCD}$. Thus we have the hierarchy

$$\Lambda_{QCD} \ll \mu \ll M_W. \tag{9.4}$$

The leading QCD quantum corrections covering the range from M_W down to μ are evaluated in terms of $\alpha_S \cdot \log(M_W/\Lambda_{QCD})$ and $\alpha_S \cdot \log(\mu/\Lambda_{QCD})$ rather than merely α_S. These logarithmically enhanced numbers can be of order unity, making quantum corrections quite sizeable. For a numerically reliable treatment we have to sum them up; a rather elaborate theoretical tool box based on the renormalization group [82] allows us to achieve this goal in what is called the 'leading log' (and 'next-leading log' etc.) approximation [86].

What we have said so far is quite general. For the particular case under study here, namely strange decays, we have to note a qualitatively new feature: the original operator $\bar{u}_L \gamma_\mu s_L \bar{d}_L \gamma^\mu u_L$ which appears in Eq. (9.2) is *non*-multiplicatively renormalized: *additional* four-quark operators with different colour, flavour and even chirality structure emerge, as can be inferred on qualitative grounds already.

- Consider the scattering of left handed quarks $s_L + u_L \rightarrow d_L + u_L$ which is given by

$$\mathcal{H} = \frac{G_F}{\sqrt{2}} \mathbf{V}_{ud} \mathbf{V}_{us}^* [O_- + O_+] + \text{h.c.}, \tag{9.5}$$

$$O_\pm = 2[\bar{s}_L \gamma_\mu u_L \bar{u}_L \gamma^\mu d_L \pm \bar{s}_L \gamma_\mu d_L \bar{u}_L \gamma^\mu u_L]. \tag{9.6}$$

$O_-[O_+]$ is odd [even] under the interchange of final state u and d quarks. Thus the final state is a pure $I_f = 0$ [$I_f = 1$] state. With s being an iso-singlet, we see that O_- drives purely $\Delta I = 1/2$ transitions, whereas O_+ drives $\Delta I = 1/2$ and $3/2$ transitions. Since the strong interactions conserve I, they preserve the isospin structure and renormalize O_- and O_+ multiplicatively, yet by a different amount:

$$O_+(M_W) \xrightarrow{QCD} C_+(\mu)O_+(\mu)$$
$$O_-(M_W) \xrightarrow{QCD} C_-(\mu)O_-(\mu). \tag{9.7}$$

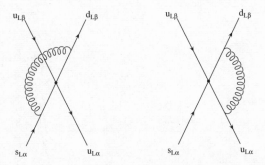

Fig. 9.1. QCD corrections to four-quark operators. The blob represents the uncorrected four-Fermi interaction $\bar{u}_{L\alpha}\gamma_\mu s_{L\alpha}\bar{d}_{L\beta}\gamma^\mu u_{L\beta}$. Exchange of a gluon shown in these diagrams generates $\bar{u}_{L\beta}\gamma_\mu s_{L\alpha}\bar{d}_{L\alpha}\gamma^\mu u_{L\beta}$ which, after Fiertz transformation, becomes $\bar{u}_{L\beta}\gamma_\mu u_{L\beta}\bar{d}_{L\alpha}\gamma^\mu s_{L\alpha}$.

Fig. 9.1 shows the radiative QCD correction to a four-quark operator. In the leading log approximation the coefficients $C_\pm$ are given by

$$C_-(\mu)|_{LL} = \left[\frac{\alpha_S(M_W^2)}{\alpha_S(\mu^2)}\right]^{-\frac{12}{33-2n_F}}, \quad C_+(\mu)|_{LL} = \left[\frac{\alpha_S(M_W^2)}{\alpha_S(\mu^2)}\right]^{\frac{6}{33-2n_F}} \quad (9.8)$$

with n_F denoting the number of active flavours. So, a new operator with different flavour structure $\bar{d}_L\gamma_\mu s_L\bar{u}_L\gamma^\mu u_L$ appears.

Since $\alpha_S(M_W^2) < \alpha_S(\mu^2)$ for $M_W > \mu$ – QCD is asymptotically free – we have

$$C_-(\mu) > 1 > C_+(\mu) ; \quad (9.9)$$

i.e., radiative QCD corrections naturally enhance $\Delta I = 1/2$ over $3/2$ transitions [87, 88].

- Even the chirality structure can be modified by QCD corrections through the so-called penguin process introduced by the ITEP group [89] and shown in Fig. 9.2(a). The penguin operator is given by

$$O_P = \sum_a \bar{s}_L\gamma_\mu t^a d_L(\bar{u}\gamma^\mu t^a u + \bar{d}\gamma^\mu t^a d). \quad (9.10)$$

The gluon field – being a vector – yields a $(V-A) \times V$ contribution; since it is an isoscalar, it contributes only to $\Delta I = 1/2$ transitions. Redrawing Fig. 9.2(a) as in Fig. 9.2(b) makes it more obvious why the name penguin has been coined for it.

- For a quantitative description we start with the three family charged

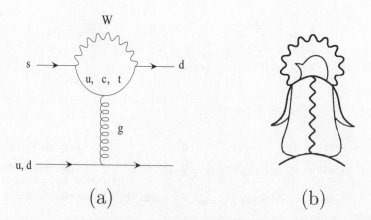

Fig. 9.2. (a) Penguin graphs generated by QCD correction to the non-leptonic Lagrangian. This graph generates O_3 to O_6. (b) If we open up the blob in (a) and expose the W boson propagator, we can guess why (a) is called a penguin graph. This figure was reproduced from Parity by permission of T. Muta and T. Morozumi.

current Hamiltonian defined at M_W:

$$\mathscr{H}(M_W) = \frac{G_F}{\sqrt{2}} \sum_{q=u,c,t} \mathbf{V}_{qs}^* \mathbf{V}_{qd}\, O_2^q + \text{h.c.}$$

$$= \frac{G_F}{\sqrt{2}} \mathbf{V}_{us}^* \mathbf{V}_{ud}[(1-\tau)(O_2^u - O_2^c) + \tau(O_2^u - O_2^t)] + \text{h.c.},$$

$$(9.11)$$

where $\tau = -\frac{\mathbf{V}_{ts}^* \mathbf{V}_{td}}{\mathbf{V}_{us}^* \mathbf{V}_{ud}}$ and $O_2^q = (\bar{s}q)_{V-A}(\bar{q}d)_{V-A}$, and we use the notation

$$(\bar{s}q)_{V-A}(\bar{q}d)_{V-A} \equiv (\bar{s}\gamma_\mu(1-\gamma_5)q)(\bar{q}\gamma^\mu(1-\gamma_5)d). \qquad (9.12)$$

Now, running the renormalization group equation down from M_W to $\mu < m_c$,

$$O_2^u - O_2^c \rightarrow \sum_{i=1}^{10} z_i(\mu)O_i$$

$$O_2^u - O_2^t \rightarrow \sum_{i=1}^{10} v_i(\mu)O_i, \qquad (9.13)$$

where we have included the electromagnetic penguin operator shown in Fig. 9.3. Putting everything together – see [89, 90, 91, 85] for numerical evaluation of $z_i(\mu)$ and $v_i(\mu)$, – we obtain as low-energy Hamiltonian:

$$\mathscr{H}(\Delta S = 1) = \frac{G_F}{\sqrt{2}} \mathbf{V}_{us}^* \mathbf{V}_{ud} \left(\sum_{i=1}^{10} (z_i(\mu) + \tau y_i(\mu))O_i \right) + \text{h.c.}, \qquad (9.14)$$

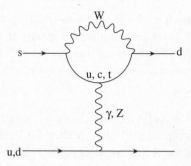

Fig. 9.3. The electroweak penguin graph, generating O_7 to O_{10}. While these diagrams are suppressed by electroweak interaction compared to gluon penguin operators, electroweak penguins generate a $\Delta I = 3/2$ amplitude. These contributions become relevant when $\Delta I = 3/2$ transitions are important – such as in ϵ'. The strength of electroweak penguins is helped by large top quark mass.

where $y_i(\mu) = v_i(\mu) - z_i(\mu)$, and

$$O_1^u = (\bar{s}d)_{V-A}(\bar{u}u)_{V-A}, \qquad O_2^u = (\bar{s}u)_{V-A}(\bar{u}d)_{V-A}$$

$$O_3 = (\bar{s}d)_{V-A}\sum_q(\bar{q}q)_{V-A}, \qquad O_4 = (\bar{s}_\alpha d_\beta)_{V-A}\sum_q(\bar{q}_\beta q_\alpha)_{V-A}$$

$$O_5 = (\bar{s}d)_{V-A}\sum_q(\bar{q}q)_{V+A}, \qquad O_6 = (\bar{s}_\alpha d_\beta)_{V-A}\sum_q(\bar{q}_\beta q_\alpha)_{V+A}$$

$$O_7 = \frac{3}{2}(\bar{s}d)_{V-A}\sum_q e_q(\bar{q}q)_{V+A}, \quad O_8 = \frac{3}{2}(\bar{s}_\alpha d_\beta)_{V-A}\sum_q e_q(\bar{q}_\beta q_\alpha)_{V+A}$$

$$O_9 = \frac{3}{2}(\bar{s}d)_{V-A}\sum_q e_q(\bar{q}q)_{V-A}, \quad O_{10} = \frac{3}{2}(\bar{s}_\alpha d_\beta)_{V-A}\sum_q e_q(\bar{q}_\beta q_\alpha)_{V-A},$$

$$(9.15)$$

with α and β being colour indicies.

9.2 Evaluating matrix elements

The $K \to (\pi\pi)_I$ amplitude is given by

$$\langle(\pi\pi)_I|\mathscr{H}(\Delta S = 1)|K\rangle = \frac{G_F}{\sqrt{2}}V_{us}^*V_{ud}\sum_{i=1}^{10}C_i(\mu)\langle(\pi\pi)_I|O_i|K\rangle_\mu, \qquad (9.16)$$

where $C_i(\mu) = z_i(\mu) + \tau y_i(\mu)$. Remember, the quantity μ was introduced to renormalize the operator $\mathscr{H}(\Delta S = 1)$ as demanded by quantum field theory; *in principle* the numerical value of such an auxiliary scale is irrelevant since the μ dependence of $C_i(\mu)$ is compensated for by a corresponding μ dependence in the matrix elements of $\langle(\pi\pi)_I|O_i|\rangle_\mu$ so that it drops out from the transition amplitudes. However *in practice* its value has to be chosen judiciously for two reasons:

- We are faced with a 'Scylla and Charybdis' problem: on one hand we need $\mu \gg \Lambda_{QCD}$ for a perturbative treatment of the radiative corrections to be meaningful; on the other hand it has to be sufficiently low so that the nonperturbative algorithm we employ in calculating matrix elements can be applied.

- The situation is actually more iffy since we have to rely mainly on models or mere phenomenological prescriptions for computing matrix elements. Typically, those tools are not sufficiently refined to allow us to track the dependence on μ. We then rely on physical intuition to argue that μ should be matched to typical momenta in the hadronic wavefunctions, i.e., $\mu \sim 0.5 - 1$ GeV.

9.3 Chiral symmetry and vacuum saturation approximation

We do not know yet how to use QCD to reliably compute matrix elements of O_i at low energies. Often symmetries come to our rescue. We know that QCD possesses chiral symmetry, i.e., if quark masses are set to zero, left-handed and right-handed quarks do not talk to each other. We shall write an effective low energy theory with $U(3) \times U(3)$ chiral symmetry which is broken only by quark mass terms [92]. Computing the Noether currents associated with chiral symmetry for both the QCD and the effective chiral Lagrangian [93], yields the following identities:

$$\bar{u}\gamma_\mu\gamma_5 u = \frac{F_\pi}{\sqrt{2}}\partial_\mu\pi^0, \qquad \bar{d}\gamma_\mu\gamma_5 u = F_\pi\partial_\mu\pi^+$$

$$\bar{s}\gamma_\mu\gamma_5 u = F_K\partial_\mu K^+, \qquad \bar{s}\gamma_\mu\gamma_5 d = F_K\partial_\mu K^0$$

$$\bar{s}\gamma_5 u = \frac{iM_K^2 F_K}{m_s + m_u}K^+, \qquad \bar{s}\gamma_5 d = \frac{iM_K^2 F_K}{m_s + m_d}K^0$$

$$\bar{s}u = \frac{M_K^2 - M_\pi^2}{m_s - m_u}\left(\pi^+ K^0 + \frac{1}{\sqrt{2}}\pi^0 K^+\right)$$

$$\bar{s}d = \frac{M_K^2 - M_\pi^2}{m_s - m_d}\left(\pi^- K^+ - \frac{1}{\sqrt{2}}\pi^0 K^0\right)$$

$$\bar{d}d = \frac{M_\pi^2}{2m_d}\left(\pi^+\pi^- + \frac{1}{2}\pi^0\pi^0\right)$$

$$\bar{s}\gamma_\mu u = -i\left(\frac{1}{\sqrt{2}}(\pi^0\partial_\mu K^+ - \partial_\mu\pi^0 K^+) + \pi^+\partial_\mu K^0 - \partial_\mu\pi^+ K^0\right)$$

$$\bar{s}\gamma_\mu d = -i\left(\frac{1}{\sqrt{2}}(K^0\partial_\mu\pi^0 - \partial_\mu K^0\pi^0) + \pi^-\partial_\mu K^+ - \partial_\mu\pi^- K^+\right).$$

$$(9.17)$$

Using these identities we can write the operators O_i to leading order in chiral perturbation theory as products of pseudo-Nambu–Goldstone bosons which accompany chiral symmetry. The evaluation of $\langle(\pi\pi)_I|O_i|K\rangle_\mu$ to this lowest order is trivial (see Problem 9.5).

$$\langle\pi^+\pi^-|O_2^u|K^0\rangle = -\langle\pi^0\pi^0|O_1^u|K^0\rangle \; = \; X$$
$$\langle\pi^+\pi^0|O_1|K^+\rangle = \langle\pi^+\pi^0|O_2|K^+\rangle \; = \; \frac{1}{\sqrt{2}}X$$
$$\langle\pi^+\pi^-|O_6|K^0\rangle = \langle\pi^0\pi^0|O_6|K^0\rangle \; = \; Y, \qquad (9.18)$$

where

$$X \; = \; iF_\pi(M_K^2 - M_\pi^2) = 0.036i \; \text{GeV}^3$$
$$Y \; = \; -2i(F_K - F_\pi)\left(\frac{M_K^2}{m_s + m_d}\right)^2$$
$$\; = \; -0.114i\left(\frac{175MeV}{m_s + m_d}\right)^2 \; \text{GeV}^3. \qquad (9.19)$$

This approximation is called *factorization* or the *vacuum saturation approximation*.

Vacuum saturation cannot be an identity. For example, in this order matrix elements are independent of μ; thus the above cancellation of μ dependences cannot happen. We can of course compute higher order terms which require renormalization, and thus introduce a μ dependence. But we do not expect this path to be fruitful for reasons which will become obvious soon. Instead, we shall introduce fudge factors $B_i^I(\mu)$ with their deviation from unity representing the error in the approximation for a given operator, where I is the isospin of the two-pion system.

For an energy scale $\mu \sim 0.7$ GeV, where we expect chiral symmetry to be reasonably intact, we expect $B_i^I(\mu) = \mathcal{O}(1)$.

The operator O_6 has a mixed chirality structure of the form $(V-A)\times(V+A)$, see Eq. (9.15), which is due to the fact that the gluon couples to left- as well as right-handed quarks. Under a Fierz transformation it changes into $(S-P)\times(S+P)$; we might then suspect that it will have enhanced matrix elements for pseudoscalar hadrons [89]. This is indeed the case – see the Problems section – and it arises explicitly in Eq. (9.19) through the factor $[M_K/(m_s + m_d)]^2 \sim 10$.

9.4 $K \to \pi\pi$ decays

We are now ready to evaluate amplitudes for $K \to \pi\pi$ decays given by

$$\langle\pi^+\pi^0|\mathcal{H}(\Delta S = 1, \mu)|K^+\rangle = \frac{G_F}{\sqrt{2}}\mathbf{V}_{us}^*\mathbf{V}_{ud}\frac{1}{\sqrt{2}}(z_1(\mu) + z_2(\mu))X. \qquad (9.20)$$

For $\mu \sim 0.7$ GeV (beyond this our chiral purturbation result is certainly questionable), $z_1 = -0.87$, $z_2 \sim 1.51$ [85]. Using these results,

$$|\langle\pi^+\pi^0|\mathcal{H}(\Delta S = 1, \mu)|K^+\rangle| \simeq 3 \times 10^{-8} \; \text{GeV}, \qquad (9.21)$$

to be compared with an experimental number 1.8×10^{-8} GeV – not bad for such a crude estimate!

Now, for the real part of the $K^0 \to \pi^+\pi^-$ decay amplitude, numerical studies show that all contributions except for those from O_2 and O_6 can be neglected:

$$\langle \pi^+\pi^- | \mathcal{H}(\Delta S = 1, \mu) | K^0 \rangle = \frac{G_F}{\sqrt{2}} V_{us}^* V_{ud} (z_2(\mu)X + z_6(\mu)Y). \tag{9.22}$$

We thus obtain, with $z_6 = -0.082$,

$$\frac{\langle \pi^+\pi^- | \mathcal{H}(\Delta S = 1, \mu) | K^0 \rangle}{\langle \pi^+\pi^0 | \mathcal{H}(\Delta S = 1, \mu) | K^+ \rangle} = \sqrt{2} \frac{z_2(\mu)X(\mu) + z_6(\mu)Y(\mu)}{(z_1(\mu) + z_2(\mu))X(\mu)} \sim 4. \tag{9.23}$$

This should be compared to the experimental value of

$$\frac{\langle \pi^+\pi^- | \mathcal{H}(\Delta S = 1, \mu) | K^0 \rangle}{\langle \pi^+\pi^0 | \mathcal{H}(\Delta S = 1, \mu) | K^+ \rangle} = 15, \tag{9.24}$$

obtained from Eq. (9.1). Do we understand the $\Delta I = 1/2$ rule? The answer is obvious and embarrassing, isn't it? While we more or less understand the $\Delta I = \frac{3}{2}$ amplitude, the precise origin of the $\Delta I = \frac{1}{2}$ enhancement still has not been identified 40 years later. Various possible culprits have been lined up; we should note that QCD renormalization accelerates quite considerably the $\Delta I = 1/2$ rate through the enhanced operator O_-, and the penguin operator through its large matrix element as inferred from chiral invariance. Yet there does not seem to exist a concise dynamical explanation for what had appeared to be an elegant rule. Instead Eq. (9.1) probably represents the cumulative result of various effects all working in the same direction [94], yet no conclusive quantitative explanation has been given; this is not surprising since we are dealing here with predominantly nonperturbative effects. On the other hand it would be unfair to claim that this rule represents more than an embarrassment for theory.

9.5 ϵ'/ϵ

Before we begin, a few general remarks might be useful:

- Since the interplay of three families is required for a **CP** asymmetry to become observable, direct **CP** violation cannot be generated from tree level diagrams – a one-loop operator is needed. Direct **CP** violation thus represents a pure quantum effect and therefore can be expected to be reduced in the KM model. This can be seen by writing

$$\frac{\epsilon'}{\epsilon} \sim \omega \frac{\xi_0 - \xi_2}{\epsilon_m} \tag{9.25}$$

  where $\epsilon_m = -\frac{\text{Im}\, M_{12}}{\Delta M}$. The phase $\xi_0 - \xi_2$ of the loop diagrams is reduced even compared to ϵ_m.

- Nevertheless the KM model does not provide an automatic realization of a superweak scenario since $\epsilon'/\epsilon \neq 0$ is expected in general.

- On the other hand we predict ϵ'/ϵ to decrease considerably the larger m_t gets. For there is a single weak phase that drives **CP** violation which drops out from the ratio ϵ'/ϵ; yet ϵ is more enhanced by a large top mass than ϵ'. This has to be kept in mind when judging older predictions on ϵ'/ϵ; they tend to be based on assuming a range for m_t that is well below the now measured value. Consider the 'old' analysis of Ref. [95]: it yields numbers very similar to Eq. (9.35) once the now known value for m_t has been adopted. Unfortunately this opens a back door for $\epsilon'/\epsilon \simeq 0$. The strong penguin is not greatly enhanced by a large top mass; however, the electroweak penguin is – due to the longitudinal component of the virtual Z boson which is the reincarnation of one of the original Higgs fields – and it contributes with the opposite sign! Even $\epsilon'/\epsilon = 0$ can then occur for m_t large enough. This happens for $m_t \geq 200$ GeV, which is not far above the observed mass.

To evaluate ϵ' using Eq. (7.39), we need to know the phases of A_0 and A_2, ξ_0 and ξ_2, respectively. Our lack of success in deriving the $\Delta I = 1/2$ rule from theory suggests that we should obtain the values of the various hadronic matrix elements from experimental data.

9.5.1 Determination of matrix elements from data

We sketch the method used to obtain matrix elements of these four quark operators to show you the complexities involved. Those who are interested in the actual determination are referred to Ref. [85].

At what value of μ shall we determine the matrix elements? It should be large enough so that the coefficient functions are reliably obtained using renormalization group equations. It will be clear below that $\mu = m_c$ is a good choice. But then we cannot rely on chiral perturbation theory to obtain the matrix element. This is just as well. For the region in which the coefficient functions are known, $\mu \gg \Lambda_{QCD}$, and the region where chiral perturbation theory determines the matrix elements, $\mu \sim m_\pi$, never overlap. Then we have nothing but experiments to help us determine the matrix elements. The experimental values we use are (see Problem 9.2):

$$\text{Re}\, A_0 = 3.38 \times 10^{-7} \text{ GeV}$$
$$\text{Re}\, A_2 = 1.49 \times 10^{-8} \text{ GeV} . \tag{9.26}$$

In the region $\mu > m_c$, the charm degree of freedom is not frozen and the Lagrangian is written as

$$\mathcal{H}(\Delta S = 1) = \frac{G_F}{\sqrt{2}} \mathbf{V}_{us}^* \mathbf{V}_{ud} \left((1-\tau) \sum_{i=1,2} z_i(\mu)(O_i^u - O_i^c) + \tau \sum_{i=1}^{10} v_i(\mu) O_i \right). \tag{9.27}$$

With ϵ' so small, we have to include higher order terms in $\mathscr{H}(\Delta S = 1)$, including, in particular, *electroweak* penguin operators. Terms multiplied by τ can safely be neglected for the real parts of the matrix elements:

$$\mathrm{Re}\, A_0 \;=\; \frac{G_F}{\sqrt{2}} \mathbf{V}_{us}^* \mathbf{V}_{ud}[z_1(\mu)\langle Q_1^u - Q_1^c\rangle_0 + z_2(\mu)\langle Q_2^u - Q_2^c\rangle_0]$$

$$\mathrm{Re}\, A_2 \;=\; \frac{G_F}{\sqrt{2}} \mathbf{V}_{us}^* \mathbf{V}_{ud}[z_1(\mu)\langle Q_1^u\rangle_2 + z_2(\mu)\langle Q_2^u\rangle_2], \tag{9.28}$$

with

$$\langle Q_i^q(\mu)\rangle_I \equiv \langle(\pi\pi)_I|O_i^q|K\rangle. \tag{9.29}$$

For illustration we use $z_1(m_c) = -0.459$ and $z_2(m_c) = 1.244$, which correspond to $\Lambda_{\overline{MS}} = 0.3$ GeV in the HV renormalization scheme. Noting that $\langle Q_1^u\rangle_2 = \langle Q_2^u\rangle_2$, and the renormalization group equation preserves this relation, we obtain

$$\langle Q_2\rangle_2 \simeq 0.010 \,(\mathrm{GeV})^3. \tag{9.30}$$

Determination of $\langle Q_i\rangle_0$ is more complicated. Putting another constraint, $\langle Q_-(m_c)\rangle_0 \geq \langle Q_+(m_c)\rangle_0 \geq 0$, where $Q_\pm$ is a matrix element of $O_\pm$ defined in Eq. (9.6), which holds in most non-perturbative approaches, we obtain

$$\langle Q_2\rangle_0 = (0.13 \pm 0.02)\,(\mathrm{GeV})^3, \;\; \langle Q_1\rangle_0 = (-0.06 \mp 0.06)\,(\mathrm{GeV})^3. \tag{9.31}$$

Note that $\langle Q_2\rangle_0$ is much larger than $\langle Q_2\rangle_2$: for $\mu \geq m_c$ the charm penguin effect is included in $\langle Q_2\rangle$ rather than appearing as a separate operator, as is the case with $\mu \ll m_c$, discussed in the previous section.

Now $\langle Q_1\rangle_{0,2}$ and $\langle Q_2\rangle_{0,2}$ are known. How about other matrix elements? $\mathrm{Im}\,A_i$ arises from terms proportional to $\mathrm{Im}\,\tau$ in Eq. (9.27). The matrix elements $\langle Q_5\rangle_0$, $\langle Q_6\rangle_0$, $\langle Q_7\rangle_2$ and $\langle Q_8\rangle_2$ have been studied in lattice QCD with the result that the vacuum saturation approximation seems to be reliable for these matrix elements. Finally, there are operator identities which allow us to write $\langle Q_4\rangle_0$, $\langle Q_9\rangle_0$, $\langle Q_{10}\rangle_0$, $\langle Q_9\rangle_2$, $\langle Q_{10}\rangle_2$ in terms of $\langle Q_1\rangle_I$, $\langle Q_2\rangle_I$, and $\langle Q_3\rangle_0$. So, if we kow $\langle Q_3\rangle_0$, we are all set. So far there is no way to improve the reliability of the vacuum saturation approximation. But we note that the vacuum saturation leads to $\langle Q_3\rangle_0 \sim X/3$, i.e., it is about a factor of five smaller than $\langle Q_2\rangle_0$. So, the result is not sensitive to $\langle Q_3\rangle_I$ and we shall use the vacuum saturation approximation.

9.5.2 Numerical estimates

It is by now quite clear that the theoretical status of the KM prediction for direct **CP** violation as expressed through ϵ'/ϵ is even more frustrating than the experimental situation used to be – and with less relief in sight. Inserting experimental numbers for ϵ and $\mathrm{Re}\,A_0$, the result for $\frac{\epsilon'}{\epsilon}$ given in Eq. (7.28) can be written as:

$$\frac{\epsilon'}{\epsilon} = -ie^{i(\delta_2 - \delta_0)} \times 10^{-4} \left(\frac{\mathrm{Im}\,\tau}{10^{-4}}\right) r \left(\sum_{i=3}^{6} y_i \langle Q_i \rangle_0 - \frac{1}{\omega} \sum_{i=7}^{10} y_i \langle Q_i \rangle_2\right) \tag{9.32}$$

where $r = \frac{G_F \omega}{|2\epsilon| \mathrm{Re}\, A_0} = 336$ GeV^{-3}, and we used the fact that $y_1 = y_2 = 0$.

The numerical value of each contribution to Eq. (9.32) is given as follows.

$$r \sum_{i=3}^{6} y_i \langle Q_i \rangle_0 \sim 0.133 - 3.93 + (-0.5 + 7.56)R_s$$

$$-\frac{r}{\omega} \sum_{i=7}^{10} y_i \langle Q_i \rangle_2 \sim (+0.19 - 3.49)R_s + 1.29 - 0.48, \tag{9.33}$$

where

$$R_s = \left(\frac{175 \text{ MeV}}{\mathrm{m_s} + \mathrm{m_d}}\right)^2, \tag{9.34}$$

and we used $\Delta S = 0$ Wilson coefficients at $\mu = m_c = 1.3$ GeV, $\Lambda_{QCD} = 325$ MeV and $m_t = 170$ GeV as given by Ref. [96].

A few comments can provide some orientation:

- The contribution from $y_3 \langle Q_3 \rangle_0$ is indeed small.

- Contributions from $y_4 \langle Q_4 \rangle_0$, and $y_6 \langle Q_6 \rangle_0$ are from charm quark and top quark penguin-type terms, respectively. For $m_s = (100 \text{ MeV} \sim 170 \text{ MeV})$ they amount to $(4 \sim 19) \times 10^{-4}$.

- Because $\frac{\epsilon'}{\epsilon}$ depends on $\frac{1}{\omega} \mathrm{Im}\, A_2$, electromagnetic penguins [91], which are suppressed by α, become relevant. For $m_s = (100 \text{ MeV} \sim 170 \text{ MeV})$ they give a contribution of order $-(2 \sim 9) \times 10^{-4}$.

- QCD and electroweak penguins contribute with the opposite sign!

There are considerable cancellations between various terms - especially between QCD and electroweak penguins. The theoretical prediction for $\frac{\epsilon'}{\epsilon}$ thus suffers from a large uncertainty. It changes for different allowed input values of μ, Λ_{QCD}, m_c, $m_s(\mu)$, and these variations are magnified due to the cancellations. In the presentation above, we have retained the leading $1/N_C$ terms only mainly to illustrate the procedure in a more transparent way. A more careful analysis can be done keeping higher order terms. Here we shall quote a typical result [85, 97]:

$$-2.1 \times 10^{-4} \leq \frac{\epsilon'}{\epsilon} \leq 13.3 \times 10^{-4} \qquad \text{for } \mathrm{m_s(m_c)} = (150 \pm 20)\text{MeV}$$

$$-0.5 \times 10^{-4} \leq \frac{\epsilon'}{\epsilon} \leq 25.2 \times 10^{-4} \qquad \text{for } \mathrm{m_s(m_c)} = (125 \pm 20)\text{MeV}.$$

$$\tag{9.35}$$

Clearly, the theoretical analysis has not reached a mature stage yet:

- The values of $\langle Q_{1,2} \rangle_0$ have to be increased considerably over expectations to reproduce the data.

- The sizes of $\langle Q_6 \rangle_0$ and $\langle Q_8 \rangle_0$ are very uncertain. For example, they may get enhanced by final state interactions. It is hoped that future lattice simulations of QCD will allow an *ab initio* determination of these matrix elements. Yet, we will need to go beyond the so-called quenched approximations to include final state interactions.

Theory folklore based on $m_s(m_c) > 150$ MeV had $\frac{\epsilon'}{\epsilon}$ actually a tad below 10^{-3} for several years, before it started to inch up recently. While it appears that direct **CP** violation – and thus a new paradigm – has definitively been established by NA31 and KTeV, the interpretation of the observed signal is far from settled: it could represent another success of the KM ansatz – or signal considerable intervention of New Physics!

9.6 $\Delta S = 2$ amplitudes

This is one of very few places in particle physics where second order weak amplitudes are accessible to experiment. It is therefore an ideal place to look for New Physics beyond the SM.

It also illustrates a point we have repeatedly made before: proper treatment of **CP** phenomenology involves applying basically the full machinery of theoretical tools available in quantum field theory and an appreciation of its subtleties; thus it can be seen as a somewhat unconventional introduction into quantum field theory.

When the **CP** violating effect is small, we have from Eq. (6.39):

$$\Delta M_K = -2\overline{M}_{12}. \tag{9.36}$$

Without an elementary $\Delta S = 2$ interaction in the SM, we obtain the $\Delta S = 2$ amplitude by iterating the basic $\Delta S = 1$ coupling, i.e., we are dealing with a second-order weak process: $\mathscr{H}_{\text{eff}}(\Delta S = 2) = \mathscr{H}_{\text{eff}}(\Delta S = 1) \otimes \mathscr{H}_{\text{eff}}(\Delta S = 1)$. In doing so we get the celebrated box diagram shown in Fig. 9.4. There are local (c and t quark) and non-local (u quark loop) contributions. We shall concentrate on local contributions first and come back to non-local ones later. The contributions that do *not* depend on the mass of the internal quark cancel against each other due to the GIM mechanism [77]. Integrating over the internal lines then yields a convergent result. With three families we arrive at [98]

$$\mathscr{H}_{\text{eff}}^{\text{box}}(\Delta S = 2, \mu) = \left(\frac{G_F}{4\pi}\right)^2 M_W^2$$

$$\cdot \left[\eta_{cc}(\mu)\lambda_c^2 E(x_c) + \eta_{tt}(\mu)\lambda_t^2 E(x_t) + 2\eta_{ct}(\mu)\lambda_c\lambda_t E(x_c, x_t)\right] [\bar{d}\gamma_\mu(1 - \gamma_5)s]^2 \tag{9.37}$$

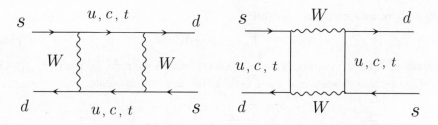

Fig. 9.4.   The box diagrams giving rise to a $\Delta S = 2$ operator.

with λ_i denoting combinations of KM parameters,

$$\lambda_i = \mathbf{V}_{is}\mathbf{V}_{id}^* , \ i = c, t , \tag{9.38}$$

$E(x_i)$ and $E(x_i, x_j)$ reflecting the box loops with equal and different internal quarks (charm or top), respectively:

$$E(x_i) = x_i \left(\frac{1}{4} + \frac{9}{4(1 - x_i)} - \frac{3}{2(1 - x_i)^2} \right) - \frac{3}{2} \left(\frac{x_i}{1 - x_i} \right)^3 \log x_i \tag{9.39}$$

$$E(x_c, x_t) = x_c x_t \left[\left(\frac{1}{4} + \frac{3}{2}\frac{1}{1 - x_t} - \frac{3}{4}\frac{1}{(1 - x_t)^2} \right) \frac{\log x_t}{x_t - x_c} + (x_c \leftrightarrow x_t) \right.$$

$$\left. - \frac{3}{4}\frac{1}{(1 - x_c)(1 - x_t)} \right] ; \ x_i = \frac{m_i^2}{M_W^2}, \tag{9.40}$$

and $\eta_{qq'}$ containing the QCD radiative corrections from evolving the effective Lagrangian from M_W down below m_c. The next-to-leading log corrections have been studied in order to understand the theoretical errors. We shall not go into the scale dependence as well as errors associated with uncertainties in Λ_{QCD}, m_t, etc. Such discussion can be found in Ref. [85]; the result is

$$\eta_{cc} \simeq 1.38 \pm 0.20 , \quad \eta_{tt} \simeq 0.57 \pm 0.01 , \quad \eta_{ct} \simeq 0.47 \pm 0.04. \tag{9.41}$$

As the last step, we have to evaluate matrix elements of the $\Delta S = 2$ transition operator. Even for a local four-fermion operator this is far from trivial since on-shell matrix elements are controlled by long-distance dynamics. Its size can be parametrized as follows (See Problem 9.6):

$$\langle K^0 | (\bar{d}\gamma_\mu(1 - \gamma_5)s)(\bar{d}\gamma_\mu(1 - \gamma_5)s) | \overline{K}^0 \rangle = -\frac{4}{3} B_K F_K^2 M_K. \tag{9.42}$$

The fudge factor B_K is the bag factor. Let us see the origin of the minus sign. From Table 4.2 note that $\mathbf{CP}\bar{s}(\vec{x})\gamma^\mu\gamma_5 d(\vec{x})\mathbf{CP}^\dagger = -\bar{d}(-\vec{x})\gamma^\mu\gamma_5 s(-\vec{x})$. But we have defined $\mathbf{CP}|K\rangle = +|\overline{K}\rangle$ by convention.

Several theoretical techniques have been employed to estimate the size of B_K. For a recent review, see Ref. [99]. The findings of several phenomenological studies can be summarized as follows:

$$B_K \simeq 0.8 \pm 0.2. \qquad (9.43)$$

At which value of μ is Eq. (9.43) evaluated? Within the error stated in Eq. (9.43), it is reasonable to set $\mu \simeq 0.5 - 1$ GeV as the appropriate scale.

9.6.1 ΔM_K

Let us start with a numerical estimate of the short distance contribution to ΔM_K. Eq. (9.36) gives

$$
\begin{aligned}
\Delta M|_{SD} &= -2\langle K | \mathscr{H}_{\text{eff}}^{\text{box}}(\Delta S = 2, \mu) | \overline{K} \rangle \\
&= \frac{G_F^2}{6\pi^2} F_K^2 B_K M_K M_W^2 \left[\lambda_c^2 E(x_c)\eta_{cc}(\mu) \right. \\
&\quad + \left. \lambda_t^2 E(x_t)\eta_{tt}(\mu) + 2\lambda_c\lambda_t E(x_c, x_t)\eta_{ct}(\mu) \right].
\end{aligned}
\qquad (9.44)
$$

For numerical estimates, we use the following numbers:

$$
\begin{aligned}
E(x_c) &\sim 3 \times 10^{-4}, \quad E(x_t) \sim 2.7, \quad E(x_c, x_t) \sim 2.7 \times 10^{-3} \\
m_t &= 180 \text{ GeV}, \quad m_c = 1.4 \text{ GeV}, \quad \lambda = 0.221.
\end{aligned}
\qquad (9.45)
$$

It is clear that since $\lambda_t \sim 6 \times 10^{-4} \ll \lambda_c$, the first term which corresponds to the $\bar{c}c$ intermediate states dominates. Numerically, we have

$$\Delta M|_{SD} = 2 \times 10^{-12} \text{ MeV}. \qquad (9.46)$$

Using the experimental value given in Eq. (5.40), we see that

$$\Delta M|_{SD} \sim \left(\frac{1}{3} \sim \frac{1}{2} \right) \cdot \Delta M_K|_{\text{exp}}. \qquad (9.47)$$

The deficit in $\Delta M|_{SD}$, viz. $\Delta M|_{\text{exp}}$, can – with the present reliability of our theoretical tools – be assigned to contributions from *non-local* operators corresponding to the u quark loop contribution which describes $K^0 \to \pi, \eta, \eta', \pi\pi, ... \to \overline{K}^0$ [100] intermediate states. These contributions can be summarized as follows:

$$
\begin{aligned}
\Delta M|_{SM} &= \Delta M|_{SD} + \Delta M|_{LD}, \\
\Delta M|_{LD} &\sim \left(\frac{1}{2} \sim \frac{2}{3} \right) \cdot \Delta M|_{\text{exp}}.
\end{aligned}
\qquad (9.48)
$$

Obviously we wish the theoretical uncertainties we have to allow for were smaller. While we can expect the parameter B_K and thus $\Delta M|_{SD}$ to be known better in the foreseeable future, there is much less reason for such optimism with respect to $\Delta M|_{LD}$. Yet for proper perspective we should note that the quark box contribution in the *absence* of charm (and top) quarks amounts to (see Problem 9.7)

$$
\begin{aligned}
\Delta M|_{u,d,t/}^{\text{box}} &\simeq \frac{G_F^2}{3\pi^2} M_W^2 \cos^2\theta_C \sin^2\theta_C B_K F_K^2 M_K \\
&\sim 10^{-8} \text{ MeV} \sim 4000 \cdot \Delta M|_{\text{exp}} \, !
\end{aligned}
\qquad (9.49)
$$

While theory gives the right order of magnitude for ΔM_K, we should not overlook the fact that the theory also implies K_L must be heavier than K_S to give the correct sign for ΔM_K. Both of these conclusions are highly nontrivial.

9.6.2 ϵ

As shown in Eq. (7.47), we have

$$\epsilon \simeq \frac{1}{\sqrt{2}} e^{i\phi_{SW}} (\epsilon_m + \xi_0) \tag{9.50}$$

where

$$\epsilon_m = -\frac{\text{Im} M_{12}}{\Delta M_K}. \tag{9.51}$$

The ξ_0 term comes from the long distance dynamics. The expression for ϵ'/ϵ given in Eq. (7.39) together with the experimental fact that $\epsilon' \ll \epsilon$ shows that the long distance dynamics *cannot* generate ϵ by themselves. Within the KM ansatz, the interplay of all three families including top quarks is essential; since $m_t, m_c > M_K$ the relevant effective $\Delta S = 2$ operator is local; i.e., $\mathcal{H}_{\text{eff}}^{\text{box}}(\Delta S = 2)$ can be used in evaluating ImM_{12}:

$$\begin{aligned}
\epsilon_{KM} &\simeq \epsilon_{KM}^{\text{box}} \\
&\simeq \frac{G_F^2}{12\sqrt{2}\pi^2} e^{i\phi_{SW}} \frac{M_W^2 M_K F_K^2 B_K}{\Delta M} \left[\text{Im}(\lambda_c^2) E(x_c) \eta_{cc} \right. \\
&\quad \left. + \text{Im}(\lambda_t^2) E(x_t) \eta_{tt} + 2\text{Im}(\lambda_c \lambda_t) E(x_c, x_t) \eta_{ct} \right] \\
&\sim 1.9 \times 10^4 \, B_K e^{i\phi_{SW}} \left[\text{Im}(\lambda_c^2) E(x_c) \eta_{cc} + \text{Im}(\lambda_t^2) E(x_t) \eta_{tt} \right. \\
&\quad \left. + 2\text{Im}(\lambda_c \lambda_t) E(x_c, x_t) \eta_{ct} \right].
\end{aligned} \tag{9.52}$$

Note that ϵ is proportional to J, defined in Eq. (8.40). Note also that the factors which depend on the KM matrix elements must provide about 10^{-8} suppression – a critical test of the KM ansatz. While ϵ has been measured very accurately, no tight bounds can be derived on these KM parameters due to the theoretical uncertainty in B_K, see Eq. (9.43). Yet, the constraint from ϵ will figure prominently in our discussion of **CP** asymmetries in B_d decays. All this will be discussed in detail later.

9.7 ♠ **Estimating** Im Γ_{12} ♠

We will show now that the dominating contribution to Im Γ_{12} is given by the 2π final state [101]. This will become important below when we analyse the information on **CPT** symmetry contained in the phases ϕ_{+-}, ϕ_{00} and ϕ_{SW}.

With the notation

$$\Gamma_{12}^f \equiv 2\pi \rho_f \langle K^0 | H_W | f \rangle \langle f | H_W | \overline{K}^0 \rangle, \tag{9.53}$$

we have

$$\Gamma_{12} = \sum_f \Gamma_{12}^f. \tag{9.54}$$

Keeping the part of ϵ in Eq. (7.41) which is proportional to Im Γ_{12}, we obtain

$$\epsilon(f) = i \cos \phi_{SW} e^{i\phi_{SW}} \frac{\text{Im}\Gamma_{12}^f}{\Delta\Gamma}. \tag{9.55}$$

We shall now analyse $|\epsilon(f)|$ for the different possible final states. Assuming **CPT** symmetry we have (see Prob. 9.10):

$$\text{Im } \Gamma_{12}^f = i\eta_f 2\pi\rho_f A(f)^2 \left(1 - \eta_f \frac{\bar{A}(f)}{A(f)}\right), \tag{9.56}$$

where η_f denotes the **CP** eigenvalue of f.

- For the isospin 2π channel, $f = (2\pi)_2$. Here $\epsilon((2\pi)_2)$ can be estimated by noting that

$$\epsilon' = \frac{1}{2\sqrt{2}} \omega \, e^{i(\delta_2 - \delta_0)} (\Delta_0 - \Delta_2), \tag{9.57}$$

where $\Delta_I = 1 - \frac{\bar{A}_I}{A_I}$. Barring unexpected cancellation between Δ_0 and Δ_2, we estimate that

$$|\epsilon((2\pi)_2)| = \omega\epsilon' \sim 3 \times 10^{-7}. \tag{9.58}$$

- $f = (3\pi)^{\text{CP}+}$ state.

$$\text{Im } \Gamma_{12}((3\pi)^{\text{CP}+}) = i\Gamma(K \rightarrow (3\pi)^{\text{CP}+}) \left(1 - \frac{\bar{A}((3\pi)^{\text{CP}+})}{A((3\pi)^{\text{CP}+})}\right) \tag{9.59}$$

$$\epsilon((3\pi)^{\text{CP}+}) = \frac{-1}{2\sqrt{2}} \text{Br}(K_S \rightarrow (3\pi)^{\text{CP}+}) \left(1 - \frac{\bar{A}((3\pi)^{\text{CP}+})}{A((3\pi)^{\text{CP}+})}\right). \tag{9.60}$$

Assuming that $\left|1 - \frac{\bar{A}((3\pi)^{\text{CP}+})}{A((3\pi)^{\text{CP}+})}\right| < 1$, and using the measured value, $\text{Br}(K_S \rightarrow (3\pi)^{\text{CP}+}) = \left(3.4^{+1.1}_{-0.9}\right) \times 10^{-7}$ [28], we obtain:

$$|\epsilon((3\pi)^{\text{CP}+})| < 2 \times 10^{-7}. \tag{9.61}$$

- $f = (3\pi)^{\text{CP}=-}$ state.

$$\epsilon((3\pi)^{\text{CP}-}) = \frac{1}{2\sqrt{2}} \text{Br}(K_L \rightarrow (3\pi)^{\text{CP}-}) \frac{\Gamma_L}{\Gamma_S} \left(1 + \frac{\bar{A}((3\pi)^{\text{CP}-})}{A((3\pi)^{\text{CP}-})}\right). \tag{9.62}$$

Using the value of η_{+-0} given in Eq. (7.91) and Eq. (7.92), and $\text{Br}(K_L \rightarrow (3\pi)^{\text{CP}-}) \sim 12\%$ [28] we are led to an estimate:

$$\left(1 + \frac{\bar{A}((\pi^+\pi^-\pi^0)^{\text{CP}-})}{A((\pi^+\pi^-\pi^0)^{\text{CP}-})}\right) \sim \left(1 + \frac{q_1}{p_1} \frac{\bar{A}((\pi^+\pi^-\pi^0)^{\text{CP}-})}{A((\pi^+\pi^-\pi^0)^{\text{CP}-})}\right)$$

$$< 6 \times 10^{-2}. \tag{9.63}$$

This gives

$$|\epsilon((\pi^+\pi^-\pi^0)^{\mathbf{CP}-})| < 10^{-5}. \tag{9.64}$$

- $f = 3\pi^0$

Using Eq. (9.56), we obtain:

$$\epsilon(3\pi^0) = \frac{\Gamma(K_L \to 3\pi^0)}{2\sqrt{2}\Gamma_S}\left(1 + \frac{\overline{A}(3\pi^0)}{A(3\pi^0)}\right). \tag{9.65}$$

The CPLEAR bound on $|\eta_{000}|$ given in Eq. (7.70) is not very useful. We have made a reasonable guess that the direct **CP** violation in $K \to 3\pi^0$ is less than ϵ:

$$\left|1 + \frac{\overline{A}(3\pi^0)}{A(3\pi^0)}\right| < \left|1 - \frac{q}{p}\frac{\overline{A}(2\pi)}{A(2\pi)}\right| < 2\epsilon \tag{9.66}$$

This gives

$$|\epsilon(3\pi^0)| < 1 \times 10^{-6}. \tag{9.67}$$

- In considering $f = \pi l \nu$ we have to allow for a violation of the $\Delta Q = \Delta S$ rule as expressed by the complex parameter x defined in Eq. (7.14). Since

$$\mathrm{Im}\,\Gamma_{12}^{\pi l \nu} \simeq \mathrm{Im}\,x\Gamma(K \to \pi\mu\nu), \tag{9.68}$$

we find

$$|\epsilon(\pi l\nu)| \leq 4 \cdot 10^{-7}, \tag{9.69}$$

where we have used the bound $\mathrm{Im}\,x = (0.5 \pm 2.5) \times 10^{-3}$ from [51], and $\mathrm{Br}(K_S \to \pi^\mp\mu^\pm\nu) \sim 5 \times 10^{-4}$.

In summary: we can conclude [101]

$$|\sum_f \epsilon^f| \leq \sum_f |\epsilon^f| \leq 10^{-5}. \tag{9.70}$$

Since $|\epsilon| \simeq 2.2 \cdot 10^{-3}$ we realize that the prediction given above for arg ϵ where we had included only the contribution of $(2\pi)_0$ to Im Γ_{12} holds indeed to within 1%. The other side of this coin is, however, that tests of **CPT** invariance based on the determination of arg ϵ are unambiguous only if the discrepancy exceeds 1% ! It should be noted that Eq. (9.66) requires an additional assumption that nothing extraordinary happens in $K_S \to 3\pi^0$ decay. This should be checked by measuring $\mathrm{Br}(K_S \to 3\pi^0)$ at the same level as $\mathrm{Br}(K_S \to \pi^+\pi^-\pi^0)$.

9.8 SM expectations for ⟨P⊥⟩ in K₁₃ decays

Let us see how a transversal polarization gets generated within the SM. The lowest order radiative correction to the $\overline{u}\gamma_\mu(1-\gamma_5)sW^\mu$ vertex is shown in Fig. 9.5. What we want to do is to investigate how we get $\mathrm{Re}\,(F_P^* F_A) - \mathrm{Im}\,(F_S^* F_V)$, where

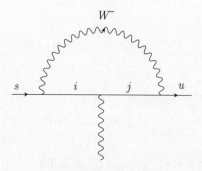

Fig. 9.5.   Feynman graph representing the vertex correction for the $\bar{u}\gamma_\mu(1-\gamma_5)sW^\mu$ vertex. Here $i = u,\ c,\ t;\ j = d,\ s,\ b$ quarks.

the formfactors F_P, F_A, F_S, and F_V are defined in Eq. (7.55). Let us analyse Fig. 9.5(a). The loop integral I has the following form in the ultraviolet region:

$$\mathcal{I}_{UV} \propto \sum_{i,j} \text{Im}\ \left(\mathbf{V}_{is}\mathbf{V}_{ij}^*\mathbf{V}_{uj}\right)$$

$$\times g^3 \int \frac{d^4l}{(2\pi)^4}\ \bar{u}\gamma_+\gamma_\alpha \frac{i}{\rlap{/}{l}-m_i}\gamma_\mu\gamma_- \frac{i}{\rlap{/}{l}-m_j}\gamma^\alpha\gamma_- s\ \frac{-i}{l^2-m_W^2}. \qquad (9.71)$$

It can be seen that the imaginary part of the integral has double GIM cancellations making the integral highly convergent. It is also seen that the most prominent part of Eq. (9.71) has exactly the $\bar{u}\gamma_\mu(1-\gamma_5)sW^\mu$ structure and it just changes the overall phase. Such terms do not contribute to the polarization. We estimate that the part which contributes to the polarization is bounded by:

$$\langle P_\perp \rangle < \frac{\alpha}{\pi \sin^2\theta_W}A^2\lambda^4\eta\frac{m_b^2}{M_W^2} \sim 10^{-7}. \qquad (9.72)$$

9.9 Hyperon decays

Let us discuss the **CP** violating parameter A, defined in Eq. (7.106). Predictions based on the KM ansatz range over a considerable interval.

The strong interaction phases for $p\pi$ final states have been measured [102], those for $\Lambda\pi$ have not. The phases of weak decays can be estimated using the Hamiltonian given in Eq. (9.27). The imaginary part of the Hamiltonian is generated by the top quark contribution in penguin graphs shown in Fig. 9.2:

$$\text{Im}z_6(\mu \sim 1\ \text{GeV}) \simeq -0.1\mathbf{V}_{td}^*\mathbf{V}_{ts} \sim -10^{-4}. \qquad (9.73)$$

At present there is no rigorous computation of the matrix elements associated with these non-leptonic decays. They have been computed using the Bag model [103]

$$A(\Lambda \to p\pi^-) \quad \sim \quad (0.05 \sim 1.2)\cdot 10^{-4} \qquad (9.74)$$

$$A(\Xi \to \Lambda\pi^-) \quad \sim \quad (0.2 \sim 3.5)\cdot 10^{-4}. \qquad (9.75)$$

There is an experiment running at FNAL during the fixed target run 1997/98 that will undertake to measure these correlations combined for Λ and Ξ decays [104],

$$A_{\Lambda\Xi} \equiv \frac{\alpha_\Lambda \alpha_\Xi - \alpha_{\overline{\Lambda}} \alpha_{\overline{\Xi}}}{\alpha_\Lambda \alpha_\Xi + \alpha_{\overline{\Lambda}} \alpha_{\overline{\Xi}}} \simeq A_\Lambda + A_\Xi \qquad (9.76)$$

with a sensitivity to be about 10^{-4} which is around (though close to the upper range of) the KM predictions listed above.

Observing direct **CP** violation in the decays of strange baryons would be a first rate discovery on the same level as establishing $\epsilon'/\epsilon \neq 0$. At the same time it will not be any easier to decide whether it is due to the KM ansatz or due to New Physics.

9.10 Résumé

The full machinery of quantum field theory has been brought to bear on the theoretical treatment of the non-leptonic decays of strange hadrons: operator renormalization, even of the non-multiplicative variety based on the renormalization group, separation of perturbative and non-perturbative dynamics, simplifying hadronic matrix elements with the help of chiral symmetry, etc. The returns for our efforts have been mixed.

- We have not found a simple elegant reason underlying the $\Delta I = 1/2$ rule. We have identified several dynamical effects all enhancing $\Delta I = 1/2$ amplitudes; yet we have not arrived at a quantitative dynamical understanding. It is a sore thumb – but not worse than that.

- We can understand why **CP** violation – and in particular its direct manifestations – are so delicate in strange decays. In addition to the required interplay of the three quark families, the $\Delta I = 1/2$ enhancement suppresses quantities like ϵ'/ϵ.

- It naturally yields a small value for ϵ'/ϵ. Predicting a reliable numerical value for it is, however, quite a different story: the history of such predictions does not represent one of the glory stories of theoretical high energy physics. Yet measuring direct **CP** violation is of crucial importance, no matter what theory might – or might not – say.

- A semi-quantitative prediction for the $K_L - K_S$ mass difference ΔM has been obtained that agrees with its measured value – including the sign! The GIM mechanism is crucial for this success.

- We have established very good theoretical control over the **CP** breaking effective $\Delta S = 2$ Lagrangian; i.e., we can reliably express

$\mathscr{L}(\Delta S = 2)$ in terms of fundamental SM parameters. Even the impact of long-distance dynamics is being brought under control, see Eq. (9.42) and Eq. (9.43).

- We will see in the next chapter that the observed value of ϵ can be reproduced without forcing any of the parameters, and that this is a highly non-trivial success that could easily have been an abject failure!

In any case, we have to be eager to tackle a new adventure where we can redeem ourselves and atone for past short-comings.

Problems

9.1   Check Eq. (9.1) and show that the interaction term Eq. (9.5) leads to

$$\frac{\Gamma(K_S \to \pi^+\pi^-)}{\Gamma(K^+ \to \pi^+\pi^0)} = 4\left(\frac{z_2}{z_1 + z_2}\right)^2. \tag{9.77}$$

Does it explain the $\Delta I = \frac{1}{2}$ rule?

9.2   First, remind yourself that

$$\Gamma(K^+ \to \pi^+\pi^0) = \frac{p_\pi}{8\pi M_K^2}|A(K^+ \to \pi^+\pi^0)|^2, \tag{9.78}$$

then from the lifetimes for K^+, K_S, and branching ratios for the decays $K_S \to \pi^+\pi^-$ and $K^+ \to \pi^+\pi^0$, obtain

$$\begin{aligned}
A(K^+ \to \pi^+\pi^0) &= 1.8 \times 10^{-8} \text{ GeV} \\
A(K^0 \to \pi^+\pi^-) &= 27.5 \times 10^{-8} \text{ GeV} \\
A(K^0 \to \pi^0\pi^0) &= 26.4 \times 10^{-8} \text{ GeV}.
\end{aligned} \tag{9.79}$$

9.3   Using identities given in Eq. (9.17), and remembering the normalization of states given in Sec. 6.1, derive

$$\langle 0|A_\mu^+(0)|P^+\rangle = -iF_P p_\mu, \quad \langle 0|A_\mu^0(0)|P^0\rangle = \frac{-iF_P}{\sqrt{2}}p_\mu,$$

$$\langle \pi^0|V_\mu^+(0)|P^+\rangle = \frac{-1}{\sqrt{2}}(p_P + p_\pi)_\mu,$$

$$\langle \pi^+|V_\mu^0(0)|P^+\rangle = -(p_P + p_\pi)_\mu. \tag{9.80}$$

In our notation $F_{\pi^+} = 130$ MeV, $F_K = 160$ MeV.

9.4   Using $\mathbf{CP}A_\mu(t, \vec{x})\mathbf{CP}^\dagger = -A^{\dagger\mu}(t, -\vec{x})$ as in Table 4.2, show that

$$\int d^4x e^{ipx}\langle 0|A_\mu(t, \vec{x})|P^0\rangle = -\int d^4x e^{ipx}\langle 0|A^{\mu\dagger}(t, -\vec{x})|\overline{P}^0\rangle. \tag{9.81}$$

for $\mathbf{CP}|P^0\rangle = |\overline{P}^0\rangle$.

Note that the minus sign comes from our definition of $\mathbf{C}|P^0\rangle = -|\overline{P}^0\rangle$ – which also influences the definition of **CP** eigenstates P_1 and P_2.

9.5 We have evaluated the matrix elements using chiral perturbation theory. Historically, matrix elements were evaluated by inserting the vacuum in all possible ways. This shows that

$$
\begin{aligned}
\langle \pi^+\pi^-|O_2|K^0\rangle &= \langle \pi^-|(\bar{s}u)_{V-A}|K^0\rangle\langle \pi^+|(\bar{u}d)_{V-A}|0\rangle \\
\langle \pi^+\pi^-|O_6|K^0\rangle &= -2[\langle \pi^-|(\bar{s}u)_S|K^0\rangle\langle \pi^+| - (\bar{u}d)_P|0\rangle \\
&\quad + \langle \pi^+\pi^-|(\bar{d}d)_S|0\rangle\langle 0|(\bar{s}u)_P|K^0\rangle].
\end{aligned} \tag{9.82}
$$

Using Eq. (9.17), derive Eq. (9.18).

9.6 To derive Eq. (9.42), show that

$$
\begin{aligned}
&\langle K^0|(\bar{s}_\alpha\gamma_\mu(1-\gamma_5)d_\alpha)(\bar{s}_\beta\gamma_\mu(1-\gamma_5)d_\beta)|\overline{K}^0\rangle \\
&= 2[\langle K^0|\bar{s}_\alpha\gamma_\mu(1-\gamma_5)d_\alpha|0\rangle\langle 0|\bar{s}_\beta\gamma_\mu(1-\gamma_5)d_\beta|\overline{K}^0\rangle \\
&\quad + \langle K^0|\bar{s}_\alpha\gamma_\mu(1-\gamma_5)d_\beta|0\rangle\langle 0|\bar{s}_\beta\gamma_\mu(1-\gamma_5)d_\alpha|\overline{K}^0\rangle].
\end{aligned} \tag{9.83}
$$

Now, using Eq. (9.80), derive Eq. (9.42).

9.7 Ignoring QCD radiative corrections, calculate the quark box contribution to ΔM assuming there are neither charm nor top quarks. Is it finite? By dimensional argument convince yourself that Eq. (9.49) gives the right order of magnitude for this contribution.

9.8 Let us simplify the situation and consider a two family case. Write an expression corresponding to the Feynman graph shown in Fig. 9.4. Show that the u and c quark propagators combine to give

$$
\frac{m_u - m_c}{(\not{p} - m_u)(\not{p} - m_c)} \tag{9.84}
$$

a GIM cancellation.

9.9 Look at Eq. (9.52): setting $\lambda_t = 0$ or $x_t = 0$ would remove the top quark contribution to $|\epsilon|_{KM}^{box}$, yet it would seem that $|\epsilon|_{KM}^{box} \neq 0$ still holds – in conflict with the result that **CP** violation à la KM requires the interplay of three families! Resolve the apparent paradox.

9.10 Set $\langle f\,;\mathrm{out}|H|K\rangle = e^{i\delta_f}A_f$ and $\langle f\,;\mathrm{out}|H|\overline{K}\rangle = e^{i\delta_f}\overline{A}_f$, where f is a **CP** eigenstate, and δ_f is the final state strong interaction phase. Show that **CPT** symmetry implies $A_f = \eta_f\overline{A}_f^*$, and that

$$
\mathrm{Im}\Gamma_{12}^f = -i\eta_f\pi\rho_f(\overline{A}_f^2 - A_f^2), \tag{9.85}
$$

where the η_f corresponds to the **CP** eigenvalue of f. Now derive Eq. (9.56).

10

Fundamentals of B physics

It's Beauty – Not Bottom!

10.1 The emerging beauty of B hadrons

In June 1997 an international conference entitled 'b20: Twenty Beautiful Years of Bottom Physics' was held at the Illinois Institute of Technology. This was a joint celebration of the 20th anniversary of the discovery of Υ – the bound state of the $b\bar{b}$ quarks – and the 75th birthday of Leon Lederman, who lead the discovery of the Υ. Following are brief excerpts from the discussions at the conference [105].

10.1.1 The discovery of beauty

Lederman and his group had performed experiments measuring

$$p + p \rightarrow \quad X \quad + \text{anything}$$
$$\hookrightarrow \quad \mu^+\mu^-. \tag{10.1}$$

This is considered to be the best way to detect any new particle decaying into $\mu^+\mu^-$ [106]. Muons can penetrate a lot more material than any other particle except neutrinos. This property makes it easy to identify muons. Iron blocks from, say, old battleships placed in front of a detector, for example, can filter out particles other than muons (and neutrinos).

Sanda in his concluding talk reminisced:

> Back in 1970, I was a post-doc at the physics department of Columbia University, where Leon was a professor. Three postdocs shared an office on the 8th floor of Pupin. Leon came in one day and showed me the muon mass spectrum form the $p\bar{p}$ collision at ISR, (Fig. 10.1). A curious thing about this spectrum is the shoulder at around 3 GeV. Leon use to tell us, jokingly, that there was a huge resonance at 6 GeV, and that the shoulder was its tail. It appears like a shoulder because there is not enough energy to actually produce a peak. He suggested that we work on computing the mass spectrum.

In the meantime, Lederman's younger competitor, Sam Ting, a professor at MIT, decided to look at this shoulder carefully. He performed an experiment

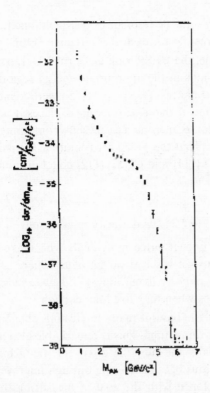

Fig. 10.1. A curious shoulder in the muon mass spectrum of $p +$ Uranium $\rightarrow$ $\mu^+\mu^- +$ anything, which turned out to be J/ψ. This figure was reproduced from *Physical Review Letters* by permission of the American Physical Society.

at Brookhaven National Laboratory with much better resolution to measure the invariant mass of e^+e^- pairs at around 3 GeV. In 1974, he and his collaborators announced the discovery of the J particle, which showed up as a big peak at an invariant mass of 3 GeV [107]. The excess events which had made up Lederman's shoulder were coming from a very narrow bound state of a new quark – the charm quark – and its antiquark. Leon had been looking at the integrated version of the peak, as his resolution was much larger than the width of the resonance. This resonance was discovered simultaneously at the e^+e^- storage ring SPEAR of the Stanford Linear Accelerator Center [108], where it was named ψ. It is now listed in the Particle Data Book as J/ψ. For this discovery, Sam Ting and Burt Richter were awarded the Nobel Prize for physics in 1976.

Lederman had to keep on looking, and he did. His persistence paid off – as it often does, at least for some people – in 1976, when he eventually bumped into yet another narrow resonance, the Υ, around 10 GeV [109].

Since the Υ had been discovered through its decay into a e^+e^- pair, it could be studied quite cleanly at an e^+e^- storage ring. The DORIS ring at DESY in Hamburg already had a successful career behind it analysing the dynamics of the charmonium family – the $\bar{c}c$ resonances just mentioned; in a perforce ride its energy was boosted threefold to reach the Υ regime [110].

Just around that time, the CESR ring at Cornell University in Ithaca was set up to study e^+e^- collisions in the appropriate energy region[1]. Now the race was on. Before long five resonances $\Upsilon(1S)$, ..., $\Upsilon(5S)$ were identified. The important discovery for us was that of the $\Upsilon(4S)$: its mass places it just slightly above the production threshold for B mesons, i.e., mesons containing a b quark together with a $\bar{d}$ or $\bar{u}$ antiquark. That the $\Upsilon(4S)$ decays into a $B\bar{B}$ pair almost all the time was observed by CLEO [111] and CUSP [112] collaborations working at CESR.

10.1.2 The longevity of B mesons

Aside from those well known gifted individuals who make major contributions in their teens, most of us get wiser as we get older. There are many things in life which can be obtained only by living longer. Likewise elementary particles can reveal interesting physics when they live long enough.

The key discovery at the base of many interesting phenomena in B physics is that B mesons live for a 'long' time. This is remarkable also in two other respects, namely what 'long' means and how the discovery was achieved. Measuring the lifetimes of beauty hadrons became feasible through microvertex detectors; those had actually been developed with the goal of measuring the lifetimes of *charm* hadrons in hadronic collisions which just happen to have rather similar decay lengths. The basic idea is to track the time through space, that is, to infer the (proper) time of a decay from locating its decay vertex – i.e., a secondary vertex separated from the production point – in a tracking device. This is easier said than done as we are dealing with flight paths of only a few hundred microns (if that much) in an environment of numerous tracks belonging to the underlying event.

It was the MAC collaboration [113] working at the PEP ring of SLAC that found the first evidence of a long beauty life time. The discovery was quickly followed by MARKII collaboration, also at PEPII [114]. Tracks emanating from a decay vertex miss the primary production vertex when extrapolated back, as illustrated in Fig. 10.2; i.e., they exhibit a non-vanishing impact parameter. The spatial resolution of the MAC detector was still too coarse for discriminating the impact parameter against zero on an event-by-event basis; yet it could be achieved statistically, as shown by the histogram of Fig. 10.2: the impact parameter distribution is shifted away from zero to a positive value.

State-of-the-art studies by ALEPH, DELPHI and OPAL at LEP and CDF at FNAL yield a very accurate value for the life time of, say, neutral B mesons

---

[1] It was actually optimized for a c.m. energy of 16 GeV, but it was never operated there.

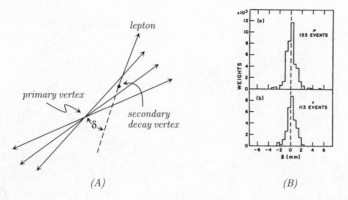

(A)   *(B)*

Fig. 10.2. A *B* hadron is created at the primary vertex. It travels to the secondary decay vertex where it decays. The lifetime can be deduced by measuring the distance between the primary and secondary vertices, but with technology available at that time it was impossible to measure this distance. So, a statistical analysis was performed. (A) δ is defined as an impact parameter of the lepton computed at the primary vertex. In (B) the distribution of δ is given. Nonvanishing δ was established first by the MAC collaboration. This figure was reproduced from *Physical Review Letters* by permission of the American Physical Society.

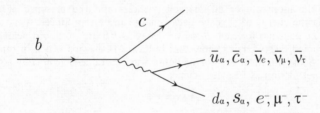

Fig. 10.3. Feynman diagram which is responsible for *b* quark decay. If the final state quark masses are neglected, each channel contributes equally to the decay, and its contribution can be computed from the μ decay. Note that there are nine channels as quarks come in three colours.

[115]:

$$\tau_{B_d} \sim (1.55 \pm 0.04) \times 10^{-12} \text{ s.} \tag{10.2}$$

To appreciate the fact that this is a long time scale for a *B* meson, let us give a naive estimate. The total width of *b* quarks can be guessed at by scaling the expression for the muon width, accounting for the larger number of decay channels with quarks coming in three colours, see Fig. 10.3:

$$\Gamma_\mu = \frac{G_F^2}{192\pi^3} m_\mu^5 \implies \Gamma_b \sim \frac{G_F^2}{192\pi^3} m_b^5 |\mathbf{V}_{bc}|^2 \times (2 \times 3 + 3). \tag{10.3}$$

The difference due to finite quark and lepton masses is ignored. For $|\mathbf{V}_{bc}| \sim 1$ we would have $\tau_b \sim \tau_\mu (m_\mu/m_b)^5/9 \simeq 10^{-15}$ s, i.e., a factor of 1000 shorter than observed. Actually at that time we typically estimated $\mathbf{V}_{cb}$ to be suppressed since

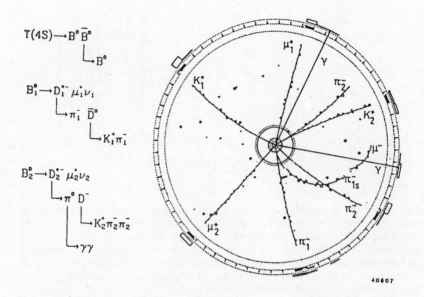

$\Upsilon(4S) \rightarrow B^0 \bar{B}^0$
$ \hookrightarrow B^0$

$B_1^0 \rightarrow D_1^{*-} \mu_1^+ \nu_1$
$ \hookrightarrow \pi_1^- \bar{D}^0$
$ \hookrightarrow K_1^+ \pi_1^-$

$B_2^0 \rightarrow D_2^{*-} \mu_2^+ \nu_2$
$ \hookrightarrow \pi^0 D^-$
$ \hookrightarrow K_2^+ \pi_2^- \pi_2^-$
$ \hookrightarrow \gamma\gamma$

40607

Fig. 10.4. The gold plated event discovered by the ARGUS collaboration. By knowing the momenta of all charged particles and the presence of π^0 from measuring the energy of $\gamma\gamma$, each decay chain can be reproduced. We can afford to have one neutrino for each B meson decay as $B\bar{B}$ are produced nearly at rest. Reconstructed decay chains show that both are B mesons which in turn shows that $B - \bar{B}$ mixing must exist. This figure was reproduced from *Physics Letters* [116] by permission of Elsevier Science.

it described an inter-family transition:

$$\tau_b \sim \tau_\mu \left(\frac{m_\mu}{m_b}\right)^5 \frac{1}{9} \frac{1}{|V_{cb}|^2} \sim 3 \cdot 10^{-14} \left|\frac{\sin\theta_C}{V_{cb}}\right|^2 \text{ s.} \qquad (10.4)$$

Compared with this expectation Eq. (10.2) represents a *long* lifetime; it shows that V_{cb} is considerably more reduced than anticipated:

$$|V_{cb}| \sim \frac{1}{30} \sim \mathcal{O}(\sin^2\theta_C) \ll \mathcal{O}(\sin\theta_C). \qquad (10.5)$$

10.1.3 The fluctuating identity of neutral B mesons

Knowing B mesons to live for a long time, we can hope that they would do something interesting while they are alive. Indeed, $B - \bar{B}$ oscillations were first discovered by the ARGUS collaboration through establishing the existence of same-sign dilepton events [116]:

$$e^+e^- \rightarrow \Upsilon(4S) \rightarrow B^0\bar{B}^0 \rightarrow \mu^\pm\mu^\pm + X, \qquad (10.6)$$

where X stands for any final state. For without oscillations the two leptons from B decays have to carry opposite charges. We must be careful, though, not to be fooled by leptons from $B^0\overline{B}^0 \to [l^+X_1][D + X_2]$ followed by $D \to l^+ + X_D$. Such secondary leptons can be separated from primary semileptonic decays $\overline{B}^0 \to l^- + X$ through various kinematical criteria like transverse momentum cuts, etc. ARGUS actually found a special beauty, namely the fully reconstructed event shown in Fig. 10.4.

To interpret oscillation data properly we have to apply quantum mechanical reasoning. The B meson pair in $\Upsilon(4S) \to B^0\overline{B}^0$ is produced into a **C** odd configuration with the two mesons flying apart from each other with momenta $\vec{k}$ and $-\vec{k}$ at time of production $t = 0$. Subsequently oscillations set in that are highly correlated for **C** = −: Bose statistics tells us that if one of the mesons is a B^0 at some time t, the other one cannot be a B^0 as well at *that* time, since the state must be odd under exchange of the two mesons. The time evolution of the pair is then simply[2]

$$|(B^0\overline{B}^0)_{\mathbf{C} = -}(t)\rangle = e^{-\Gamma_B t}\frac{1}{\sqrt{2}}\left[|B^0(\vec{k})\overline{B}^0(-\vec{k})\rangle - |B^0(-\vec{k})\overline{B}^0(\vec{k})\rangle\right], \quad (10.7)$$

and we can never have like-sign primary dileptons emerge at a single time t. Once one of the B hadrons has decayed, the coherence is lost, and the surviving $\overline{B}$ meson will oscillate without production constraint. From Eq. (10.7) it is straightforward to compute the time (t_2) dependence of the remaining B once the first one decays at t_1 see (Problem 10.1). The probability that one charged lepton emerges at time t_1 and another one with the same charge at t_2 is given by [117]:

$$\Gamma((B^0\overline{B}^0)_{\mathbf{C}=-} \to l^+_{t_1} l^+_{t_2} X) \propto e^{-\Gamma(t_1+t_2)}\left|\frac{p}{q}\right|^2 \sin^2\frac{\Delta M_B}{2}(t_1 - t_2) ;$$

$$\Gamma((B^0\overline{B}^0)_{\mathbf{C}=-} \to l^-_{t_1} l^-_{t_2} X) \propto e^{-\Gamma(t_1+t_2)}\left|\frac{q}{p}\right|^2 \sin^2\frac{\Delta M_B}{2}(t_1 - t_2) ; \quad (10.8)$$

it indeed vanishes for $t_1 = t_2$. Integrating over times of decay t_1, t_2 we obtain for the general case of a $B^0\overline{B}^0$ pair (Problem 10.2):

$$N_{++} = N([B^0\overline{B}^0]_{\mathbf{C}=\mp} \to l^+l^+ + X) \sim \left|\frac{p}{q}\right|^2\left(1 - \frac{1 \pm x^2}{(1 + x^2)^2}\right)$$

$$N_{--} = N([B^0\overline{B}^0]_{\mathbf{C}=\mp} \to l^-l^- + X) \sim \left|\frac{q}{p}\right|^2\left(1 - \frac{1 \pm x^2}{(1 + x^2)^2}\right)$$

$$N_{+-} = N([B^0\overline{B}^0]_{\mathbf{C}=\mp} \to l^\pm l^\mp + X) \sim \left(1 + \frac{1 \pm x^2}{(1 + x^2)^2}\right), \quad (10.9)$$

with $x = \Delta M(B)_B/\Gamma_B$; we have taken $\left|\frac{q}{p}\right|^2 = 1, y \equiv \frac{\Delta\Gamma}{\Gamma} \ll 1$, which will be shown later to be an excellent approximation. The rate for production of same sign

---

[2] It will be shown later that $\Delta\Gamma(B) \ll \Delta M(B)$. To spotlight the essence of our argument, we set $\Delta\Gamma = 0$.

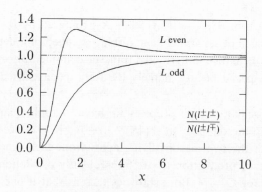

Fig. 10.5. Rate for producing dilepton events for even and odd angular momentum $B^0 \bar{B}^0$ states. The fact that there are fewer same sign dilepton events when the $B\bar{B}$ state is an odd angular momentum state than an even one can be understood by the fact that Bose statistics does not allow BB or $\bar{B}\bar{B}$ to exist at the same time.

leptons compared to opposite sign leptons is shown in Fig. 10.5. Interpretating the difference in the number of same-sign dilepton events for **C** even and odd is left as an exercise.

A **CP** violating observable which is bound to be discussed extensively among experimentalists is:

$$\frac{N_{--} - N_{++}}{N_{--} + N_{++}} = \frac{1 - \left|\frac{p}{q}\right|^4}{1 + \left|\frac{p}{q}\right|^4} \tag{10.10}$$

which is, not surprisingly, the same as Eq. (6.60).

Now let us look at the data. Starting with

$$\left.\frac{N_{++} + N_{--}}{N_{+-}}\right|_{C=-} = \frac{x^2}{2 + x^2}, \tag{10.11}$$

present data yield for B_d mesons

$$x_d = 0.71 \pm 0.06, \qquad \Delta M_{B_d} \simeq 3 \cdot 10^{-10} \text{ MeV}. \tag{10.12}$$

The physical meaning of x is very simple. Write:

$$\Delta M_B \cdot t = \frac{\Delta M_B}{\Gamma} \cdot \frac{t}{\tau}, \tag{10.13}$$

where Γ and τ are the total width and lifetime, respectively. What is important for the *observability* (rather than the occasional occurrence) of oscillations is that x be not too small, since

$$\frac{\Delta M_B}{\Gamma} \sim \frac{\text{lifetime}}{\text{oscillation time}}. \tag{10.14}$$

10.2 Theoretical tool kit

10.2.1 The KM paradigm of huge **CP** asymmetries

It was found early on that beauty had a preference to decay into charm rather than u quarks, thus establishing the hierarchy

$$|\mathbf{V}_{ub}|^2 \ll |\mathbf{V}_{cb}|^2 \ll |\mathbf{V}_{us}|^2 \ll 1, \tag{10.15}$$

which places tight unitarity constraints on the CKM matrix if there are three families only. As pointed out by Wolfenstein [118], we can express the CKM matrix through an expansion in powers of $\sin \theta_C = \lambda$; through $\mathcal{O}(\lambda^4)$ we have:

$$\mathbf{V} = \begin{pmatrix} 1 - \frac{1}{2}\lambda^2 & \lambda & A\lambda^3(\rho - i\eta + \frac{i}{2}\eta\lambda^2) \\ -\lambda & 1 - \frac{1}{2}\lambda^2 - i\eta A^2\lambda^4 & A\lambda^2(1 + i\eta\lambda^2) \\ A\lambda^3(1 - \rho - i\eta) & -A\lambda^2 & 1 \end{pmatrix}. \tag{10.16}$$

The three Euler angles and one KM phase of the PDG representation, Eq. (8.35), are replaced by the four real quantities λ, A, ρ and η. For such an expansion to be self-consistent, we have to require $|A|$, $|\rho|$ and $|\eta|$ to be all of order unity. This will be shown below.

Let us look at the CKM matrix in a semi-quantitative way: Eq. (10.16) suggests one general observation while at the same time leading to a global prediction:

- This pattern goes well beyond the requirements of merely being unitary: the matrix is almost diagonal and symmetric, and its elements get smaller the more we move away from the diagonal. Nature most certainly has encoded a profound message in this peculiar pattern. Alas – we have not (yet) succeeded in deciphering it!

- The six unitarity triangles [119, 120, 121] can now be characterized through their dependence on λ:

$$\underset{\mathcal{O}(\lambda)}{\mathbf{V}_{ud}^*\mathbf{V}_{us}} + \underset{\mathcal{O}(\lambda)}{\mathbf{V}_{cd}^*\mathbf{V}_{cs}} + \underset{\mathcal{O}(\lambda^5)}{\mathbf{V}_{td}^*\mathbf{V}_{ts}} = \delta_{ds} = 0 \tag{10.17}$$

$$\underset{\mathcal{O}(\lambda)}{\mathbf{V}_{ud}\mathbf{V}_{cd}^*} + \underset{\mathcal{O}(\lambda)}{\mathbf{V}_{us}\mathbf{V}_{cs}^*} + \underset{\mathcal{O}(\lambda^5)}{\mathbf{V}_{ub}\mathbf{V}_{cb}^*} = \delta_{uc} = 0 \tag{10.18}$$

$$\underset{\mathcal{O}(\lambda^4)}{\mathbf{V}_{us}^*\mathbf{V}_{ub}} + \underset{\mathcal{O}(\lambda^2)}{\mathbf{V}_{cs}^*\mathbf{V}_{cb}} + \underset{\mathcal{O}(\lambda^2)}{\mathbf{V}_{ts}^*\mathbf{V}_{tb}} = \delta_{sb} = 0 \tag{10.19}$$

$$\underset{\mathcal{O}(\lambda^4)}{\mathbf{V}_{td}\mathbf{V}_{cd}^*} + \underset{\mathcal{O}(\lambda^2)}{\mathbf{V}_{ts}\mathbf{V}_{cs}^*} + \underset{\mathcal{O}(\lambda^2)}{\mathbf{V}_{tb}\mathbf{V}_{cb}^*} = \delta_{ct} = 0 \tag{10.20}$$

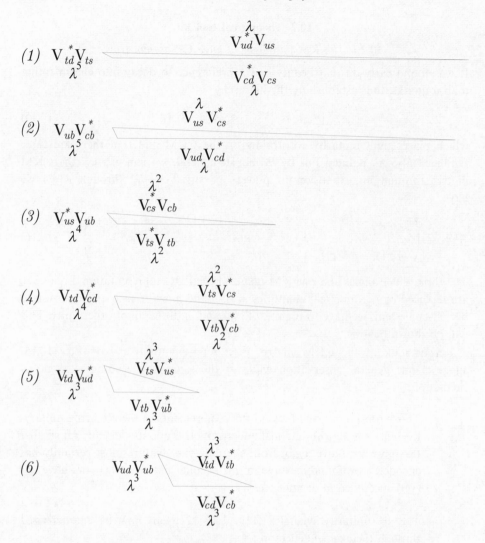

Fig. 10.6.   Each of six unitarity relations Eq. (10.17)–Eq. (10.22) can be represented by a triangle on the complex plane with equal area. The one corresponding to **CP** violation in K decay is given by (1). It is a spear rather than a triangle – this leads to small **CP** violation. In contrast, the one corresponding to B decay, (6), is expected to have three sides of $\mathcal{O}(\lambda^3)$ and decent angles – leading to large **CP** violation.

$$\begin{array}{cccc} \mathbf{V}_{td}\mathbf{V}_{ud}^* + & \mathbf{V}_{ts}\mathbf{V}_{us}^* + & \mathbf{V}_{tb}\mathbf{V}_{ub}^* = & \delta_{ut} = 0 \\ \mathcal{O}(\lambda^3) & \mathcal{O}(\lambda^3) & \mathcal{O}(\lambda^3) & \end{array} \qquad (10.21)$$

$$\begin{array}{cccc} \mathbf{V}_{ud}\mathbf{V}_{ub}^* + & \mathbf{V}_{cd}\mathbf{V}_{cb}^* + & \mathbf{V}_{td}\mathbf{V}_{tb}^* = & \delta_{db} = 0, \\ \mathcal{O}(\lambda^3) & \mathcal{O}(\lambda^3) & \mathcal{O}(\lambda^3) & \end{array} \qquad (10.22)$$

where below each product of matrix elements we have noted their size in powers of λ.

We see that the six triangles fall into three categories, as illustrated in Fig. 10.6:

1 The first two triangles are extremely 'squashed': two sides are of order λ, the third one of order λ^5 and their ratio of order $\lambda^4 \simeq 2.3 \cdot 10^{-3}$; Eq. (10.17) and Eq. (10.18) control the situation in strange and charm decays; the effective weak phases there are obviously tiny.

2 The third and fourth triangles are still rather squashed, yet less so: two sides are of order λ^2 and the third one of order λ^4.

3 The last two triangles have sides that are all of the same order, namely λ^3. All their angles are therefore naturally large, i.e. $\sim$ several $\times$ 10 degrees, since to leading order in λ we have

$$\mathbf{V}_{ud} \simeq \mathbf{V}_{tb} \; , \; \mathbf{V}_{cd} \simeq -\mathbf{V}_{us} \; , \; \mathbf{V}_{ts} \simeq -\mathbf{V}_{cb} \; . \tag{10.23}$$

The triangles of Eq. (10.22) and Eq. (10.21) thus coincide to that order.

The sides of this triangle, having naturally large angles, are given by $\lambda \cdot \mathbf{V}_{cb}$, $\mathbf{V}_{ub}$ and $\mathbf{V}_{td}^*$; these are all quantities that control important aspects of B decays, namely CKM favoured and disfavoured B decays and $B_d - \overline{B}_d$ oscillations! Hence we can conclude: *the KM ansatz unequivocally predicts beauty transitions to contain relative weak phases of order unity!* To repeat once more: the crucial element in reaching this conclusion is the 'long' B life time.

We can actually go considerably further. With the $B^0 - \overline{B}^0$ oscillation rate being quite similar to its decay rate, we have two different, yet *coherent* amplitudes of *comparable* size available in B^0 transitions. *Therefore there have to be large – or even huge – CP asymmetries in some decays of beauty hadrons!* The challenge for theory is to figure out in which specific channels such asymmetries will surface and how reliably – parametrically as well as numerically – we can predict their size.

10.2.2 The unitarity triangle

We shall see that all six parameters of the unitarity triangle, three angles and three sides, defined by the unitarity relation Eq. (10.22), can be measured. The sides of the triangle are obtained by measuring decay rates, and the angles are measured by various asymmetries. When angles are determined from asymmetries, we must be careful about their signs. For this reason, we shall record here the relationships between angles of the unitarity triangle and the KM phases. The

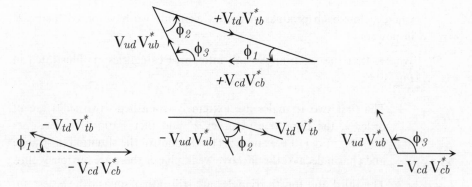

Fig. 10.7.   Angles of the unitarity triangle are related to the phases of the KM matrix. The right hand rule gives the positive direction of the angle between two vectors.

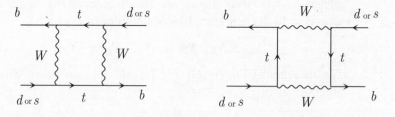

Fig. 10.8.   Feynman diagrams which are responsible for $\Delta B = 2$ transition.

angles are defined by the right hand rule.

$$\phi_1 = \pi - \arg\left(\frac{-\mathbf{V}_{tb}^* \mathbf{V}_{td}}{-\mathbf{V}_{cb}^* \mathbf{V}_{cd}}\right),$$

$$\phi_2 = \arg\left(\frac{\mathbf{V}_{tb}^* \mathbf{V}_{td}}{-\mathbf{V}_{ub}^* \mathbf{V}_{ud}}\right),$$

$$\phi_3 = \arg\left(\frac{\mathbf{V}_{ub}^* \mathbf{V}_{ud}}{-\mathbf{V}_{cb}^* \mathbf{V}_{cd}}\right). \tag{10.24}$$

These results are obvious from Fig. 10.7.

10.2.3 $\Delta B = 2$ transitions

Physical observables ΔM_B and $\Delta \Gamma_B$ have been computed in Eq. (6.42). On rather general grounds we are confident that ΔM_B – in contrast to ΔM_K – can reliably be calculated in the 'femto-universe', i.e., in the world of quarks and gluons – as opposed to that of hadrons. In such a world, the Feynman diagrams which generate $B^0 \to \overline{B}^0$ oscillation are shown in Fig. 10.8. Integrating out the internal fields – top quarks and W bosons – leads to, as already stated, a *local* operator of the form $(\overline{b}q)_{V-A}(\overline{b}q)_{V-A}$.

The QCD corrected effective Hamiltonian is given by

$$\mathcal{H}_{\text{eff}}(\Delta B = 2; \mu) = \left(\frac{G_F}{4\pi}\right)^2 M_W^2 \left[(\bar{b}q)_{V-A}(\bar{b}q)_{V-A}\right]$$

$$\times \left[\eta_c^B(\mu)\xi_{cq}^{*2}E(x_c) + \eta_t^B(\mu)\xi_{tq}^{*2}E(x_t) + 2\eta_{ct}^B(\mu)\xi_{cq}^*\xi_{tq}^*E(x_c, x_t)\right] \quad (10.25)$$

where $\eta_i^B(\mu)$ denotes QCD radiative corrections evaluated at an energy scale μ, $x_q = m_q^2/M_W^2$, $\xi_{Qq} = \mathbf{V}_{Qb}\mathbf{V}_{Qq}^*$, and $q = d, s$ for $B^0 = B_d, B_s$, respectively.

$$E(x_i) = \frac{4x_i - 11x_i^2 + x_i^3}{4(1-x_i)^2} - \frac{3x_i^3 \ln x_i}{2(1-x_i)^3}, \quad i = c, t,$$

$$E(x_c, x_t) = x_c \ln \frac{x_t}{x_c} - \frac{3x_c x_t}{4(1-x_t)} - \frac{3x_c x_t^2 \ln x_t}{4(1-x_t)^2}, \quad x_c \ll 1 \quad (10.26)$$

where $E(0) = 0 = E(0,0)$ is a reflection of the GIM mechanism.

10.2.4 Calculating ΔM_B

The off-diagonal element of the mass matrix is given by

$$\langle B^0|\mathcal{H}_{\text{eff}}(\Delta B = 2; \mu)^\dagger|\overline{B}^0\rangle_{(\mu)} \equiv M_{12} - \frac{i}{2}\Gamma_{12}. \quad (10.27)$$

While there are many similarities in the computation for ΔM_{B^0} and $\Delta\Gamma_{B^0}$, it is better to present them separately. The major contribution to ΔM_{B^0} or $\Delta M_{B_s^0}$ is given by the top quark contribution to the box graph:

$$M_{12} = \left(\frac{G_F}{4\pi}\right)^2 \xi_{tq}^2 M_W^2 E\left(\frac{m_t^2(\mu)}{M_W^2}\right) \eta_t^B(\mu)\langle B^0|(\bar{q}b)_{V-A}(\bar{q}b)_{V-A}|\overline{B}^0\rangle_{(\mu)}, \quad (10.28)$$

where $q = d$ or s. The observable ΔM_{B^0} of course cannot depend on μ, since the latter's introduction merely reflects how we arrange our computation and not how nature goes about its business. Nevertheless, the value adopted for μ is not of purely academic interest: we must be able to choose it such that perturbative calculations of the coefficient functions as well as non-perturbative determinations of the matrix element can be performed in a meaningful way. Applying these general observations to the problem at hand, we realize the following: a choice like $\mu \simeq m_b$ which might seem reasonable at first sight is actually far from that! For the quantity m_b is much larger than characteristic scales for the strong interactions. We don't know how to compute the matrix element in Eq. (10.28) at $\mu = m_b$. If we had full computational control over the non-perturbative effects in the matrix elements, we could express the μ dependence as follows:

$$\langle B^0|(\bar{q}b)_{V-A}(\bar{q}b)_{V-A}|\overline{B}^0\rangle_{(\mu)} = -\frac{4}{3}B_B(\mu)F_B^2(\mu)M_B. \quad (10.29)$$

$B_B = 1$ is again referred to as 'vacuum saturation' (VS) or factorization with F_B representing the decay constant for B mesons. It is natural to expect VS to hold with fair accuracy for $\mu \sim 1\,\text{GeV} \ll m_b$. While these considerations do not fix μ uniquely, they put non-trivial constraints on the range over which it can reasonably vary.

The mixing parameter x_d can now be computed:

$$x_d \simeq -\frac{2\overline{M}_{12}}{\Gamma_{B_d}} \simeq \frac{G_F^2 F_B^2 M_B M_W^2}{\Gamma_{B_d}} \frac{\eta_t^B E(x_t)\overline{\xi}_{td}^2 B_B}{6\pi^2}$$

$$\simeq 0.63 \left(\frac{F_B}{150 MeV}\right)^2 B_B \left(\frac{E(x_t)}{E(5)}\right) A^2[(1-\rho)^2 + \eta^2]$$

$$\times \text{ sign of Re}\xi_{td}^2 \tag{10.30}$$

where $\overline{\xi}_{td}^2 = |\xi_{td}^2| \times \text{ sign}(\text{Re } \xi_{td}^2)$; we have calibrated the expression to $x_t = 5$ which corresponds to $m_t = 180$ GeV. Also, $\tau_B \sim 1.5$ ps and $\eta_t^B = .55$ have been used.

Note that the probability of observing like-sign dilepton events is proportional to x_d^2, see Eq. (10.11), and thus nearly proportional to $E(x_t)^2 \sim (m_t/M_W)^4$. While it is trivial to extend the oscillation analysis from the K to the B meson system, it takes considerable daring to entertain the possibility of the top quark being as heavy as 180 GeV – which is needed to generate $x_d \simeq 1$ – when data yielded a lower bound of a mere 20 GeV. Premature claims of the discovery of top quarks with $m_t \sim 40$ GeV did not help theorists much, either, at that time. The attentive reader will discern a slightly defensive tone here; yet in fairness this should be kept in mind when wondering why theorists did not predict $x_d \sim \mathcal{O}(1)$. On the other hand the ARGUS [116] discovery, confirmed speedily by CLEO [122], was readily perceived as the first clear evidence for an 'unusually heavy' top with m_t exceeding 100 GeV.

With a top quark mass of 20 GeV – and no New Physics – the B_d meson would decay before it could exhibit a significant oscillation; likewise if τ_B had been of order 10^{-14} s – in either case we would not have written this book.

For B_s mesons we obtain in an analogous way:

$$x_s \simeq \frac{G_F^2 M_W^2}{6\pi^2} \frac{M_{B_s}}{\Gamma_{B_s}} \eta_t^B E(x_t)\overline{\xi}_{ts}^2 B_{B_s} F_{B_s}^2$$

$$\simeq 19 \left(\frac{\sqrt{B_{B_s}}F_{B_s}}{180 \text{ MeV}}\right)^2 A^2 \cdot \frac{E(x_t)}{E(5)}. \tag{10.31}$$

The new feature is that here all microscopic parameters are known, i.e., also $\xi_{ts}^2 \simeq A^2\lambda^4$ in addition to $\eta_t^B E(x_t)$. The only unknown is the size of the hadronic matrix element expressed by $B_{B_s} F_{B_s}^2$.

10.2.5 On the sign of ΔM_B

In considering the sign of ΔM_B we have to remember a few subtleties explained in detail in Chapter 6:

- Whereas $\bar{\xi}^2_{ts}$ is necessarily positive within SM, $\mathrm{Re}\,\xi^2_{td}$ a priori could be either positive or negative; yet constraints from existing data tell us that it has to be positive as well.

- Thus we have $\bar{M}_{12} < 0$ for both B_d and B_s mesons with the convention $\mathbf{CP}|B^0\rangle = |\bar{B}^0\rangle$.

- However, the sign of ΔM_B by itself becomes an observable only if the two mass eigenstates can be distinguished by a difference in lifetimes or by their **CP** quantum numbers (if **CP** symmetry holds approximately).

- Neither of these criteria can be utilized in the B_d complex:

  - As stated below we predict the lifetimes of the two B_d mass eigenstates to differ by no more than about 1%. It is quite unlikely that such a small difference can be identified in the foreseeable future.

  - The KM ansatz unequivocally predicts large **CP** violation here.

- The situation is quite different for the B_s system:

  - As explained below lifetime differences here might amount to about 20% or so and become observable.

  - The $\Delta B = 2$ and the CKM favoured $\Delta B = 1$ effective operators for B_s transitions are **CP** conserving.

- Even if the sign of ΔM_B is not observable, the sign of the **CP** asymmetry $\sin\Delta M_B \cdot \mathrm{Im}\frac{q}{p}\bar{\rho}(f)$ is and can be predicted within a given theory of $\Delta B = 2$ dynamics, see Chapter 6.

10.2.6 The search for physics beyond the SM

Top quarks have been discovered with $m_t \sim 180$ GeV; their mass is expected to be measured with an accuracy of $\mathcal{O}(1\%)$ in the foreseeable future. Yet even then no accurate SM prediction can be given for ΔM_B since neither $|V_{td}|$ nor $B_B F_B^2$ are well known numerically. Fortunately we can go beyond this agnostic statement in four respects:

(i) Nature has been generous enough to provide us with two kinds of neutral B states that can exhibit oscillations, namely B_d and B_s, i.e., beauty mesons without and with strangeness, respectively. The dependence on m_t drops out from the ratio of their oscillation rates[3]

$$\frac{\Delta M_{B_s}}{\Delta M_{B_d}} = \frac{[B_B F_B^2]_{B_s}}{[B_B F_B^2]_{B_d}} \frac{\overline{\mathbf{V}}^2_{ts}}{\overline{\mathbf{V}}^2_{td}}; \qquad (10.32)$$

---

[3] We ignore here small and calculable differences due to $M_{B_d} \neq M_{B_s}$.

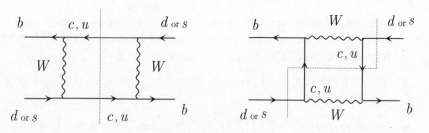

Fig. 10.9.   Diagrams contributing to Γ_{12}.

with $|\overline{\mathbf{V}}_{ts}^2/\overline{\mathbf{V}}_{td}^2| \sim \mathcal{O}(1/\lambda^2) \gg 1$ we predict $\Delta M_{B_s} \gg \Delta M_{B_d}$. We can actually say more. The ratio of the hadronic parameters $B_B F_B^2$ represents a quantity characterising flavour $SU(3)$ breaking; therefore we should be able to predict it with decent accuracy. The best available theoretical technologies, at present, yield:

$$\frac{F_{B_s}}{F_{B_d}} \simeq 1.1 \sim 1.2, \tag{10.33}$$

and therefore

$$\frac{\Delta M_{B_s}}{\Delta M_{B_d}} = (1.2 \sim 1.5) \times \frac{|\mathbf{V}_{ts}|^2}{|\mathbf{V}_{td}|^2}. \tag{10.34}$$

A measurement of $\Delta M_{B_s}/\Delta M_{B_d}$ will then allow us to extract $|\mathbf{V}_{td}|$ since $|\mathbf{V}_{ts}| \simeq |\mathbf{V}_{cb}|$ holds (for three families). Eq. (10.32) holds, of course, only as long as New Physics does not intervene to cause a shift in $\Delta M_{B_s}/\Delta M_{B_d}$; this could easily happen, as will be discussed in Secs. 18.4, 19.3 and 20.4.

(ii) Once a theoretical prediction for F_D, the decay constant for D mesons, has been confirmed in $D^+, D_s^+ \to \mu^+ \nu$ and the underlying method thus authenticated, we can be confident of arriving at a reliable estimate for F_B by extrapolating from m_c to m_b. Measuring both F_{D_s} and F_D with, say, about 10 per cent accuracy would obviously be quite beneficial in this context. We can also hold out the hope of measuring F_B directly through $\Gamma(B \to \tau \nu)$.

(iii) We might succeed in extracting $|\mathbf{V}_{td}|$ also from other decay rates – either directly from $\mathrm{Br}(K^+ \to \pi^+ \nu \bar{\nu})$ or as $|\mathbf{V}_{td}/\mathbf{V}_{ts}|$ from $\mathrm{Br}(B \to \gamma \rho/\omega)/\mathrm{Br}(B \to \gamma K^*)$. Once this has been achieved we can – limited by the numerical accuracy to which $|\mathbf{V}_{td}|$ is known –

• extract F_B from ΔM_{B_d} and compare that value with other determinations of it;
• scrutinize $\Delta M_{B_s}/\Delta M_{B_d}$ for manifestations of New Physics.

(iv) Measuring **CP** asymmetries in B^0 decays will shed further light on F_B, $|\mathbf{V}_{td}|$ and $|\mathbf{V}_{ts}|$, as discussed in detail later.

10.2.7 Calculating $\Delta\Gamma_{B^0}$

We expect the quark box diagram to provide us with a decent approximation to $\Delta\Gamma_B$ as well[4]. Yet the internal quarks are now charm and up quarks which are considerably lighter than M_B; it is then the latter that sets the scale for $\Delta\Gamma_B$: $\Delta\Gamma_B \propto M_B$. Since $\Delta M_{B_q} \propto m_t$, we expect on rather general grounds

$$\Delta\Gamma_B \ll \Delta M_B. \tag{10.35}$$

Even so, $\Delta\Gamma_{B_s}$ could still be sizeable compared to the average width

$$\overline{\Gamma}_{B_s} \equiv [\Gamma(B_{s,\text{short}}) + \Gamma(B_{s,\text{long}})]/2, \tag{10.36}$$

since $\overline{\Gamma}_{B_s} \ll \Delta M_{B_s}$, as discussed above. We arrive at a first guestimate of $\Delta\Gamma_{B^0}$ in the following handwaving way: the transition amplitude $T(B_s \to \overline{D}_s^{(*)} D_s^{(*)} \to \overline{B}_s)$ which contributes prominently to $\Delta\Gamma_{B_s}$ is of second order in $\mathbf{V}_{cb}$ – as is $\overline{\Gamma}(B_s \to [c\bar{c}s\bar{s}])$. With the latter making up a smallish yet significant part of $\overline{\Gamma}_{B_s}$, we can state an order of magnitude estimate: $\Delta\Gamma_{B_s}/\overline{\Gamma}_{B_s} \sim \mathcal{O}(10\%)$ [123]. On the other hand, channels common to B_d and $\overline{B}_d$ decays are KM suppressed; e.g., $B_d \to \overline{D}^{(*)} D^{(*)} / \pi\pi / \rho\rho \to \overline{B}_d$; or $B_d \to \overline{D}^{(*)}\pi + D^{(*)}\pi \to \overline{B}_d$ which are KM favoured for B_d and doubly KM suppressed for $\overline{B}_d$ decays or vice versa. Accordingly, we guestimate $\Delta\Gamma_{B_d}/\overline{\Gamma}_{B_d} \sim \mathcal{O}(\text{few} \times 10^{-3})$, where cancellations can take place between the two classes of common channels exemplified above.

In evaluating the quark box diagram we have to keep in mind that here it does *not* yield a *local* operator: while the W boson fields can be integrated out as for ΔM_B, the internal charm quark fields *cannot* since m_c and even $2m_c$ fall below m_b; this gives rise to absorptive parts at $m_c + m_u$ and $2m_c$. Evaluating the quark box diagram [124], [125],

$$(\Gamma_{B_q})_{12} \simeq -\frac{G_F^2 M_{B_q}^3}{8\pi} B_B F_B^2\big|_{B_q} \left(\xi_{cq}^2 P(cc) + \xi_{uq}^2 P(uu) + 2\xi_{u,q}\xi_{cq} P(uc) \right). \tag{10.37}$$

The functions $P(cc)$, $P(uu)$ and $P(uc)$ denote the weight of the configurations with $\bar{c}c$, $\bar{u}u$ and $\bar{u}c$ or $\bar{c}u$ intermediate states, respectively, see Fig. 10.9. With $m_u^2 \ll m_c^2 \ll m_b^2$ they are given by very simple expressions:

$$P(uu) \simeq 1 , \ P(uc) \simeq 1 - \frac{4}{3}\frac{m_c^2}{m_b^2} , \ P(cc) \simeq 1 - \frac{8}{3}\frac{m_c^2}{m_b^2}. \tag{10.38}$$

Hence

$$(\Gamma_{B_q})_{12} \simeq -\frac{G_F^2 M_{B_q}^3}{8\pi} B_B F_B^2\big|_{B_q} \left[(\xi_{uq} + \xi_{cq})^2 - \frac{8}{3}\frac{m_c^2}{m_b^2}\xi_{cq}(\xi_{uq} + \xi_{cq}) \right]. \tag{10.39}$$

Unitarity of the 3x3 KM matrix requires

$$\xi_{uq} + \xi_{cq} = -\xi_{tq}, \tag{10.40}$$

---

[4] Such an expectation would be utterly absurd for $\Delta\Gamma_K$ and highly miraculous for $\Delta\Gamma_D$.

and thus

$$\left(\frac{(\Gamma_{B_q})_{12}}{(M_{B_q})_{12}}\right)_{B_q} \simeq \frac{3\pi}{2}\frac{M_{B_q}^2}{M_W^2}\frac{1}{\eta_t^B(\mu)E(x_t)}\left(1+\frac{8}{3}\frac{m_c^2}{m_b^2}\frac{\xi_{cq}}{\xi_{tq}}\right) \sim \mathcal{O}(\text{few}\times 10^{-3}). \quad (10.41)$$

This is in line with the general expectations stated above, although we have to take the number in Eq. (10.41) with a grain of salt: for it reflects cancellations taking place between $P(uu)$, $P(cc)$ and $P(ud)$.

The above result can be turned into

$$\frac{\Delta\Gamma}{\Gamma} = \left(\frac{\Gamma_{12}}{M_{12}}\right)_{B_q}\frac{x_q}{2}$$

$$= \mathcal{O}(10^{-3})x_q. \quad (10.42)$$

Depending on the numerical value for x_s, we might expect

$$\frac{\Delta\Gamma}{\Gamma}|_{B_s} \simeq (1\sim 10)\%. \quad (10.43)$$

Similar numbers are obtained when we employ phenomenological models to describe $B_s \to \overline{D}_s^{(*)}D_s^{(*)} \to \overline{B}_s$.

10.3 The quest for the ultimate prize – CP violation in B decays

Around 1976, A. Pais gave a seminar at Rockefeller University entitled 'CP Violation in Charmed Particle Decays' [126]. Sanda's recollection of how Pais started out his seminar is as follows:

> There is good news and bad news. The good news is that **CP** violation in a heavy meson system is quite similar to that of the K meson system. The bad news is that there is little distinction like K_L and K_S mass eigenstates. For heavy meson systems, both lifetimes are short.

This paper stimulated the search for large **CP** violation in B decays.

Not too long afterwards came the paper by Bander, Silverman and Soni [127], where they discussed **CP** asymmetries in b quark decay generated by penguin amplitudes[5]. For anyone familiar with kaon physics in the context of the SM, it did not take much imagination to extend it to the B meson system. We were all convinced that $B^0 - \overline{B}^0$ mixing existed at some level, although it could be too feeble to be observable in the conventional way. Instead its effects might surface in other ways – like in **CP** asymmetries that depended on mixing!

Consider a final state f that is common to B^0 and $\overline{B}^0$ decays with $\overline{\rho}(f) \equiv A(\overline{B}^0 \to f)/A(B^0 \to f)$; for simplicity we assume $|\overline{\rho}(f)| = 1$, and also $\Delta\Gamma = 0$. We

---

[5] This effect will be discussed later.

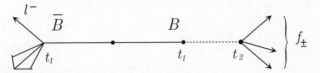

Fig. 10.10. Configuration for a $B\overline{B}|_{C=-}$ event in which $\overline{B}$ is identified at t_1 through its semileptonic decay with l^-, and its partner, a B at t_1, decays to a **CP** eigenstate f at $t_2 > t_1$.

then find from Eq. (6.64) and Eq. (6.65):

$$\Gamma(B^0(t) \to f) \propto e^{-\Gamma_B t}|A(B^0 \to f)|^2[1 - \mathrm{Im}\,\frac{q}{p}\overline{\rho}(f)\sin(\Delta M_B t)]$$

$$\Gamma(\overline{B}^0(t) \to f) \propto e^{-\Gamma_B t}|A(B^0 \to f)|^2[1 + \mathrm{Im}\,\frac{q}{p}\overline{\rho}(f)\sin(\Delta M_B t)]. \quad (10.44)$$

In $e^+e^- \to B^0\overline{B}^0$ the situation is more involved since both mesons oscillate – in a correlated fashion!

Consider the event shown in Fig. 10.10, where a $B^0\overline{B}^0$ pair is produced in a C odd configuration, as in $\Upsilon(4S) \to B^0\overline{B}^0$. The time evolution of the *pair* is given by Eq. (10.7). Once one of the beauty mesons is tagged as a $\overline{B}$ by, for example, its semileptonic decay $\overline{B} \to l^- + X$ at time t, we know that its pair produced partner has to be a B at that time; its amplitude for decaying an interval Δt later into a **CP** eigenstate f can then depend only on Δt – the time elapsed since the first decay – and is given by [133]:

$$A(f, t_2, t_1) = e^{-\frac{1}{2}\Gamma(t+\Delta t)}e^{-i\frac{1}{2}(M_1+M_2)(t+\Delta t)}$$
$$\times \left[A(f)\cos(\frac{1}{2}\Delta M \Delta t) + i\frac{q}{p}\overline{A}(f)\sin\left(\frac{1}{2}\Delta M_B \Delta t\right)\right]. \quad (10.45)$$

The same expression applies when $B \to f_{\pm}$ actually occurs before $\overline{B} \to l^- + X$ by an amount Δt; i.e., Eq. (10.45) holds for Δt both positive and negative. The probabilities for $(B^0\overline{B}^0)_{C=-} \to [l^{\pm} + X]_t + [f]_{t+\Delta t}$ is then given by (see Problem 10.1):

$$\Gamma(B^0\overline{B}^0|_{C=-} \to [l^+ X]_{t_l} f_{t_f}) \propto e^{-\Gamma(t_l+t_f)}|A_{SL}|^2|A(f)|^2$$
$$\cdot \left[1 + \mathrm{Im}\left(\frac{q}{p}\overline{\rho}(f)\right)\sin\Delta M_B(t_f - t_l)\right]$$

$$\Gamma(B^0\overline{B}^0|_{C=-} \to [l^- X]_{t_l} f_{t_f}) \propto e^{-\Gamma(t_l+t_f)}|A_{SL}|^2|A(f)|^2$$
$$\cdot \left[1 - \mathrm{Im}\left(\frac{q}{p}\overline{\rho}(f)\right)\sin\Delta M_B(t_f - t_l)\right], \quad (10.46)$$

where we have assumed $|\overline{\rho}(f)| = 1$. This correlated time evolution is shown in Fig. 10.11 for $f = \psi K_S$ with $\mathrm{Im}\,\frac{q}{p}\overline{\rho}(\psi K_S) \simeq \sin 2\phi_1 = +0.6$.

The authors started their collaboration in 1980. We realized [134] that

$$B \to \psi K_S \quad (10.47)$$

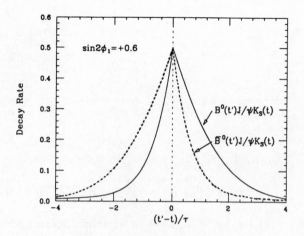

Fig. 10.11. Calculated proper time distribution of $J/\psi K_s$ decay for a **CP** asymmetry corresponding to $\sin 2\phi_1 = +0.6$.

is an optimal channel for detecting a **CP** asymmetry:

- The decays $\psi \to l^+l^-$ provide a striking signature; the accompanying $K_S \to \pi^+\pi^-$ is also relatively easy to detect.

- This mode is predicted to exhibit a **CP** asymmetry that is reliably expressed in terms of fundamental KM parameters and is numerically large[6].

- It is not KM suppressed. Yet due to its two-body nature it commands a small branching ratio:

$$\text{Br}(B \to \psi K^0) = (7.5 \pm 2.1) \cdot 10^{-4}. \tag{10.48}$$

The often repeated comment that all channels with large **CP** asymmetries suffer from small branching ratios, however, overstates the facts: since a 5 GeV hadron has so many exclusive (nonleptonic) channels available for decay, they will all have smallish branching ratios. Semi-inclusive modes such as $B \to \psi K_S + X$ will occur more frequently; yet the asymmetry will get washed out, unless X is chosen wisely, in particular if the final state is not a pure **CP** eigenstate. Note that the statistical error on an asymmetry is

$$\text{Error}\left(\frac{N_+ - N_-}{N_+ + N_-}\right) \propto \frac{1}{\sqrt{N}}, \tag{10.49}$$

---

[6] Fig. 2 of [134] contains a prediction for the asymmetry that could be close to 100 % for $|V_{td}| \ll |V_{cb}| \ll |V_{us}|$ together with vacuum saturation.

where N is the smaller of $N_\pm$. To establish an asymmetry in a perfect detector, we need the ratio of asymmetry to the error to be large.

$$\text{Asym} \times \sqrt{N} \gg 1. \tag{10.50}$$

By choosing a decay mode with large branching ratio, we gain in N but lose in the size of the asymmetry. The fact that the N dependence enters through a square root means that it is often much more effective to choose a decay channel with a large asymmetry at the expense of a smaller branching ratio than to choose a decay with larger branching ratio but smaller asymmetry.

10.4 From sweatshops to beauty factories

Due to a high signal-to-noise ratio and low associated multiplicities, e^+e^- annihilation provides a clean laboratory for studying beauty physics; in addition, $\Upsilon(4S) \to B\overline{B}$ offers the considerable advantage that beam-energy constraints are particularly powerful for this two-body final state. Yet for it to be hailed as a factory rather than a sweatshop we have to create a large sample of B mesons. The cross section for B production just above threshold is shown in Fig. 10.12.

At the $\Upsilon(4S)$ we have 1 nb available. Let us aim for 100 events of the type $B\overline{B} \to [l^\pm X][\psi K_S]$ to establish the asymmetry during one year of running. With

$$
\begin{aligned}
\text{Br}(K_S \to \pi^+\pi^-) &= (68.61 \pm 0.28)\% \\
\text{Br}(\psi \to \mu^+\mu^-) &= (12.03 \pm 0.27)\% \\
\text{Br}(B^0 \to l^+ + X) &= (10.3 \pm 1.0)\%,
\end{aligned}
\tag{10.51}
$$

we need a luminosity of

$$L = \frac{3 \times 10^7 \text{ events}}{10^7 \text{ s } 10^{-33} \text{ cm}^2} = 3 \times 10^{33} \text{ s}^{-1} \text{ cm}^2 \tag{10.52}$$

for a perfect detector, where we have assumed that we can run 10^7 s in a calendar year with $\pi \cdot 10^7$ s. This is ten times the highest luminosity achieved so far at CESR, the world record holder.

There are further pieces of good news and bad news. The good news is the fact that the asymmetry depends on $t_2 - t_1$. It is practically impossible to determine the e^+e^- collision point and time with sufficient accuracy. Thus the magnitude of t_1 and t_2 separately cannot be determined. On the other hand, $t_2 - t_1$ depends only on the spacial separation of the two decay vertices irrespective of the production point.

The bad news is – that the asymmetry depends on $t_2 - t_1$! That means we have to be able to track the decay vertices routinely – otherwise the sought-after asymmetry gets integrated out. Unfortunately the B mesons travel a mere 30 μm in the $\Upsilon(4S)$ rest-frame before they decay. Presently available technology for microvertex detectors does not provide us with the necessary resolution.

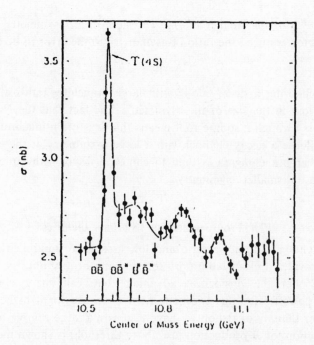

Fig. 10.12. The hadronic cross section for $e^+e^- \to$ hadrons near $\Upsilon(4S)$. By sitting on top of $\Upsilon(4S)$ the signal to noise ratio for $B\overline{B}$ production can be increased to 1 : 2.5. Not only that, $B\overline{B}$ is nearly at rest in the rest frame of $\Upsilon(4S)$ and also there is no $B\overline{B}^*$ production as it is below threshold. This is a blessing as $B\overline{B}^*$ results in an even angular momentum $B\overline{B}$ state. This figure was reproduced from *Physical Review Letters* by permission of the American Physical Society.

For a while there was hope that we could harness the seemingly magic powers of quantum mechanics rather than detector technology to overcome this hurdle. Consider

$$e^+e^- \to B^{0*}\overline{B}^0 \to B^0\overline{B}^0 + \gamma \; ; \qquad (10.53)$$

the $B\overline{B}$ pair is now produced in a **C** *even* configuration[7]. For a **C** even $B\overline{B}$ state, the **CP** asymmetry depends on $\sin \Delta M_B(t_f + t_l)$ rather than $\sin \Delta M_B(t_f - t_l)$; in that case integrating over all times of decay would *not* average out the asymmetry. CLEO and CUSP have searched for the reaction of Eq. (10.53) at $\Upsilon(4S)$ alas, without success!

It was a seemingly crazy idea that provided the breakthrough out of this impasse, namely to build an *asymmetric* e^+e^- collider, as first suggested by P. Oddone during our conversations with him on the difficulty mentioned above. If

---

[7] The $B\overline{B}$ pair produced in $\Upsilon(4S)$ decays is equivalently labelled as a P wave or a **C** odd pair; like in this case, the **C** label often allows for a more concise argument.

$\Upsilon(4S)$ was moving so that B mesons have a large enough boost, they would leave a track that could be measured by existing vertex detectors.

At first, and for some time, accelerator physicists at SLAC, KEK and other places let it be known that an asymmetric collider with a luminosity $\sim 10^{33} \mathrm{cm}^{-2} \, \mathrm{s}^{-1}$ could not be built. Yet at the moment of the writing of this book both KEK and SLAC are building e^+e^- rings to collide positrons with about 3 GeV on electrons of about 9 GeV. Those machines are scheduled to become operational before the turn of the millenium! Which is to show that truly formidable obstacles can be overcome through human ingenuity coupled with persistence – if the prize is attractive enough. This is certainly the case here.

10.5 Snatching victory from the jaws of defeat – the 'unreasonable' success of the CKM description

Before discussing B decays in more detail we want to briefly recapitulate salient features of the theory of flavour-changing neutral currents and emphasize the major successes scored by the SM with the KM ansatz added. Our argument is based on four findings:

1 Effective strangeness-changing neutral currents do exist, albeit with a highly suppressed strength

$$0 < \Delta M_K |_{\exp} \simeq 3.5 \cdot 10^{-12} \text{ MeV} \ll 10^{-8} \text{ MeV.} \qquad (10.54)$$

The GIM mechanism can generate the required suppression, as discussed in Sec. 9.6. While it is true that the theoretical uncertainty in the SM postdiction of ΔM_K is about 50%, we should not lose sight of the big picture: with ΔM_K being smaller by a factor of ~ 4000 than a priori expected, it is highly non-trivial for even a postdiction to come within a factor of two of the observation!

2 Deriving charged current dynamics from a non-abelian gauge theory naturally leads to the implementation of the GIM mechanism with a unitary CKM matrix reflecting weak universality. .

3 The observed value of ΔM_{B_d} is reproduced without forcing any of the parameters. While this is not further evidence for the GIM mechanism per se, it is a nontrivial consequence of the pattern in the mass related SM parameters: the large size of m_t offsets the smallness of $\mathbf{V}_{td}$. From $\Delta M_{B_d}|_{\exp}$ we infer for the Wolfenstein parameters ρ and η (A is fixed by $|\mathbf{V}_{cb}|$.)

$$0.7 \leq \sqrt{(1-\rho)^2 + \eta^2} \leq 1.4, \qquad (10.55)$$

the main uncertainty being that of $B_B f_B^2$.

4 The argument presented above to illustrate the fact that we have been very lucky to have such a long living B meson is too naive and handwaving

to obtain a reliable number for V_{cb}. To get a quantitative result for $\mathbf{V}_{cb}$ we resort to heavy quark expansions [128]. A rigorous description of $B \rightarrow D^* + l^+ v$ and $B \rightarrow lvX$ decays has been developed within the framework of QCD. We infer from the data [129] that

$$|\mathbf{V}_{cb}| = 0.040 \pm 0.004. \tag{10.56}$$

This leads to

$$A = 0.83 \pm 0.09. \tag{10.57}$$

5 We have derived in Eq. (9.52) the KM prediction for ϵ_K. This can be used to obtain a constraint among KM parameters. Inserting numerical values given in Eq. (9.45), we obtain [121]

$$
\begin{aligned}
\eta &\simeq \frac{\epsilon}{4.4 A^2 B_K e^{i\phi_{sw}}} \left[-\eta_{cc} E(x_c) + \eta_{ct} E(x_c, x_t) + \eta_t E(x_t) \right]^{-1} \\
&\sim \frac{.3}{B_K(1.4 - \rho)}.
\end{aligned}
\tag{10.58}
$$

6 Summarizing ARGUS [130], CLEO I [131] and CLEO II [132] data, we obtain: $|\mathbf{V}_{ub}/\mathbf{V}_{cb}| \sim 0.08 \pm 0.03$, where the listed error represents a guestimate only[8]. This result leads to

$$\sqrt{\rho^2 + \eta^2} \simeq 0.38 \pm 0.11, \tag{10.59}$$

as inferred from $|\mathbf{V}_{ub}|$, which leads to Fig. 10.13.

At first sight we might attribute little weight to the observation that the bands representing the three constraints do intersect, since they appear fairly broad, mainly due to theoretical uncertainties. However, such an evaluation misses the profound message as we read it:

- The observables ΔM_K, ϵ_K, ΔM_{B_d} and $\Gamma(B \rightarrow lX_u)$ which get connected by the KM framework represent quite different dynamical regimes that proceed on very different time scales. A priori it would seem that the extracted value of parameters could have been quite different – by orders of magnitude! – from what they turned out to be. We are tempted to say that it borders on the miraculous that these quantities can be described in terms of a unitary CKM matrix with $|\rho|, |\eta| \leq 1$. An uncertainty by a factor of two in the values of ρ and η should be viewed as puny if the related observables could have been different by orders of magnitude. In other words: the KM phenomenology can accommodate only a tiny corner of a general parameter space – yet nature has put the observables exactly into that 'neighbourhood'!

---

[8] In the future an uncertainty reliably estimated to be below 10 % appears achievable through a combination of methods and cross checks.

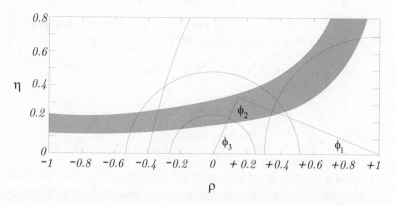

Fig. 10.13. The allowed region of (ρ, η) of the unitarity triangle. The semicircular band with centre at $(0,0)$ represents Eq. (10.59), the constraint from $|\mathbf{V}_{ub}/\mathbf{V}_{cb}|$. The semicircle with centre at $(1,0)$ represents Eq. (10.55), the constraint from $B - \bar{B}$ mixing. Finally, the meshed band represents Eq. (10.58), the constraint from ϵ_K.

- Even after committing ourselves to the KM 'neighbourhood', phenomenological success is far from guaranteed. Keep in mind that in the 1970s and early 1980s values like $|\mathbf{V}_{cb}| \sim 0.04$ and $|\mathbf{V}_{ub}| \sim 0.004$ would have seemed quite unnatural; claiming the top mass having to be ~ 180 GeV would have been perceived as preposterous even in the 1980s! Consider a scenario with $|\mathbf{V}_{cb}| \simeq 0.04$ and $|\mathbf{V}_{ub}| \simeq 0.003$, yet $m_t = 40$ GeV; in the mid 1980s this appeared to be quite natural (in particular in view of the claimed 'discovery' – since withdrawn – of top quarks with $m_t = 40 \pm 10$ GeV). In that case we would need

$$\eta \simeq \frac{1.2}{B_K} \frac{1}{1 - \frac{1}{3}\rho} \qquad (10.60)$$

to reproduce $|\epsilon_K|$. Such a large value of η is incompatible with Eq. (10.59) since best estimates at present yield $B_K \simeq 0.5 \sim 1$.

The point of this numerological exercise is to illustrate that the CKM description could easily have suffered a phenomenological collapse over the last ten years – but it did not!

We humbly submit to the reader that the CKM ansatz is quite unlikely to be merely a coincidence or – worse still – mirage: we see it as reflecting a significant aspect of **CP** violation – though presumably not the only one – even if we are ignorant about its dynamical foundation.

Problems

10.1 Suppose that a $B^0\bar{B}^0$ pair is created at $t = 0$, with momenta $\vec{k}$ and $-\vec{k}$ in the centre of mass system with given **C** parity. Assuming $\Delta\Gamma = 0$, show that a state with B^0 or $\bar{B}^0$ at time t_1 with momentum $\vec{k}$ and B^0 or $\bar{B}^0$ at time t_2 with momentum $-\vec{k}$ can be written as:

$$|(P^0\bar{P}^0)_{C=\pm}(t_1,k),(t_2,-k)\rangle = \frac{1}{\sqrt{2}}e^{-\frac{1}{2}\Gamma(t_1+t_2)}$$

$$\left(-i\sin\frac{\Delta M(t_{-k}\pm t_k)}{2}\left[\frac{p}{q}|P_k^0 P_{-k}^0\rangle \pm \frac{q}{p}|\bar{P}_k^0\bar{P}_{-k}^0\rangle\right]\right.$$

$$\left.+\cos\frac{\Delta M(t_{-k}\pm t_k)}{2}\left[|P_k^0\bar{P}_{-k}^0\rangle \pm |\bar{P}_k^0 P_{-k}^0\rangle\right]\right). \qquad (10.61)$$

Consider the case in which one of the particles decays in a flavour-specific way, while the other one generates a flavour-nonspecific final state:

$$\bar{P}^0 \rightarrow l^- + X \nleftarrow P^0$$
$$\bar{P}^0 \nrightarrow l^+ + X \leftarrow P^0$$
$$\bar{P}^0 \rightarrow f \leftarrow P^0, \qquad (10.62)$$

where for simplicity we have assumed the final state f to be a CP eigenstate.

Derive the decay rates:

$$\Gamma\left(P^0\bar{P}^0\Big|_{C=\pm} \rightarrow (l^- X)_k(t_k)f_{-k}(t_{-k})\right)$$

$$\sim e^{-\Gamma(t_k+t_{-k})}|A_{SL}|^2|A(f)|^2\left[(1-\cos\Delta M(t_k\pm t_{-k}))\left|\frac{q}{p}\right|^2|\bar{\rho}(f)|^2\right.$$

$$\left.+(1+\cos\Delta M(t_k\pm t_{-k}))-2\sin\Delta M(t_{-k}\pm t_k)\mathrm{Im}\frac{q}{p}\bar{\rho}(f)\right], \qquad (10.63)$$

where

$$\bar{\rho}(f) \equiv \frac{\bar{A}(f)}{A(f)} \equiv \frac{1}{\rho(f)}. \qquad (10.64)$$

Derive a similar expression for $\Gamma\left(P^0\bar{P}^0\Big|_{C=\pm} \rightarrow (l^+ X)_k(t_k)f_{-k}(t_{-k})\right)$.

Now for B mesons with $\left|\frac{p}{q}\right|=1$ and for special final states with $|\rho(f)|=1$ derive Eq. (10.8).

10.2   Using Eq. (10.61), find the amplitude for same- and opposite-sign dilepton production

$$(B^0\bar{B}^0)_{C=\pm} \rightarrow l^\pm l^\mp + X$$
$$(B^0\bar{B}^0)_{C=\pm} \rightarrow l^\pm l^\pm + X, \qquad (10.65)$$

and derive Eq. (10.9).

10.3 Under special circumstances the mere *existence* of a transition rather than an asymmetry between two reactions can establish **CP** violation. Consider

$$e^+e^- \to \gamma^* \to P^0\overline{P}^0 \to f_+^a f_+^b, \; f_-^a f_-^b \qquad (10.66)$$

with

$$\mathbf{CP}|f_\pm^i\rangle = \pm|f_\pm^i\rangle,$$

i.e., both P^0 and $\overline{P}^0$ decay into **CP** eigenstates with the same **CP** parity. Convince yourself that this is a **CP** violating reaction.

For $|q/p| = 1$ and $\Delta\Gamma = 0$ verify the time integrated decay rate

$$\text{rate}\left(\left. P^0\overline{P}^0\right|_{C=-} \to f_\pm^a f_\pm^b\right) \propto \left|A(f_\pm^a)A(f_\pm^b)\right|^2$$

$$\cdot \left[\left(1 + \frac{1}{1+x^2}\right)\left|\overline{\rho}(f_\pm^a) - \overline{\rho}(f_\pm^b)\right|^2\right.$$

$$\left. + \frac{x^2}{1+x^2}\left|1 - \left(\frac{q}{p}\right)^2 \overline{\rho}(f_\pm^a)\overline{\rho}(f_\pm^b)\right|^2\right]. \qquad (10.67)$$

10.4 If the two final states are identical

$$f_\pm^a = f_\pm^b \equiv f_\pm,$$

then show that

$$\text{Br}\left(\left. P^0\overline{P}^0\right|_{C=-} \to f_\pm f_\pm\right) = [\text{Br}(P^0 \to f_\pm)]^2 \cdot \frac{x^2}{1+x^2} \qquad (10.68)$$

$$\cdot \left(2\text{Im}\frac{q}{p}\overline{\rho}(f_\pm)\right)^2. \qquad (10.69)$$

Explain why

$$e^+e^- \to \gamma^* \to P^0\overline{P}^0 \to f_\pm f_\pm \qquad (10.70)$$

is not forbidden by Bose statistics. Hint: look at this decay in terms of mass eigenstates P_1 and P_2,

$$\gamma^* \;\not\to\; P_1 P_1, \; P_2 P_2$$
$$\gamma^* \;\to\; P_1 P_2 \to (f_\pm)_1 (f_\pm)_2. \qquad (10.71)$$

10.5 Consider a decay in which the amplitude ratios for $P \to f_\pm^a$ and $P \to f_\pm^b$ exhibit a relative phase

$$\frac{\overline{\rho}(f_\pm^b)}{\overline{\rho}(f_\pm^a)} = e^{i\alpha_{ab}}, \quad \alpha_{ab} \neq 0, \pi, \qquad (10.72)$$

which means that $f_\pm^a$ and $f_\pm^b$ definitely have to be different states. Then show that $\gamma^* \rightarrow P^0\overline{P}^0 \rightarrow f_\pm^a f_\pm^b$ can proceed even in the absence of oscillations, and the decay rate is given by:

$$\text{Br}\left(P^0\overline{P}^0\Big|_{C=-} \rightarrow f_\pm^a f_\pm^b\right) \sim \text{Br}(P^0 \rightarrow f_\pm^a)\text{Br}(P^0 \rightarrow f_\pm^b)\left(\sin\frac{\alpha_{ab}}{2}\right)^2 \tag{10.73}$$

for $x = 0$.

10.6   Discuss why you expect more same-sign dilepton events from semileptonic decays of **C** even than **C** odd $B^0\overline{B}^0$ pairs.

10.7   Considering the fact that $M(\Upsilon(4S)) = 10.5800 \pm .0035$ and $m(B^0) = 5.2792 \pm .0018$ GeV, and $c\tau = 468\mu m$, compute the distance a B^0 travels before it decays when the $B^0\overline{B}^0$ pair is produced at rest. Compute the typical track length for a 3 GeV $\times$ 9 GeV asymmetric e^+e^- collider.

10.8   Derive Eq. (10.24) from Fig. 10.7, paying particular attention to the sign, as the SM will predict the sign as well as the magnitude of the angles. Verify that these three angles add up to π.

11

Basics of the **CP** phenomenology in *B* decays

CP violation in B decays:
exactly like in K decays –
only different!

Based on an Austrian saying

Let us turn the firm, albeit qualitative, expectation that somewhere in the decays of beauty hadrons large **CP** asymmetries are waiting to be discovered into specific predictions of the KM ansatz. First we survey the general landscape.

11.1 The six paths to CP violation in *B* decays

The first gratifying feature is that $B^0 - \overline{B}^0$ oscillations do occur with considerable speed. For good measure nature has actually provided us with two such systems, namely B_d and B_s mesons. There are then two mass eigenstates, defined in Eq. (6.19), with different lifetimes for neutral beauty mesons without and with strangeness: $B_{d,2}$ vs $B_{d,1}$ and $B_{s,2}$ vs $B_{s,1}$. Discussing **CP** violation in B^0 and K^0 decays along the same lines – while possible in principle – would be quite impractical. This is not immediately apparent when considering the measurements $\Delta M_K / \Gamma_{K_S} \simeq 0.47$ vs $\Delta M_{B_d} / \Gamma_{B_d} \simeq 0.71 \pm 0.06$. What is obscured in this representation of the data is a basic difference in the lifetime pattern. In the $K^0 - \overline{K}^0$ system we have

$$\Delta \Gamma_K \equiv \Gamma_{K_S} - \Gamma_{K_L} \simeq \Gamma_{K_S} + \Gamma_{K_L} \simeq \Gamma_{K_S}, \qquad (11.1)$$

since $\Gamma_{K_S} \gg \Gamma_{K_L}$. For B_d mesons, on the other hand, we predict

$$\Delta \Gamma_{B_d} \equiv \Gamma_{B_1} - \Gamma_{B_2} \ll \Gamma_{B_1} + \Gamma_{B_2}, \qquad (11.2)$$

i.e., $\Gamma_{B_1} \simeq \Gamma_{B_2}$. Accordingly, a better measure for the situation is provided by

$$x_K \equiv \frac{\Delta M_K}{\overline{\Gamma}_K} \equiv \frac{\Delta M_K}{\frac{1}{2}(\Gamma_{K_S} + \Gamma_{K_L})} \equiv 0.95, \qquad (11.3)$$

$$y_K \equiv \frac{\Delta \Gamma_K}{2\overline{\Gamma}_K} \simeq 1, \qquad (11.4)$$

179

whereas

$$x_d \equiv \frac{\Delta M_{B_d}}{\overline{\Gamma}_{B_d}} \simeq 0.71 \pm 0.06, \tag{11.5}$$

$$y_d \equiv \frac{\Delta \Gamma_B}{2\overline{\Gamma}_{B_d}} \sim \mathcal{O}(\%) \ll 1. \tag{11.6}$$

Eq. (11.4) and Eq. (11.6) have a very practical impact: while an initial beam of K^0 and $\overline{K}^0$ mesons is transformed through patience – i.e., waiting for the K_S component to decay away – and care – i.e., no regeneration of K_S – into a practically pure K_L beam, this does not happen with $B_d/\overline{B}_d$ beams in any appreciable way. However it should be stressed that $\Delta \Gamma \ll \Gamma$ represents the typical situation; $\Gamma_{K_S} \gg \Gamma_{K_L}$, on the other hand, is caused by the fact that only a handful of channels make up K^0 decays and the 'freak' accident of nature posting the mass of K_L meson barely above the 3π threshold.

The situation is somewhat ambiguous for B_s mesons, where we expect

$$x_s \gg 1 , \tag{11.7}$$

$$y_s \sim \mathcal{O}(10 - 20\%) . \tag{11.8}$$

This may lead to a certain preponderance of $B_{s,2}$ mesons in a $B_s/\overline{B}_s$ beam which might be exploitable through modern detection devices.

The fact that $|(\Gamma_K)_{12}| \sim |(M_K)_{12}|$ whereas $|(\Gamma_B)_{12}| \ll |(M_B)_{12}|$ has also a direct impact on **CP** observables. For ϵ_K it generates a complex phase of $\simeq 45^o$; concentrating on the observable $\text{Re}\,\epsilon_K$ rather than $|\epsilon_K|$ thus leads to only a slight reduction in sensitivity. In the $B_d - \overline{B}_d$ system on the other hand with $|\Gamma_{12}| \ll |M_{12}|$ we have $\text{Re}\,\epsilon_B \ll |\epsilon_B|$ (see Problem 11.2).

These differences have a dramatic impact on how to go about the business of searching for **CP** asymmetries. Let us explain by going through the phenomenology of six classes given in Sections 7.1.1 to 7.1.6.

11.1.1 Class (i)

The analogy to $K_L \to \pi^+\pi^-$ – i.e., establishing **CP** violation through existence of a reaction – is provided by

$$e^+e^- \to \Upsilon(4S) \to B_d\overline{B}_d \to f_+^{(i)}f_+^{(j)} \text{ or } f_-^{(i)}f_-^{(j)}, \tag{11.9}$$

with $f_+^{(i)}$ and $f_-^{(i)}$ denoting an even and odd **CP** eigenstate, respectively. Since $\mathbf{CP}[\Upsilon(4S)] = +1$ and $\mathbf{CP}[f_\pm^{(i)}f_\pm^{(j)}]_{l=1} = -1$, this reaction can proceed only due to **CP** violation[1]. The analogy to $K_L \to \pi\pi$ can be seen more clearly as follows:

---

[1] The **CP** conserving reaction $e^+e^- \to [B_d\overline{B}_d]_{l=0} + \gamma \to [f_\pm^{(i)}f_\pm^{(j)}]_{l=0} + \gamma$ provides a negligible background.

Bose statistics imposes the selection rules

$$\Upsilon(4S) \rightarrow [B_{d,1}B_{d,2}]_{l=1}, \tag{11.10}$$

and the $[B\overline{B}]_{l=1}$ state cannot be $[B_{d,1}B_{d,1}]_{l=1}$ or $[B_{d,2}B_{d,2}]_{l=1}$. With **CP** conserved, the B^0 mass eigenstates must be **CP** eigenstates as well and only $\Upsilon(4S) \rightarrow [f_+^{(i)}f_-^{(j)}]_{l=1}$ can occur. Rather than wait for the short-lived component to die away, which would not be practical for B_d mesons, we employ an EPR-like correlation: the first decay tells us what the **CP** parity of the *surviving* B meson and its final state should be, namely opposite – if **CP** is conserved! The two states $f_\pm^{(i)}f_\pm^{(j)}$ do not have to be identical; yet even so the drawback in this study is the heavy penalty we have to pay statistics-wise, namely the tiny combined branching ratios $\sum_{i,j}$ Br$(B_d \rightarrow f_\pm^{(i)}) \cdot$ Br$(B_d \rightarrow f_\pm^{(j)})$ since both exclusive channels have to be reconstructed.

11.1.2 Class (ii)

An analogy to **CP** asymmetry in semileptonic K_L decays might be obtained by exploiting EPR correlations:

$$e^+e^- \rightarrow \Upsilon(4S) \rightarrow \quad B_{d,1} \quad B_{d,2} \rightarrow l^\pm + X.$$
$$\hookrightarrow \quad f_+ \tag{11.11}$$

Here the **CP** $= +1$ state $B_{d,1}$ is tagged by detecting $B_{d,1} \rightarrow f_+$. Any difference between two decay widths $\Gamma(B_{d,2} \rightarrow l^+ + X)$ and $\Gamma(B_{d,2} \rightarrow l^- + X)$ is a sign of **CP** violation. For this procedure to work, we cannot make a mistake in tagging, i.e., $B_{d,2} \rightarrow f_+$ must be absent. But the whole point, as we shall see below, is that we expect large **CP** violation in this decay. So, the above procedure of 'CP tagging' does not work, and we do not know how to observe Class (ii) asymmetry in B decays.

11.1.3 Class (iii)

There are nonleptonic final states *common* to B^0 and $\bar{B}^0$ decays. This applies in particular to **CP** eigenstates. Transitions like $B_d(\rightsquigarrow \bar{B}_d) \rightarrow \psi K_S$ or $B_d(\rightsquigarrow \bar{B}_d) \rightarrow \pi\pi$ can be discussed in close analogy to $K^0(\rightsquigarrow \bar{K}^0) \rightarrow \pi\pi$.

11.1.4 Class (iv)

Tagging B $[\overline{B}]$ by its leptonic decay at time t allows us to construct a pure $\overline{B}$ $[B]$ beam at that time. So, we have

$$e^+e^- \rightarrow B_d\overline{B}_d \quad \rightarrow \quad (l^+ + X)_B(l^+ + X)_B$$
$$\text{vs} \quad \rightarrow \quad (l^- + X)_{\overline{B}}(l^- + X)_{\overline{B}}. \tag{11.12}$$

11.1.5 Class (v)

Direct **CP** violation can be established by comparing the **CP** asymmetries[2] in $B_d \to \psi K_S$ and $B_d \to \pi^+\pi^-$

$$\frac{\Gamma(B_d \to \psi K_S) - \Gamma(\overline{B}_d \to \psi K_S)}{\Gamma(B_d \to \psi K_S) + \Gamma(\overline{B}_d \to \psi K_S)} \neq -\frac{\Gamma(B_d \to \pi^+\pi^-) - \Gamma(\overline{B}_d \to \pi^+\pi^-)}{\Gamma(B_d \to \pi^+\pi^-) + \Gamma(\overline{B}_d \to \pi^+\pi^-)} \quad (11.13)$$

$$\Longrightarrow \quad direct \; \textbf{CP} \; \text{violation.} \quad (11.14)$$

CP asymmetries in the decays of *charged B* mesons (and of beauty baryons) are quite likely to be larger than in strange decays since the weak phases are large here, and there is no $\Delta I = 1/2$ rule reducing the interference between two different amplitudes that lies at the bottom of these effects. The expression for the asymmetry is given in Eq. (4.148).

Two conditions thus have to be satisfied to obtain a non-zero result:

- Two different quark diagrams with different weak phases contribute to the same decay.

- There is a relative strong interaction phase.

Eq. (4.148) also shows that the effect is largest for $|\mathscr{A}_1| \simeq |\mathscr{A}_2|$.

The definition of a **CP** asymmetry requires distinguishing particles and antiparticles in the *initial* state, which is called 'flavour tagging'. For *neutral B* mesons this often is a highly non-trivial task. *Charged B* decays on the other hand are 'self-tagging': the final state specifies the flavour of the initial state.

On the other hand, the analysis of charged *B* decays tends to suffer from one significant drawback: the asymmetry depends crucially on other factors besides the *KM* matrix elements, namely $|\mathscr{A}_i|$ and the final state strong interaction phase, δ_i, which are shaped by strong dynamics and therefore are known only within considerable theoretical uncertainties. This makes it very difficult

- to interpret the absence of a signal or

- to extract the weak parameters from an observed asymmetry.

11.1.6 Class (vi)

Consider, for example, $B \to \psi K^*$ decay, where ψ and K^* are detected through their decay products $\psi \to l^+l^-$ and $K^* \to K\pi$. T odd correlation, for example $\langle \vec{p}_K \cdot (\vec{p}_{l^+} \times \vec{p}_{l^-}) \rangle$, can be generated from the **CP** violating phase in $B \to \psi K^*$ decay. The problem of the final state interaction phase mimicing the asymmetry must be dealt with carefully.

---

[2] Without direct **CP** violation the two asymmetries must be equal in size and opposite in sign since the two final states carry opposite **CP** parity: $\textbf{CP}|\psi K_S\rangle \simeq -|\psi K_S\rangle$ vs $\textbf{CP}|\pi\pi\rangle = |\pi\pi\rangle$.

11.1.7 Summary

To summarize this qualitative discussion: in principle we can analyse the **CP** phenomenology in beauty decays with its large effects in close analogy to the situation in strange decays. On the other hand, some distinctions have to be kept in mind in treating particle–antiparticle oscillations: for while the descriptions in terms of *flavour* or *interaction* eigenstates – $K^0/\overline{K}^0$ or $B^0/\overline{B}^0$ – and of mass eigenstates – K_L/K_S or B_2/B_1 are equivalent, they are not always equally appropriate. The main difference is represented by $|(\Gamma_K)_{12}| \sim |(M_K)_{12}|$ vs $|(\Gamma_B)_{12}| \ll |(M_B)_{12}|$; hence

- it becomes impractical to prepare a beam of neutral B_2 mesons and

- the observable **CP** *violation in* $B_d - \overline{B}_d$ *oscillations*, as probed in semileptonic B decays is much smaller than **CP** asymmetries involving $B_d - \overline{B}_d$ oscillations in nonleptonic B_d decays (although not necessarily smaller than in K_L decays!).

In the remainder of this chapter we introduce the basic kit of theoretical tools for describing **CP** asymmetries in beauty decays, and we illustrate their use through some simple examples. We want to elucidate here the basic strategies we have available for identifying and dealing with promising channels; this should enable the serious reader to come up with many more examples on her or his own. In the subsequent chapter we will address several complications due to the interventions of penguin transitions, etc.

11.2 Master equations

Particle–antiparticle oscillations provide for the presence of two different amplitudes contributing coherently, with their relative weight varying with the time of decay. This is illustrated in Fig. 11.1.

We shall record here the essential results for reference, which are derived in Sec. (6.6).

It is convenient to separate out the main exponential time dependence of the decay rate [125]:

$$\Gamma(B^0(t) \to f) \propto \frac{1}{2} e^{-\Gamma_1 t} \cdot G_f(t) \,,$$

$$G_f(t) = a + b e^{\Delta \Gamma_B t} + c e^{\Delta \Gamma_B t/2} \cos \Delta M_B t + d e^{\Delta \Gamma_B t/2} \sin \Delta M_B t \,, \tag{11.15}$$

$$a = |A(f)|^2 \left[\frac{1}{2} \left(1 + \left| \frac{q}{p} \overline{\rho}(f) \right|^2 \right) + \mathrm{Re} \left(\frac{q}{p} \overline{\rho}(f) \right) \right] , \tag{11.16}$$

$$b = |A(f)|^2 \left[\frac{1}{2} \left(1 + \left| \frac{q}{p} \overline{\rho}(f) \right|^2 \right) - \mathrm{Re} \left(\frac{q}{p} \overline{\rho}(f) \right) \right] , \tag{11.17}$$

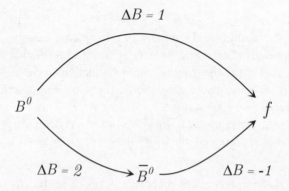

Fig. 11.1. With $B - \overline{B}$ mixing and non vanishing amplitudes for both $B \to f$ and $\overline{B} \to f$ decays there are two decay chains $B \to f$ and $B \to \overline{B} \to f$. This is another version of Fig. 6.1(c). Just as in the two-slit interference pattern observed in Young's experiment in optics, these two decay chains interfere according to the principle of quantum mechanics.

$$c = |A(f)|^2 \left[1 - \left| \frac{q}{p} \overline{\rho}(f) \right|^2 \right] \quad , \quad d = -2|A(f)|^2 \text{Im} \left(\frac{q}{p} \overline{\rho}(f) \right) ; \qquad (11.18)$$

$$\Gamma(\overline{B}^0(t) \to \overline{f}) \propto \frac{1}{2} e^{-\Gamma_1 t} \cdot \overline{G}_{\overline{f}}(t) ,$$

$$\overline{G}_{\overline{f}}(t) = \overline{a} + \overline{b} e^{\Delta \Gamma_B t} + \overline{c} e^{\Delta \Gamma_B t/2} \cos \Delta M_B t + \overline{d} e^{\Delta \Gamma_B t/2} \sin \Delta M_B t, \qquad (11.19)$$

$$\overline{a} = |\overline{A}(\overline{f})|^2 \left[\frac{1}{2} \left(1 + \left| \frac{p}{q} \rho(\overline{f}) \right|^2 \right) + \text{Re} \left(\frac{p}{q} \rho(\overline{f}) \right) \right], \qquad (11.20)$$

$$\overline{b} = |\overline{A}(\overline{f})|^2 \left[\frac{1}{2} \left(1 + \left| \frac{p}{q} \rho(\overline{f}) \right|^2 \right) - \text{Re} \left(\frac{p}{q} \rho(\overline{f}) \right) \right], \qquad (11.21)$$

$$\overline{c} = |\overline{A}(\overline{f})|^2 \left[1 - \left| \frac{p}{q} \rho(\overline{f}) \right|^2 \right] \quad , \quad \overline{d} = -2|\overline{A}(\overline{f})|^2 \text{Im} \left(\frac{p}{q} \rho(\overline{f}) \right). \qquad (11.22)$$

Obviously **CP** invariance is violated if

$$G_f(t) \neq \overline{G}_{\overline{f}}(t). \qquad (11.23)$$

To illuminate the underlying physics for the subsequent discussion, we distinguish two basic categories of final states f exhibiting quite a different phenomenology, namely

- 'Flavour-*specific*' modes emerging from either B^0 or $\overline{B}^0$ decays, i.e., either $A(f)$ or $\overline{A}(f)$ vanish. Here are some examples:

$$\overline{B}_d \equiv (b\overline{d}) \to l^- \nu X, \, D^+ \pi^-, \, \psi K^- \pi^+, \, K^- \pi^+. \qquad (11.24)$$

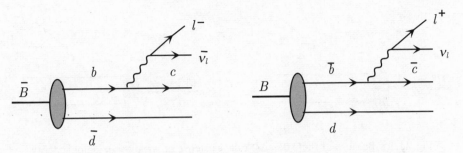

Fig. 11.2. Feynman diagrams for right-sign semileptonic decays. These decays obey the $\Delta B = \Delta Q$ rule, i.e., only $\overline{B} \to l^- + X$ and $B \to l^+ + X$ occur. Thus leptonic decays can be used to identify the decaying objects as either B or $\overline{B}$.

- 'Flavour-*non*specific' modes which can be fed by B^0 as well as $\overline{B}^0$ decays; for example,

$$\overset{(-)}{B}_d \to \psi K_S, \, D\overline{D}, \, \pi\pi \,. \qquad (11.25)$$

Flavour-nonspecific final states can arise only in nonleptonic transitions due to the $\Delta Q_l = +\Delta B$ selection rule.

Staring at the most general and therefore most complex case of an equation is rarely illuminating. Instead we employ Eq. (11.15) and Eq. (11.19) as master equations to derive and discuss complementary special cases.

11.3 Case A: flavour-specific decays with oscillations

Two examples can illustrate this scenario:

1 *Right*-sign semileptonic decays: $\overline{B}^0(t) \to l^- \nu X$ vs $B^0(t) \to l^+ \nu X$. These decays proceed through Fig. 11.2. With $A(l^- X) = 0 = \overline{A}(l^+ X)$ we have in the notation of Eq. (11.15) and Eq. (11.19):

$$a = b = \frac{1}{2}c = \frac{1}{2}|A(l^+ X)|^2 \,, \, d = 0 \,, \qquad (11.26)$$

$$\overline{a} = \overline{b} = \frac{1}{2}\overline{c} = \frac{1}{2}|\overline{A}(l^- X)|^2 \,, \, \overline{d} = 0 \,, \qquad (11.27)$$

and therefore

$$G_{l^+ X}(t) = |A(l^+ X)|^2 \cdot \left[\frac{1}{2}(1 + e^{\Delta \Gamma_B t}) + e^{\Delta \Gamma_B t/2} \cos \Delta M_B t\right] \,, \qquad (11.28)$$

$$\overline{G}_{l^- X}(t) = |\overline{A}(l^- X)|^2 \cdot \left[\frac{1}{2}(1 + e^{\Delta \Gamma_B t}) + e^{\Delta \Gamma_B t/2} \cos \Delta M_B t\right] \,. \qquad (11.29)$$

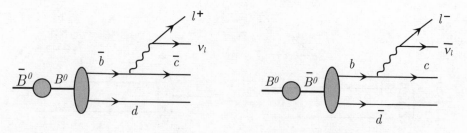

Fig. 11.3. Because of the $\Delta B = \Delta Q$ rule, existence of wrong-sign leptonic decay implies $B - \overline{B}$ mixing.

Although both $G_{l^+X}(t)$ and $\overline{G}_{l^-X}(t)$ depend on t – the time of decay – their ratio does not:

$$\frac{d}{dt}\left(\frac{G_{l^+X}}{\overline{G}_{l^-X}}\right) = \frac{d}{dt}\left(\frac{|A(l^+X)|^2}{|\overline{A}(l^-X)|^2}\right) \equiv 0. \tag{11.30}$$

A **CP** asymmetry can exist here only if

$$|A(l^+X)| \neq |\overline{A}(l^-X)|, \tag{11.31}$$

i.e., if there were *direct* **CP** violation in semileptonic beauty decays. This is not the case in the SM.

2 *Wrong-sign semileptonic decays:* $\overline{B}^0(t) \rightarrow l^+\nu X$ vs $B^0(t) \rightarrow l^-\nu X$. These decays proceed through Fig. 11.3; i.e., they require the intervention of $B^0 - \overline{B}^0$ oscillations. With

$$a = b = -\frac{1}{2}c = \frac{1}{2}\left|\frac{q}{p}\right|^2 |\overline{A}(l^-X)|^2, \; d = 0, \tag{11.32}$$

$$\overline{a} = \overline{b} = -\frac{1}{2}\overline{c} = \frac{1}{2}\left|\frac{p}{q}\right| |A(l^+X)|^2, \; \overline{d} = 0, \tag{11.33}$$

we arrive at

$$G_{l^-X}(t) = \left|\frac{q}{p}\right|^2 |\overline{A}(l^-X)|^2 \cdot \left[\frac{1}{2}(1 + e^{\Delta\Gamma_B t}) - e^{\Delta\Gamma_B t/2} \cos\Delta M_B t\right] \tag{11.34}$$

$$\overline{G}_{l^+X}(t) = \left|\frac{p}{q}\right|^2 |A(l^+X)|^2 \cdot \left[\frac{1}{2}(1 + e^{\Delta\Gamma_B t}) - e^{\Delta\Gamma_B t/2} \cos\Delta M_B t\right]. \tag{11.35}$$

Their *ratio* is independent of t:

$$\frac{d}{dt}\left(\frac{G_{l^-X}(t)}{\overline{G}_{l^+X}(t)}\right) = \frac{d}{dt}\left(\left|\frac{q}{p}\right|^4 \frac{|\overline{A}(l^-X)|^2}{|A(l^+X)|^2}\right) \equiv 0. \tag{11.36}$$

For the time integrated asymmetry defined by Eq. (6.60), we obtain:

$$a_{SL}(B_q) = \frac{1 - |p/q|^4}{1 + |p/q|^4} = -r \sin\zeta_B \tag{11.37}$$

where $r \sin \zeta_B = \operatorname{Im} \Gamma_{12}/M_{12}$ as defined in Eq. (6.37). The quantity $r \sin \zeta_B$ can be predicted in the KM ansatz within some uncertainties. Using Eq. (10.41), we have

$$r \simeq \frac{3\pi}{2} \frac{M_B^2}{M_W^2} \frac{1}{\eta_t^B(\mu)E(x_t)} \tag{11.38}$$

$$\sin \zeta_{B_q} \simeq \tan \zeta_{B_q} \simeq \frac{8}{3} \frac{m_c^2}{m_b^2} \operatorname{Im} \frac{\mathbf{V}_{cb}\mathbf{V}_{cq}^*}{\mathbf{V}_{tb}\mathbf{V}_{tq}^*} . \tag{11.39}$$

Thus

$$\sin \zeta_{B_d} \simeq -\frac{8}{3} \frac{m_c^2}{m_b^2} \frac{\eta}{(1-\rho)^2 + \eta^2} , \quad \sin \zeta_{B_s} \simeq -\frac{8}{3} \frac{m_c^2}{m_b^2} \lambda^2 \eta. \tag{11.40}$$

Several features of Eq. (11.39) and Eq. (11.40) are to be noted:

(a) On the positive side:

⊕ The quantity $\sin \zeta_B$ measures the intrinsic strength of **CP** violation in $B^0 \leftrightarrow \overline{B}^0$ oscillations in an unambiguous way.

⊕ It can be expressed with reasonable accuracy within the KM scheme in terms of its basic parameters in an intuitively understandable way:

• In addition to the suppression from $r = \mathcal{O}(\text{few} \times 10^{-3})$, there is a suppression from $\sin \zeta_B$. For $\sin \zeta_B$, the reduction factor m_c^2/m_b^2 is produced by the GIM mechanism (and actually reads $(m_c^2 - m_u^2)/m_b^2$, for there can be no **CP** violation in the KM scheme when $m_c = m_u$).

We have to remember, however, that the extent of this cancellation depends on taking the quark diagrams seriously. See the discussion leading to Eq. (10.41).

• We find $\zeta_{B_d} \gg \zeta_{B_s}$ since ζ_{B_s} is CKM suppressed: for on the leading level, only quarks of the second and third family contribute here, namely t, b, c and s; accordingly no **CP** violation can arise to leading order in the KM parameters.

(b) On the negative side:

⊖ The observable $1 - |q/p|$ provides a numerically poor handle on $\sin \zeta_{B_s}$. For the latter's coefficient $- (\Gamma_B)_{12}/(M_B)_{12} \simeq \Delta\Gamma_B/\Delta M_B$ – is, as discussed in the preceding section, expected to be quite small on rather general grounds and independently of the strength of **CP** violation. To be more specific: $\Delta\Gamma_B/\Delta M_B \leq 0.01$ and thus

$$\left| 1 - \left| \frac{q}{p} \right|^2 \right| < 0.01 . \tag{11.41}$$

The situation is quite different for neutral kaons where – due to purely accidental reasons – we have $\Gamma_{12} \simeq M_{12}$!

⊖ The prediction for $\Delta\Gamma_B/\Delta M_B$ is not 'gold-plated'; it could conceivably represent an underestimate. In principle $\Delta\Gamma_B$ could be measured in addition to ΔM_B; yet this would go beyond a merely academic proposition only if $\Delta\Gamma_B$ were much larger than anticipated – at least for B_d mesons.

⊖ Altogether we arrive at the predictions

$$a_{SL}(B_d)|_{KM} \sim \mathcal{O}(10^{-3}),$$

(11.42)

$$a_{SL}(B_s)|_{KM} \sim \mathcal{O}(10^{-4}).$$

(11.43)

These are very small asymmetries, in particular for $B_s \leftrightarrow \overline{B}_s$ oscillations, and searching for them clearly poses challenging demands on the control over the systematics an experiment has to maintain. Nevertheless such searches should be and will be performed because:

• they can be executed once we obtain a sample of 'wrong-sign' leptons from B^0 decays;

• establishing an unambiguous signal would constitute a discovery of the very first rank;

• New Physics could manifest itself here in unequivocal ways:

1 It is conceivable that data would reveal $a_{SL}(B_d)$ to be around 0.01, i.e., considerably larger than suggested by Eq. (11.42) while at the same time the SM expectations for $\Delta\Gamma_B/\Delta M_B$ were satisfied. If, on the other hand, both $a_{SL}(B_d)$ and $\Delta\Gamma_B/\Delta M_B$ were found to be enhanced over the KM and SM estimates, respectively, then no clear conclusion could be drawn – beyond conceding that we had presumably overestimated our computational powers!

2 The prediction for $a_{SL}(B_s)$ in Eq. (11.43) is suppressed relative to the natural scale of $\sim$ few $\cdot 10^{-3}$, as set by $\Delta\Gamma_B/\Delta M_B$, due to the very specific mechanism in which **CP** violation is implemented in the KM scheme: it makes the *relative* phase between $(M_{B_s})_{12}$ and $(\Gamma_{B_s})_{12}$ CKM suppressed, Eq. (11.40). New Physics can quite conceivably contribute to M_{12}, but is unlikely to do so for Γ_{12}; it would then vitiate the CKM suppression of ζ_{B_s}, quite possibly leading to

$$a_{SL}(B_s)|_{NP} \sim \mathcal{O}(0.01).$$

(11.44)

11.4 Case B: decays to CP eigenstates with oscillations

In spite of the fact that q/p can have a large phase, at least for B_d mesons,

$$\left.\frac{q}{p}\right|_{B_d} = \left.\sqrt{\frac{M_{12}^*}{M_{12}} + \mathcal{O}\left(\frac{\Gamma_{12}}{M_{12}}\right)}\right|_{B_d} \simeq \frac{V_{tb}^* V_{td}}{V_{tb} V_{td}^*} \simeq \frac{(1-\rho)^2 - \eta^2 - 2i\eta(1-\rho)}{(1-\rho)^2 + \eta^2} ,$$

(11.45)

see Eq. (6.43) and Eq. (10.28), the effects just discussed, Eq. (11.42) and Eq. (11.43), are disappointingly small. There is no mystery involved here, though: the asymmetry in semileptonic B decays depends on $1 - |q/p|^2$. We need a novel idea to find an observable depending on q/p rather than its modulus. This was the impetus behind analysing flavour-nonspecific modes [133]. It lead to the

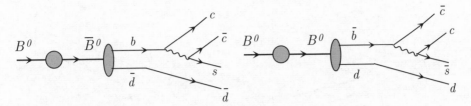

Fig. 11.4. Two quark diagrams which contribute to B decays if there is $B^0 - \overline{B}^0$ mixing. Note that the first diagram contains ($s\overline{d}$) and the second contains ($\overline{s}d$). Thus they do not interfere. Special care must be given to make them interfere.

realization that $(q/p)\overline{\rho}(f)$ is such a quantity; let us recall a few of its general properties as pointed out in Sec. 6.12:

- This quantity does not depend on the phase convention adopted for the antiparticles;

- it combines the effects of $\Delta B = 2$ and $\Delta B = 1$ dynamics;

- measuring it in a *single* class of decays does not allow us – as a matter of principle – to decide whether the observed **CP** violation is of the *superweak* or *direct* type (or a combination of the two). For that purpose we have to compare the asymmetry (or lack thereof) in two distinct classes of decays. We will return to this point more explicitly.

We still have to overcome an apparent stumbling block, namely that $b \rightarrow c$ transitions do not generate final states common to B_d and $\overline{B}_d$ decays on the *quark* level. Let us consider the two graphs shown in Fig. 11.4. It would seem that the difference between ($\overline{s}d$) and ($\overline{d}s$) rules out interference. Yet once you realize that they can lead to the same *hadronic* state

$$(\overline{s}d) \rightarrow K_S \leftarrow (\overline{d}s) \tag{11.46}$$

you have opened up a whole new territory to explore.

The bulk of B decays leads to multibody final states, yet the required coherence decreases for them. Thus they are expected to exhibit **CP** asymmetries that are significantly reduced relative to what we predict on the quark level; they are further suppressed by cancellations among the subprocesses. In two-body final states, on the other hand, the kinematics enforce coherence leading to larger asymmetries at the price of smaller statistics. Cognizant of this point, we [134] have emphasized from the beginning the importance of two-body decays of B mesons in general and of **CP** eigenstates in particular. In this section we will discuss those. The goal here is *not* to list all interesting modes, which would take up too much space and actually would be impractical since so few of the relevant branching ratios are known experimentally. Instead we will concentrate on a few 'typical' examples illustrating various dynamical issues; our expectation is that

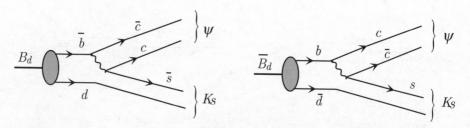

Fig. 11.5.   This is one of the very few places in which we can take advantage of the fact that we can only detect hadrons instead of quarks. Quark pairs $(\bar{s}d)$ and $(\bar{d}s)$ can be considered identical if we detect hadronic final states $K_S + X$ or $K_L + X$.

their treatment will provide the determined reader with the tools necessary to deal with any channel that might become interesting in this context. In the next section we will address some more technical issues and also discuss **CP** asymmetries in (semi)inclusive transitions.

11.4.1 $B_d \to \psi K_S$

This mode is described by the Feynman graphs of Fig. 11.5; it has been observed by ARGUS, CLEO, LEP and the Tevatron through its striking signature $B_d \to (l^+l^-)_\psi (\pi^+\pi^-)_{K_S}$ with

$$\mathrm{Br}(B_d \to \psi K_S) \sim 5 \times 10^{-4} \,, \tag{11.47}$$

roughly as expected theoretically. With this transition being driven by the isoscalar coupling $b \to c\bar{c}s$, the final state is described by a *single* isospin amplitude. Therefore it is quite unlikely even in the presence of New Physics to exhibit *direct* **CP** violation; thus

$$|A(\psi K_S)| = |\bar{A}(\psi K_S)|; \quad |\bar{\rho}(\psi K_S)| = 1. \tag{11.48}$$

Taken together with $|q|^2 \simeq |p|^2$, we see that $(q/p)\bar{\rho}(\psi K_S)$ represents practically a unit vector in the complex plane. Using $\Delta\Gamma_{B_d} \ll \Delta M_{B_d} \sim \Gamma_{B_d}$, we arrive at a simple expression:

$$G_{\psi K_S}(t) = 2|A(\psi K_S)|^2 \cdot \left[1 - \mathrm{Im}\left(\frac{q}{p}\bar{\rho}(\psi K_S) \right) \sin \Delta M_{B_d} t \right], \tag{11.49}$$

$$\bar{G}_{\psi K_S}(t) = 2|A(\psi K_S)|^2 \cdot \left[1 + \mathrm{Im}\left(\frac{q}{p}\bar{\rho}(\psi K_S) \right) \sin \Delta M_{B_d} t \right], \tag{11.50}$$

i.e., the **CP** violation represented by $\mathrm{Im}((q/p)\bar{\rho}(\psi K_S)$ becomes observable due to $\Delta M_{B_d} \neq 0$. Furthermore, the observable **CP** asymmetry now depends on the time of decay:

$$\frac{\mathrm{d}}{\mathrm{d}t}\frac{G_{\psi K_S}(t)}{\overline{G}_{\psi K_S}(t)} \neq 0. \tag{11.51}$$

Having already stated the KM prediction for q/p in Eq. (11.45), we have to evaluate the specific decay amplitudes. With the effective *isoscalar* Hamiltonian

$$
\begin{aligned}
\mathscr{H}_{\mathrm{eff}}(\Delta B = 1) \;=\; & \frac{G_F}{\sqrt{2}}\mathbf{V}_{cb}\mathbf{V}_{cs}^* \cdot (C_1 \bar{s}\gamma_\mu(1-\gamma_5)b\bar{c}\gamma_\mu(1-\gamma_5)c \\
& +C_2 \bar{c}\gamma_\mu(1-\gamma_5)b\bar{s}\gamma_\mu(1-\gamma_5)c) + \mathrm{h.c.},
\end{aligned}
\tag{11.52}
$$

we find

$$\frac{q}{p}\overline{\rho}(\psi K_S) \simeq -\frac{\mathbf{V}_{tb}^*\mathbf{V}_{td}}{\mathbf{V}_{tb}\mathbf{V}_{td}^*} \cdot \frac{\mathbf{V}_{cb}\mathbf{V}_{cs}^*}{\mathbf{V}_{cb}^*\mathbf{V}_{cs}} = -e^{-2i\phi_1}, \tag{11.53}$$

where the minus sign is due to ψK_S being a **CP** odd state. There are two important points to Eq. (11.53):

- The quantity $(q/p)\overline{\rho}(\psi K_S)$ and thus the asymmetry can be be predicted with great parametric precision; i.e., it is reliably expressed purely in terms of ratios of CKM matrix elements. The hadronic matrix element (including its strong phase shift) drops out from the ratio since only a single isospin amplitude contributes in this channel [134].

- The asymmetry is predicted to be naturally large:

$$\mathrm{Im}\frac{q}{p}\overline{\rho}(\psi K_S) \simeq \frac{2(1-\rho)\eta}{(1-\rho)^2+\eta^2} = \sin 2\phi_1 \tag{11.54}$$

with ϕ_1 denoting one of the angles of the unitarity triangle, see Fig. 10.6(6).

- The last statement can be transformed into a semi-quantitative one. From Fig. 10.13, showing the phenomenological constraints on the ρ and η, we can read off the KM prediction:

$$0.20 \;\leq\; \sin 2\phi_1 \;\leq\; 0.85. \tag{11.55}$$

- Our discussion in Secs. 6 and 10 shows that the sign of ϕ_1 can also be extracted irrespective of the sign of ΔM_B. This can be expressed as follows:

$$
\begin{aligned}
A_{\psi K_S} \;\equiv\; & \frac{\Gamma(\overline{B}_d(t) \to \psi K_S) - \Gamma(B_d(t) \to \psi K_S)}{\Gamma(\overline{B}_d(t) \to \psi K_S) + \Gamma(B_d(t) \to \psi K_S)} \\
=\; & \sin(\Delta M_{B_d}t)\mathrm{Im}\left(\frac{q}{p}\overline{\rho}(\psi K_S)\right) \\
=\; & \sin(|\Delta M_{B_d}|t)\sin(2\phi_1). \tag{11.56}
\end{aligned}
$$

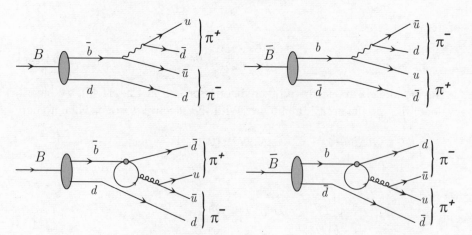

Fig. 11.6.   Tree and penguin quark diagrams which contribute to $B \to \pi\pi$ decay. We will find that the penguin diagrams give us considerable problem in predicting **CP** violation in $B \to \pi\pi$ decay.

Two groups have undertaken pilot studies for measuring this asymmetry: They found

$$\sin(2\phi_1) = \begin{cases} 3.2^{+1.8}_{-2.0} \pm 0.5 & \text{OPAL Collaboration} \quad [135] \\ 1.8 \pm 1.1 \pm 0.3 & \text{CDF Collaboration} \quad [136] \\ .79^{+0.41}_{-0.44} & \text{CDF Collaboration} \quad [137]. \end{cases}$$

As the book was going to press, a new result from the CDF collaboration became available. It indicates that a negative value of $\sin(2\phi_1)$ is disfavoured.

11.4.2 $B_d \to \pi^+\pi^-$

At present this decay mode has not been established experimentally. Its branching ratio is certainly not significantly larger than what is expected, i.e., $\text{Br}(B_d \to \pi^+\pi^-) \sim (1 \sim 2) \times 10^{-5}$. The relevant diagrams are shown in Fig. 11.6. In contrast to the previous example we have

$$|A(\pi^+\pi^-)| \neq |\bar{A}(\pi^+\pi^-)|; \quad |\bar{\rho}(\pi\pi)| \neq 1, \tag{11.57}$$

due to the concurrence of two facts:

- The final state, which happens to be **CP** even, is made up of a combination of $I = 0$ and $I = 2$ configurations which can conceivably possess significantly different strong phase shifts.

- There are two quark-level diagrams with different topologies and different CKM parameters that contribute, namely the spectator process, and the Cabibbo reduced penguin reaction, as shown in Fig. 11.6; the latter affects the $I = 0$ final state only.

There are quite a few subtleties involved here, which will be addressed in detail in Chapter 12. For now we just state

$$G_{\pi^+\pi^-}(t) = |A(\pi^+\pi^-)|^2 \left[1 + |\overline{\rho}(\pi^+\pi^-)|^2 + \left(1 - |\overline{\rho}(\pi^+\pi^-)|^2\right) \cos \Delta M_{B_d} t \right.$$
$$\left. - 2\mathrm{Im}\left(\frac{q}{p}\overline{\rho}(\pi^+\pi^-)\right) \sin \Delta M_{B_d} t \right], \tag{11.58}$$

$$\overline{G}_{\pi^+\pi^-}(t) = |\overline{A}(\pi^+\pi^-)|^2 \left[1 + |\rho(\pi^+\pi^-)|^2 + \left(1 - |\rho(\pi^+\pi^-)|^2\right) \cos \Delta M_{B_d} t \right.$$
$$\left. - 2\mathrm{Im}\left(\frac{p}{q}\rho(\pi^+\pi^-)\right) \sin \Delta M_{B_d} t \right]. \tag{11.59}$$

The **CP** asymmetry can then be expressed as follows:

$$\frac{G_{\pi^+\pi^-}(t) - \overline{G}_{\pi^+\pi^-}(t)}{G_{\pi^+\pi^-}(t) + \overline{G}_{\pi^+\pi^-}(t)} \tag{11.60}$$

$$= \frac{(1 - |\overline{\rho}(\pi^+\pi^-)|^2) \cos \Delta M_{B_d} t - 2\mathrm{Im}((q/p)\overline{\rho}(\pi^+\pi^-)) \sin \Delta M_{B_d} t}{1 + |\overline{\rho}(\pi^+\pi^-)|^2}. \tag{11.61}$$

Eq. (11.61) shows explicitly that the observable **CP** asymmetry in this case may have two sources: namely *direct* **CP** violation residing in $\mathscr{H}(\Delta B = 1)$ – $|\overline{\rho}(\pi^+\pi^-)|^2 \neq 1$; and *indirect* **CP** violation in $\mathscr{H}(\Delta B = 2)$ – $\mathrm{Im}((q/p)\overline{\rho}(\pi^+\pi^-)) \neq 0$. The time dependence allows a clear separation of the two sources.

If we could ignore the penguin contribution and consider only the tree level graph shown in Fig. 11.6, then we could proceed exactly as for $B \to \psi K_S$, with the matrix elements being:

$$\overline{A}(\pi\pi) = \frac{4G_F}{\sqrt{2}} \mathbf{V}_{ub}\mathbf{V}_{ud}^* \cdot \left[C_1 \langle \pi\pi | \overline{d}_L \gamma_\mu b_L \overline{u}_L \gamma_\mu u_L | \overline{B} \rangle + C_2 \langle \pi\pi | \overline{u}_L \gamma_\mu b_L \overline{d}_L \gamma_\mu u_L | \overline{B} \rangle \right], \tag{11.62}$$

$$A(\pi\pi) = \frac{4G_F}{\sqrt{2}} \mathbf{V}_{ub}^*\mathbf{V}_{ud} \cdot \left[C_1 \langle \pi\pi | (\overline{d}_L \gamma_\mu b_L \overline{u}_L \gamma_\mu u_L)^\dagger | B \rangle + C_2 \langle \pi\pi | (\overline{u}_L \gamma_\mu b_L \overline{d}_L \gamma_\mu u_L)^\dagger | B \rangle \right]. \tag{11.63}$$

In that simple scenario the hadronic matrix element still drops out from the ratio $\overline{\rho}(\pi^+\pi^-)$; i.e.,

$$\frac{q}{p}\overline{\rho}(\pi\pi) \simeq \frac{\mathbf{V}_{tb}^*\mathbf{V}_{td}}{\mathbf{V}_{tb}\mathbf{V}_{td}^*} \frac{\mathbf{V}_{ub}\mathbf{V}_{ud}^*}{\mathbf{V}_{ub}^*\mathbf{V}_{ud}} = e^{2i\phi_2}, \quad \mathrm{Im}\frac{q}{p}\overline{\rho}(\pi\pi) \simeq \sin 2\phi_2. \tag{11.64}$$

Yet even in the general case we can conclude:

• $\mathrm{Im}\,(q/p)\overline{\rho}(\psi K_S) \neq -\mathrm{Im}(q/p)\overline{\rho}(\pi\pi)$ establishes direct **CP** violation.

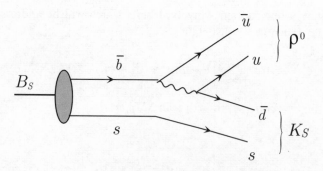

Fig. 11.7.   Tree diagram responsible for $B_s \rightarrow \rho K_S$. In addition, there is a penguin graph which makes precise prediction of this asymmetry difficult.

11.4.3 $B_s \rightarrow K_S \rho^0$

The tree level quark diagram shown in Fig. 11.7 generates this transition. If it were the only underlying mechanism $(q/p)\overline{\rho}$ could be expressed merely in terms of CKM parameters:

$$\frac{q}{p}\overline{\rho}(B_s \rightarrow K_S \rho) \simeq -\frac{\mathbf{V}^*_{tb}\mathbf{V}_{ts}}{\mathbf{V}_{tb}\mathbf{V}^*_{ts}}\frac{\mathbf{V}_{ub}\mathbf{V}^*_{ud}}{\mathbf{V}^*_{ub}\mathbf{V}_{ud}} \equiv -e^{-2i\phi_3}. \tag{11.65}$$

Yet this channel also receives contributions from a CKM reduced penguin operator which introduces additional weak phases, giving rise to direct **CP** violation and making the extraction of ϕ_3 murky. This difficulty is compounded by $B_s - \overline{B}_s$ oscillations being very speedy and the transition being CKM suppressed with a less than outstanding signature for the final state!

11.4.4   *Seeking New Physics: $B_s \rightarrow \psi\phi$, $\psi\eta$, $D^+_s D^-_s$*

It will be a repeated theme that measuring ϕ_3 will be far from straight-forward. It then behooves us to reflect why we want to determine its size. The two angles ϕ_1 and ϕ_2 will probably be measured with decent or even good accuracy over the next several years, and in any case much better than any direct extraction of ϕ_3. The numerically most precise value for it follows from simple trigonometry

$$\phi_3 = 180^o - \phi_1 - \phi_2 \tag{11.66}$$

– within the KM ansatz! The real goal behind any direct determination of ϕ_3 is thus to uncover the intervention of New Physics. It then makes eminent sense to search for it in a reaction where 'Known Physics' generates a practically zero result. The channels $B_s \rightarrow \psi\phi$, $\psi\eta$, $D^+_s D^-_s$ occurring on the *leading* KM level fit the bill. Since the width difference $\Delta\Gamma_B$ might not be insignificant for B_s mesons, we retain it while still using $|p| = |q|$. Thus for $f = \psi\phi$, $\psi\eta$,

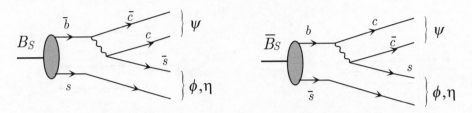

Fig. 11.8. Feynman diagrams for $B \to \psi\phi$ and $\overline{B} \to \psi\phi$ decays. Note that there are only second and third generation quarks involved in these graphs. Thus, in the SM, we do not expect any asymmetry. Observation of sizeable asymmetry implies existence of New Physics.

$D_s^+ D_s^-$:

$$G_f(t) = |A(B \to f)|^2 \cdot \left[1 + e^{\Delta\Gamma_{B_s}t} + (1 - e^{\Delta\Gamma_{B_s}t})\text{Re}\left(\frac{q}{p}\overline{\rho}(f)\right)\right. \tag{11.67}$$

$$\left. -2\text{Im}\left(\frac{q}{p}\overline{\rho}(f)\right) e^{\Delta\Gamma_{B_s}t/2} \sin\Delta M_{B_s}\right], \tag{11.68}$$

$$\overline{G}_f(t) = |\overline{A}(B \to f)|^2 \cdot \left[1 + e^{\Delta\Gamma_{B_s}t} + (1 - e^{\Delta\Gamma_{B_s}t})\text{Re}\left(\frac{p}{q}\overline{\rho}(f)\right)\right. \tag{11.69}$$

$$\left. -2\text{Im}\left(\frac{p}{q}\rho(f)\right) e^{\Delta\Gamma_{B_s}t/2} \sin\Delta M_{B_s}\right], \tag{11.70}$$

The diagrams responsible for $B_s \to \psi\phi$ (or $B_s \to \psi\eta$) are shown in Fig. 11.8; those for $B_s \to D_s^+ D_s^-$ are obtained by re-arranging the quark lines. The crucial point here is the observation that to leading order in λ only four quarks and antiquarks participate in $B_s \to \psi\phi$ and $B_s \to \overline{B}_s \to \psi\phi$: t, b, c and s, i.e., members of the third and second family only. It is the peculiar feature of the KM ansatz that no **CP** violation can arise with only two active families. That means that the **CP** asymmetry in this channel has to be CKM suppressed, since it enters through the effective violation of weak universality due to the presence of the first family.

To be more specific, we note that these transitions are described by single isospin amplitudes each; therefore – as for $B_d \to \psi K_S$ – no *direct* **CP** violation can arise: $|A(B_s \to f)| = |A(\overline{B}_s \to f)|$ and within KM we find

$$\frac{q}{p}\overline{\rho}(B_s \to f) \simeq \text{Im}\left[\frac{(\mathbf{V}_{tb}^* \mathbf{V}_{ts})^2}{|\mathbf{V}_{tb}\mathbf{V}_{ts}|^2} \frac{(\mathbf{V}_{cb}\mathbf{V}_{cs}^*)^2}{|\mathbf{V}_{cb}\mathbf{V}_{cs}|^2}\right] \simeq 2\lambda^2\eta < 0.05. \tag{11.71}$$

In calling such asymmetries small, we might note that they are still an order of magnitude larger than what has been observed in K_L decays. On the other hand New Physics scenarios to be sketched in later chapters will generate asymmetries that can well reach the few $\times 10$ % level.

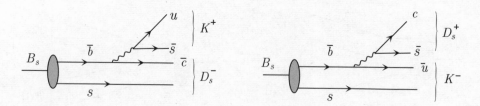

Fig. 11.9. These two decay channels and their charge conjugate decays allow us to determine ϕ_3.

At least the channel $B_s \to \psi\phi$ possesses a striking signature and has been observed at LEP and Tevatron. Its branching ratio has so far not been determined, yet is expected (by analogy with $B \to \psi K^*$) to exceed that for $B_d \to \psi K_S$, namely $\mathrm{Br}(B_s \to \psi\phi) \sim$ few $\times 10^{-3}$. Of course we have to be able to resolve the rapid $B_s \to \overline{B}_s$ oscillations predicted.

11.5 ♠ Case C: other flavour-nonspecific modes with oscillations ♠

The condition *sine qua non* for the flavour-nonspecific channels is that they are common to B^0 and $\overline{B}^0$ decays; final states that are **CP** eigenstates are just one of the possible realizations of that scenario with some built-in simplifying features. For flavour-nonspecific final states that are not **CP** eigenstates the situation becomes more complex. Since $\overline{f} \neq f$, we have to deal with four possibly distinct decay modes, namely $\overline{B}^0 \to f$, $\overline{f}$ and $B^0 \to f$, $\overline{f}$. Yet they are of great practical value, as our subsequent discussion will show. In this chapter we will describe only one such example; others will be given later.

$$11.5.1 \quad B_s \to D_s^{\pm} K^{\mp}$$

The four channels

$$
\begin{aligned}
\overline{B}_s &\to D_s^+ K^-, D_s^- K^+ \\
B_s &\to D_s^- K^+, D_s^+ K^-
\end{aligned}
\tag{11.72}
$$

might allow us a relatively decent direct extraction of ϕ_3. They proceed through the diagrams shown in Fig. 11.9 and their **CP** conjugates.

In $\overline{B}_s \to D_s^+ K^-$ $[D_s^- K^+]$ driven by $b \to c(\overline{u}s)$ $[b \to u(\overline{c}s)]$ isospin changes by $I = (1/2, -1/2)$ $[I = (1/2, 1/2)]$; i.e., both transitions are described by a single isospin amplitude:

$$
\begin{aligned}
A(\overline{B}_s \to D_s^+ K^-) &= e^{i\delta_-} \mathbf{V}_{cb} \mathbf{V}_{us}^* \mathscr{A}_- \\
A(\overline{B}_s \to D_s^- K^+) &= e^{i\delta_+} \mathbf{V}_{ub} \mathbf{V}_{cs}^* \mathscr{A}_+ \\
A(B_s \to D_s^- K^+) &= e^{i\delta_-} \mathbf{V}_{cb}^* \mathbf{V}_{us} \mathscr{A}_- \\
A(B_s \to D_s^+ K^-) &= e^{i\delta_+} \mathbf{V}_{ub}^* \mathbf{V}_{cs} \mathscr{A}_+.
\end{aligned}
\tag{11.73}
$$

Therefore

$$|A(\overline{B}_s \to D_s^+ K^-)| = |A(B_s \to D_s^- K^+)|$$
$$|A(\overline{B}_s \to D_s^- K^+)| = |A(B_s \to D_s^+ K^-)|. \tag{11.74}$$

Yet $|A(\overline{B}_s \to D_s^+ K^-)| = |A(\overline{B}_s \to D_s^- K^+)|$ obviously does *not* hold as an identity. Setting $\Delta\Gamma_B = 0$ for simplicity we find

$$G_{D_s^+ K^-} = |A(D_s^+ K^-)|^2 \Big[1 + |\overline{\rho}(D_s^+ K^-)|^2$$

$$+ (1 - |\overline{\rho}(D_s^+ K^-)|^2)\cos\Delta M_B t - 2\mathrm{Im}\frac{q}{p}\overline{\rho}(D_s^+ K^-)\sin\Delta M_B t \Big] \tag{11.75}$$

$$\overline{G}_{D_s^- K^+} = |\overline{A}(D_s^- K^+)|^2 \Big[1 + |\rho(D_s^- K^+)|^2$$

$$+ (1 - |\rho(D_s^- K^+)|^2)\cos\Delta M_B t - 2\mathrm{Im}\left(\frac{p}{q}\rho(D_s^- K^+) \right)\sin\Delta M_B t \Big] \tag{11.76}$$

$$G_{D_s^- K^+} = |A(D_s^- K^+)|^2 \Big[1 + |\overline{\rho}(D_s^- K^+)|^2$$

$$+ (1 - |\overline{\rho}(D_s^- K^+)|^2)\cos\Delta M_B t - 2\mathrm{Im}\frac{q}{p}\overline{\rho}(D_s^- K^+)\sin\Delta M_B t \Big] \tag{11.77}$$

$$\overline{G}_{D_s^+ K^-} = |\overline{A}(D_s^+ K^-)|^2 \Big[1 + |\rho(D_s^+ K^-)|^2$$

$$+ (1 - |\rho(D_s^+ K^-)|^2)\cos\Delta M_B t - 2\mathrm{Im}\left(\frac{p}{q}\rho(D_s^+ K^-) \right)\sin\Delta M_B t \Big] \tag{11.78}$$

$$\overline{\rho}(D_s^+ K^-) = e^{i(\delta_- - \delta_+)}\frac{\mathbf{V}_{cb}\mathbf{V}_{us}^*}{\mathbf{V}_{ub}^*\mathbf{V}_{cs}}\frac{\mathscr{A}_-}{\mathscr{A}_+}$$

$$\simeq e^{i(\delta_- - \delta_+)}e^{-i\phi_3}\frac{1}{\mathscr{R}}\frac{\mathscr{A}_-}{\mathscr{A}_+}$$

$$\overline{\rho}(D_s^- K^+) = e^{-i(\delta_- - \delta_+)}\frac{\mathbf{V}_{ub}\mathbf{V}_{cs}^*}{\mathbf{V}_{cb}^*\mathbf{V}_{us}}\frac{\mathscr{A}_+}{\mathscr{A}_-}$$

$$\simeq e^{-i(\delta_- - \delta_+)}e^{-i\phi_3}\mathscr{R}\frac{\mathscr{A}_+}{\mathscr{A}_-}, \tag{11.79}$$

where $\mathscr{R} = \sqrt{\rho^2 + \eta^2}$. From the four observables $G_{D_s^\pm K^\mp}$, $\overline{G}_{D_s^\pm K^\mp}$ we can reliably extract the magnitudes and phases of $\overline{\rho}(D_s^+ K^-)$ and $\overline{\rho}(D_s^- k^+)$. This leads to the determination of ϕ_3 and $\delta_- - \delta_+$. It is curious to note that the **CP** asymmetries in these channels depend on $\sin\phi_3$ rather than $\sin 2\phi_3$ as in $B_s \to K_S \rho^0$; it is thus maximal for $\phi_3 = 90°$, whereas $\sin 2\phi_3$ would vanish then.

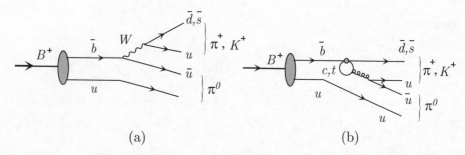

Fig. 11.10.   Quark diagrams for $B^+ \to K^+\pi^0$ and $B^+ \to \pi^+\pi^0$. (a) The tree graph contribution. (b) Penguin contribution.

11.6 Case D: no $B - \overline{B}$ oscillations

In the absence of oscillations the expressions simplify considerably:

$$\Gamma(B \to f) \quad \propto \quad e^{-\Gamma t}|A(f)|^2$$
$$\Gamma(\overline{B} \to \overline{f}) \quad \propto \quad e^{-\Gamma t}|\overline{A}(\overline{f})|^2. \tag{11.80}$$

We have dropped here the restriction to neutral B mesons since this case applies also to charged mesons and baryons. Since $G_f(t) = |A(f)|^2$, $\overline{G}_{\overline{f}}(t) = |\overline{A}(\overline{f})|^2$, obviously $dG_f(t)/dt = 0 = d\overline{G}_{\overline{f}}(t)/dt$ holds and an asymmetry, i.e., $|A(f)|^2 \neq |\overline{A}(\overline{f})|^2$, would represent *direct* **CP** violation.

11.6.1 $B^\pm \to K^\pm\pi^0$

A priori all the dynamical ingredients are there for a sizeable direct **CP** asymmetry to surface in these modes [133, 138].

- Two isospin amplitudes drive $B^\pm \to K^\pm\pi^0$, one with $\Delta I = 0$ and one with $\Delta I = 1$; their strong phase shifts have no reason to be the same.

- In addition to the spectator reaction there are penguin operators containing different CKM parameters.

Let us denote the spectator amplitude, Fig. 11.10(a), as tree amplitude and write

$$T(K\pi) = \mathbf{V}_{ub}\mathbf{V}_{us}^* T. \tag{11.81}$$

Penguin amplitudes, Fig. 11.10(b), with top and charm contribution can be written with their CKM factors:

$$P^c(K\pi) = \mathbf{V}_{cb}\mathbf{V}_{cs}^* P^c, \qquad P^t(K\pi) = \mathbf{V}_{tb}\mathbf{V}_{ts}^* P^t. \tag{11.82}$$

A **CP** asymmetry is thus proportional to $\sin\phi_3$.

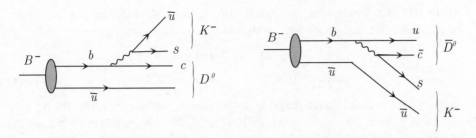

Fig. 11.11. Feynman diagrams for $B^- \to D^0 K^-$ and $B^- \to \overline{D}^0 K^-$.

- The $b \to u$ spectator contribution is CKM suppressed, making it roughly comparable to the penguin term. This enhances the interference between them and thus improves the prospects for a sizeable **CP** asymmetry.

- The penguin process with an internal charm quark provides at least a qualitative model for final state interactions.

Thus we may expect sizeable direct **CP** asymmetries to emerge in some of the $B \to K + \pi$ channels. A more careful evaluation is, however, needed to make these qualitative remarks more specific; this will be presented in the next chapter.

11.6.2 $B^- \to D^{\mathrm{neut}} + K^-$

Consider the two modes [134, 139, 140]

$$B^- \to D^0 K^- \ , \quad B^- \to \overline{D}^0 K^- \tag{11.83}$$

for which the quark level diagrams are shown in Fig. 11.11.

The transition amplitudes with the CKM parameters factored out are

$$
\begin{aligned}
A(B^- \to D^0 K^-) &= \mathbf{V}_{cb} \mathbf{V}^*_{us} e^{i\delta_D} \mathscr{A}_D \\
A(B^- \to \overline{D}^0 K^-) &= \mathbf{V}_{ub} \mathbf{V}^*_{cs} e^{i\delta_{\overline{D}}} \mathscr{A}_{\overline{D}}.
\end{aligned}
\tag{11.84}
$$

At first we might think that they are clearly distinct and they do not interfere. However, that is not necessarily the case. For the flavour identity of the neutral D meson has to reveal itself through its weak decay. Semileptonic transitions, being flavour-specific, tell us unambigously whether the decaying meson is a D^0 or $\overline{D}^0$. Yet final states that are **CP** eigenstates reveal nothing about the flavour of the D meson:

$$D^0 \to K^+ K^-, \pi^+ \pi^-, K_S \pi^0, K_S \rho^0, K_S \omega, K_S \eta \leftarrow \overline{D}^0. \tag{11.85}$$

Let us consider six types of channel:

$$
\begin{aligned}
B^- &\to D^0 K^-, \overline{D}^0 K^-, (D^0/\overline{D}^0) K^- \\
B^+ &\to \overline{D}^0 K^+, D^0 K^+, (D^0/\overline{D}^0) K^+,
\end{aligned}
\tag{11.86}
$$

where $(D^0/\overline{D}^0)$ means that the flavour of the neutral D meson has *not* been established. **CPT** invariance tells us that

$$|A(B^- \to D^0 K^-)| = |A(B^+ \to \overline{D}^0 K^+)|$$
$$|A(B^- \to \overline{D}^0 K^-)| = |A(B^+ \to D^0 K^+)|, \tag{11.87}$$

since a single quark level diagram drives these, see Eq. (11.84). In $B^\pm \to (D^0/\overline{D}^0)K^\pm$ both diagrams shown in Fig. 11.11 contribute, and they do it coherently. Their interference will generate a direct **CP** asymmetry that depends on $\sin(\delta_D - \delta_{\overline{D}})$ and $\sin\phi_3$:

$$\frac{|A(B^- \to (D^0/\overline{D}^0)K^-)|^2 - |A(B^+ \to (D^0/\overline{D}^0)K^+)|^2}{|A(B^- \to (D^0/\overline{D}^0)K^-)|^2 + |A(B^+ \to (D^0/\overline{D}^0)K^+)|^2}$$

$$= \frac{2|R|\sin(\delta_D - \delta_{\overline{D}})}{1 + |R|^2 + 2|R|\cos\phi_3\cos(\delta_D - \delta_{\overline{D}})} \cdot \sin\phi_3, \tag{11.88}$$

where $R = A(B^- \to \overline{D}^0 K^-)/A(B^- \to D^0 K^-)$. The exciting new element here is that the final state phase shift $(\delta_D - \delta_{\overline{D}})$ – instrumental for the asymmetry to materialize in the first place – can be extracted from the data. We shall come back to this point in the next chapter.

11.7 Résumé

The goal of this chapter was to illuminate how rich a **CP** phenomenology we can expect in the weak decays of beauty hadrons. While we can use the same classification as in kaon decays, the numbers and even the pattern are quite different, which is to a large degree due to $\Gamma_{K_L} \ll \Gamma_{K_S}$ vs $\Gamma_{B_L} \sim \Gamma_{B_S}$, $\Delta\Gamma_K \sim \Delta M_K$ vs $\Delta\Gamma_B \ll \Delta M_B$. The situation can be summarized as follows:

- The flavour-specific semileptonic B^0 decays unambiguously probe for **CP** violation in $B^0 - \overline{B}^0$ oscillations. The asymmetry expressed through $|q/p| \neq 1$ is small, namely 0.1 % or less. While in absolute terms this is not smaller than what happens in K_L decays, it is a poor measure of the strength of **CP** violation in B decays.

- We can summarize the landscape for nonleptonic transitions as shown in Table 11.1. The third column might require one comment: only if the final state f is a **CP** eigenstate, does $|\overline{\rho}_{B^0 \to f}| \equiv |A(\overline{B}^0 \to f)/A(B^0 \to f)| \neq 1$ necessarily constitute a direct **CP** asymmetry.

- The predictions for the asymmetries in $B_d \to \psi K_S$ and $B_s \to \psi\phi$ are done with *high parametric* reliability. This is however not true for those in the other decay classes: apart from stating that they can exhibit direct **CP** asymmetries, we have ignored here the impact of penguin operators

Table 11.1. KM predictions on **CP** asymmetries involving $B^0 - \overline{B}^0$ oscillations. Here $\mathscr{R} = \sqrt{\rho^2 + \eta^2}$. For the analysis of $B \rightarrow D^0 K_S$ and $B \rightarrow \overline{D}^0 K_S$ decays see Problem 11.9.

Quark level transition	Example of hadronic channel f	$|\bar{\rho}_{B^0 \rightarrow f}|$	KM prediction for **CP** asymmetry	**CP** parity
$b \rightarrow c\bar{c}s$	$B_d \rightarrow \psi K_S$	1	$\sin 2\phi_1 + \mathcal{O}(\lambda^2)$	odd
	$B_s \rightarrow \psi\phi$	1	$2\lambda^2\eta$	$\sim$ even
$b \rightarrow u\bar{u}d$	$B_d \rightarrow \pi^+\pi^-$	$\simeq 1$	$\sim \sin 2\phi_2$	even
	$B_s \rightarrow K_S\rho^0$	$\simeq 1$	$\sim \sin 2\phi_3$	odd
	$B_d \rightarrow \rho^\pm\pi^\mp$	$\mathcal{O}(1)$	$\sim \sin 2\phi_2$	none
$b \rightarrow c(\bar{u}s)$	$B_d \rightarrow D^0 K_S$	$\mathcal{O}(1)$	$\sim -\frac{1}{\mathscr{R}}\sin(2\phi_1 + \phi_3)$	none
$b \rightarrow u(\bar{c}s)$	$B_d \rightarrow \overline{D}^0 K_S$	$\mathcal{O}(1)$	$\sim -\mathscr{R}\sin(2\phi_1 + \phi_3)$	none
	$B_s \rightarrow D_s^+ K^-$	$\mathcal{O}(1)$	$\sim -\frac{1}{\mathscr{R}}\sin\phi_3$	none
	$B_s \rightarrow D_s^- K^+$	$\mathcal{O}(1)$	$\sim -\mathscr{R}\sin\phi_3$	none

that can muddle the extraction of ϕ_1 and ϕ_2. This is merely indicated by the symbol $\sim$.

- While in these channels penguin operators are thus a nuisance we could easily do without, they become an essential agent in other modes, for they can provide the second weak amplitude required for a direct **CP** asymmetry to become observable. Of course they remain annoying in the sense that their evaluation is beset with considerable theoretical uncertainties.

In this chapter we have presented a painting in rather broad (theoretical) brush strokes. In the next chapter we will address more precisely various elements of our theoretical treatment, their subtleties and uncertainties and how to deal with them.

Problems

11.1 We have repeatedly emphasized that strong final state interaction phases are important in exhibiting a **CP** asymmetry in the width. At the same time, we have insisted that the asymmetry in $B \rightarrow \psi K_S$ can be expected entirely with CKM parameters. Resolve the apparent paradox.

11.2 What do we mean by ϵ_B and is this phase convention dependent?

11.3 Show that $|q/p| \simeq 1$ for $\Delta M \gg \Delta\Gamma$. Under what condition can we write

$$\text{Im}\left(\frac{q}{p}\bar{\rho}\right) = -\text{Im}\left(\frac{p}{q}\rho\right)? \tag{11.89}$$

11.4   What is the relationship between $\phi_{\psi K_S}$ and $\phi_{\pi\pi}$ if the phase enters only through q/p? This is a case for the superweak theory.

11.5   Consider $B \to f_i$ decay where f_i is a **CP** eigenstate, and assume that there is only one amplitude contributing. Define

$$\rho(f) = \frac{\langle f|\mathcal{H}|B\rangle}{\langle \overline{f}|\mathcal{H}|\overline{B}\rangle}. \tag{11.90}$$

If **CP** is conserved, $\rho(f) = $ **CP** eigenvalue of f. Now if **CP** is not conserved, $\mathcal{H}$ contains a complex phase. If decays to f_1 and f_2 are both caused by the same Feynman graph, show that $\rho(f_1) = \pm\rho(f_2)$ where the $\pm$ sign is given by the relative **CP** eigenvalue.

11.6   Show that

$$\operatorname{Im}\frac{q}{p}\overline{\rho}(\psi K_S) = \sin 2\phi_1$$

$$\operatorname{Im}\frac{q}{p}\overline{\rho}(\pi\pi) = \sin 2\phi_2. \tag{11.91}$$

11.7   Show that **CP** symmetry implies:

$$\langle \psi K_S|\overline{b}_L\gamma_\mu c_L\overline{c}_L\gamma^\mu s_L|B\rangle = -\langle \psi K_S|(\overline{b}_L\gamma_\mu c_L\overline{c}_L\gamma^\mu s_L)^\dagger|\overline{B}\rangle$$

$$\langle \pi\pi|\overline{b}_L\gamma_\mu c_L\overline{c}_L\gamma^\mu s_L|B\rangle = \langle \pi\pi|(\overline{b}_L\gamma_\mu c_L\overline{c}_L\gamma^\mu s_L)^\dagger|\overline{B}\rangle. \tag{11.92}$$

11.8   Show Eq. (11.54) and Eq. (11.64) by going back to the definition of ϕ_i given in Eq. (10.24).

11.9   Consider $B \to D^0 K_S$ and $B \to \overline{D}^0 K_S$ decays with strong final state interaction phases δ_{DK_S} and $\delta_{\overline{D}K_S}$, respectively. Show that

$$\rho(D^0 K_S) = e^{i(\delta_{\overline{D}K_S} - \delta_{DK_S})}\frac{\mathcal{A}_{\overline{D}}}{\mathcal{A}_D}e^{i\phi_3}$$

$$\rho(\overline{D}^0 K_S) = e^{-i(\delta_{\overline{D}K_S} - \delta_{DK_S})}\frac{\mathcal{A}_D}{\mathcal{A}_{\overline{D}}}e^{i\phi_3}. \tag{11.93}$$

Can we use these to extract ϕ_3?

12

♠ The nitty-gritty of B physics ♠

Per ardua
Ad astra

We have warned the reader repeatedly that the treatment given in the preceding chapter was oversimplified and incomplete in several respects:

- Our discussion of **CP** asymmetries in the context of $B^0 - \overline{B}^0$ oscillations assumed the presence of a *single* weak transition operator leading to $|\overline{\rho}(f)| = 1$. However, most nonleptonic transitions are driven by more than one operator with different combinations of weak parameters. The quantity $(q/p)\overline{\rho}(f)$ is then not a unit vector in the complex plane and – even more importantly – its phase is no longer purely that of a product of CKM parameters:

$$\mathrm{Im}\frac{q}{p}\overline{\rho}(f) \neq \sin 2\phi_i. \tag{12.1}$$

  We have to ask how these complications can be unfolded and the angles extracted from observed asymmetries in a reliable way.

- The intervention of a second weak transition operator (incorporating final state interactions as well) is essential – rather than a nuisance – for *direct* **CP** asymmetries to emerge. To relate measurements with CKM parameters again requires us to come to grips with these dynamical features in a quantitative way.

- Penguin operators figure prominently in these contexts. Therefore they require and deserve a more careful and detailed treatment than was just given.

- Final states like $\pi^{\pm}\rho^{\mp}$ or $\rho^+\rho^-$ are common to B_d and $\overline{B}_d$ decays without being **CP** eigenstates (or not pure ones). Yet they will play an important role in extracting ϕ_2.

- The question of how summing over nonleptonic decay channels will affect a **CP** asymmetry is one of obvious practical value needing attention.

This chapter will address these issues.

12.1 Pollution from water fowls and others

Consider two operators, differing in their CKM parameters, driving $B \to f$:

$$A(B \to f) = e^{i\xi_1} e^{i\delta_1} |\mathscr{A}_1| + e^{i\xi_2} e^{i\delta_2} |\mathscr{A}_2|, \qquad (12.2)$$

where δ_i and ξ_i are the strong interaction and weak phases, respectively; the moduli of the CKM parameters have been incorporated into $|\mathscr{A}_i|$. We then find

$$\bar{\rho}(f) = e^{-2i\xi_1} \frac{1 + e^{-i\Delta\xi} e^{i\Delta\delta} \left|\frac{\mathscr{A}_2}{\mathscr{A}_1}\right|}{1 + e^{i\Delta\xi} e^{i\Delta\delta} \left|\frac{\mathscr{A}_2}{\mathscr{A}_1}\right|} \quad ; \quad \Delta\xi = \xi_2 - \xi_1, \, \Delta\delta = \delta_2 - \delta_1, \qquad (12.3)$$

leading to

$$\bar{\rho}(f) \sim e^{-2i\xi_1} \cdot \left(1 - 2i \left|\frac{\mathscr{A}_2}{\mathscr{A}_1}\right| e^{i\Delta\delta} \sin\Delta\xi\right), \qquad (12.4)$$

where the last approximation holds for $|\mathscr{A}_2/\mathscr{A}_1| \ll 1$. $|\bar{\rho}(f)| \neq 1$ is an unambiguous sign for the presence of a second weak operator; yet $|\bar{\rho}(f)| = 1$ does not rule against it since it follows, for example, for $\Delta\delta = 0$ (or $\Delta\xi = 0$). From Eq. (12.4) we obtain [141, 142]:

$$\mathrm{Im}\frac{q}{p}\bar{\rho}(f) \sim \sin 2(\Phi_m - \xi_1) - 2\left|\frac{\mathscr{A}_2}{\mathscr{A}_1}\right| \sin\Delta\xi \cos(2\Phi_m - 2\xi_1 + \Delta\delta), \qquad (12.5)$$

where $2\Phi_m = \arg\left(\frac{q}{p}\right)$. This situation is often referred to as penguin pollution since the penguin mechanism typically generates such an operator. Yet we find it unfair to single out penguins in such a negative way. It would be more appropriate to state that they enrich the phenomenology even when making it more complex. The presence of a second weak operator poses a challenge to our ability to extract weak parameters from the data. How this can best be achieved depends on the specifics of the channel under study.

One more general remark is in order: we have so far employed the terms *tree* and *penguin* operators in a somewhat unreflected way, guided by simple Feynman diagrams. Yet we should keep in mind that if two operators contribute to the same final state they obviously mix under QCD renormalization; their relative weight depends on the scale at which the operators are evaluated (which in turn is compensated by the scale dependence of their matrix elements). *They should then be distinguished by their dependence on the CKM parameters, together with their colour and chiral structure.*

Before we go on to discuss the effects of pollution, we need to fix our notation for the $\Delta B = 1$ weak Hamiltonian, because at the energy scale comparable to M_B, the charm degree of freedom is not frozen. So, we have four types of operator: $(\bar{b}c)_{V-A}(\bar{c}q')_{V-A}$, $(\bar{b}u)_{V-A}(\bar{u}q')_{V-A}$, where $q' = s, d$. Let us define four-Fermi operators following similar notation as in Eq. (9.15) which defined $O_1 - O_{10}$.

$$O^q_{q'1} = (\bar{b}q')_{V-A}(\bar{q}q)_{V-A}, \qquad O^q_{q'2} = (\bar{b}q)_{V-A}(\bar{q}q')_{V-A},$$

$$O_{q'3} = (\bar{b}q')_{V-A}\sum_q (\bar{q}q)_{V-A}, \qquad O_{q'4} = (\bar{b}_\alpha q'_\beta)_{V-A}\sum_q (\bar{q}_\beta q_\alpha)_{V-A},$$

$$O_{q'5} = (\bar{b}q')_{V-A}\sum_q (\bar{q}q)_{V+A}, \qquad O_{q'6} = (\bar{b}_\alpha q'_\beta)_{V-A}\sum_q (\bar{q}_\beta q_\alpha)_{V+A},$$

$$O_{q'7} = \tfrac{3}{2}(\bar{b}q')_{V-A}\sum_q e_q(\bar{q}q)_{V+A}, \quad O_{q'8} = \tfrac{3}{2}(\bar{b}_\alpha q'_\beta)_{V-A}\sum_q e_q(\bar{q}_\beta q_\alpha)_{V+A},$$

$$O_{q'9} = \tfrac{3}{2}(\bar{b}q')_{V-A}\sum_q e_q(\bar{q}q)_{V-A}, \quad O_{q'10} = \tfrac{3}{2}(\bar{b}_\alpha q'_\beta)_{V-A}\sum_q e_q(\bar{q}_\beta q_\alpha)_{V-A},$$

where α and β are colour indicies.

The $\Delta B = 1$ Hamiltonian is then given by

$$
\begin{aligned}
\mathcal{H}(\Delta B = 1) \;=\; & \frac{G_F}{\sqrt{2}}(\mathbf{V}^*_{cb}\mathbf{V}_{cq'}[C_1(\mu)O^c_{q'1} + C_2(\mu)O^c_{q'2}] \\
+\; & \mathbf{V}^*_{ub}\mathbf{V}_{uq'}[C_1(\mu)O^u_{q'1} + C_2(\mu)O^u_{q'2}] \\
-\; & \mathbf{V}^*_{tb}\mathbf{V}_{tq'}\sum_{i=3}^{10} C_i(\mu)O_{q'i} + \text{h.c.}
\end{aligned}
\tag{12.6}
$$

Finally, we shall denote hadronic matrix elements as

$$Q^q_{q'i}(F) = \langle F|O^q_{q'i}|B\rangle. \tag{12.7}$$

12.1.1 How big are penguins?

Recently the CLEO collaboration has established [143] two-prong B decays: $B \to h^+h^-$. More specifically, their findings can be summarized as follows:

$$\text{Br}(B \to K\bar{K}) < \text{Br}(B \to \pi\pi) \le \text{Br}(B \to K\pi) \simeq (1 \sim 2) \times 10^{-5}. \tag{12.8}$$

These decays are generated by the Feynman graphs shown in Fig. 11.10.

The amplitudes for tree and penguin contributions for $K\pi$ decay mode are:

$$
\begin{aligned}
T(K\pi) \;=\; & \frac{G_F}{\sqrt{2}}\mathbf{V}^*_{ub}\mathbf{V}_{us}[C_1(\mu)Q^u_{s1}(K\pi) + C_2(\mu)Q^u_{s2}(K\pi)] \\
P(K\pi)_c \;=\; & \frac{G_F}{\sqrt{2}}\mathbf{V}^*_{cb}\mathbf{V}_{cs}[C_1(\mu)Q^c_{s1}(K\pi) + C_2(\mu)Q^c_{s2}(K\pi)] \\
P(K\pi)_t \;=\; & \frac{G_F}{\sqrt{2}}(-\mathbf{V}^*_{tb}\mathbf{V}_{ts})\sum_{i=3}^{10} C_i(\mu)Q_{si}(K\pi).
\end{aligned}
\tag{12.9}
$$

For $B \to K\pi$, $P(K\pi)$ is $\mathcal{O}(\lambda^2)$ and $T(K\pi)$ is $\mathcal{O}(\lambda^4)$. For $B \to \pi\pi$, these diagrams give $T(\pi\pi) = \lambda^3 T$, $P(\pi\pi)_t = \lambda^3 P_t$, and $P(\pi\pi)_c = \lambda^3 P_c$. Since $T(K\pi)/T(\pi\pi) \sim \lambda$, if the tree graph matrix elements dominate, we would expect $\text{Br}(B \to K\pi)/\text{Br}(B \to$

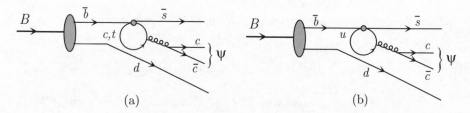

Fig. 12.1. Penguin graphs for $B \to \psi K_S$ decay. (a) has the same weak phase as the tree graph. Thus this diagram does not change the asymmetry. (b) introduces some theoretical uncertainties, but this contribution is expected to be less than 1%.

$\pi\pi) \sim \mathcal{O}(\lambda^2)$. Experimentally this is not so. This indicates that the $P(K\pi)$ amplitude is at least as large as the $T(\pi\pi)$. If $P(K\pi)_{c,t} \simeq T(\pi\pi)$, this suggests

$$\frac{[C_1(\mu)Q^c_{s1}(K\pi) + C_2(\mu)Q^c_{s2}(K\pi)]}{[C_1(\mu)Q^u_{s1}(\pi\pi) + C_2(\mu)Q^u_{s2}(\pi\pi)]} = \mathcal{O}(\lambda), \qquad (12.10)$$

and similar relation for $P(K\pi)_t$, i.e., the penguin contribution is considerably larger than what a naive estimate of the loop graph would suggest:

$$\frac{[C_1(\mu)Q^c_{s1}(K\pi) + C_2(\mu)Q^c_{s2}(K\pi) + \sum_{i=3}^{10} C_i(\mu)Q_{si}(K\pi)]}{[C_1(\mu)O^u_{s1}(\pi\pi) + C_2(\mu)Q^u_{s2}(\pi\pi)]}$$

$$\sim \frac{\alpha_S}{12\pi^3}\log\frac{m_t}{m_c} \sim \mathcal{O}(0.01). \qquad (12.11)$$

Since we now have evidence that loop graphs compete with tree graphs, we have to be prepared for a substantially more complex situation.

12.1.2 $B_d \to \psi K_S$

In addition to the diagram shown in Fig. 11.5, the decay $B \to \psi K_S$ gets contributions from the penguin diagrams shown in Fig. 12.1. The resulting operators are

$$T(\psi K_S) = \frac{G_F}{\sqrt{2}}V^*_{cb}V_{cs}[C_1(\mu)Q^c_{s1}(\psi K_S)) + C_2(\mu)Q^c_{s2}(\psi K_S)]$$

$$P_u(\psi K_S) = \frac{G_F}{\sqrt{2}}V^*_{ub}V_{us}[C_1(\mu)Q^u_{s1}(\psi K_S) + C_2(\mu)Q^u_{s2}(\psi K_S)]$$

$$P_t(\psi K_S) = \frac{G_F}{\sqrt{2}}(-V^*_{tb}V_{ts})\sum_{i=3}^{10} C_i(\mu)Q_{si}(\psi K_S), \qquad (12.12)$$

where the letters $T(\psi K_S)$, $P_u(\psi K_S)$ and $P_t(\psi K_S)$ obviously refer to the diagrammatic origin of these operators.

First note that the weak phases of $T(\psi K_S)$ and $P_t(\psi K_S)$ are the same to leading order in λ. So, we see from Eq. (12.5) that the presence of $P_t(\psi K_S)$ does not change the asymmetry. It is different for $P_u(\psi K_S)$ – yet that operator is considerably suppressed by CKM parameters and the coefficient:

$$\left| \frac{P_u(\psi K_S)}{T(\psi K_S)} \right| \sim \mathcal{O}\left(\lambda^2 \cdot \frac{Q^u_{s1,s2}(\psi K_S)}{Q^c_{s1,s2}(\psi K_S)} \right) \leq 1\%, \tag{12.13}$$

hence the influence on our theoretical prediction of the penguins shown in Fig. 12.1, while it may be large in magnitude, is limited to less than 1%.

$$\left| \text{Im} \frac{q}{p} \overline{\rho}(\psi K_S) - \sin 2\phi_1 \right| < 1\%. \tag{12.14}$$

12.1.3 $B_d \to \pi\pi$

From the diagrams of Fig. 11.6 we infer the presence of the following transition operators[1]:

$$T(\pi\pi) = \frac{G_F}{\sqrt{2}} V^*_{ub} V_{ud} [C_1(\mu) Q^u_{d1}(\pi\pi) + C_2(\mu) Q^u_{d2}(\pi\pi)]$$

$$P_c(\pi\pi) = \frac{G_F}{\sqrt{2}} V^*_{cb} V_{cd} [C_1(\mu) Q^c_{d1}(\pi\pi) + C_2(\mu) Q^c_{d2}(\pi\pi)]$$

$$P_t(\pi\pi) = \frac{G_F}{\sqrt{2}} (-V^*_{tb} V_{td}) \sum_{i=3}^{10} C_i(\mu) Q_{di}(\pi\pi). \tag{12.15}$$

The qualitatively new element is that now $T(\pi\pi)$, $P_c(\pi\pi)$ and $P_t(\pi\pi)$ are of roughly comparable strength in terms of their CKM parameters:

$$|V^*_{ub} V_{ud}| \sim |V^*_{cb} V_{cd}| \sim |V^*_{tb} V_{td}| \sim \mathcal{O}(\lambda^3), \tag{12.16}$$

yet with different weak phases. The contributions from the $P_{t,c}(\pi\pi)$ amplitudes are expected to be reduced relative to those from $T(\pi\pi)$, due to either their small coefficients reflecting them being a pure quantum effect, or suppressed matrix elements:

$$|C_1(\mu) Q^c_{d1}(\pi\pi) + C_2(\mu) Q^c_{d2}(\pi\pi)| < |C_1(\mu) Q^u_{d1}(\pi\pi) + C_2(\mu) Q^u_{d2}(\pi\pi)|. \tag{12.17}$$

Using Eq. (11.61), the effect of direct **CP** violation – $|\overline{\rho}(\pi^+\pi^-)| \neq 1$ – can unambiguously be separated from that involving $B_d - \overline{B}_d$ oscillations, but we cannot immediately extract $\sin 2\phi_2$. How big is the deviation of $\text{Im} \frac{q}{p} \overline{\rho}(\pi^+\pi^-)$ from $\sin 2\phi_2$? Let's write:

$$\text{Im} \frac{q}{p} \overline{\rho}(\pi^+\pi^-) = \sin 2\phi_2 + \Delta_{\pi\pi}. \tag{12.18}$$

For purpose of illustration, we set $|\mathcal{A}_2/\mathcal{A}_1| \sim \frac{P(\pi\pi)_c}{T(\pi\pi)} \sim \lambda$ and $\sin \Delta\delta = 1$ in Eq. (12.4). The difference $\Delta_{\pi\pi}$ can be quite sizeable, as illustrated in Fig. 12.2, and such a $\Delta_{\pi\pi}$ is not acceptable as an *uncertainty*!

---

[1] As before it should be noted that while the penguin diagram with an internal *top* quark generates a *local* operator, the other two do not.

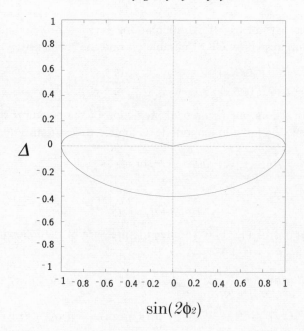

Δ

$$\sin(2\phi_2)$$

Fig. 12.2. $\Delta_{\pi\pi}$ as a function of $\sin 2\phi_2$ with $\sin \Delta\delta = 1$.

12.1.4 $B_s \to K_S \rho^0$

Comparing Fig. 11.7 and Fig. 11.6, we see that this decay differs from $B \to \pi\pi$ by a spectator quark. The analysis therefore proceeds in complete analogy to that just described for $B \to \pi\pi$.

12.2 Overcoming pollution

Nothing in our preceding discussion diminishes the firm general expectation of truly large **CP** asymmetries in B decays. What is unsatisfactory, though, is the lack of parametric reliability, namely the considerable uncertainty in relating the observable **CP** asymmetry in, say, $B_d \to \pi\pi$ to the microscopic quantities, namely the CKM parameters. While our understanding of nonleptonic B decays will certainly improve, we think an ab initio theoretical calculation of the relevant transition amplitudes will not become available soon. Instead we have to harness additional experimental information in a judicious way. The concrete procedure depends on the specifics of the mode and will involve a learning curve. It is therefore not our intent to discuss it in exhaustive detail; instead we want to outline important features of possible strategies. An essential element is employing isospin and – to a lesser degree – $SU(3)_{Fl}$ relations [138, 144, 146]. We will focus on the task of extracting ϕ_2 from $B_d \to \pi's$; application of these ideas to determining ϕ_3 from $B_s \to K\pi's$ is quite analogous, though probably much more difficult quantitatively.

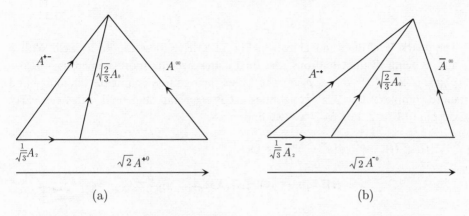

Fig. 12.3. (a) Isospin relations between $B^+ \to \pi^+\pi^0$, $B^0 \to \pi^+\pi^-$, $B^0 \to \pi^0\pi^0$ modes. We see that both the magnitudes of A_0, A_2 and the relative phase between these two amplitudes can be obtained. (b) Isospin relations between the charge conjugate decays.

12.2.1 $B \to \pi\pi$

There are six channels, namely $B^+ \to \pi^+\pi^0$, $B^0 \to \pi^+\pi^-$, $\pi^0\pi^0$ plus their charge conjugate ones. The isospin decomposition of these amplitudes is exactly the same as that of $K \to \pi\pi$ decays given in Eq. (7.23). The only difference is that we can no longer use Watson's theorem to relate the final state interaction phase to the $\pi\pi$ phase shift. We shall absorb the final state interaction phase in A_i. With two isospin amplitudes, A_0 and A_2, and three measurable rates for B mesons (and likewise for $\overline{B}$), there is a constraint among the latter [147]

$$A^{+-} = A^{00} + \sqrt{2}A^{+0}. \tag{12.19}$$

Now we can write

$$\rho(\pi^+\pi^-) = \frac{A(\pi^+\pi^-)}{\overline{A}(\pi^+\pi^-)} = \frac{A_2}{\overline{A}_2}\frac{1+z}{1+\overline{z}}, \tag{12.20}$$

where $z = \sqrt{2}A_0/A_2$, $\overline{z} = \sqrt{2}\overline{A}_0/\overline{A}_2$ and $A(\overline{B} \to \pi\pi) \equiv \overline{A}(\pi\pi)$. Since the $P_{c,t}(\pi\pi)$ operators for $b \to d$ can generate $\Delta I = 1/2$ modes only, the $I = 2$ amplitude is given by the T operator alone, and therefore:

$$\mathrm{Im}\frac{q}{p}\overline{\rho}(\pi^+\pi^-) = \left|\frac{1+\overline{z}}{1+z}\right| \sin\left[2\phi_2 + \arg\left(\frac{1+\overline{z}}{1+z}\right)\right]. \tag{12.21}$$

Fig. 12.3(a) and (b) illustrate how we can obtain z and $\overline{z}$ from branching ratios. We can thus extract $\sin 2\phi_2$. Theoretically, it is a clean method; its drawback is experimental, namely the difficulty of measuring tiny $\mathrm{Br}(B \to \pi^0\pi^0)$ well. Information on the different amplitudes can be gleaned also from $B \to K\pi$ invoking $SU(3)_{Fl}$ relations.

12.2.2 $B \to \pi\rho$

The quark level diagrams shown in Fig. 11.6 drive these modes as well. With ρ and π being distinct hadrons, the final states become more numerous – $B_d \to \rho^+\pi^-,\ \rho^-\pi^+,\ \rho^0\pi^0,\ B^- \to \rho^-\pi^0,\ \rho^0\pi^-$ plus their charge conjugate versions – and more complex: only $\rho^0\pi^0$ constitutes a **CP** eigenstate and final states can carry isospin 0, 1 or 2. For $\Delta\Gamma_B \simeq 0$ we find

$$
G_{\pi^+\rho^-}(t) = |A(\pi^+\rho^-)|^2 \Big[1 + |\bar{\rho}(\pi^+\rho^-)|^2 +
$$
$$
+ (1 - |\bar{\rho}(\pi^+\rho^-)|^2)\cos\Delta M_B t - 2\mathrm{Im}\frac{q}{p}\bar{\rho}(\pi^+\rho^-)\sin\Delta M_B t \Big]
$$

$$
\overline{G}_{\pi^-\rho^+}(t) = |\overline{A}(\pi^-\rho^+)|^2 \Big[1 + |\rho(\pi^-\rho^+)|^2 +
$$
$$
+ (1 - |\rho(\pi^-\rho^+)|^2)\cos\Delta M_B t - 2\mathrm{Im}\frac{p}{q}\rho(\pi^-\rho^+)\sin\Delta M_B t \Big]
$$

$$
G_{\pi^-\rho^+}(t) = |A(\pi^-\rho^+)|^2 \Big[1 + |\bar{\rho}(\pi^-\rho^+)|^2
$$
$$
+ (1 - |\bar{\rho}(\pi^-\rho^+)|^2)\cos\Delta M_B t - 2\mathrm{Im}\frac{q}{p}\bar{\rho}(\pi^-\rho^+)\sin\Delta M_B t \Big]
$$

$$
\overline{G}_{\pi^+\rho^-}(t) = |\overline{A}(\pi^+\rho^-)|^2 \Big[1 + |\rho(\pi^+\rho^-)|^2
$$
$$
+ (1 - |\rho(\pi^+\rho^-)|^2)\cos\Delta M_B t - 2\mathrm{Im}\frac{p}{q}\rho(\pi^+\rho^-)\sin\Delta M_B t \Big]
$$

with, in general, $|\bar{\rho}(\pi^+\rho^-)| \neq 1 \neq |\bar{\rho}(\pi^-\rho^+)|$. There are two reasons why we look at these transitions, nevertheless:

- The branching ratios for the $B \to \pi\rho$ modes might be significantly larger and easier to measure than for $B \to \pi\pi$.

- The CKM parameters are of course the same for $B \to \pi\pi$ and $B \to \rho\pi$; analysing $B \to \pi\rho$, including their time evolution, can then provide us with important information on the impact of the strong forces.

Details can be found in Refs. [148, 149].

12.2.3 $B^0 \to \psi K_S^*$

With a branching ratio for $B \to \psi K_S$ as small as 5×10^{-4}, we can't just sit back and wait for this decay to happen. We must look for other decay modes which can help in measuring a **CP** asymmetry in B decays [150, 151, 152]. $B \to \psi K^*$ might be useful along this direction. Note that this decay has a branching ratio

$$
Br(B \to \psi K^*) = (1.58 \pm 0.27) \times 10^{-3}. \tag{12.22}
$$

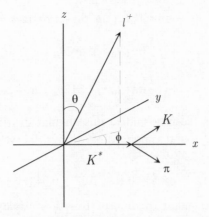

Fig. 12.4.   Let ψ be at rest. The x-axis is defined by the K^* momentum and the x–y plane is defined by the momentum of K and π in the K^* decay; θ and ϕ are defined by the momentum of l^+ in $\psi \to l^+l^-$.

Let us denote by K_S^* those K^* decaying to $K_S\pi^0$. In $B \to \psi K_S^*$ we encounter a qualitatively new feature. Amplitudes for $B \to PP, PV$, with P and V denoting a pseudoscalar and vector meson, respectively, are described by a *single* number since only one partial wave contributes: $l = 0$ [1] for $B \to PP$ [PV]. $B \to VV$, on the other hand, is made up by $l = 0, 1, 2$ configurations; i.e., its amplitudes contain more dynamical information than can be derived from their widths.

While this feature can profitably be exploited in searches for *direct* **CP** violation, it acts as a nuisance here; for it means that $B_d \to \psi K^*$ is not a **CP** eigenstate. To be more precise: $(\psi K_S^*)_{J=0;l=0,2}$ is **CP** even while $(\psi K_S^*)_{J=0;l=1}$ is **CP** odd. This implies that summing over l even and odd combinations will tend to wash out the total asymmetry. Fortunately, final states with different **CP** quantum numbers can be separated by measuring angular distributions. Define θ and ϕ as in Fig. 12.4, and let θ_K be the direction of K relative to $-\vec{p}_\psi$ in the rest frame of K^*. Denote

$$A_0 = -\sqrt{\frac{1}{3}}S + \sqrt{\frac{2}{3}}D; \qquad A_\| = \sqrt{\frac{2}{3}}S + \sqrt{\frac{1}{3}}D; \qquad A_\perp = P, \qquad (12.23)$$

where S, D and P denote the orbital angular momentum of the ψK^* state with $l = 0, 2$ and 1 respectively. So, $A_\|$ and A_0 are **CP** even amplitudes and $A_\perp$ is a **CP** odd amplitude.

In the ψ rest frame defined by Fig. 12.4, the projection of the ψ spin along the z-direction directly reveals the intrinsic parity of the ψK^* system. The angular distribution is given by [153]:

$$\frac{d\Gamma}{d\cos\theta\, d\cos\theta_K\, d\phi} = \frac{9}{32\pi}\left[2|A_0|^2\cos^2\theta_K(1-\sin^2\theta\cos^2\phi)\right.$$
$$\left.+|A_\||^2\sin^2\theta_K(1-\sin^2\theta\sin^2\phi)+|A_\perp|^2\sin^2\theta_K\sin^2\theta\right.$$

$$-\mathrm{Im}\,(A_\parallel^* A_\perp)\sin^2\theta_K\sin 2\theta\sin\phi$$

$$+\sqrt{\frac{1}{2}}\mathrm{Re}\,(A_0^* A_\parallel)\sin 2\theta_K\sin^2\theta\sin 2\phi$$

$$+\sqrt{\frac{1}{2}}\mathrm{Im}\,(A_0^* A_\perp)\sin 2\theta_K\sin 2\theta\cos\phi\Big]. \qquad (12.24)$$

The CLEO group has recently studied this angular distribution for $B^0\to\psi K^*$ and $B^+\to\psi K^{*+}$ and found, among other things, the very important fact that [154]

$$|P|^2 = 0.16\pm 0.08\pm 0.04, \qquad (12.25)$$

which implies that the major fraction of $B\to\psi K^*$ decay goes via a **CP** even final state. As more data accumulates, this channel should prove itself to be very useful for **CP** violation studies of B decays.

As the book is going to press, CDF collaboration has measured these amplitudes [155]:

$$
\begin{aligned}
A_0 &= 0.07\pm 0.039\pm 0.012 \\
A_\parallel &= (0.530\pm 0.106\pm 0.034)e^{i(2.16\pm 0.46\pm 0.10)} \\
A_\perp &= (0.355\pm 0.156\pm 0.039)e^{i(-0.56\pm 0.53\pm 0.12)} \\
|P|^2 &= 0.126{}^{+0.121}_{-0.093}\pm 0.028.
\end{aligned}
\qquad (12.26)
$$

Again, these numbers imply that $B\to\psi K^*$ decay is useful for **CP** studies.

12.3 The ϕ_3 saga

Measuring ϕ_3 directly and comparing it with the value inferred from CKM trigonometry represents a high sensitivity probe for the intervention of New Physics. Many observables depend on ϕ_3 – and, alas, on other and ill-known quantities as well, making an accurate determination of ϕ_3 a formidable task theoretically as well as experimentally. Here we want to address some of the nitty-gritty details in such endeavours.

12.3.1 $B^\pm\to D^{\mathrm{neut}}K^\pm$, revisited

In the preceding chapter we have outlined the basic strategy [139, 140]. We have four hadronic matrix elements containing weak and strong phases to contend with:

$$
\begin{aligned}
A(D^0 K^-) &= |\mathscr{A}_D|\,e^{i[\arg(\mathbf{V}_{cb}\mathbf{V}_{us}^*)+\delta_D]} \\
A(\overline{D}^0 K^-) &= |\mathscr{A}_{\overline{D}}|\,e^{i[\arg(\mathbf{V}_{ub}\mathbf{V}_{cs}^*)+\delta_{\overline{D}}]}
\end{aligned}
\qquad (12.27)
$$

$$
\begin{aligned}
\overline{A}(\overline{D}^0 K^+) &= |\mathscr{A}_D|\,e^{i[\arg(\mathbf{V}_{cb}^*\mathbf{V}_{us})+\delta_D]} \\
\overline{A}(D^0 K^+) &= |\mathscr{A}_{\overline{D}}|\,e^{i[\arg(\mathbf{V}_{ub}^*\mathbf{V}_{cs})+\delta_{\overline{D}}]}
\end{aligned}
\qquad (12.28)
$$

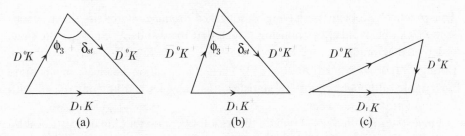

Fig. 12.5. (a) Triangle representing the relationship between $A(D_1K^-)$, $A(D^0K^-)$ and $A(\overline{D}^0K^-)$. (b) The triangle representing the charged conjugate version. (c) A more realistic triangle (spear). Here we have denoted $\delta_{st} = \delta_{\overline{D}} - \delta_D$.

If the neutral D meson decays into a **CP** eigenstate, its flavour remains completely undefined and the transition amplitude is given by

$$A(D_{1,2}K^-) = \frac{1}{\sqrt{2}} \left[A(D^0K^-) \pm A(\overline{D}^0K^-) \right], \tag{12.29}$$

which defines a triangle in the complex plane [139] as shown in Fig. 12.5 (a). Measuring the widths for $B^- \to D^0K^-$, $\overline{D}^0K^-$ and $D_{1,2}K^-$ yields the sides of the triangle and hence allows us to determine the phases. Likewise for the triangle formed by the **CP** conjugate amplitudes, see Fig. 12.5(b):

$$\angle[A(D^0K^-), A(\overline{D}^0K^-)] \equiv |-\phi_3 + \delta_{\overline{D}} - \delta_D|$$

$$\angle[\overline{A}(\overline{D}^0K^+), \overline{A}(D^0K^+)] \equiv |+\phi_3 + \delta_{\overline{D}} - \delta_D|. \tag{12.30}$$

Comparing those two angles allows us to deduce the size of ϕ_3 – up to a binary ambiguity. We should note that ϕ_3 can be determined this way even if $\delta_{\overline{D}} - \delta_D = 0$! There is, however, a major practical drawback to this method: we expect Br($B^- \to \overline{D}^0K^-$) to be tiny since it represents a colour suppressed mode, see Fig. 12.5.

$$\frac{\text{Br}(B^- \to \overline{D}^0K^-)}{\text{Br}(B^- \to D^0K^-)} \simeq \left| \frac{V_{ub}V_{cs}^*}{V_{cb}V_{us}^*} \right|^2 \left| \frac{a_2}{a_1} \right|^2 \sim \mathcal{O}(0.01), \tag{12.31}$$

where a_2/a_1 is the usual phenomenological colour suppression factor. The triangle under discussion will be quite squashed, with one of its sides being a mere $\sim 10\%$ of the others. The resulting error on ϕ_3 may be too large for it to be useful. It is possible that we are too pessimistic here; yet we had better have an alternative plan.

12.3.2 Using doubly Cabibbo suppressed decays

Consider again the overall reaction chain [156]:

$$B^- \to \begin{pmatrix} K^-D^0 \\ K^-\overline{D}^0 \end{pmatrix} \to K^-f. \tag{12.32}$$

The practical problem of having a squashed triangle, stated in the previous section, can be avoided by choosing f such that the two decay rates are the same order of magnitude. i.e., we compensate $|A(B^- \to D^0 K^-)| \gg |A(B^- \to \overline{D}^0 K^-)|$ through $|A(D^0 \to f)| \ll |A(\overline{D}^0 \to f)$. Then the situation becomes much more favourable: although we lose in statistics, the triangles can be constructed with more quantitative precision.

Final states f with $S = +1$ will do the trick for us since they are doubly Cabibbo suppressed for D^0, yet Cabibbo allowed for $\overline{D}^0$: $|A(D^0 \to [S = +1])/A(\overline{D}^0 \to [S = +1])| \sim \mathcal{O}(\theta_C^2)$. More specifically we observe

$$\frac{\mathrm{Br}(D^0 \to K^+\pi^-)}{\mathrm{Br}(\overline{D}^0 \to K^+\pi^-)} = 0.0077 \pm 0.0025 \pm 0.0025. \qquad (12.33)$$

Altogether we estimate

$$\frac{\mathrm{Br}(B^- \to K^- D^0)}{\mathrm{Br}(B^- \to K^- \overline{D}^0)} \cdot \frac{\mathrm{Br}(D^0 \to [S = +1])}{\mathrm{Br}(\overline{D}^0 \to [S = +1])} \sim 100 \cdot \mathcal{O}(\tan^4\theta_C) \sim 1 \qquad (12.34)$$

as desired! For details we refer you to Ref. [156].

12.3.3 $B \to K\pi$

As pointed out before, all ingredients should be there for a sizeable or even large **CP** asymmetry to become observable in, say, $B_d \to K\pi$ modes. The question is: can we go beyond such qualitative pronouncements?

The theoretical uncertainties enter mainly through the size and the strong phases of the hadronic matrix elements of the operators $T(K\pi)$ and $P(K\pi)_{c,t}$. In close analogy to the discussion given above, we can use the data set of all $B \to K\pi$, $\pi\pi$ and $K\overline{K}$ transition rates to infer the size of hadronic matrix elements. Consider

$$A(B^+ \to K^+\pi^0) = T(K\pi) + P(K\pi)_c + P(K\pi)_t. \qquad (12.35)$$

We have seen that the experimental result $\mathrm{Br}(B \to K\pi) > \mathrm{Br}(B \to \pi\pi)$ implies $P(K\pi)_c + P(K\pi)_t$ cannot be neglected compared to $T(K\pi)$. So, the tree graph contribution and penguin contribution may interfere. In addition, $P(K\pi)_c$ represents an amplitude for a process

$$B = (\overline{b}u) \to c\overline{c}sd \to K^+\pi^0, \qquad (12.36)$$

and physical intermediate states $D_s^+ D^-$, $D_s^+ D^- \pi$, $D_s^{+*} D^-$, $D_s^{+*} D^- \pi$ may contribute. These intermediate states will generate a final state interaction strong phase. The size of the asymmetry depends on the ratio of tree to penguin amplitudes as well as on the strong interaction phases.

Theoretical technologies developed to deal with non-perturbative dynamics – QCD sum rules and simulations of QCD on a lattice – are, for good reasons, defined in Euclidean rather than Minkowski space. Such treatments are quite insensitive to final state interactions for exclusive channels which are intimately connected with Minkowski space. It seems that a conceptual breakthrough is required for establishing theoretical – in contrast to phenomenological – control over final state interactions.

12.4 Time-averaged B_s transitions

Resolving the rapid oscillation rate expected for B_s mesons with x_s being conceivably much larger than 10 might pose a very stiff experimental challenge. Yet, even failing to do so, we can recover equivalent information on **CP** violation – if the the lifetime difference between the two B_s mass eigenstates is not too small [157].

Assume that the oscillations driven by ΔM_{B_s} indeed cannot be resolved. The $\cos\Delta M_B t$ and $\sin\Delta M_B t$ terms in the master equations of Eq. (11.15) and Eq. (11.19) have to be dropped then and we find:

$$\Gamma(B_s(t) \to f) \propto \frac{1}{4}e^{-\Gamma_1 t}|A(f)|^2$$
$$\cdot \left[\left(1 + \left|\frac{q}{p}\overline{\rho}(f)\right|^2\right)\left(1 + e^{\Delta\Gamma_{B_s}t}\right) + 2\mathrm{Re}\frac{q}{p}\overline{\rho}(f)\left(1 - e^{\Delta\Gamma_{B_s}t}\right)\right], \qquad (12.37)$$

$$\Gamma(\overline{B}_s(t) \to \overline{f}) \propto \frac{1}{4}e^{-\Gamma_1 t}|\overline{A}(\overline{f})|^2$$
$$\cdot \left[\left(1 + \left|\frac{p}{q}\rho(\overline{f})\right|^2\right)\left(1 + e^{\Delta\Gamma_{B_s}t}\right) + 2\mathrm{Re}\frac{p}{q}\rho(\overline{f})\left(1 - e^{\Delta\Gamma_{B_s}t}\right)\right]. \qquad (12.38)$$

Two general comments might help in elucidating these expressions:

- If f is flavour-specific – $A(f) \cdot \overline{A}(f) = 0$ – then the two rates obviously coincide.

- If f is flavour-nonspecific, yet without direct **CP** violation – $|A(f)| = |\overline{A}(\overline{f})|$ and thus $|\rho(\overline{f})| = |\overline{\rho}(f)|$ in addition to $|p| = |q|$ – we find

$$\Gamma(B_s(t) \to f) - \Gamma(\overline{B}_s(t) \to \overline{f}) \propto$$

$$\frac{1}{4}e^{-\Gamma_1 t}|A(f)\overline{A}(f)| \cdot \left(1 - e^{\Delta\Gamma_{B_s}t}\right)\left(\mathrm{Re}\frac{q}{p}e^{i\arg\overline{\rho}(f)} - \mathrm{Re}\frac{p}{q}e^{-i\arg\overline{\rho}(f)}\right), \qquad (12.39)$$

i.e., the two rates in Eq. (12.37) and Eq. (12.38) can differ only if

$$\Delta\Gamma_{B_s} \neq 0 \quad \mathrm{Im}\frac{q}{p}\overline{\rho}(f) \neq 0 \; ! \qquad (12.40)$$

This type of analysis can be undertaken for any flavour-nonspecific channel, in particular also for $f = \psi\phi$, $D_s^+ D_s^-$ or $f = D_s^\pm K^\mp$.

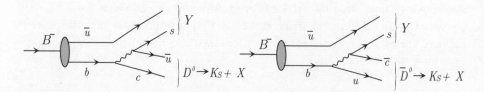

Fig. 12.6. Two Feynman graphs give amplitudes for inclusive decays which can interfere to produce **CP** asymmetries. Branching ratios for these decays should be considerable. It does not require decay time information. We expect a non-vanishing final state interaction phase. So, with some luck **CP** violation may be seen. When we sum over many channels, however, the sign of the asymmetry for various channels may be such that these asymmetries may cancel – making it hard to establish.

12.5 (Semi-)inclusive decays

It would be highly desirable to increase statistics by comparing the rates for sums of channels rather than single modes – if it could be done without significantly jeopardizing the size of the asymmetry. The ultimate limit to which this could be pushed is of course set by **CPT** invariance enforcing the equality of total rates and of subclasses thereof for particles and antiparticles. Consider the inclusive transitions [133]

$$B^+ \to [K_S X]_{D^{\text{neut}}} + K^+ + Y \quad \text{vs} \quad B^- \to [K_S \overline{X}]_{D^{\text{neut}}} + K^- + \overline{Y}. \tag{12.41}$$

A difference can arise because two different amplitudes – one via D^0 and the other via $\overline{D}^0$ – can contribute coherently, see Fig. 12.6. Yet an inspection of these quark level diagrams makes it obvious that for a multi-body final state (with additional $q\bar{q}$ pairs materializing during hadronization) the interference can be highly reduced: there may be little overlap in phase space between the $u\bar{s}$ pairs in the two diagrams!

Taking note of the hadrons' transformation under **CP** leads to the same general conclusion, namely that summing over a sizeable number of channels will lead to a significant reduction in the size of an asymmetry:

- The *relative* sign between the asymmetry in two different decay modes is determined by $\bar{\rho}(f)$. When f is self-conjugate, then its **CP** parity determines the sign of $\bar{\rho}(f)$:

$$\bar{\rho}(f_{\pm}) = \frac{\langle f_{\pm}|\mathscr{H}(\Delta B = 1)|\overline{B}^0\rangle}{\langle f_{\pm}|\mathscr{H}(\Delta B = 1)|B^0\rangle} = \pm \frac{\langle f_{\pm}|\mathscr{H}^{\mathbf{CP}}(\Delta B = 1)|B^0\rangle}{\langle f_{\pm}|\mathscr{H}(\Delta B = 1)|B^0\rangle}, \tag{12.42}$$

with $\mathscr{H}^{\mathbf{CP}}$ denoting the **CP** transformed version of $\mathscr{H}$.

- Similar relations hold also when f is not a **CP** eigenstate. Since

$$\mathbf{CP}|\pi^0\rangle = -|\pi^0\rangle, \tag{12.43}$$

just adding a pion to a final state will have the tendency to flip the sign of the asymmetry. In the limit of states with even and odd numbers of pions contributing equally, the asymmetry would average out to zero!

12.6 Measuring the same angle more than once

According to the three-family KM ansatz, there are different classes of channels from which we can extract the same angle.

- The channels $B_d \to \psi K_S$ driven by $b \to c\bar{c}s$ and the Cabibbo suppressed ones, $B_d \to \psi\pi^+\pi^-, D^+D^-$, generated by $b \to c\bar{c}d$ are predicted to exhibit the same **CP** asymmetry given by $\sin 2\phi_1$. Any difference would reflect New Physics in the $\Delta B = 1$ dynamics.

- Likewise for $B_d \to \pi^+\pi^-$ and $B_s \to K_S\pi^0, K_S\eta, K_S\eta', K_S\pi^+\pi^-$ – if penguin contributions in particular to $B_d \to \pi\pi$ can be unfolded: their asymmetries are predicted to be given by $\sin 2\phi_2$.

- We had discussed before several ways – none too appealing – to extract ϕ_3; they involved analysing $B^+ \to DK^+$ vs $B^- \to DK^-$. New Physics producing a difference could reside in the $\Delta B = 1$ or $\Delta B = 2$ sector.

12.7 Impact of the first CP asymmetry in B decays

Any observation of **CP** violation outside the neutral kaon system – no matter where – will constitute a breakthrough discovery – no matter what! A measurement of a **CP** asymmetry in B^0 decays will change the landscape in a particularly significant way. Presumably $B_d \to \psi K_S$ will first be analyzed with an accuracy sufficient to allow a determination of $\sin 2\phi_1$. This measurement would then enable us to construct the unitarity triangle from two of its sides – the normalized baseline and the second side of length $|V_{ub}/\lambda V_{cb}|$ as sketched in Sec. 10.5 – and one angle, namely ϕ_1. Elementary trigonometry tells us that in general two solutions will emerge, see Fig. 10.13. Yet apart from this binary ambiguity in the shape of the triangle we can expect an accuracy of around 10% (or even better) to be achievable in the long run for the triangle parameters from these measurements; even an incomplete quantitative understanding of $B_d - \bar{B}_d$ oscillations and of $|V_{td}|$ might suffice to resolve the ambiguity mentioned above.

This would represent a very important qualitative landmark: it would mean that *the unitarity triangle had been determined purely from B physics, which had thus become 'autonomous'* with the umbilical cord to K physics being severed.

12.7.1 Indirect manifestations of New Physics

Most physicists in our community – and certainly Yours Truly – suspect that the SM of high energy physics is incomplete, that New Physics has to exist, and quite possibly at energy scales that are not hugely higher than the 1 TeV level. Various conjectures have been made concerning the nature of the New Dynamics with SUSY figuring prominently among them; some of them will be described in Chapters 18–20. The rich **CP** phenomenology in beauty decays provides us with a high sensitivity probe for such new dynamical elements, based on a careful quantitative analysis of the KM triangle. Above, we have expressed our optimism that this triangle will be constructed with an uncertainty not exceeding 10%; we should also keep in mind that New Physics can affect the asymmetries *linearly* in *amplitude* rather than quadratically. Not only can the existence of some kind of New Physics be established, but we can also infer some of its specific features.

New Physics will in general introduce new weak phases and thus shift **CP** asymmetries away from their KM values. However, it is possible that *no* appreciable new phases emerge: the minimal supersymmetric Standard Model (MSSM) is one natural example for such a scenario. Nevertheless, the intervention of even such New Physics can be identified through a comprehensive analysis of the triangle. In all of this we have to keep in mind that we are not engaged in a mathematical exercise where three elements define a triangle and any difference between expectation and measurement is a significant deviation. Some data that from a mathematical perspective would be redundant or an over-constraint might at first be employed to obtain an empirically stable form of the triangle; likewise we will initially have to insist that deviations are quite sizeable before they can be perceived as significant.

To recall our previous discussion in Sec. 10.5: the baseline of the triangle is normalized to unity and the second side determined from semileptonic B decays, very probably irrespective of New Physics. The two observables $|\Delta M_{B_d}|$ and $|\epsilon|$ allow us to express the third side as well as the first angle ϕ_1 as a function of $B_B f_B^2$ (and B_K in the latter case):

$$\sin 2\phi_1 = g\left(B_B f_B^2, B_K, \epsilon, m_t\right) \tag{12.44}$$

$$|\mathbf{V}_{td}| = h\left(B_B f_B^2, \Delta M_{B_d}, m_t\right). \tag{12.45}$$

Determining $|\mathbf{V}_{td}|$ as outlined above or measuring ϕ_1 will then not only enable us to construct the unitarity triangle, but give us already a meaningful self-consistency test. For from either quantity we can infer the required size of $B_B f_B^2$; inserting this value into into the theoretical expression for the other quantity provides us with a non-trivial test of the triangle. A non-closure by a significant amount unmasks the intervention of New Physics either through a new weak phase affecting the **CP** asymmetry in $B \to \psi K_S$ or through a contribution to ΔM_{B_d}.

12.7.2 New Physics without new weak phases

It will be shown in Chapter 20 that some scenarios of New Physics, like MSSM, do not introduce new weak phases:

$$\arg\left[\frac{q}{p}\overline{\rho}(B_s \to \psi\phi)\right]\Bigg|_{MSSM} \simeq \arg\left[\frac{q}{p}\overline{\rho}(B_s \to \psi\phi)\right]\Bigg|_{KM} \simeq 0 \qquad (12.46)$$

$$\phi_2|_{MSSM} \simeq \phi_2|_{KM} , \ \phi_1|_{MSSM} \simeq \phi_1|_{KM} , \ \phi_3|_{MSSM} \simeq \phi_3|_{KM}. \qquad (12.47)$$

In that case a measurement of all three angles would not reveal non-closure:

$$\phi_1 + \phi_2 + \phi_3 = \pi. \qquad (12.48)$$

The MSSM is even more atypical (though still not unnatural) in the sense that the quark–squark–gluino couplings are controlled by the same KM parameters as the quark charged current couplings. The third side of the triangle would therefore reflect properly the KM value of $|V_{td}/V_{ts}|$, even in the presence of MSSM. Yet with the value inferred for $B_B f_B^2$ from $\sin 2\phi_1$ we would – with the given top quark mass – fail to reproduce ΔM_{B_d} to the degree that MSSM contributes to this quantity!

In other New Physics scenarios without new weak phases $|\epsilon|$ and ΔM_{B_d} will probably be affected differently. While Eq. (12.48) would still hold, we would find that the value inferred for the angle ϕ_1 from constructing the triangle through its three sides would differ from its observed size:

$$\sin 2\phi_1|_{\text{inferred}} \neq \sin 2\phi_1|_{\text{measured}} \ \text{ or } \ \sin 2\phi_2|_{\text{inferred}} \neq \sin 2\phi_2|_{\text{measured}}$$

$$\Longrightarrow \text{New Physics !} \qquad (12.49)$$

12.7.3 New Physics with new weak phases

Next we would undertake to measure directly the other angles ϕ_2 and ϕ_3 and compare them to their values as inferred from the triangle. If New Physics enters – as would be quite natural – through $\langle B_d|\mathscr{H}(\Delta B = 2)|\overline{B}_d\rangle$, it would change both ϕ_2 and ϕ_1 relative to their KM values

$$(\phi_2)_{\text{measured}} = (\phi_2)_{KM} - \phi_{NP} \qquad (12.50)$$

$$(\phi_1)_{\text{measured}} = (\phi_1)_{KM} + \phi_{NP} . \qquad (12.51)$$

Yet the contribution ϕ_{NP} drops out from their sum – as long as the New Physics indeed enters only through $B_d - \overline{B}_d$ oscillations and does not differentiate between the two final states (ψK_S and $\pi\pi$ in the example).

While a measurement of the third angle ϕ_3 would certainly be desirable – and every effort should be made in that direction – it is not obvious that this could be done with great accuracy. We should keep the following in mind then:

once the angles ϕ_2 and ϕ_1 have been measured with good accuracy we know the value of ϕ_3 from Eq. (12.48) within the KM scheme with an accuracy that will be superior to any direct measurement. The motivation behind measuring ϕ_3 is of course to discover the manifestation of New Physics! Yet in that case it makes even more sense to search for a **CP** asymmetry in $B_s \to \psi\phi$ or $D_s\overline{D}_s$, since that channel possesses a more striking signature and a cleaner theoretical interpretation, in the sense that the KM scheme predicts a small **CP** asymmetry here: $\mathrm{Im}[(q/p)\overline{\rho}(B_s \to \psi\phi)] < \tan^2\theta_C \simeq 0.05$, which might well be too small to be observable. However if New Physics entered $B_s \to K_S\rho^0$ through $B_s - \overline{B}_s$ oscillations – as is quite conceivable – it would do likewise of course for $B_s \to \psi\phi$ and it would be 'undisguised' by a KM 'background':

$$\arg\left[\frac{q}{p}\overline{\rho}(B_s \to \psi\phi)\right] \simeq \phi_{NP} \tag{12.52}$$

$$\arg\left[\frac{q}{p}\overline{\rho}(B_s \to K_S\rho^0)\right] \simeq \phi_3 + \phi_{NP}. \tag{12.53}$$

12.8 Résumé

- The unitarity triangle can be constructed with a considerable amount of redundancy: its three sides can be extracted irrespective of **CP** asymmetries in B decays; one, maybe two and possibly even three angles can be determined through observations.

- This redundancy can be harnessed to obtain the shape of the triangle with confidence and to probe for the presence of New Physics with high sensitivity.

- A comprehensive analysis will reveal whether New Physics enters through new weak phases or through contributions to $B^0 - \overline{B}^0$ oscillations and rare B decays that do not contain new phases or both. This will provide us with precious information about the nature of the intervening New Physics.

- Such a program of analysis will be intriguing, exciting – but neither easy nor quick!

Problems

12.1   Derive Eq. (12.5).

12.2   From Eq. (12.27) and Eq. (12.28) we see that the imaginary part of these amplitudes do not vanish in the limit of the CKM matrix elements being real. Does this mean that **CP** violation exists even for a real CKM matrix?

12.3 Show that in Eq. (12.27) the strong final state interaction phase for $A(DK^-)$ is equal to that of $A(\overline{D}K^+)$.

12.4 Consider

$$B_s \to D^0\phi, \overline{D}^0\phi, D_\pm\phi \qquad (12.54)$$

with $|D_\pm\rangle = \frac{1}{\sqrt{2}}(|D^0\rangle \pm |\overline{D}^0\rangle)$. Show that

$$\frac{q}{p}\overline{\rho}(D_\pm\phi) = \frac{(q/p)\overline{\rho}(D^0\phi) \pm 1}{1 \pm (p/q)\overline{\rho}(\overline{D}^0\phi)^{-1}} \qquad (12.55)$$

holds in the KM ansatz. Check $(q/p)\overline{\rho}(D^0\phi)$ and $(q/p)\overline{\rho}(\overline{D}^0\phi)$ are rephasing invariant.

Hint: first show that

$$\frac{p}{q}\rho(D^0_{1,2}\phi) = \frac{\frac{q}{p}\rho(D^0\phi) \pm \frac{p}{q}e^{2i\arg[V^*_{cb}V_{us}]}}{1 \pm \left(\frac{q}{p}\rho(\overline{D}^0\phi)\right)^{-1}\frac{p}{q}e^{2i\arg[V^*_{cb}V_{us}]}}. \qquad (12.56)$$

Then show that $\lambda(D^0_{1,2}\phi)$ is rephasing invariant. If you concluded that $\frac{p}{q}e^{-2i\arg[V^*_{cb}V_{us}]}$ is not rephasing invariant, you are correct. Then think about the definition of $D_{1,2}$. How does this definition change as we change the phases of the u and c quarks?

13

Rare K and B decays – almost perfect laboratories

If a meson lives long enough, it has a chance to show us many interesting phenomena. With ordinary transitions slowed down, quantum effects due to virtual intermediate states become relevant, and those can involve dynamical entities beyond the SM operating at much higher energy scales than M_K or M_B. Search for new phenomena where uninteresting ordinary decays are suppressed has yielded many hints in the past, and we can expect it to do so again in the future.

A voluminous tool chest has been created over the years. This field was pioneered by Gaillard and Lee [158]; the renormalization group analysis including penguins was introduced by Shifman, Vainshtein, Voloshin and Zakharov [89]; Gilman and Wise [90, 159] and also Visotsky [160] refined the analysis; electromagnetic penguins were introduced by Flynn and Randall [91]; the heavy quark mass corrections were computed by Inami and Lim [98]; finally the QCD corrections to two loops have been evaluated by Buras and co-workers [85, 86, 96, 161]. In order not to divert too much from our main theme we just sketch the procedure and quote numerical results, and refer the truly committed student to the aforementioned literature.

13.1 Rare K decays

13.1.1 $K_L \to \mu^+\mu^-$ and $K^+ \to \pi^+ e^+ e^-$

The phenomenological landscape can be characterized by two tiny ratios:

$$\frac{\Gamma(K^+ \to \pi^+ e^+ e^-)}{\Gamma(K^+ \to \pi^0 e^+ \nu)} \sim 6 \times 10^{-6}, \tag{13.1}$$

$$\frac{\Gamma(K_L \to \mu^+ \mu^-)}{\Gamma(K^+ \to \mu^+ \nu)} \sim 3 \times 10^{-9}. \tag{13.2}$$

These ratios concern the strength of strangeness-changing neutral currents. They are greatly suppressed; this near-absence of strangeness-changing currents provided one of the guiding principles in constructing the SM. Yet the suppression is of quite different strength – by roughly three orders of magnitude – in leptonic and in semileptonic transitions. Such a huge numerical difference cannot be laid at the doorstep of nonperturbative corrections due to the strong interations. There

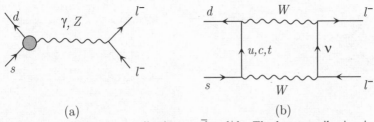

Fig. 13.1. The diagrams contributing to $s\bar{d} \to l^+l^-$. The key contribution is the $sd\gamma$ vertex in (a) which has a GIM suppression of the form $\alpha_s \log(m_c/m_u)$, and it contributes only to $K \to \pi e^+e^-$.

has to be a structural explanation. Let us see how the SM addresses this challenge. Both $K_L \to \mu^+\mu^-$ and $K^+ \to \pi^+e^+e^-$ are driven by the process depicted in Fig. 13.1. The effective interaction $\mathscr{H}(s \to d\gamma)$ which is responsible for Fig. 13.1(a) is of special interest [162]:

$$\mathscr{H}(s \to d\gamma) = \frac{1}{6}\langle r_{sd}\rangle \bar{s}(-\Box g_{\mu\nu} + \partial_\mu\partial_\nu)\gamma^\nu(1 - \gamma_5)dA^\mu$$
$$+ \langle \mu_{sd}\rangle \bar{s}\sigma_{\mu\nu}[(m_s + m_d) + \gamma_5(m_s - m_d)]dF^{\mu\nu},$$
$$\frac{1}{6}\langle r_{sd}\rangle = \frac{G_F}{\sqrt{2}}\sin\theta_c \cos\theta_c\left(-\frac{e}{9\pi^2}\right)\ln\frac{m_c}{m_u},$$
$$\langle \mu_{sd}\rangle = \frac{G_F}{\sqrt{2}}\sin\theta_c \cos\theta_c\left(\frac{7e}{192\pi^2}\right)\frac{m_c^2 - m_u^2}{M_W^2}, \tag{13.3}$$

where A^μ denotes the photon field and $F^{\mu\nu}$ its field strength; the top quark contribution has been ignored for simplicity. The first term in Eq. (13.3) is called the charge radius term and the second one the magnetic moment term.

The GIM mechanism requires these amplitudes to vanish if $m_u = m_c$. This is realized in two different ways: while the charge radius term has a GIM suppression $\sim \log m_c/m_u$ – which actually provides no suppression – it is of the form $(m_c^2 - m_u^2)/M_W^2$ for the magnetic moment term as well as the other diagram, Fig. 13.1(b). The fact that the charge radius interaction cannot contribute to $K \to \mu^+\mu^-$ (see Problem 13.2) explains the factor ~ 1000 difference shown in Eq. (13.2). A word of caution: QCD corrections change the effective Hamiltonian considerably. For example, the magnetic moment term gets an additional $\alpha_s \log(m_c/m_u)$ term [89].

13.1.2 $K_L \to \pi^0l^+l^-$

The final state $\pi^0l^+l^-$ is **CP** even to lowest order in the electroweak coupling. (Problem 13.3). It can be reached from the **CP** even component in K_L

$$K_L \to K_S \to \pi^0l^+l^- \tag{13.4}$$

or through direct **CP** violation in $\mathscr{H}(\Delta S)$, see Fig. 13.1. In the case of $K_L \to \pi\pi$, the direct **CP** violation parameter, $\frac{\epsilon'}{\epsilon}$, is suppressed by an additional factor of ω due to the $\Delta I = \frac{1}{2}$ rule. For many rare decays, such non-leptonic dynamical effects

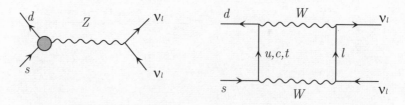

Fig. 13.2.   The diagrams driving $K \to l\bar{l}$ and $s\bar{d} \to \nu\bar{\nu}$.

are not expected to play a role. So, we expect the strength of *direct* **CP** violation to be roughly comparable to that via $K_L \to K_S$. Detailed calculations indeed yield [161]

$$\text{Br}(K_L \to \pi^0 e^+ e^-) = (4.5 \pm 2.6) \times 10^{-12}$$
$$\text{Br}(K_L \to K_S \to \pi^0 e^+ e^-) < 1.5 \times 10^{-12}. \tag{13.5}$$

Unfortunately, there are other contributions [159], the chain reaction

$$K_L \to \pi^0 \gamma^* \gamma^* \to \pi^0 l^+ l^-, \tag{13.6}$$

which is neither necessarily short distance dominated nor **CP** violating. Using chiral perturbation theory [163], we deduce from the data on $K_L \to \pi^0 \gamma\gamma$ [164, 165] that

$$\text{Br}(K_L \to \pi^0 e^+ e^-) = (0.3 - 1.8) \times 10^{-12}, \tag{13.7}$$

which is not much smaller than the **CP** breaking contributions.

A Dalitz plot analysis would enable us to extract the **CP** violating component unambigously [166]. Since the experimental limit for this decay is

$$\text{Br}(K_L \to \pi^0 e^+ e^-) < 4.3 \times 10^{-9}, \quad 90\% \text{ CL}, \tag{13.8}$$

this is not likely to happen soon – unless unexpected New Physics generates a signal to be observed by those who dare venture in this direction.

13.1.3 $K \to \pi\nu\bar{\nu}$

The matrix element for this decay is given by [98]:

$$\langle \pi\nu\bar{\nu}|H(d\bar{s} \to \nu\bar{\nu})|K \rangle = \frac{G_F}{\sqrt{2}} \frac{\alpha}{2\pi \sin^2 \theta_W} \sum_{i=e,\mu\tau} D\langle \pi|(\bar{s}d)_{V-A}|K \rangle (\bar{\nu}\nu)_{V-A}, \tag{13.9}$$

where

$$D = \mathbf{V}_{td}\mathbf{V}_{ts}^* X(x_t) + \mathbf{V}_{cd}\mathbf{V}_{cs}^* X(x_c),$$
$$X(x) = \frac{x}{8}\left[3\frac{x-2}{(1-x)^2}\ln x + \frac{x+2}{x-1} \right], \quad x_i = \frac{m_i^2}{M_W^2}. \tag{13.10}$$

In evaluating the hadronic matrix element $\langle \pi|J_\mu^{\text{had}}|K \rangle$ we can no longer ignore long-distance dynamics. What saves the day is the following: isospin invariance implies the equality of the quantities $\langle \pi^+ | J_\mu^{\text{had,neut}} | K^+ \rangle$ and $\langle \pi^0 | J_\mu^{\text{had,neut}} | K^0 \rangle$ with $\langle \pi^0 | J_\mu^{\text{had,ch}} | K^+ \rangle$; the latter is known from the data on the ordinary semileptonic transition $K^+ \rightarrow \pi^0 l^+ \nu$.

• $K^+ \rightarrow \pi^+ \nu \bar{\nu}$

The SM prediction is [161]:

$$\text{Br}(K^+ \rightarrow \pi^+ \nu_i \bar{\nu}_i) = (9.1 \pm 3.2) \times 10^{-11}, \tag{13.11}$$

where the uncertainty comes from higher order QCD corrections and the value of $|\mathbf{V}_{td} \mathbf{V}_{ts}|$ in particular.

Experimentally, the way to measure the branching ratio is to look for events of the type

$$K^+ \rightarrow \pi^+ + \text{nothing}. \tag{13.12}$$

The AGS experiment E787 at Brookhaven National Laboratory has been searching for events of this type. Just to give you some idea how formidable the search is, note that what we are searching for is an event – with less than sterling signature! – occurring only once in about 10^{10} decays. For example, we must make sure that other K^+ decays

$$K^+ \rightarrow \mu^+ \nu_\mu, \; \mu^+ \nu_\mu \gamma, \; \pi^+ \pi^0, \tag{13.13}$$

do not mimic the signal at the level of once every 10^{10} decay. The presence of π^0 must be excluded by vetoing γ. The misidentification of μ^+ for π^+ is avoided by observing the chain decay

$$\begin{aligned} \pi^+ \rightarrow \quad & \mu^+ \quad \nu \\ & \hookrightarrow \quad e^+ \nu_e \bar{\nu}_\mu. \end{aligned} \tag{13.14}$$

E787 has found a candidate event for $K^+ \rightarrow \pi^+ \nu \bar{\nu}$ [167]. If this is a signal, the branching ratio is

$$\text{Br}(K^+ \rightarrow \pi^+ \nu_i \bar{\nu}_i) = 4.2 \, ^{+9.7}_{-3.5} \times 10^{-10}. \tag{13.15}$$

Comparing Eq. (13.11) and Eq. (13.15), we see that the SM will be tested in the near future.

What good is this decay once it is found? The main theoretical uncertainty enters in the QCD correction. Determination of this branching ratio at the level of 30% should yield a quite reliable determination of $\mathbf{V}_{td} \mathbf{V}_{td}^*$ – yet another constraint on the unitarity triangle.

Physics beyond the SM may affect this process. For example, a decay

$$K^+ \rightarrow \pi^+ + X^0, \tag{13.16}$$

where X^0 is a massless particle, will lead to a striking signature. The final state π^+ is monoenergetic in the rest frame of K^+. The 90% confidence limit for such a decay is [167]

$$\text{Br}(K^+ \to \pi^+ + X^0) < 5.2 \times 10^{-10}. \tag{13.17}$$

Also, in Ref. [167], the upper limit is given as a function of X^0 mass. Less drastic extensions of the SM are discussed in Ref. [67]. Indeed, the existence of additional Higgs bosons, right handed gauge bosons and SUSY particles changes the branching ratio significantly.

• $K_L \to \pi^0 \nu \bar\nu$

This decay can occur only due to **CP** violation since there is no two-photon and practically no two-Z^0 intermediate state. To go beyond these qualitative remarks we follow the usual and by now familiar procedure, namely to first derive the effective $\Delta S = 1$, $\Delta Q = 0$ coupling on the quark level and then evaluate the appropriate matrix elements.

It can formally be expressed through an effective low energy Hamiltonian involving hadronic fields:

$$\mathcal{H}_{\text{eff}} = c_{CPV} \phi_\pi \partial_\mu \phi_{K_L} \bar\nu_L \gamma^\mu \nu_L. \tag{13.18}$$

Under **CP** we have $\bar\nu \gamma^\mu \nu \to -\bar\nu \gamma_\mu \nu$, $\partial_\mu \phi_{K_L} \to -\partial^\mu \phi_{K_L}$ and $\phi_{\pi^0} \to -\phi_{\pi^0}$, i.e., $\mathcal{H}_{\text{eff}}$ is odd under **CP**; therefore c_{CPV} has to be proportional to $\text{Im}V_{td}V_{ts}^*$. Putting everything together, we arrive at

$$\frac{\text{Br}(K_L \to \pi^0 \nu \bar\nu)}{\text{Br}(K^+ \to \pi^0 e^+ \nu_e)} = \frac{3}{2} \frac{\tau(K_L)}{\tau(K^+)} \frac{\alpha^2}{\pi^2 \sin^4 \theta_W} \left| \frac{\text{Im}(V_{td} V_{ts}^*)}{V_{us}} X(x_t) \right|^2. \tag{13.19}$$

Again using presently available information on the KM parameters, we arrive at the numerical prediction [168]:

$$\text{Br}(K_L \to \pi^0 \nu \bar\nu) = (2.8 \pm 1.7) \times 10^{-11}. \tag{13.20}$$

This decay is dominated by *direct* **CP** violation. If for some reason the direct **CP** violation computed above is absent, the **CP** even component in the K_L wavefunction can give rise to this decay. The decay rate for $K_L \to \pi^0 \nu \bar\nu$ is then reduced relative to that for $K^+ \to \pi^+ \nu \bar\nu$ by the impurity parameter ϵ_K^2.

Of course, the present experimental bound [169] is above the expectation by several orders of magnitude;

$$\text{Br}(K_L \to \pi^0 \nu \bar\nu) < 5.8 \times 10^{-5}, \quad 90\% \text{ CL.} \tag{13.21}$$

Yet it appears that a branching ratio as low as 10^{-12} might be achievable.

13.1.4 $K \to \pi\pi\gamma^{(*)}$

• $K^\pm \to \pi^\pm\pi^0\gamma$

CPT symmetry enforces $\Gamma(K^+ \to \pi^+\pi^0) = \Gamma(K^- \to \pi^-\pi^0)$ to the degree that electromagnetic corrections can be ignored; the latter allow a tiny asymmetry to sneak in. It would then appear more promising to search for direct **CP** violation in radiative K decays – $K^+ \to \pi^+\pi^0\gamma$ vs. $K^- \to \pi^-\pi^0\gamma$ – thus trading in a larger branching ratio for a larger asymmetry. QCD corrections actually enhance its rate. Upon closer examination, however, we have encountered [170] a strong kinematical suppresion. The decay is dominated by the bremsstrahlung amplitude and a magnetic multipole amplitude. **CP** violation is generated by the interference of the bremsstrahlung and a small electric multipole amplitude. Theoretical estimates yield

$$a(\pi\pi\gamma) = \frac{\Gamma(K^+ \to \pi^+\pi^0\gamma) - \Gamma(K^- \to \pi^-\pi^0\gamma)}{\Gamma(K^+ \to \pi^+\pi^0\gamma) + \Gamma(K^- \to \pi^-\pi^0\gamma)} \sim 10^{-4} - 10^{-5}, \qquad (13.22)$$

at best, which is discouraging.

The difficulty can be traced to the fact that the dominating magnetic multipole amplitude cannot interfere with the bremsstrahlung amplitude, once we sum over the photon polarization.

• $K \to \pi^+\pi^-\gamma^{(*)}$

The modes $K_{L,S} \to \pi^+\pi^-\gamma$ have been observed with [171]

$$\text{Br}(K_L \to \pi^+\pi^-\gamma) = (4.66 \pm 0.15) \cdot 10^{-5}, \qquad (13.23)$$
$$\text{Br}(K_S \to \pi^+\pi^-\gamma) = (4.87 \pm 0.11) \cdot 10^{-3}, \qquad (13.24)$$

for $E_\gamma > 20$ MeV. Two mechanisms can drive these channels and an analysis of the photon spectra indeed reveals the intervention of both:

* bremsstrahlung off the pions through an E1 transition:

$$K_L \xrightarrow{\Delta S=1} \pi^+\pi^- \xrightarrow{E1} \pi^+\pi^-\gamma \ , \ K_S \xrightarrow{\Delta S=1} \pi^+\pi^- \xrightarrow{E1} \pi^+\pi^-\gamma \ , \qquad (13.25)$$

where only the first step in the K_L decay is **CP** violating.

* Direct photon emission of the M1 type

$$K_L \xrightarrow{M1 \& \Delta S=1} \pi^+\pi^-\gamma \ , \ K_S \xrightarrow{M1 \& \Delta S=1} \pi^+\pi^-\gamma \ , \qquad (13.26)$$

which is **CP** conserving [violating] for the K_L [K_S] process.

In analogy to ϵ_K we define a ratio of E1 amplitudes

$$\eta_{+-\gamma} = \frac{A(K_L \to \pi^+\pi^-\gamma, E1)}{A(K_S \to \pi^+\pi^-\gamma, E1)}, \qquad (13.27)$$

that measures **CP** violation. In the absence of direct **CP** violation we have $\eta_{+-\gamma} = \eta_{+-}$.

We can proceed in close analogy to the $K_L \to \pi^+\pi^-$, $\pi^0\pi^0$ case and through the use of a regenerator in a K_L beam compare the decay rate evolution of $K_L \to \pi^+\pi^-\gamma$ and $K_S \to \pi^+\pi^-\gamma$ as a function of the time of decay. This has been done in Ref. [171] where it has been demonstrated that $K_L - K_S$ interference occurs in $K \to \pi^+\pi^-\gamma$ and that the **CP** violating parameters of these modes are quite consistent with those of $K_L \to \pi\pi$.

We want to focus on a special aspect of it. The interference of the **CP** violating E1 and conserving M1 amplitudes will yield a circularly polarized photon. This polarization represents **CP** violation. To be more specific: the *interference* yields a triple correlation between the pion momenta and the photon polarization

$$P_\perp^\gamma = \langle \vec{\epsilon}_\gamma \cdot (\vec{p}_{\pi^+} \times \vec{p}_{\pi^-}) \rangle, \tag{13.28}$$

which is **CP** odd; its leading contribution is proportional to η_{+-} entering in the E1 amplitude.

- $K_L \to \pi^+\pi^- e^+ e^-$

The photon polarization which constitutes the **CP** signal can be probed best for off-shell photons

$$K_L \to \pi^+\pi^-\gamma^* \to \pi^+\pi^- e^+ e^-. \tag{13.29}$$

A general discussion which includes direct E1 transition as well as the charge radius term $K_L \to K_S e^+ e^- \to \pi^+\pi^- e^+ e^-$ can be performed [172, 173]. Actually these give new direct **CP** violating effects, but as discussed for the asymmetry in $K \to \pi\pi\gamma$ decay, these new effects are rather small. Here we restrict ourselves to the leading contributions for the asymmetry. The amplitude for this decay can be written as

$$T(K_L \to \pi^+\pi^- e^+ e^-) = e|T(K_S \to \pi^+\pi^-)|\cdot$$

$$\cdot \left[\frac{g_{BR}}{M_K^4} \left(\frac{p_+^\mu}{p_+ \cdot k} - \frac{p_-^\mu}{p_- \cdot k} \right) + \frac{g_{M1}}{M_K^4} \epsilon_{\mu\nu\alpha\beta} k^\nu p_+^\alpha p_-^\beta \right] \frac{e}{k^2} \bar{u}(k_-)\gamma_\mu v(k_+) \tag{13.30}$$

(with $k = k_+ + k_-$) in terms of two couplings $g_{BR,M1}$ for the bremsstrahlung and M1 transitions, respectively.

From the observed M1 rate, we infer

$$g_{BR} = \eta_{+-} e^{i\delta_0(m_K^2)}, \quad g_{M1} = 0.76 i e^{i\delta_1(s_\pi)}, \tag{13.31}$$

where $\delta_{0,1}$ denote the S- and P-wave $\pi\pi$ phase shifts.

The **CP** violating effect appears as a correlation between the $e^+ e^-$ and $\pi^+\pi^-$ planes. Denoting by Φ the angle between the $e^+ e^-$ and $\pi^+\pi^-$ planes, we obtain

$$\frac{d}{d\Phi} \Gamma(K_L \to \pi^+\pi^- e^+ e^-) = \Gamma_1 \cos^2\Phi + \Gamma_2 \sin^2\Phi + \Gamma_3 \cos\Phi \sin\Phi. \tag{13.32}$$

It is easy to see that the term $\cos\Phi\sin\Phi$ changes sign under **CP** and **T** (see Prob. 13.4); Γ_3 thus represents **CP** violation. It can be projected out by comparing the Φ distribution integrated over two quadrants:

$$A = \frac{\int_0^{\pi/2} \mathrm{d}\Phi \frac{\mathrm{d}\Gamma}{\mathrm{d}\Phi} - \int_{\pi/2}^{\pi} \mathrm{d}\Phi \frac{\mathrm{d}\Gamma}{\mathrm{d}\Phi}}{\int_0^{\pi} \mathrm{d}\Phi \frac{\mathrm{d}\Gamma}{\mathrm{d}\Phi}} = \frac{2\Gamma_3}{\pi(\Gamma_1 + \Gamma_2)}. \tag{13.33}$$

Including all the terms, the result is [172]:

$$A \simeq (15\cos\Theta_1)\% + \left(38 \left|\frac{g_{E1}}{g_{M1}}\right| \cos\Theta_2\right)\%, \tag{13.34}$$

where $\Theta_1 = \phi_{+-} + \delta_0 - \bar{\delta}_1 - \frac{\pi}{2}$ (mod π); $\Theta_2 = \phi_{+-} + \delta_0 - \frac{\pi}{2}$ (mod π); and g_{E1} is a coupling for the E1 transition. Here $\bar{\delta}_1$ is the P wave $\pi\pi$ phase shift averaged over the kinematical region. Experimentally, $\delta_0 - \bar{\delta}_1 \sim 30°$. For $\phi_{+-} = \phi_{SW}$, and $g_{E1}/g_{M1} = 0.05$, the asymmetry becomes:

$$A = (14.3 \pm 1.3)\%. \tag{13.35}$$

The main theoretical uncertainty resides in what we assume for the hadronic form factors. In the above analysis a phenomenological ansatz was employed for them; evaluating them in chiral perturbation theory yields similar numbers [174].

The KTeV experiment at FNAL has established the existence of this channel confirming both predictions [175]:

$$\mathrm{Br}(K_L \to \pi^+\pi^- e^+ e^-) = (3.32 \pm 0.14 \pm 0.28) \cdot 10^{-7}$$

$$A = (13.5 \pm 2.5 \pm 3.0)\%, \text{ preliminary.} \tag{13.36}$$

The discovery of such a spectacularly large **CP** asymmetry is a significant result, be it only to show that **CP** violation is not uniformly tiny in K_L decays. We should note, though, that A is driven by η_{+-} entering through $K_L \to \pi^+\pi^- \to \pi^+\pi^-\gamma^* \to \pi^+\pi^- e^+ e^-$ and not the prediction of a specific model.

Some comments are in order:

- This **CP** asymmertry is so large because the **CP** violating amplitude is enhanced by kinematical bremsstrahlung factors.

- It is often stated that this asymmetry represents the observation of *direct* **T** violation. It is tempting to think that for the first time **T** violation is clearly seen, independent of **CPT** conservation. Yet with the time reversal operator being *anti*unitary a **T** odd correlation can arise, as discussed in Sec. 4.10.2, even with **T** invariant dynamics, if complex phases are present; final state interactions can generate such phases.

- *Direct* **CP** violation can contribute as well to A and its size depends on the details of the dynamics underlying **CP** violation. Yet its contributions to the observable A averaged over all final states are tiny, namely $< 10^{-3}$

for the KM ansatz [176]; it is hard to see how they could be significantly larger for other models. While they can be significantly larger in certain parts of phase space, no promising avenue has been pointed out yet.

13.2 Beauty decays

Rare beauty decays consistent with SM one-loop transitions have been observed in the form of radiative B decays. One exclusive channel has been found [177]

$$\text{Br}(B^+ \to K^* \gamma) = (4.2 \pm 0.8 \pm 0.9) \times 10^{-5}. \qquad (13.37)$$

This is a flavour changing neutral current effect which occurs only through quantum corrections, namely a penguin graph. Yet, it is comparable to $\text{Br}(B \to \psi K)$ which occurs through a tree graph. Thus we are tempted to think that B decays are a good place to study quantum corrections to the gauge theory where New Physics might be more prominent. Also, because the b quark is heavy, we have better control over QCD corrections. This gives us a handle on the background to New Physics contribution – the SM contribution. In this section we investigate B decays which only occur through loop effects: $B \to \gamma X$, $B \to l^+l^-X$, and $B \to l^+l^-$. The form of the effective Hamiltonian without QCD corrections is quite similar to that of corresponding reactions in K decays. However, the top quark contributions enter in full strength, and the GIM mechanism no longer leads to simple suppressions.

13.2.1 $B \to X_s \gamma$

This is an inclusive process in which we sum over all possible final states. Using a heavy quark expansion we can show that the leading term is given by the decay of a b quark, with the non-perturbative corrections suppressed by factors which are $\sim \mathcal{O}(\frac{1}{m_b^2}, \frac{1}{m_c^2})$.

As for the weak interaction, the Hamiltonian for B decay is given by

$$\mathcal{H} = -\frac{G_F}{\sqrt{2}} \mathbf{V}_{ts}^* \mathbf{V}_{tb} C_{7\gamma}(\mu) Q_{7\gamma}(\mu), \qquad (13.38)$$

where

$$Q_{7\gamma} = \frac{e}{8\pi^2} m_b \bar{s} \sigma^{\mu\nu} (1 + \gamma_5) b F_{\mu\nu}. \qquad (13.39)$$

The value for $C_{7\gamma}(M_W)$ can be read off from Ref. [98], and $C_{7\gamma}(\mu)$ with two loop QCD correction can be found in Ref. [96]. We normalize this decay to the inclusive semileptonic decay:

$$\frac{\text{Br}(B \to X_s \gamma)}{\text{Br}(B \to X_c e^+ \nu_e)} = \frac{\text{Br}(\bar{b} \to \bar{s} \gamma)}{\text{Br}(\bar{b} \to \bar{c} e \bar{\nu}_e)} = \frac{|\mathbf{V}_{ts}^* \mathbf{V}_{tb}|^2}{|\mathbf{V}_{cb}|^2} \frac{6\alpha}{\pi f(z)} |C_{7\gamma}(\mu)|^2, \qquad (13.40)$$

where $f(z) = 1 - 8z^2 + 8z^6 - z^8 - 24z^4 \ln z$, and $z = m_c/m_b$ is the phase space factor. Numerically [178],

$$\text{Br}(B \to X_s \gamma) = (3.28 \pm 0.33) \times 10^{-4}, \tag{13.41}$$

where the error is dominated by the uncertainty in μ. Similar results have also been obtained in Refs. [179, 180]. This is in good agreement with the experimental value [181, 182]

$$\text{Br}(B \to X_s \gamma) = \left(2.6 \pm 0.6 \text{ exp.} + \left(^{+0.37}_{-0.30}\right)_{\text{th}} \right) \times 10^{-4}. \tag{13.42}$$

There is room for improvement in the theoretical prediction by including higher loop corrections. Also, experiment will improve. This is a promising place for hunting New Physics in the near future.

13.2.2 $B \to \mu^+ \mu^-$

This decay may not add much to our fundamental understanding of **CP** violation. It has, however, a nice signature which can be used to pick up this decay in a hadronic environment. A measurement of the branching ratio for this decay leads to a determination of $|V_{tb}^* V_{td} f_B|$. The effective interaction Hamiltonian is given by [98]

$$\mathcal{H} = -\frac{G_F}{\sqrt{2}} \alpha 2\pi \sin \theta_w V_{tb}^* V_{ts} C(x_t)(\bar{b}s)_{V-A}(\bar{\mu}\mu)_{V-A}, \tag{13.43}$$

with

$$C(x) = \frac{x}{8} \left[\frac{3x}{(1-x)^2} \ln x + \frac{4-x}{1-x} \right]. \tag{13.44}$$

This leads to [96]

$$\text{Br}(B_d \to \mu^+ \mu^-) = 1.7 \times 10^{-10} \left(\frac{\tau(B_d)}{1.6\text{ps}} \right) \left(\frac{F_B}{230\text{MeV}} \right) \left(\frac{|V_{td}|^2}{0.008} \right) \left(\frac{m_t(m_t)}{170\text{MeV}} \right)^{3.12}. \tag{13.45}$$

A similar result is obtained for $B_s \to \bar{\mu}\mu$. Taking the ratio of branching ratios yields

$$\frac{\text{Br}(B_d \to \bar{\mu}\mu)}{\text{Br}(B_s \to \bar{\mu}\mu)} = \frac{\tau(B_d)M_{B_d}F_{B_d}^2|V_{td}|^2}{\tau(B_s)M_{B_s}F_{B_s}^2|V_{ts}|^2}, \tag{13.46}$$

which allows for a determination of $|V_{td}|^2/|V_{ts}|^2$. The branching ration $B_d \to \bar{\mu}\mu$ is expected to be $\mathcal{O}(10^{-10})$ and we need about 10^{14} B_d for 1% measurement of the branching ratio!

13.2.3 $B \to X + \nu \bar{\nu}$

Compared to $B \to \bar{\mu}\mu$ decay this decay is more abundant – if we can only identify them cleanly! Also, theoretical prediction is clean. Computing corresponding

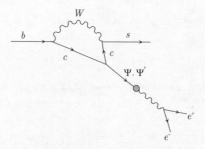

Fig. 13.3.   The long distance contribution to $b \to se^+e^-$.

diagrams shown in Fig. 13.2, where we make replacements $s \to b$ and $d \to s$, the decay rate is given by [96]:

$$\frac{\mathrm{Br}(B \to X_s v \bar{v})}{\mathrm{Br}(B \to X_c e \bar{v})} = \frac{3\alpha^2}{4\pi^2 \sin^4 \theta_W} \frac{|V_{ts}|^2}{|V_{cb}|^2} \frac{X^2(x_t)}{f(z)} \frac{\bar{\eta}}{\kappa(z)}, \qquad (13.47)$$

$\bar{\eta} \simeq 0.83$ and $\kappa(z) \simeq 0.88$ are QCD corrections.

Putting the numbers in, we obtain

$$\mathrm{Br}(B \to X_s v \bar{v}) = 4.1 \times 10^{-5} \frac{|V_{ts}|^2}{|V_{cb}|^2} \left(\frac{m_t(m_t)}{170 MeV} \right)^{2.3}. \qquad (13.48)$$

The ALEPH collaboration has searched for this decay and obtained an upper limit [183]:

$$\mathrm{Br}(B \to X_s + v\bar{v}) < 7.7 \times 10^{-4} \quad (90\% \text{ CL}). \qquad (13.49)$$

Let us wish lots of good luck to experimentalists taking this challenge.

13.2.4 $B \to X_s + \mu^+\mu^-$

This decay gets a contribution from the Feynman diagram in Fig. 13.3. Note that the final state in the KM favoured $b \to c\bar{c}s$ contains a $(c\bar{c})$ pair which may form a resonance J/ψ, ψ',... [184, 185]. These nonperturbative corrections will introduce inherent uncertainties. An estimate of the branching ratio as a function of $\hat{s} = Q^2/m_b^2$, where Q^2 is the invariant mass square of the l^+l^- system, is shown in Fig. 13.4.

In this figure, QCD corrections are not fully considered. For detailed QCD corrections to this decay, see Ref. [186]. Note that the width of J/ψ and ψ' is much larger than their actual width. This is because the diagram shown in Fig. 13.3 interferes with the penguin graph with internal top quark. Similarly, if there is New Physics, such a contribution can also interfere with the diagram shown in Fig. 13.3. Such interference will also affect the angular distributions [187]. It is a fertile searching ground for New Physics.

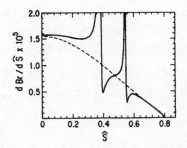

Fig. 13.4. We see that due to the interference of long distance and short distance contributions, the shapes of Ψ and Ψ' resonances are altered. New Physics contributions may change the interference pattern. This figure was reproduced from *Physics Letters* by permission of Elsevier Science.

13.3 Résumé

Various K and B decays allow us to test the SM, including quantum corrections. When many rare decay branching ratios agree, even to within 20% or so, we gain much confidence that the SM is correct. With improvement in experimental results, these studies will allow us to fix the KM parameters. In some rare decays such as $K_L \to \pi\bar{\nu}\nu$, direct **CP** violation dominates indirect **CP** violation. So, it offers an oportunity to test the KM scheme. Of course, if a significant disagreement between experiment and the SM prediction is found, then we hit the jackpot – a discovery of New Physics!

Study of K and B physics along this line will no doubt continue for many decades. Don't forget, rare decays of the K meson have been under scrutiny forover 50 years. Who knows, some bright young experimentalist might come up with a way to do a dedicated experiment to collect large number of B decays (say 10^{14}?).

Problems

13.1 Show that the charge radius term can only contribute to a virtual photon process $s \to de^+e^-$ and not to a real photon emission process $s \to d\gamma$.

13.2 Show that $\langle \mu\bar{\mu}|\mathscr{H}|K\rangle = 0$, where $\mathscr{H}$ is a charge radius contribution to the effective Hamiltonian for $s \to d\mu\bar{\mu}$. Show also that the magnetic moment term cannot contribute to $K \to \mu\bar{\mu}$ either. Hint: consider $\langle 0|\sigma_{\mu\nu}|K\rangle$.

13.3 Show that the contribution from Fig. 13.1 to $K_L \to Z^*\pi \to l^+l^-\pi$ and $K_L \to \gamma^*\pi \to l^+l^-\pi$ must violate **CP** symmetry. First, a physical argument: consider the **CP** quantum number of the final states $(Z^*\pi)$ and $(\gamma^*\pi)$. Noting that $\mathbf{CP}|Z^*\rangle = |Z^*\rangle$, $\mathbf{CP}|\gamma^*\rangle = |\gamma^*\rangle$ and two particles are in a relative P wave state, deduce the **CP** eigenvalues of the final states.

Second, a more formal argument: note that Fig. 13.1 gives an operator of the form $\bar{d}\gamma_\mu\gamma_- \bar{s}l\gamma^\mu\gamma_- l$. Show that the matrix element of this operator between the initial and the final states vanishes.

13.4 Consider $K_L \to \pi^+\pi^- e^+ e^-$; let p_+, p_-, k_+ and k_- denote the four-momenta of π^+, π^-, e^+ and e^-, respectively.

(a) How does the triple correlation

$$P = (\vec{p}_+ \times \vec{p}_-) \cdot (\vec{k}_+ - \vec{k}_-), \tag{13.50}$$

transform under **CP** and **T**? What does a non-vanishing expectation value of P in K_L decays teach you?

(b) Define the normal vectors of the $\pi^+ - \pi^-$ and $e^+ - e^-$ planes:

$$\vec{n}_\pi \equiv \frac{\vec{p}_+ \times \vec{p}_-}{|\vec{p}_+ \times \vec{p}_-|}, \quad \vec{n}_e \equiv \frac{\vec{k}_+ \times \vec{k}_-}{|\vec{k}_+ \times \vec{k}_-|}. \tag{13.51}$$

Show that $\vec{n}_\pi$ and $\vec{n}_e$ change sign under **CP**. How does

$$\vec{n}_e \times \vec{n}_\pi \cdot \frac{(\vec{p}_+ + \vec{p}_-)}{|\vec{p}_+ + \vec{p}_-|} \tag{13.52}$$

transform under **CP** and **T**? Denote by Φ the angle between $\vec{n}_\pi$ and $\vec{n}_e$; derive the identity

$$\sin\Phi \cos\Phi = \vec{n}_e \times \vec{n}_\pi \cdot \frac{(\vec{p}_+ + \vec{p}_-)}{|\vec{p}_+ + \vec{p}_-|} \vec{n}_e \cdot \vec{n}_\pi. \tag{13.53}$$

How does $\sin\Phi \cos\Phi$ behave under **CP** and **T**?

(c) Show that the interference between the bremsstrahlung and M1 amplitudes generates a term in the angular distribution which is proportional to $\sin\Phi \cos\Phi$.

13.5 With 10^{14} identified Bs, what is the sensitivity on the branching ratio of rare B decay? To what extent is it sensitive to New Physics?

14

♠ CPT invariance in K and B decays ♠

Peccate Fortiter[1]

As we have seen, **CPT** symmetry is a very general property of quantum field theories: it can be derived by invoking little more than locality and Lorentz invariance [188]. Despite this impeccable pedigree, it makes sense to ask whether limitations exist:

- Precisely because the **CPT** theorem rests on such essential pillars of our present paradigm, we have to make every reasonable effort to probe its universal validity.

- Although no *appealing* theory of **CPT** violation exists, we should keep in mind that superstring theories – suggested as more fundamental than quantum field theories and by some seen as TOE, the 'Theory of Everything' – are intrinsically *non*-local and thus do not satisfy one of the basic axioms of the **CPT** theorem.

- An intriguing phenomenon has been suggested by Hawking [189]. Near a black hole *pure* quantum states can evolve into *mixed* ones, since some of the information carried by them gets funnelled into the black hole due to the latter's overpowering gravitational pull, and thus is lost for the 'outside world'. This sequence violates both conventional quantum mechanics and **CPT** invariance. Hawking used the density formalism to specifically discuss $K^0 - \overline{K}^0$ oscillations in the presence of a black hole. An experimental test of Hawking's idea was proposed in Ref. [190].

While the intrinsic non-locality of superstring theories allows for limitations to **CPT** invariance, it does *not* demand it. Our attitude can then be described as one of educated curiosity: while we know of no benchmark for where **CPT** invariance might be violated, we like to emphasize two points: (i) **CPT** studies have in the past played a role in strengthening the community's belief in the relevance of quantum field theories. (ii) Particle–antiparticle oscillations allow measurements of spectacular sensitivity by tracking coherent effects building up over *macroscopic* distances. Thus they enable us to probe **CPT** invariance – and the foundations it

---

[1] Sin strongly!

235

is based on, namely Lorentz invariance and linear quantum mechanics – with un-parallelled accuracy. In the future, the scope of such tests will be greatly extended through detailed studies of $B^0 - \overline{B}^0$ oscillations and by analyzing EPR correlations in the $K^0 \overline{K}^0$ and $B^0 \overline{B}^0$ states produced at the Φ and $\Upsilon(4S)$ factories, respectively.

For this purpose we have to apply the expressions derived in Chapter 6 from linear quantum mechanics, yet *without* imposing **CPT** invariance. First we sketch relevant existing data and then make some brief remarks concerning possible theoretical scenarios for limitations to **CPT** and Lorentz invariance and linear quantum mechanics.

14.1 Equality of masses and lifetimes

The 'classical' tests of **CPT** symmetry concern the equality of masses and lifetimes for particles and antiparticles (see Chapter 4). The experimental situation can be characterized by the following set of upper bounds[2]:

$$\frac{m_{e^+} - m_{e^-}}{m_e} < 4 \times 10^{-8}, \quad 90\% \text{ CL}, \qquad \frac{\frac{|q_{\overline{p}}|}{M_{\overline{p}}} - \frac{q_p}{M_p}}{\frac{q_p}{M_p}} = (1.5 \pm 1.1) \times 10^{-9},$$

$$\frac{M_{\overline{n}} - M_n}{M_n} = (9 \pm 5) \times 10^{-5}, \qquad \frac{M_{\pi^+} - M_{\pi^-}}{M_\pi} = (2 \pm 5) \times 10^{-4},$$

$$\frac{M_{K^+} - M_{K^-}}{M_K} = (-0.6 \pm 1.8) \times 10^{-4}, \quad \frac{M_\Lambda - M_{\overline{\Lambda}}}{M_\Lambda} = (-1.0 \pm 0.9) \times 10^{-5},$$

$$\frac{M_{W^+} - M_{W^-}}{M_W} = (-2 \pm 7) \times 10^{-3}, \qquad \frac{\tau_{\mu^+} - \tau_{\mu^-}}{\tau_{\mu^+}} < 2 \times 10^{-4},$$

$$\frac{\tau_{\pi^+} - \tau_{\pi^-}}{\tau_{\pi^+}} < 2 \times 10^{-3}, \qquad \frac{\tau_{K^+} - \tau_{K^-}}{\tau_{K^+}} < 2 \times 10^{-4}.$$

$$(14.1)$$

For proper evaluation we should keep the following in mind:

- The masses provide a very ad-hoc calibration stick for a *difference* in particle–antiparticle masses, since the (overwhelming bulk of the) masses is not generated by the weak forces.

- While the decay widths in Eq. (14.1) are of weak origin, bounds $\sim \mathcal{O}(\text{few} \times 10^{-4})$ can hardly be called decisive *empirical* verifications of a fundamental symmetry.

Another bound is usually quoted as best test:

$$\frac{|M_{\overline{K}^0} - M_{K^0}|}{M_K} < 9 \times 10^{-19} \triangleq \frac{|M_{\overline{K}^0} - M_{K^0}|}{\Gamma_{K_S}} < 7 \times 10^{-5}. \qquad (14.2)$$

---

[2] **CPT** invariance requires the charge q of particle and antiparticle to be the same in magnitude and opposite in sign.

Later in this chapter we will derive it and state the assumptions that go into it. For now we want to state that even with the caveat just mentioned, Eq. (14.2) appears impressive – as also indicated by the more meaningful comparison to Γ_{K_S}. It demonstrates that the interferometry in particle–antiparticle oscillations allows the measurement of truly tiny energy differences.

14.2 Theoretical scenarios

Proofs of the **CPT** theorem are based on some very natural axioms like Lorentz invariance, the existence of a unique ground state and local fields obeying the 'correct' spin-statistics connection, see Chapter 4. In addition they involve some more technical assumptions, like the fields containing a *finite* number of components only – i.e., belonging to a *finite* dimensional representation of the Lorentz group. In 1968 Oksak and Todorov gave two explicit examples for a Lorentz invariant free field theory that does *not* obey **CPT** invariance. The fields form unitary representations of $SL(2, C)$ – the special linear group in two dimensions with complex coefficients – and contain an *infinite* number of components[3].

While this might be viewed as an academic curiosity, we can face the foundations of the **CPT** theorem more squarely by casting doubts on the absolute validity of *Lorentz invariance*. This had been done in Ref. [191] and was revived in the context of so-called tumbling gauge theories in Ref. [192]: Lorentz breaking is engineered by postulating a *vector* condensate to form in the vacuum: $\langle 0|\overline{\psi}\gamma_\mu\psi|0\rangle \neq 0$. This apparent Lorentz violation could still get relegated to an unphysical sector if $\langle 0|\overline{\psi}\gamma_\mu\psi|0\rangle$ takes on the role of a gauge fixing vector [191, 192]. Alternatively, it could lead to physical effects [193]. Since $(\mathbf{CPT})^{-1}\overline{\psi}\gamma_\mu\psi(x)\mathbf{CPT} = -\overline{\psi}\gamma_\mu\psi(-x)$, its VEV would break **CPT** symmetry. Fermion masses can then obtain a contribution that is not a Lorentz scalar, but the zeroth component of a vector, which would lift the degeneracy in the masses of particles and antiparticles [194].

The even more radical suggestion has been made that Lorentz symmetry is merely an approximate infrared phenomenon [195]. Unlike the previous scenario, the invariance is now broken at the highest energy scales like the Planck mass. Yet there is an attractive infrared fixed point in the renormalization flow that is Lorentz invariant; i.e., the coefficients of *non*-covariant operators, which are assumed to be of order unity at the Planck scale, are reduced when the scales are lowered and get renormalized down to zero in the infrared limit.

As mentioned above, recent theoretical developments have provided some fresh impetus to raising these seemingly esoteric questions. Superstring theories are defined in 10 dimensions in a (generalized) Lorentz invariant way. It has been argued [196] that in the compactification down to four space-time dimensions *spontaneous* violation of Lorentz invariance can arise, due to intrinsically non-

---

[3] One of the examples satisfies the spin-statistics theorem, the other one does not. We should note that this theorem does in general not hold for infinite-component fields.

local string interactions which lead to the emergence of VEVs for some non-scalar Lorentz operators. This is somewhat similar to the scenario sketched before and generates **CPT** violation.

These speculations are certainly on the more exotic side, yet should not be dismissed completely out-of-hand. The basic idea behind them can be formulated as follows: it is quite conceivable that at the Planck scale – the domain of quantum gravity – **CPT** and Lorentz invariance are of limited validity. It is likewise conceivable that such effects can get communicated to lower energies. The real and most difficult question – to which we do not know the answer at all – is whether and where they can retain an observable magnitude. In the absence of any reliable guidance from theory, we will concentrate on those processes that

- promise us the highest experimental sensitivity and

- differentiate between different flavour sectors.

14.3 CPT phenomenology for neutral kaons

Without **CPT** constraint, the two neutral kaon mass eigenstates are still described by linear combinations of K^0 and $\overline{K}^0$, yet with two coefficients q_1/p_1 and q_2/p_2 rather than the single term q/p [58, 197, 198, 199], as described in Chapter 6. Consequently, even a perfect determination of the K_L state would *not* tell us the exact composition of the K_S state; i.e., there are more independent ways for **CP** violation to manifest itself.

14.3.1 Semileptonic decays

Rather than q_1/p_1 and q_2/p_2, we will use here two other parameters to describe the kaon states, namely $\cos\theta$ and $i\phi$ [197]:

$$|K^0(t)\rangle = \tfrac{1}{2}\left[(1+\cos\theta)|K^0\rangle + (1+i\phi)|\overline{K}^0\rangle\right]e^{-iM_st}e^{-\frac{1}{2}\Gamma_st}$$
$$+\tfrac{1}{2}\left[(1-\cos\theta)|K^0\rangle - (1+i\phi)|\overline{K}^0\rangle\right]e^{-iM_Lt}e^{-\frac{1}{2}\Gamma_Lt},$$
$$|\overline{K}^0(t)\rangle = \tfrac{1}{2}\left[(1-\cos\theta)|\overline{K}^0\rangle + (1-i\phi)|K^0\rangle\right]e^{-iM_st}e^{-\frac{1}{2}\Gamma_st}$$
$$+\tfrac{1}{2}\left[(1+\cos\theta)|\overline{K}^0\rangle - (1-i\phi)|K^0\rangle\right]e^{-iM_Lt}e^{-\frac{1}{2}\Gamma_Lt}. \tag{14.3}$$

In the notation of Eq. (7.14) for the semileptonic decay amplitudes, we can compute the following decay rates:

$$\Gamma(K^0(t) \to \pi^-l^+v) = \left|\frac{F_l}{2}\right|^2\Big[\{1 - 2Re\,y_l - 2Re\,(\cos\theta + x_l)\}\,e^{-\Gamma_Lt}$$
$$+\{1 - 2\mathrm{Re}\,y_l + 2Re\,(\cos\theta + x_l)\}\,e^{-\Gamma_st}$$
$$+2e^{-\overline{\Gamma}t}\{(1 - 2\mathrm{Re}\,y_l)\cos(\Delta M\,t)$$
$$-2\mathrm{Im}\,(\cos\theta + x_l)\sin(\Delta M\,t)\}\Big], \tag{14.4}$$

$$\Gamma(\overline{K}^0(t) \to \pi^- l^+ v) = \left|\frac{F_l}{2}\right|^2 \left[\{1 + 2\text{Re}\,(-i\phi - y_l) - 2\text{Re}\,x_l\}\, e^{-\Gamma_L t}\right.$$
$$+ \{1 + 2\text{Re}\,(-i\phi - y_l) + 2\text{Re}\,x_l\}\, e^{-\Gamma_S t}$$
$$- 2e^{-\overline{\Gamma} t}\{[1 + 2\text{Re}\,(-i\phi - y_l)]\cos(\Delta M t)$$
$$\left. - 2\text{Im}\,x_l \sin(\Delta M t)\}\right], \tag{14.5}$$

$$\Gamma(K^0(t) \to \pi^+ l^- \overline{v}) = \left|\frac{F_l}{2}\right|^2 \left[\{1 - 2\text{Re}\,(-i\phi - y_l) - 2\text{Re}\,\overline{x}_l\}\, e^{-\Gamma_L t}\right.$$
$$+ \{1 - 2\text{Re}\,(-i\phi - y_l) + 2\text{Re}\,\overline{x}_l\}\, e^{-\Gamma_S t}$$
$$- 2e^{-\overline{\Gamma} t}\{[1 - 2\text{Re}\,(-i\phi - y_l)]\cos(\Delta M t)$$
$$\left. + 2\text{Im}\,\overline{x}_l \sin(\Delta M t)\}\right], \tag{14.6}$$

$$\Gamma(\overline{K}^0(t) \to \pi^+ l^- \overline{v}) = \left|\frac{F_l}{2}\right|^2 \left[\{1 + 2\text{Re}\,y_l + 2\text{Re}\,(\cos\theta - \overline{x}_l)\}\, e^{-\Gamma_L t}\right.$$
$$+ \{1 + 2\text{Re}\,y_l - 2\text{Re}\,(\cos\theta - \overline{x}_l)\}\, e^{-\Gamma_S t}$$
$$+ 2e^{-\overline{\Gamma} t}\{[1 + 2\text{Re}\,y_l]\cos(\Delta M t)$$
$$\left. + 2\text{Im}\,(\cos\theta + \overline{x}_l)\sin(\Delta M t)\}\right], \tag{14.7}$$

with $\overline{\Gamma} = (\Gamma_L + \Gamma_S)/2$ and where we recall the following constraints:

$\Delta S = \Delta Q$ rule:	$x_l = \overline{x}_l = 0$
CP invariance:	$x_l = \overline{x}_l^*$; $F_l = F_l^*$; $y_l = -y_l^*$; $\cos\theta = 0$; $\phi = 0$
T invariance:	$\text{Im}\,F = \text{Im}\,y_l = \text{Im}\,x_l = \text{Im}\,\overline{x}_l = 0$; $\phi = 0$
CPT invariance:	$y_l = 0$; $x_l = \overline{x}_l$; $\cos\theta = 0$.

Eq. (14.4)–Eq. (14.7), which hold to first order in $\cos\theta, \phi, x_l$, and y_l, are the master equations from which we form various combinations to test these different selection rules. For very late decay times, $t \gg \Gamma_S^{-1}$, we can write down much simpler expressions:

$$\Gamma(K^0(t) \to \pi^- l^+ v) \simeq \left|\frac{F_l}{2}\right|^2 e^{-\Gamma_L t}[1 - 2\text{Re}\,y_l - 2\text{Re}(\cos\theta + x_l)]$$

$$\Gamma(\overline{K}^0(t) \to \pi^- l^+ v) \simeq \left|\frac{F_l}{2}\right|^2 e^{-\Gamma_L t}[1 + 2\text{Re}(-i\phi - y_l) - 2\text{Re}\,x_l]$$

$$\Gamma(K^0(t) \to \pi^+ l^- \overline{v}) \simeq \left|\frac{F_l}{2}\right|^2 e^{-\Gamma_L t}[1 - 2\text{Re}(-i\phi - y_l) - 2\text{Re}\,\overline{x}_l]$$

$$\Gamma(\overline{K}^0(t) \to \pi^+ l^- \overline{v}) \simeq \left|\frac{F_l}{2}\right|^2 e^{-\Gamma_L t}[1 + 2\text{Re}\,y_l + 2\text{Re}(\cos\theta - \overline{x}_l)]. \tag{14.8}$$

•*Test of the* $\Delta S = \Delta Q$ *rule and measurement of* ΔM

Semileptonic decays allow us to track the strangeness of the meson *at the time of decay* – *if* the $\Delta Q = \Delta S$ rule holds. This can be analysed through the following ratio:

$$A_{\Delta M} = \frac{\Gamma(K^0(t) \to e^+) + \Gamma(\overline{K}^0(t) \to e^-) - \Gamma(K^0(t) \to e^-) - \Gamma(\overline{K}^0(t) \to e^+)}{\Gamma(K^0(t) \to e^+) + \Gamma(\overline{K}^0(t) \to e^-) + \Gamma(K^0(t) \to e^-) + \Gamma(\overline{K}^0(t) \to e^+)}$$

$$= \frac{2e^{-\overline{\Gamma}t}[\cos \Delta Mt + \mathrm{Im}\,(\overline{x}_l - x_l) \sin \Delta Mt]}{[1 + \mathrm{Re}\,(x_l + \overline{x}_l)]e^{-\Gamma_S t} + [1 - \mathrm{Re}\,(x_l + \overline{x}_l)]e^{-\Gamma_L t}}, \tag{14.9}$$

where the short-hand notation $\Gamma(K^0(t) \to e^+) \equiv \Gamma(K^0(t) \to e^+ \nu \pi^-)$ etc. has been employed. Given enough statistics, we can separate out $\mathrm{Re}\,(x_l + \overline{x}_l)$, $\mathrm{Im}\,(x_l - \overline{x}_l)$ and ΔM.

After a long hiatus with no experimental activity on this topic, the CPLEAR collaboration has performed sensitive studies of Eq. (14.9). With their limited statistics, they chose to assume **CPT** invariance for the semileptonic $\Delta S = 1$ (but not for the $\Delta S = 2$) dynamics: $x_l = \overline{x}_l$. Tracking the evolution in time t of Eq. (14.9) thus simplified, they have obtained [201]:

$$\mathrm{Re}\,x_l = (-1.8 \pm 4.1 \pm 4.5) \times 10^{-3}$$
$$\Delta M = (529.5 \pm 2.0 \pm 0.3) \times 10^7 \hbar/s\,. \tag{14.10}$$

•*Tests of* **T** *and* **CPT** *invariance*

Since

$$\Gamma(\overline{K}^0 \to K^0) \overset{\mathbf{T}}{\Longrightarrow} \Gamma(K^0 \to \overline{K}^0)$$
$$\Gamma(\overline{K}^0 \to \overline{K}^0) \overset{\mathbf{CPT}}{\Longrightarrow} \Gamma(K^0 \to K^0), \tag{14.11}$$

we can probe **T** and **CPT** symmetry through the following ratios:

$$\mathscr{A}_T(t) = \frac{\Gamma(\overline{K}^0 \to K^0(t)) - \Gamma(K^0 \to \overline{K}^0(t))}{\Gamma(\overline{K}^0 \to K^0(t)) + \Gamma(K^0 \to \overline{K}^0(t))}$$

$$\mathscr{A}_{CPT}(t) = \frac{\Gamma(\overline{K}^0 \to \overline{K}^0(t)) - \Gamma(K^0 \to K^0(t))}{\Gamma(\overline{K}^0 \to \overline{K}^0(t)) + \Gamma(K^0 \to K^0(t))}. \tag{14.12}$$

We now define asymmetries which can be measured experimentally. The above two asymmetries can be re-expressed as follows:

$$A_T(t) = \frac{\Gamma(\overline{K}^0(t) \to e^+) - \Gamma(K^0(t) \to e^-)}{\Gamma(\overline{K}^0 \to e^+) + \Gamma(K^0 \to e^-)}$$

$$= \frac{ae^{-\Gamma_L t} + be^{-\Gamma_S t} + ce^{-\overline{\Gamma}t}\cos \Delta Mt + de^{-\overline{\Gamma}t}\sin \Delta Mt}{e^{-\Gamma_L t} + e^{-\Gamma_S t} - 2e^{-\overline{\Gamma}t}\cos \Delta Mt}$$

$$\simeq a \quad \text{for } t \gg \Gamma_S^{-1}. \tag{14.13}$$

$$
\begin{aligned}
A_{CPT}(t) &= \frac{\Gamma(\overline{K}^0(t) \to e^-) - \Gamma(K^0(t) \to e^+)}{\Gamma(\overline{K}^0 \to e^-) + \Gamma(K^0 \to e^+)} \\
&= \frac{\tilde{a}e^{-\Gamma_L t} + \tilde{b}e^{-\Gamma_S t} + \tilde{c}e^{-\overline{\Gamma}t}\cos\Delta Mt + \tilde{d}e^{-\overline{\Gamma}t}\sin\Delta Mt}{e^{-\Gamma_L t} + e^{-\Gamma_S t} + 2e^{-\overline{\Gamma}t}\cos\Delta Mt} \\
&\simeq \tilde{a} \quad \text{for } t \gg \Gamma_S^{-1},
\end{aligned}
\tag{14.14}
$$

with the following notation:

$$
\begin{aligned}
a &= 2\text{Im}\,\phi - 2\text{Re}\,y_l - \text{Re}\,(x_l - \overline{x}_l) \\
b &= 2\text{Im}\,\phi - 2\text{Re}\,y_l + \text{Re}\,(x_l - \overline{x}_l) \\
c &= -4(\text{Im}\,\phi - \text{Re}\,y_l) \\
d &= 2\text{Im}\,(x_l + \overline{x}_l)
\end{aligned}
\tag{14.15}
$$

$$
\begin{aligned}
\tilde{a} &= 2\text{Re}\,y_l + 2\text{Re}\cos\theta + \text{Re}\,(x_l - \overline{x}_l) \\
\tilde{b} &= 2\text{Re}\,y_l - 2\text{Re}\cos\theta - \text{Re}\,(x_l - \overline{x}_l) \\
\tilde{c} &= 4\text{Re}\,y_l \\
\tilde{d} &= 4\text{Im}\cos\theta + 2\text{Im}\,(x_l + \overline{x}_l).
\end{aligned}
\tag{14.16}
$$

In going from Eq. (14.12) to Eq. (14.13), the interpretation of A_T has changed significantly: $A_T \neq 0$ *now does not establish* **T** *violation!* If $\phi = \text{Im}x_l = \text{Im}\overline{x}_l = \text{Im}y_l = 0$ – as required by **T** invariance (see above) – we still have

$$
A_T(t) \simeq -2\text{Re}y_l - \text{Re}(x_l - \overline{x}_l) = -2y_l - (x_l - \overline{x}_l) \quad \text{for } t \gg \Gamma_S^{-1}; \tag{14.17}
$$

i.e., $A_T \neq 0$ can be due to **CPT** not being conserved in two variants: (i) $y_l \neq 0$, which in turn corresponds to **CP** violation as well, since $y_l = -y_l^*$ cannot be satisfied for Re $y_l \neq 0$. This is of course consistent with **T** invariance implying **CP** invariance unless **CPT** is violated. (ii)$x_l \neq \overline{x}_l$, i.e., a limitation of the $\Delta S = \Delta Q$ rule. Since this rule is used to track the oscillating strangeness of the kaons through their semileptonic decays, it is easy to understand how its violation can *fake* a **T** and **CPT** asymmetry.

We obviously have

$$
A_T(t) + A_{CPT}(t) \simeq 2(\text{Im}\,\phi + \text{Re}\cos\theta) \quad \text{for } t \gg \Gamma_S^{-1}; \tag{14.18}
$$

i.e., the sum of these two asymmetries is independent of y_l, x_l and $\overline{x}_l$; if it does not vanish, then **T** and/or **CPT** are violated.

The CPLEAR collaboration has also studied the combination:

$$
\begin{aligned}
A_\delta(t) &= \frac{\Gamma(\overline{K}(t) \to l^+) - \Gamma(K(t) \to l^-)\alpha}{\Gamma(\overline{K}(t) \to l^+) + \Gamma(K(t) \to l^-)\alpha} \\
&+ \frac{\Gamma(\overline{K}(t) \to l^-) - \Gamma(K(t) \to l^+)\alpha}{\Gamma(\overline{K}(t) \to l^-) + \Gamma(K(t) \to l^+)\alpha},
\end{aligned}
\tag{14.19}
$$

where $\alpha = \left|\frac{p_2}{q_2}\right|^2$.

For large times t, $A_\delta(t) \to 4\text{Re} \cos \theta$ and they found [200]

$$\text{Re} \cos \theta = (6.0 \pm 6.6 \pm 1.2) \times 10^{-4}; \qquad (14.20)$$

studying the time dependence, they obtained

$$\text{Im} \cos \theta = (-3.0 \pm 4.6 \pm 0.6) \times 10^{-2},$$

$$\frac{1}{2}\text{Re}\,(x_l - \bar{x}_l) = (0.2 \pm 1.3 \pm 0.3) \times 10^{-2},$$

$$\frac{1}{2}\text{Im}\,(x_l + \bar{x}_l) = (1.2 \pm 2.2 \pm 0.3) \times 10^{-2}. \qquad (14.21)$$

Now that we know something about **CPT** violation, consider:

$$A_T(t) + A_{CPT}(t) \simeq 2(\text{Im}\,\phi + \text{Re}\cos\theta) \quad \text{for } t \gg \Gamma_S^{-1}\,; \qquad (14.22)$$

i.e., the sum of these two asymmetries is independent of y_l, x_l and $\bar{x}_l$; if it does not vanish, then **T** and/or **CPT** are violated.

Averaging $A_T(t)$ and $A_{CPT}(t)$ over the range $\tau_S \leq t \leq 20\,\tau_S$, CPLEAR has found [202, 203]:

$$\langle A_T \rangle = (6.6 \pm 1.3 \pm 1.0) \times 10^{-3}$$

$$\langle A_{CPT} \rangle = (0.07 \pm 0.50 \pm 0.45) \times 10^{-3}. \qquad (14.23)$$

Then we have

$$\text{Im}\,\phi \simeq (3.3 \pm 0.1 \pm 0.5) \times 10^{-3}. \qquad (14.24)$$

These data are fully consistent with the observed **CP** violation in K_L decays being matched by a corresponding **T** violation as required by **CPT** symmetry.

As the book was being sent to press, CPLEAR [204] announced a new result for the right hand side of Eq. (14.17). They obtained $(-0.4 \pm 0.6) \times 10^{-3}$. This together with the value for $\langle A_T \rangle$ given in Eq. (14.23) allow us to conclude that again $\text{Im}\,\phi \neq 0$. This is a new observation of **T** violation.

•*Charge asymmetry in* $K_L \to l^\pm \nu \pi^\mp$

Another relevant observable is the charge asymmetry in semileptonic K_L decays:

$$\delta_{\text{Lept}} = \frac{\Gamma(K_L \to l^+\nu\pi^-) - \Gamma(K_L \to l^-\nu\pi^+)}{\Gamma(K_L \to l^+\nu\pi^-) + \Gamma(K_L \to l^-\nu\pi^+)}$$

$$= \text{Im}\,\phi - \text{Re}\cos\theta - \text{Re}\,(x_l - \bar{x}_l) - 2\text{Re}\,y_l. \qquad (14.25)$$

Unlike the asymmetries A_T and A_{CPT}, $\delta_l \neq 0$ does not distinguish between **CP** and **CPT** violation. Data yield [28]:

$$\delta_{\text{Lept}} = (3.27 \pm 0.12) \times 10^{-3}. \qquad (14.26)$$

14.3.2   Nonleptonic K^{neut} decays

We proceed in close analogy (including the notation) to the derivation for ϵ and ϵ' given in Chapter 7 for the **CPT** symmetric case. We obtain (see Problem 14.2):

$$\epsilon \;\simeq\; \frac{1}{2}\left(1 - \frac{q_2}{p_2}\frac{\overline{A}_0}{A_0}\right)$$

$$\epsilon' \;\simeq\; \frac{1}{2\sqrt{2}}\omega e^{i(\delta_2-\delta_0)}\frac{q_2}{p_2}\left(\frac{\overline{A}_0}{A_0} - \frac{\overline{A}_2}{A_2}\right), \tag{14.27}$$

which illustrate some important general features:

- As said before, **CPT** violation in the $\Delta S = 2$ sector leads to the two mass eigenstates being described by two different mixing parameters q_1/p_1 and q_2/p_2.

- ϵ and ϵ' measured in $K_L \to \pi\pi$ depend on $\Delta_I = 1-(q_2\overline{A}_I)/(p_2A_I)$, yet are *not* sensitive to q_1/p_1. In Sect. 7.7 we have discussed $K^0, \overline{K}^0 \to \pi^+\pi^-\pi^0$ decays; from Eq. (7.76) we read off

$$A_{+-0}(\infty) = |A_L(3\pi)|^2\left(1 - \left|\frac{q_1}{p_1}\right|\right); \tag{14.28}$$

  i.e., we can extract $|q_1/p_1|$ from the **CP** asymmetry in $K_L \to \pi^+\pi^-\pi^0$ and compare it with $|q_2/p_2|$ from $K_L \to \pi^+\pi^-$. Nature seems to be kind to us. By looking at 2π and 3π decay modes of K_L we can extract *both* $\frac{q_1}{p_1}$ and $\frac{q_2}{p_2}$.

- *Direct* **CPT** violation in $\Delta S = 1$ amplitudes leads to $|\overline{A}_i| \neq |A_i|$.

  The amount of **CPT** – and **CP** violation as well – can thus depend on the isospin I of the final state. Eq. (14.27) shows there are then three gateways through which a direct asymmetry can emerge in $K \to \pi\pi$: $|A_0| \neq |\overline{A}_0|$ and $|A_2| \neq |\overline{A}_2|$ in addition to $\arg(A_0\overline{A}_2/\overline{A}_0A_2) \neq 0$, which can occur already within **CPT** symmetry.

- As discussed in Chapter 7, within **CPT** symmetry, since the phase of ϵ – $\arg\epsilon \simeq \phi_{SW} = \tan^{-1}(\Delta M/\Delta\Gamma) = 43.46° \pm 0.08°$ – and the strong phase shift basically coincide – $\delta_0 - \delta_2 = 42° \pm 4°$, the relative phase between ϵ and ϵ' is either close to $0°$ or $180°$. A component of ϵ' *orthogonal* to ϵ thus breaks **CPT** invariance.

●**CPT** *Tests in the Phases of* η_{+-} *and* η_{00}

The most stringent test of **CPT** is provided by the phases of η_{+-} and η_{00}, ϕ_{+-} and ϕ_{00}, respectively. In Problem 14.4 we are going to derive [206]

$$\frac{|\eta_{+-}|\left(\frac{2}{3}\phi_{+-} + \frac{1}{3}\phi_{00} - \phi_{SW}\right)}{\sin\phi_{SW}} = \left(\frac{M_{\overline{K}} - M_K}{2\Delta M_K} + R_{\text{direct}}\right). \tag{14.29}$$

Here

$$R_{\text{direct}} = \frac{-ie^{-i\phi_{SW}}}{\sin \phi_{SW}} \sum_{f \neq (2\pi)_{I=0}} \epsilon(f) + \frac{1}{2}(\text{Re} \frac{\overline{A}_0}{A_0} - 1) \qquad (14.30)$$

where $\epsilon(f)$ is defined by Eq. (9.55).

With the data yielding $\phi_{+-} = (43.5 \pm 0.6)°$ and $\Delta\phi = \phi_{00} - \phi_{+-} = (-0.1 \pm 0.8)°$ [28, 205] we obtain

$$\frac{2}{3}\phi_{+-} + \frac{1}{3}\phi_{00} - \phi_{SW} \simeq (0.01 \pm 0.66)° \qquad (14.31)$$

we find [206]

$$\left(\frac{M_{\overline{K}} - M_K}{2\Delta M_K} + R_{\text{direct}}\right) = (0.06 \pm 4.0) \times 10^{-5}, \qquad (14.32)$$

which is completely consistent with being zero. A few comments might help in evaluating Eq. (14.32).

- Assuming $R_{\text{direct}} = 0$ – Eq. (14.32) can be re-expressed as follows:

$$\frac{M_{\overline{K}} - M_K}{M_K} = (0.08 \pm 5.6) \times 10^{-19}, \qquad (14.33)$$

  which is the updated version of an often quoted bound.

- Eq. (14.33) looks numerically truly spectacular – yet it overstates the case since M_K is largely generated by the strong interactions! A more reasonable scale for the $\Delta S = 2$ difference $M_{\overline{K}} - M_K$ is provided noting that **CP** invariance would yield $M_{\overline{K}} - M_K = 0$ even if **CPT** symmetry were invalid. So the most natural calibrator for $M_{\overline{K}} - M_K$ might be $|\eta|\Gamma_S$; Eq. (14.33) is then re-expressed as follows:

$$\frac{M_{\overline{K}} - M_K}{|\eta|\Gamma_S} = (0.02 \pm 1.6) \times 10^{-2}. \qquad (14.34)$$

  The numerical accuracy here is good, though not overwhelming!

- Using the previously obtained bound $\sum_f \epsilon(f) < 10^{-5}$ derived in Eq. (9.70), we have

$$R_{\text{direct}} < 10^{-5} \qquad (14.35)$$

  to arrive at

$$\left(\frac{M_{\overline{K}} - M_K}{2\Delta M_K}\right) = (0.06 \pm 4.0_{\text{exp.}} \pm 1_{\text{theory}}) \times 10^{-5}, \qquad (14.36)$$

  which is completely consistent with being zero.

- Casting this to a test of **CPT** symmetry,

$$\left|\frac{2}{3}\phi_{+-} + \frac{1}{3}\phi_{00} - \phi_{SW}\right| = (0.01 \pm 0.66_{\text{exp.}} \pm 0.2_{\text{th.}})°. \qquad (14.37)$$

So, comparing the right hand side to ϕ_{SW}, we have tested **CPT** at the level of 1.5 % and if the statistics improves, we have a potential to test **CPT** to the level of 4×10^{-3}.

- Strictly speaking, in deriving $\sum_f \epsilon(f) < 10^{-5}$, we assumed that direct **CP** violation in the $3\pi^0$ channel is bounded by ϵ. While it is probably a very good assumption, it is an assumption, nevertheless. More work in the $K_L \to 3\pi^0$ channel is needed to relax this assumption.

14.4 Harnessing EPR correlations

In Chapter 10 we saw that the reaction

$$e^+e^- \to \Upsilon(4S) \to B_d \overline{B}_d \tag{14.38}$$

enables us to perform novel measurements of high sensitivity:

- Both beauty mesons can oscillate into each other.

- Since the $B\overline{B}$ pair is produced into a *single coherent quantum mechanical state* their oscillations are highly *correlated* with each other. We have actually an unusual EPR situation at hand [207]: irrespective of **CPT** invariance and of how fast the $B_d - \overline{B}_d$ oscillations proceed, the **C** *odd* $B_d \overline{B}_d$ pair in Eq. (14.38) can never transmogrify itself into a $B_d B_d$ or $\overline{B}_d \overline{B}_d$ pair; this is a consequence of Bose statistics. Equivalently, we can say that the $\Upsilon(4S)$ resonance always decays into two *different* neutral *mass* eigenstates B_1 and B_2:

$$e^+e^- \to \Upsilon(4S) \to B_1 B_2 \; ; \; \not\to B_1 B_1, \; B_2 B_2. \tag{14.39}$$

Yet we cannot predict whether the particular neutral meson reaching the detector will turn out to be B_1 or B_2 any more than whether it will be a B_d or $\overline{B}_d$.

- Studying

$$e^+e^- \to \Upsilon(4S) \to B_d \overline{B}_d \to f_1 f_2 \tag{14.40}$$

for various channels $B_d/\overline{B}_d \to f_{1,2}$ thus represents double interferometry, which typically allows us to measure tiny quantities like $\Delta M(B_d)$ and phases that otherwise are unaccessible. Some examples are discussed in the Problems section of Chapter 10.

Now we want to go beyond these statements:

- Qualitatively, we encounter an analogous quantum mechanical situation at ϕ factories[4]

$$e^+e^- \rightarrow \phi(1020) \rightarrow K_S K_L. \tag{14.41}$$

Yet big quantitative differences arise due to $\Gamma(K_S) \gg \Gamma(K_L)$ vs $\Gamma(B_1) \simeq \Gamma(B_2)$.

- The high sensitivity of the double interferometry referred to above can be harnessed to probe for **CPT** violation.

- The specific form predicted for the EPR correlations is based on the superposition principle. A detailed analysis of the observed correlations will probe for the presence of *nonlinear* effects beyond ordinary quantum mechanics.

The formalism presented in Chapter 6 is completely general and applicable even without **CPT** symmetry. The amplitude for finding a final state f_1 emerging at time t_k and f_2 at time t_{-k} is given by [208]

$$
\begin{aligned}
A &\left(P^0 \overline{P}^0 |_{C=\pm} \rightarrow f_1(t_k) f_2(t_{-k}) \right) \\
&= \left(g_+(t_k) \overline{g}_-(t_{-k}) + (-1)^C \overline{g}_-(t_k) g_+(t_{-k}) \right) A(f_1) A(f_2) \\
&+ \left(g_+(t_k) \overline{g}_+(t_{-k}) + (-1)^C \overline{g}_-(t_k) g_-(t_{-k}) \right) A(f_1) \overline{A}(f_2) \\
&+ \left(g_-(t_k) \overline{g}_-(t_{-k}) + (-1)^C \overline{g}_+(t_k) g_+(t_{-k}) \right) \overline{A}(f_1) A(f_2) \\
&+ \left(g_-(t_k) \overline{g}_+(t_{-k}) + (-1)^C \overline{g}_+(t_k) g_-(t_{-k}) \right) \overline{A}(f_1) \overline{A}(f_2),
\end{aligned} \tag{14.42}
$$

where

$$
\begin{aligned}
g_+(t) &= f_+(t) + \cos\theta f_-(t) \\
g_-(t) &= \sin\theta e^{i\phi} f_-(t) \\
\overline{g}_+(t) &= f_+(t) - \cos\theta f_-(t) \\
\overline{g}_-(t) &= \sin\theta e^{-i\phi} f_-(t),
\end{aligned} \tag{14.43}
$$

and the functions $f_\pm(t)$ are as defined by Eq. (6.48):

$$f_\pm(t) = \frac{1}{2} e^{-iM_1 t} e^{-\frac{1}{2}\Gamma_1 t} \left[1 \pm e^{-i\Delta M t} e^{\frac{1}{2}\Delta\Gamma t} \right].$$

As stated in Eq. (6.18), $\cos\theta \neq 0$ [$\phi \neq 0$] requires simultaneous **CPT** [**T**] and **CP** violation.

It is obvious from Eq. (14.42) that by studying the rate for a certain final state f_1 appearing at some time and f_2 an interval Δt later, we can – at least in principle – extract $\cos\theta$, ϕ and the transformation properties of the matrix elements. We will sketch the procedure for ϕ and beauty factories.

---

[4] Likewise for the reaction $e^+e^- \rightarrow \psi'' \rightarrow D^0 \overline{D}^0 / D_1 D_2$.

14.4.1 φ factory

It might be of help to repeat some statements on the quantum mechanical analysis of

$$e^+e^- \to \phi(1020) \to K^0\overline{K}^0. \tag{14.44}$$

The kaon pair is produced in a coherent **C** *odd* quantum state[5], the evolution of which is described in terms of a single time variable – until one of the kaons decays at time $t = t_1$. We infer from Bose statistics that *up to that moment* the two kaons have to remain *distinct*; this is expressed most concisely through their mass eigenstates:

$$|K^0\overline{K}^0; t = 0\rangle = |K_SK_L; t = 0\rangle \rightsquigarrow |K_SK_L; t > 0\rangle \quad \text{for } t \le t_1. \tag{14.45}$$

Once the first kaon decays as, say, a K^0 or K_S, the other keeps evolving as a $\overline{K}^0$ or K_L, respectively, till it meets its decay at time t_2.

Even at a symmetric ϕ factory like DAΦNE [209] the vertices for the kaon decays can be resolved with existing technology. Yet the *primary* vertex where the kaons were produced is known very poorly. We can therefore deduce quite reliably the *difference* in the (proper) time of decay – $\Delta t = t_2 - t_1$ – but not t_1 and t_2 themselves. In consequence we have to integrated the predicted distributions over $t_1 + t_2$ to obtain observable effects:

$$|A(\Delta t; f_1; f_2)|^2 \equiv \frac{1}{2} \int_{|\Delta t|}^{\infty} d(t_1 + t_2)|A(f_1, t_1; f_2, t_2)|^2. \tag{14.46}$$

We can then consider two types of asymmetries, namely

- those where we compare the rates *integrated* over $\Delta t > 0$ and $\Delta t < 0$, respectively, and

- those that are given as a function of Δt, see Eq. (14.46).

As final states we can employ $f = \pi\pi, l^\pm \nu \pi^\mp$. In this way all the quantities of interest – $\cos\theta$, ϕ, y_l, x_l and $\overline{x}_l$ can be measured. Since a detailed discussion can be found in Ref. [197], we limit ourselves to stating the main results.

●*Time-integrated asymmetries*

With **CPT** symmetry requiring $|A(\Delta t; \pi^- l^+ \nu_l, \pi^+ l^- \overline{\nu}_l)|^2$ to be an *even* function of Δt, the following asymmetry measures **CPT** breaking with or without the $\Delta S = \Delta Q$ rule:

$$\langle \mathscr{A}_{l^+l^-} \rangle \equiv \frac{\int_0^\infty d(\Delta t)|A(\Delta t; \pi^- l^+ \nu_l, \pi^+ l^- \overline{\nu}_l)|^2 - \int_{-\infty}^0 d(\Delta t)|A(\Delta t; \pi^- l^+ \nu_l, \pi^+ l^- \overline{\nu}_l)|^2}{\int_0^\infty d(\Delta t)|A(\Delta t; \pi^- l^+ \nu_l, \pi^+ l^- \overline{\nu}_l)|^2 + \int_{-\infty}^0 d(\Delta t)|A(\Delta t; \pi^- l^+ \nu_l, \pi^+ l^- \overline{\nu}_l)|^2}$$

$$= -2\text{Re}(\cos\theta + x_l - \overline{x}_l^*). \tag{14.47}$$

---

[5] The radiative process $\phi \to K\overline{K}\gamma$ provides a negligible background.

If both kaons decay through the same semileptonic mode, we can compare *like*-sign dilepton final states

$$\langle \mathscr{A}_{l\pm l\pm} \rangle \equiv \frac{\int_{-\infty}^{\infty} \mathrm{d}(\Delta t) |A(\Delta t; \pi^+ l^- \overline{\nu}_l, \pi^+ l^- \overline{\nu}_l)|^2 - \int_{-\infty}^{\infty} \mathrm{d}(\Delta t) |A(\Delta t; \pi^- l^+ \nu_l, \pi^- l^+ \nu_l)|^2}{\int_{-\infty}^{\infty} \mathrm{d}(\Delta t) |A(\Delta t; \pi^+ l^- \overline{\nu}_l, \pi^+ l^- \overline{\nu}_l)|^2 + \int_{-\infty}^{\infty} \mathrm{d}(\Delta t) |A(\Delta t; \pi^- l^+ \nu_l, \pi^- l^+ \nu_l)|^2}$$

$$= 4\mathrm{Re}\, y_l - 2\mathrm{Im}\, \phi \; ; \tag{14.48}$$

i.e., $\langle \mathscr{A}_{l\pm l\pm} \rangle \neq 0$ reveals **T** and **CP** breaking – $\phi \neq 0$ – and/or **CPT** and **CP** violation – Re $y_l \neq 0$.

•Δt-dependent asymmetries

Assuming both decays to be semileptonic, we find

$$\mathscr{A}_{l+l-}(\Delta t) \equiv \frac{|A(\Delta t > 0; \pi^- l^+ \nu_l, \pi^+ l^- \overline{\nu}_l)|^2 - |A(\Delta t < 0; \pi^- l^+ \nu_l, \pi^+ l^- \overline{\nu}_l)|^2}{|A(\Delta t > 0; \pi^- l^+ \nu_l, \pi^+ l^- \overline{\nu}_l)|^2 + |A(\Delta t < 0; \pi^- l^+ \nu_l, \pi^+ l^- \overline{\nu}_l)|^2}$$

$$= -\frac{2\mathrm{Re}(\Delta_l)\left(e^{-\Gamma_L \Delta t} - e^{-\Gamma_S \Delta t}\right) + 4\mathrm{Im}(\Delta_l)e^{-\frac{\overline{\Gamma}}{2}\Delta t}\sin(\Delta M \Delta t)}{e^{-\Gamma_L \Delta t} + e^{-\Gamma_S \Delta t} + 2e^{-\frac{\overline{\Gamma}}{2}\Delta t}\cos(\Delta M \Delta t)}, \tag{14.49}$$

where $\Delta_l \equiv \cos\theta + x_l - \overline{x}_l^*$. Both the real and the imaginary part of Δ_l can be extracted by analysing this asymmetry for small and large time intervals, namely $\Delta t = (0 \div 15)\tau_S$ and $\Delta t \geq 20\tau_S$, respectively [197].

The corresponding ratio for like-sign dileptons is actually *independent* of Δt:

$$\mathscr{A}_{l\pm l\pm} \equiv \frac{|A(\Delta t; \pi^+ l^- \overline{\nu}_l, \pi^+ l^- \overline{\nu}_l)|^2 - |A(\Delta t; \pi^- l^+ \nu_l, \pi^- l^+ \nu_l)|^2}{|A(\Delta t; \pi^+ l^- \overline{\nu}_l, \pi^+ l^- \overline{\nu}_l)|^2 + |A(\Delta t; \pi^- l^+ \nu_l, \pi^- l^+ \nu_l)|^2}$$

$$= 4\mathrm{Re}\, y_l - 2\mathrm{Im}\, \phi. \tag{14.50}$$

14.4.2 Tests of **CPT** symmetry in B decays

The completion of the program for a ϕ factory outlined above does not free us from the obligation to perform an analogous detailed study for beauty decays:

- 'Exotic' physics could have a different effect on the decays of beauty than strange hadrons.

- The time-scale characterizing decays and oscillations is quite different in the two systems.

- Asymmetric beauty factories allowing one to track the time evolution of $B_d \overline{B}_d$ pairs are available!

We start from the same master equations as for the $K\overline{K}$ case, yet with $\Delta\Gamma_B \simeq 0$. Again we aim at illustrating salient features rather than giving an exhaustive discussion.

Defining

$$s = \cot \theta = \frac{1}{2} \left(\frac{q_2}{p_2} - \frac{q_1}{p_1} \right) e^{-i\phi}, \qquad (14.51)$$

we can express the time-integrated ratio of like-sign to opposite-sign dileptons and the charge asymmetry in it as follows:

$$R_{(C=-1)} = \frac{\Gamma(B^0 \overline{B}^0 |_{C=-} \rightarrow l^\pm l^\pm X)}{\Gamma(B^0 \overline{B}^0 |_{C=-} \rightarrow l^\pm l^\mp X)} = \frac{1}{2} \frac{(|e^{-i\phi}|^2 + |e^{-i\phi}|^2)x^2}{|1 + s^2|(2 + x^2) + |s|^2 x^2} \qquad (14.52)$$

$$\begin{aligned}
a &= \frac{\Gamma(B^0 \overline{B}^0 |_{C=-} \rightarrow l^+ l^+ X) - \Gamma(B^0 \overline{B}^0 |_{C=-} \rightarrow l^- l^- X)}{\Gamma(B^0 \overline{B}^0 |_{C=-} \rightarrow l^\pm l^\mp X)} \\[2mm]
&= \frac{|e^{-i\phi}|^2 - |e^{-i\phi}|^2}{|e^{-i\phi}|^2 + |e^{-i\phi}|^2}, \qquad (14.53)
\end{aligned}$$

where $x = \Delta M(B^0)/\Gamma(B^0)$ is to be distinguished from x_l. Note that while a is independent of s, its measurement will lead to a determination of Im ϕ. For the reasons discussed in Sec. 10.2.7, we expect Im ϕ to be small. Then R gives only a constraint on s and x. To determine x in the presence of **CPT** violation, we need to measure the time dependence of the dilepton rates:

$$\begin{aligned}
N_{C=-}^{\pm\pm} &\sim e^{-\Gamma(t_1+t_2)}[1 - \cos \Delta M(t_1 - t_2)] \\
N_{C=-}^{+-} &\sim e^{-\Gamma(t_1+t_2)}[1 + 2s^2 + \cos \Delta M(t_1 - t_2)]. \qquad (14.54)
\end{aligned}$$

Finally, let us now discuss how **CPT** violation might influence the ψK_S **CP** asymmetry. In the presence of small **CPT** violation, Eq. (10.46) becomes:

$$\begin{aligned}
\Gamma([l^\pm X]_{t_l} f_{t_f}) &\propto e^{-\Gamma(t_l+t_f)} |A_{SL}|^2 |A(f)|^2 \left[1 \pm \text{Im} \left(e^{i\phi} \rho(f) \right) \sin\Delta M(t_f - t_l) \right. \\
&\quad \left. \pm s\text{Re} \left(e^{i\phi} \rho(f) \right) [\cos \Delta M(t_l + t_f) - 1] \right]. \qquad (14.55)
\end{aligned}$$

where we have kept only the leading term in s. There is one non-trivial modification. **CPT** violation leads to a time integrated asymmetry,

$$\frac{\Gamma([l^+ X]_{t_l} f_{t_f}) - \Gamma([l^- X]_{t_l} f_{t_f})}{\Gamma([l^+ X]_{t_l} f_{t_f}) + \Gamma([l^- X]_{t_l} f_{t_f})} = -s\text{Re} \left[e^{i\phi} \rho(f) \right] \frac{x^2}{1 + x^2}. \qquad (14.56)$$

Note that this asymmetry depends linearly on s. So, for small s, it might be more sensitive to **CPT** violation than like-sign dilepton decays.

14.5 The moralist's view

We often hear the claim that **CPT** invariance has been tested at the $\mathcal{O}(10^{-18})$ level. However, this refers to a very specific observable only, namely $(M_{\overline{K}^0} - M_{K^0})/M_K$, and it has been inflated by relating the mass difference to the overall kaon mass. Calibrating this mass difference by the weak width Γ_S, we arrive at a bound $\sim \mathcal{O}(10^{-4})$. Final results from CPLEAR will improve this situation somewhat.

Of course, the **CPT** *theorem* is based on noble assumptions about nature's basic structure. Yet these are assumptions none-the-less, and it behooves us to subject them to empirical scrutiny. We are not claiming that such an obligation would justify the construction of new facilities, in particular in the absence of meanigful benchmarks about the size and form of violations. However, the arrival of ϕ and beauty factories enables us to perform novel tests of **CPT** invariance; the high sensitivity that can be achieved by harnessing EPR correlations makes it even relevant as a probe for the possible presence of *non*-linear terms in the Schrödinger equation. It is conceivable that **CP** asymmetries in $e^+e^- \to B_d\bar{B}_d$ decays emerge, as predicted, but with a dependence on the time difference Δt that differs from expectations.

Problems

14.1   Ignore the effect of **CPT** and **CP** violation and consider a violation of the $\Delta S = \Delta Q$ rule. Then we can set $\phi = \cos\theta = y_l = 0$. Compute the time dependence of the decay:

$$
\begin{aligned}
\Gamma(K \to \pi^- l^+ v) &= \left|\frac{F_l}{2}\right| [(1 - 2\mathrm{Re}\ x_l)e^{-\Gamma_L t} + (1 + 2\mathrm{Re}\ x_l)e^{-\Gamma_S t} \\
&\quad + 2e^{\frac{1}{2}\Gamma t}\cos(\Delta M t) - 4\mathrm{Im}\ x_l e^{\frac{1}{2}\Gamma t}\sin(\Delta M t), \\
\Gamma(K \to \pi^+ l^- \bar{v}) &= \left|\frac{F_l}{2}\right| [(1 - 2\mathrm{Re}\ x_l)e^{-\Gamma_L t} + (1 + 2\mathrm{Re}\ x_l)e^{-\Gamma_S t} \\
&\quad - 2e^{\frac{1}{2}\Gamma t}\cos(\Delta M t) - 4\mathrm{Im}\ x_l e^{\frac{1}{2}\Gamma t}\sin(\Delta M t).
\end{aligned} \quad (14.57)
$$

14.2   Repeat the derivation of ϵ and ϵ' given in Sec. 7.3 without assuming **CPT** symmetry, i.e., keeping track of label i on p_i and q_i.

14.3   What is the **CP** eigenvalue of the $n\pi^0$ state? Show that the **CP** violating $K_S \to \pi^+\pi^-\pi^0$ decay is kinematically enhanced over the **CP** conserving $K_S \to \pi^+\pi^-\pi^0$ decay.

14.4   Here is a problem to guide you through the derivation of Eq. (14.29).

(a)   Using the expression for ϵ from Eq. (14.27) and for q_2/p_2 from Eq. (6.20) and Eq. (7.28), and assuming that **CP** and **CPT** violation is small, show that

$$
\begin{aligned}
\epsilon &= \frac{1}{2}[1 - (1 + i\phi)(1 + \cos\theta)(1 + r_A)] \\
&= -\frac{1}{2}\left(i\frac{E_2}{E_1} + \frac{E_3}{E} + r_A\right)
\end{aligned} \quad (14.58)
$$

where $r_A = \frac{\bar{A}_0}{A_0} - 1$.

(b) Derive:

$$\epsilon = \frac{-\text{Im}\,M_{12} + \frac{i}{2}\text{Im}\,\Gamma_{12}}{\sqrt{(\Delta M)^2 + (\frac{1}{2}\Delta\Gamma)^2}} e^{i\phi_{sw}}$$

$$-\frac{i(M_{11} - M_{22}) + \frac{1}{2}(\Gamma_{11} - \Gamma_{22})}{2\sqrt{(\Delta M)^2 + (\frac{1}{2}\Delta\Gamma)^2}} e^{i\phi_{sw}} - \frac{1}{2}r_A. \quad (14.59)$$

(c) Starting from

$$r_A = \text{Re}\,r_A + 2i\frac{\text{Im}\,\Gamma_{12}((2\pi)_0)}{\Delta\Gamma}, \quad (14.60)$$

show that

$$-\frac{1}{2}r_A e^{-i\phi_{sw}} = -\frac{i}{2}\frac{\text{Im}\,\Gamma_{12}((2\pi)_0)}{\sqrt{(\Delta M)^2 + (\frac{1}{2}\Delta\Gamma)^2}}$$

$$-\frac{\text{Im}\,\Gamma_{12}((2\pi)_0)}{\Delta\Gamma}\sin\phi_{sw} - \frac{1}{2}\text{Re}\,r_A e^{-i\phi_{sw}}. \quad (14.61)$$

(d) Now, inserting the expression for r_A into that of ϵ, show that the $(2\pi)_0$ contribution to $\text{Im}\,\Gamma_{12}$ cancels and derive

$$\epsilon = e^{i\phi_{sw}}\sin\phi_{sw}\frac{1}{\Delta M}\left(-\text{Im}\,M_{12} + \frac{i}{2}\sum_{f\neq(2\pi)_0}\text{Im}\,\Gamma_{12}(f)\right.$$

$$-\frac{i}{2}(M_{11} - M_{22}) - \frac{1}{4}(\Gamma_{11} - \Gamma_{22})$$

$$\left. -\frac{1}{2}\text{Im}\,\Gamma_{12}((2\pi)_0)\tan\phi_{sw}\right) - \frac{1}{2}\text{Re}\,r_A. \quad (14.62)$$

(e) Multiplying by a phase factor $e^{-i\phi_{sw}}$ and taking the imaginary part, derive:

$$|\eta_{+-}|\left(\frac{2}{3}\phi_{+-} + \frac{1}{3}\phi_{00} - \phi_{sw}\right)$$

$$= \left(\frac{M_{\bar{K}} - M_K}{2\Delta M_K} + R_{\text{direct}}\right)\sin\phi_{sw}, \quad (14.63)$$

where

$$R_{\text{direct}} = \frac{1}{2}\left(\frac{\text{Im}\,\Gamma_{12}}{\Delta M} + \text{Re}\,r_A\right)$$

$$= \frac{-ie^{-i\phi_{sw}}}{\sin\phi_{sw}}\sum_{f\neq(2\pi)_0}\epsilon(f) + \frac{1}{2}\text{Re}\,r_A \quad (14.64)$$

and $\epsilon(f)$ is defined in Eq. (9.55).

15

♠CP violation in other heavy systems♠

In this chapter we analyse the dynamics of charm hadrons, top quarks and τ leptons. They share a few features:

- Within the KM ansatz we predict no or only marginally observable **CP** asymmetries there.

- The intervention of New Physics can improve those prospects dramatically.

- More detailed theoretical model building and theoretical engineering is required and possible there.

15.1 The decays of charm hadrons

As far as experimental issues are concerned, charm decays are almost optimal for **CP** studies. (i) Charm states are being produced in large numbers, not only at e^+e^- machines – even 'parasitically' at B factories – but also in hadroproduction and in the rather clean environment of photoproduction. (ii) Their branching ratios into pions and kaons are large. (iii) They reside in the resonance region where direct **CP** asymmetries can be enhanced by prominent final state interactions; the latter can be determined through analyzing $\pi - \pi$, $\pi - K$ and $K - K$ rescattering. (iv) Flavour tagging can be achieved very efficiently through $D^{*+} \to D^0 \pi^+$ vs $D^{*-} \to \overline{D}^0 \pi^-$ decays.

On the theoretical side, we can entertain the hope that Monte Carlo simulations of QCD on the lattice will one day enable us to establish computational control over all these hadronization effects. For that purpose we have to go beyond the 'quenched approximation', and that will be achievable for the charm decays much earlier than for beauty transitions.

Of course there is a fly in this ointment making the outlook less rosy! We expect the weak phases to be quite small, for reasons that are partly germane to the SM and partly quite specific to the KM ansatz. In general we can say that within the SM the intrinsically weak aspects of charm decays are decidedly on the dull side: hugely suppressed charm changing neutral currents and small **CP** asymmetries. There is, however, another more up-beat perspective to this: **CP** studies in charm transitions represent an almost zero-background search for

252

New Physics, since the KM ansatz within the SM predicts small effects only. Of course we have to quantify what we mean by small.

In the remainder of this section we will first describe the **CP** phenomenology specific for charm decays, then we will analyze the corresponding KM predictions and how New Physics can intervene.

The general **CP** phenomenology discussed above for beauty decays applies here as well; yet the specifics make a big difference.

15.1.1 **CP** *violation involving* $D^0 - \overline{D}^0$ *oscillations*

$D^0 - \overline{D}^0$ oscillations proceed rather slowly, as expressed by the present experimental bound [210]:

$$r_D \equiv \frac{\Gamma(D^0 \to l^- X)}{\Gamma(D^0 \to l^+ X)} = \frac{x_D^2 + y_D^2}{2 + x_D^2 + y_D^2} \simeq \frac{x_D^2 + y_D^2}{2} \leq 0.0056, \qquad (15.1)$$

where $x_D = \Delta M_D / \Gamma_D, y_D = \Delta \Gamma_D / 2\Gamma_D$.

In the SM we expect r_D to be considerably smaller, as will be discussed below. Searching for an asymmetry of the unambiguously superweak type (see Eq. (6.60)),

$$a_{SL}(D^0) \equiv \frac{\Gamma(D^0 \to l^- X) - \Gamma(\overline{D}^0 \to l^+ X)}{\Gamma(D^0 \to l^- X) + \Gamma(\overline{D}^0 \to l^+ X)} = \frac{\left|\frac{q}{p}\right|^2 - \left|\frac{p}{q}\right|^2}{\left|\frac{q}{p}\right|^2 + \left|\frac{p}{q}\right|^2}, \qquad (15.2)$$

involving only 'wrong-sign' leptons then has hardly a practical value.

Yet that is not the end of it. To properly evaluate the impact of Eq. (15.1) we have to keep in mind that the bound of Eq. (15.1) is translated into

$$x_D \leq \mathcal{O}(10\%), \quad y_D \leq \mathcal{O}(10\%), \qquad (15.3)$$

which is considerably less impressive numerically, once we realize that certain **CP** asymmetries depend *linearly* on x_D rather than quadratically as r_D [211]. This enhances considerably the odds for **CP** violation to surface in some flavour-nonspecific modes. We illustrate the general method by two relevant examples.

A. $D^0 \to K_S \pi^0, K^+ K^-, \pi^+ \pi^-$. These channels have all been well studied and their branching ratios measured [28]:

$$\mathrm{Br}(D^0 \to K_S \pi^0) \simeq 1\%,$$

$$\mathrm{Br}(D^0 \to K^+ K^-) \simeq 0.5\%, \quad \mathrm{Br}(D^0 \to \pi^+ \pi^-) \simeq 0.2\%. \qquad (15.4)$$

They lead to **CP** eigenstates – with $K\overline{K}$ or $\pi\pi$ being even and $K_S \pi^0$ mainly odd – and thus can be discussed [211] in close analogy to $B_d \to \psi K_S, \pi^+ \pi^-$ (see Eq. (6.65)):

$$\Gamma(D^0(t) \to f) \propto e^{-\Gamma_D t}\left(1 - \mathrm{Im}\frac{q}{p}\overline{\rho}(D \to f)\sin(\Delta M_D t)\right)$$

$$\simeq e^{-\Gamma_D t}\left(1 - x_D \tau \mathrm{Im}\frac{q}{p}\overline{\rho}(D \to f)\right), \qquad (15.5)$$

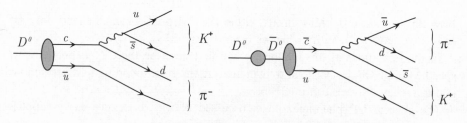

Fig. 15.1. Feynman diagram which gives $D^0 \to K^+\pi^-$.

$$\Gamma(\overline{D}^0(t) \to f) \propto e^{-\Gamma_D t}\left(1 - \text{Im}\frac{p}{q}\rho(D \to f)\sin(\Delta M_D t)\right)$$

$$\simeq e^{-\Gamma_D t}\left(1 + x_D\tau\text{Im}\left[\frac{q}{p}\overline{\rho}(D \to f)\right]^{-1}\right), \tag{15.6}$$

with $f = K_S\pi^0, K^+K^-$ or $\pi^+\pi^-$, and $\tau = t\Gamma$, the time in units of D lifetime. We have retained here only terms at most linear in the small quantity x_D. As stated above, the bound $r_D \leq \mathcal{O}(1\%)$ still allows for $x_D \sim 0.1$ – which is not that small – and *we could have an asymmetry of a few percent if* $\text{Im}(q/p)\overline{\rho}(D \to f)$ *were not too much smaller than unity*. This cannot happen for the KM ansatz, but quite conceivably with New Physics. In any case, we might discover $D^0 - \overline{D}^0$ oscillations first through a **CP** asymmetry of the type illustrated in this example or the next.

B. $D^0 \to K^+\pi^-$

The final state here is obviously not a **CP** eigenstate, yet it is flavour-nonspecific: it can occur directly through a $\Delta C = -\Delta S$ doubly Cabibbo suppressed transition (DCSD) or as a two-step process via a $D^0 - \overline{D}^0$ oscillation $D^0 \to \overline{D}^0 \to K^+\pi^-$.

The important point here is that the non-oscillation DCSD amplitude is quite considerably suppressed as well, giving the oscillation amplitude a much better chance to be competitive. We find [212, 213]

$$\Gamma(D^0(t) \to K^+\pi^-) \propto e^{-\Gamma_{D^0}t}|A(D^0 \to K^+\pi^-)|^2 \times \left[1 + y_D\tau + \frac{(x_D\tau)^2}{4}\left|\overline{\rho}(K^+\pi^-)\right|^2\right.$$

$$\left. -y_D\tau\text{Re}\left(\frac{q}{p}\overline{\rho}(K^+\pi^-)\right) - x_D\tau\text{Im}\left(\frac{q}{p}\overline{\rho}(K^+\pi^-)\right)\right] \tag{15.7}$$

and for the **CP** conjugate channel

$$\Gamma(\overline{D}^0(t) \to K^-\pi^+) \propto e^{-\Gamma_{D^0}t}|A(\overline{D} \to K^-\pi^+)|^2 \times [1 + y_D\tau + \frac{(x_D\tau)^2}{4}\left|\rho(K^-\pi^+)\right|^2$$

$$-y_D\tau\text{Re}\left(\frac{p}{q}\rho(K^-\pi^+)\right) - x_D\tau\text{Im}\left(\frac{p}{q}\rho(K^-\pi^+)\right)] \tag{15.8}$$

where

$$\overline{\rho}(K^+\pi^-) = \frac{A(\overline{D} \to K^+\pi^-)}{A(D \to K^+\pi^-)}; \; \rho(K^-\pi^+) = \frac{A(D \to K^-\pi^+)}{A(\overline{D} \to K^-\pi^+)}$$

$$\sim \mathcal{O}(\tan^{-2}\theta_C) \sim 20. \tag{15.9}$$

There are several observations to be noted here:

- The dependence on the time of decay t has a rather complex structure, with terms linear as well as quadratic in t coming from expanding the $\sin\Delta M_D t$ and $\cos\Delta M_D t$ terms.

- The last term in the square brackets is in general present even if **CP** is conserved [213]: for while the isospin is changed by $(I, I_3) = (1, 1)$ in Cabibbo allowed modes, we have a combination of $(\Delta I, \Delta I_3) = (1, 0)$ and $(0, 0)$ in DCSD; therefore the two amplitudes will undergo different phase shifts, yielding a relative phase leading to an imaginary part.

- While the **CP** asymmetry is suppressed by the smallness of x_D, it gets a boost from the fact that unmixed decays $D^0 \to K^+\pi^-$ and $\overline{D}^0 \to K^-\pi^+$ are Cabibbo suppressed, i.e., large $\overline{\rho}(K^+\pi^-) \sim \rho(K^-\pi^+) \sim 20$.

- However, a very sizeable **CP** asymmetry can arise if New Physics drives $D^0 - \overline{D}^0$ oscillations in a way that generates a weak phase in $\langle D^0 | \mathcal{H}(\Delta C = 2) | \overline{D}^0 \rangle$. Possible scenarios for this happening will be given below.

15.1.2 Direct **CP** violation

A. $D \to K\overline{K}, \pi\pi$

Unambiguously Cabibbo allowed (and DCSD) channels are fed by a *single* weak amplitude and thus cannot exhibit *direct* **CP** violation. Singly Cabibbo suppressed modes, on the other hand, can. Channels like $D^0 \to K^+K^-, \pi^+\pi^-$ or $D^+ \to K_S K^+, \eta\pi^+$ are prime examples[1].

With D decays proceeding in an environment with many hadronic resonances nearby ($m_{\text{Resonance}} \leq M_D$), there is no reason why the required strong final state interactions would be absent or particularly small. So far they are not calculable; on the other hand, we can at least hope that lattice simulations of QCD will overcome that in the not-too-distant future. Also, an accurate measurement of a comprehensive set of two-body branching ratios for D^+, D^0 and D_s mesons will provide useful constraints on these phase shifts [214]. Yet at present we can make at best order-of-magnitude predictions for **CP** asymmetries in a given model. An observed signal will therefore in general allow more than one interpretation.

B. $D^\pm \to K_S \pi^\pm$

At first sight we might think that these modes are Cabibbo allowed and thus cannot exhibit direct **CP** violation, apart from a quite unlikely scenario where New Physics affects even Cabibbo allowed amplitudes significantly. However, as seen in Fig. 15.2, we realize [215] that (a) $D \to K_S \pi$ receives a contribution both from $D \to \overline{K}^0 \pi$ and the DCSD $D \to K^0 \pi$ reaction and that (b) both proceed

---

[1] Even $D^0 \to K_S K_S$ is a legitimate, though experimentally unappealing candidate, while $D^+ \to \pi^0\pi^+$ described by a single isospin amplitude is not even that.

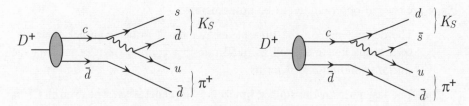

Fig. 15.2.   Feynman diagrams describing $D^+ \rightarrow K_S\pi^+$. The doubly Cabibbo suppressed decay amplitude (b) can be made to interfere with the Cabibbo allowed amplitude (a) by observing the $K_S\pi^+$ channel.

coherently! That means that two different weak amplitudes contribute coherently; the rate then depends linearly on the DCSD amplitude, making the weight of the latter of order $\tan^2\theta_C$ rather than $\tan^4\theta_C$. Furthermore, with the isospin content of the two amplitudes being different, we can count on a non-trivial phase-shift difference to arise here. Finally, we can measure these phase shifts and the relevant hadronic matrix elements through an isospin analysis of the various $D \rightarrow K\pi$ and $D \rightarrow \overline{K}\pi$ branching ratios. The phenomenology of this example is actually the charm analogue of $B^\pm \rightarrow D^{\mathrm{neut}}K^\pm$, discussed before, as a way to determine the third KM angle ϕ_3. There is an intriguing feature to be kept in mind here. The observed **CP** violation in $K_L \rightarrow \pi\pi$ decays induces a difference of known magnitude in $\Gamma(D^+ \rightarrow K_S\pi^+)$ vs $\Gamma(D^- \rightarrow K_S\pi^-)$:

$$A(D^+ \rightarrow K_S\pi^+) = A(D^+ \rightarrow \overline{K}^0\pi^+)\langle K_S|\overline{K}^0\rangle + A(D^+ \rightarrow K^0\pi^+)\langle K_S|K^0\rangle. \quad (15.10)$$

With the **CPT** invariant version of Eq. (6.19)

$$\langle K^0|K_S\rangle = p_K, \quad \langle \overline{K}^0|K_S\rangle = q_K, \quad (15.11)$$

we obtain

$$A(D^+ \rightarrow K_S\pi^+) = A(D^+ \rightarrow \overline{K}^0\pi^+)q_K^*\left(1 + \frac{p_K^*}{q_K^*}\rho_{DCSD}\right),$$

$$A(D^- \rightarrow K_S\pi^-) = A(D^- \rightarrow K^0\pi^-)p_K^*\left(1 + \frac{q_K^*}{p_K^*}\overline{\rho}_{DCSD}\right), \quad (15.12)$$

where

$$\rho_{DCSD} \equiv \frac{A(D^+ \rightarrow K^0\pi^+)}{A(D^+ \rightarrow \overline{K}^0\pi^+)}, \quad \overline{\rho}_{DCSD} \equiv \frac{A(D^- \rightarrow \overline{K}^0\pi^-)}{A(D^- \rightarrow K^0\pi^-)}. \quad (15.13)$$

Ignoring terms $\sim \mathcal{O}(|\rho_{DCSD}|^2)$, we obtain for the decay widths

$$\Gamma(D^+ \rightarrow K_{S,L} + \pi^+) \propto \Gamma(D^+ \rightarrow \overline{K}^0 + \pi^+)|q_K|^2\left[1 \pm 2\mathrm{Re}\left(\frac{p_K^*}{q_K^*}\rho_{DCSD}\right)\right],$$

$$\Gamma(D^- \rightarrow K_{S,L} + \pi^-) \propto \Gamma(D^- \rightarrow K^0 + \pi^-)|p_K|^2\left[1 \pm 2\mathrm{Re}\left(\frac{q_K^*}{p_K^*}\overline{\rho}_{DCSD}\right)\right].$$

$$(15.14)$$

We see there are two sources for a difference in $D^+ \to K_S \pi^+$ vs $D^- \to K_S \pi^-$ or $D^+ \to K_L \pi^+$ vs $D^- \to K_L \pi^-$:

- $q_K \neq p_K$, i.e., the well-known **CP** violation in the $K^0 - \overline{K}^0$ complex irrespective of its dynamical origin;

- $\rho_{DCSD} \neq \overline{\rho}_{DCSD}$, i.e., a **CP** asymmetry in $\Delta S \neq 0$ *charm* decays, most likely in its DCSD modes.

15.1.3 KM predictions on asymmetries involving $D^0 - \overline{D}^0$ oscillations

It has often been stated that **CP** asymmetries in charm decays are small within the KM scheme. Yet – how small is small? And when do we have to invoke New Physics to explain an observed signal? This has to be addressed on a case-by-case basis.

CP asymmetries with $D^0 - \overline{D}^0$ oscillations are controlled by the product of two quantities, $x_D = \Delta M_D / \Gamma_D$ and $\mathrm{Im}(q/p)\overline{\rho}(f)$ – see Eq. (15.5) and Eq. (15.8).
(i) As far as ΔM_D is concerned, we are faced with a dichotomy: the *local* and *short-distance* contributions, which we know how to calculate, are unnaturally tiny, whereas *long-distance* dynamics, which dominates ΔM_D by a large margin, is beyond our computational control. However, it seems safe to say that x_D (and similarly y_D) cannot exceed 0.01 [211, 216, 217] which results in $r_D|_{SM} \leq 10^{-4}$.
(ii) From the KM matrix given in Eq. (10.16), we read off that:

$$\mathrm{Im}\left(\frac{q}{p}\overline{\rho}(f)\right) \sim 20 \times \mathcal{O}(\lambda^4) \leq 0.05, \qquad (15.15)$$

for the decay shown in Fig. 15.1 holds. Combining the two estimates, we conclude that

$$\frac{\Delta M_D}{\Gamma_D} \cdot \frac{t}{\tau_D} \cdot \mathrm{Im}\frac{q}{p}\overline{\rho}_f \Big|_{KM} \leq \mathcal{O}(10^{-4}), \qquad (15.16)$$

and a **CP** asymmetry within the context of the KM ansatz will not exceed the $\mathcal{O}(10^{-4})$ level; it is so tiny since it represents the product of two small effects, namely $D^0 - \overline{D}^0$ oscillations and the size of the KM phases in charm transitions.

15.1.4 KM prediction – direct CP violation

The KM ansatz provides the ingredients for *direct* **CP** *asymmetries* to arise in *singly Cabibbo suppressed* channels. (i) On the quark level there are transition operators with different weak couplings, namely $c_L \to s_L \overline{s}_R u_L$, $c_L \to d_L \overline{d}_R u_L$ and $c_L \to u_L \sum_q \overline{q}q$. The last one is induced by the penguin process and also reflects $s\overline{s} \leftrightarrow d\overline{d}$ re-scattering.

That means, the final state in $D \to \pi\pi$ can be reached also through re-scattering from an intermediate $K\overline{K}$ state [218]; likewise for $D \to K\overline{K}$; i.e., these channels receive coherent contributions from different quark-level operators. (ii) With $c_L \to s_L \bar{s}_R u_L$ and $c_L \to u_L \sum_q \bar{q}q$ changing isospin by $(I, I_3) = (\frac{1}{2}, +\frac{1}{2})$, yet $c_L \to d_L \bar{d}_R u_L$ by $(I, I_3) = (\frac{1}{2}, +\frac{1}{2})$ or $(\frac{3}{2}, +\frac{1}{2})$, we should encounter sizeable phase shifts in these modes. (iii) Since $\arg \frac{V_{cs} V_{us}^*}{V_{cd} V_{ud}^*} \sim \eta A^2 \lambda^4 \sim \mathcal{O}(10^{-3})$, we expect the typical size for such asymmetries to be of order 0.1% [219, 220]. This is of course only a rough order of magnitude estimate, since the magnitude of a possible asymmetry is greatly affected by the presence of matrix element enhancements due to, say, nearby resonances and by phase shifts [221]. To refine the KM prediction we can apply some 'theoretical engineering'. The basic procedure is the following: we express the amplitude for two-body modes in terms of a finite number of hadronic matrix elements, where we make allowance for final state interactions producing absorption and phase shifts. Matching up those expressions with a host of well-measured branching ratios, we extract the size of the matrix elements and phase shifts. Overconstraining them through additional data provides a gauge for the self-consistency and completeness of such a phenomenological treatment. Once this test is passed, we can use these hadronic parameters to make at least semi-quantitative KM predictions on **CP** asymmetries.

No *direct* **CP** asymmetry can arise in pure Cabibbo allowed or doubly suppressed modes since they are driven by a single weak amplitude. We have used the term 'pure' since in $D \to K_S + \pi$'s **CP** asymmetries can arise in two different ways, as pointed out above:

- through the **CP** impurity in K_S,

- through interference of two weak $\Delta C \neq 0$ amplitudes.

As for the second effect, the relative weak phase between them is very tiny, of order

$$\left| \text{Im} \left(\frac{V_{cd} V_{us}^*}{V_{cs} V_{ud}^*} \right) \right| \simeq \eta A^2 \lambda^6 \sim \text{few} \times 10^{-5}, \quad (15.17)$$

and thus can be ignored [215]. Accordingly,

$$\frac{\Gamma(D^+ \to K_S \pi^+) - \Gamma(D^- \to K_S \pi^-)}{\Gamma(D^+ \to K_S \pi^+) + \Gamma(D^- \to K_S \pi^-)}$$
$$\simeq \frac{\Gamma(D^+ \to K_L \pi^+) - \Gamma(D^- \to K_L \pi^-)}{\Gamma(D^+ \to K_L \pi^+) + \Gamma(D^- \to K_L \pi^-)}$$
$$\simeq \frac{|q_k|^2 - |p_K|^2}{|q_K|^2 + |p_K|^2} \simeq -3.3 \times 10^{-3}. \quad (15.18)$$

We expect a non-vanishing asymmetry in $D^\pm \to K_S \pi^\pm$ and in $D^\pm \to K_L \pi^\pm$ of practically the same size and sign, which is due to the **CP** impurity in the K_S and K_L state.

15.1.5 Manifestations of New Physics

New Physics could conceivably enter in several different ways; yet it has the best chance to make its impact felt in suppressed transitions, namely in $D^0 - \overline{D}^0$ oscillations in the first place and secondly in DCSD, rather than in Cabibbo allowed or even once suppressed transitions.

$D^0 - \overline{D}^0$ *oscillations* are somewhat unnaturally slow in the SM with $x_D|_{SM} \leq 0.01$. New Physics could easily enhance its strength up to its present bound. Once that happens it would be natural that **CP** invariance is violated in the $D^0 - \overline{D}^0$ mass matrix as well. The size of an asymmetry involving $D^0 - \overline{D}^0$ oscillations depends on $x_D \cdot \text{Im}[(q/p)\overline{\rho}(f)]$. With $x_D \leq 0.1$, as inferred from present data, the possible size of an asymmetry is reduced by at least a factor of ten, but conceivably not by much more. For $\text{Im}[(q/p)\overline{\rho}_f] \sim 0.1$ – as pointed out above, a natural scenario if $D^0 - \overline{D}^0$ oscillations are enhanced by New Physics – we have then a benchmark asymmetry of 1% for $D^0 \to K^+K^-$, $\pi^+\pi^-$ and of even 20% for $D^0 \to K^+\pi^-$! The observation of even considerably reduced asymmetries, namely 0.1% in the first example and 1% in the second, would represent clear manifestations of New Physics [211, 212, 213, 222].

If New Physics intervened through DCSD the asymmetry in $D^0 \to K^+\pi^-$ would of course be effected, yet in a way that could not be disentangled from it making its entry through $D^0 - \overline{D}^0$ oscillations. On the other hand, it would have the cleanest impact on $D^+ \to K_{S,L}\pi^+$ vs $D^- \to K_{S,L}\pi^-$ if it generates, as it would be natural, a weak phase in DCSD. For example, for $A(\Delta C = -\Delta S)|_{\text{New Physics}} = 0.2e^{i\pi/6}A(\Delta C = \Delta S)|_{KM}$, i.e., if New Physics contributed to the DCSD amplitude with a mere 20% of the KM amplitude, albeit with a relative weak phase of 30°, we would have an asymmetry of around 1%!

There are two ways in which such effects can be distinguished against the KM expectations. (a) The KM asymmetry in $D^+ \to K_S\pi^+$ vs $D^- \to K_S\pi^-$ is given by Eq. (15.18) with good accuracy[2]. Observing a larger effect, say a 1% difference, or a smaller one – including zero – manifests the presence of New Physics. (b) We can undertake to measure also $D^+ \to K_L\pi^+$ vs $D^- \to K_L\pi^-$ and study whether the **CP** asymmetry in this channel is equal, including the sign, to that in $D^+ \to K_S\pi^+$ vs $D^- \to K_S\pi^-$, or not.

With respect to *direct CP violation in singly Cabibbo suppressed channels*, the interpretation of an observed **CP** asymmetry is much less clear: for within the KM ansatz we expect **CP** asymmetries in the few$\times 10^{-3}$ range to arise in some channels there; in addition we have to allow for larger effects due to possible hadronic enhancements. We can only hope that the 'theoretical engineering' outlined above will lead to fairly reliable KM predictions.

---

[2] The size of the correction factor in Eq. (15.18) can be probed independently.

15.2 Production and decay of top quarks

The existence of all the postulated members of three quark families has been established now with the discovery of the top quark. Even before that time it had been realized [223] that the top quark, once it becomes sufficiently massive, will decay (semi-)weakly – $t \to bW$ – *before* it can hadronize; i.e., top states decay as quarks rather than hadrons. This transition occurs around the 110–130 GeV region (for $|V(tb)| \sim 1$), i.e., well below the mass value now observed. *Non-perturbative* dynamics thus plays hardly any role in top decays, and the strong forces can be treated perturbatively. While this implies that top decays are under good theoretical control, it carries also a negative message concerning the observability of **CP** asymmetries:

- If top *hadrons* can hardly form before top decays, $T^0 - \overline{T}^0$ oscillations become a moot point.

- Due to the huge phase space that is available in top decays, there is little chance that interference can occur between different sub-processes.

Thus we see that most of the **CP** sensitive observables we have discussed before do *not* work for top states, because the strong interactions are no longer sufficiently efficient in 'cooling' different transitions into a coherent state. Top quarks in many ways behave like charged leptons since the residual strong interactions can be treated perturbatively. Accordingly, we consider similar phenomenological categories when analysing possible **CP** violation in top and in τ physics: we are dealing with not only the production, but also the decay of spin-1/2 objects whose dynamics is not prone to 'non-perturbative voodoo'; they can be polarized and their polarization is revealed in their decays.

From today's perspective it appears that the prospects for finding **CP** violation are better in top *production* than in decay. This will be illustrated through two examples.

15.2.1 $\sigma(t_L \bar{t}_L)$ vs $\sigma(t_R \bar{t}_R)$

Consider the reaction

$$e^+ e^- \to \gamma^*, Z^* \to t\bar{t}. \tag{15.19}$$

The coupling of both the photon and the Z boson to top quarks conserves chirality. Yet chirality and helicity coincide only for fast particles; close to threshold there is a significant component when t and $\bar{t}$ both have either helicity $+1$ or -1: $e^+ e^- \to t_R \bar{t}_R, t_L \bar{t}_L$. With t_L and $\bar{t}_R$ being conjugate to each other, i.e.,

$$t_R \overset{\text{CP}}{\Longrightarrow} \bar{t}_L, \tag{15.20}$$

a difference in $t_R \bar{t}_R$ vs $t_L \bar{t}_L$ production,

$$\sigma(e^+ e^- \to t_R \bar{t}_R) \neq \sigma(e^+ e^- \to t_L \bar{t}_L), \tag{15.21}$$

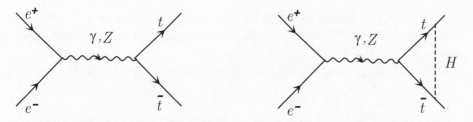

Fig. 15.3. Feynman diagrams which interfere to generate **CP** violation in $e^+e^- \to$ $t\bar{t}$ pair production.

implies **CP** violation [224]; for it to arise the usual two conditions have to be satisfied: (i) Two amplitudes have to contribute coherently to the production process that involve a **CP** *non*invariant interaction. (ii) An absorptive component has to be generated.

Exchanges of Higgs fields between the quark and antiquark pair can certainly satisfy the second condition, and also the first one if the Higgs fields form part of a non-minimal Higgs sector. The leading diagrams are shown in Fig. 15.3. The asymmetry is then produced by the interference between the tree diagram and the one-loop diagram.

While all of this would be true for any quark, it first becomes relevant for top quarks for two reasons:

- It is natural for top quarks to possess large Yukawa couplings to Higgs fields in general. Higgs exchanges thus can generate significant absorption and **CP** violation as well.

- Since top states decay weakly as quarks rather than *pseudoscalar* hadrons (as it would be for beauty, charm etc. production) we can employ the parity violating decay

$$t \to W^+ + b_L \tag{15.22}$$

  to analyze the *polarization* of the decaying top quark and thus probe the production process where the asymmetry resides. The lack of hadronization then becomes an asset.

This scenario has been analyzed in some detail in Ref. [224]. It was found that asymmetries as 'large' as $\mathcal{O}(10^{-3})$ could arise here. It remains to be seen whether an effect of this size could be observed at an e^+e^- or $\gamma\gamma$ collider or even in hadronic collisions that promise higher rates, yet present a less 'clean' environment.

15.2.2 Final state distributions in $e^+e^- \to t\bar{t}H^0$

We can probe the limits of **CP** symmetry also through analyzing **T** odd correlations, as it is done in $K_{\mu 3}$ decays. The intervention of final state interactions is not required to generate a signal here although it can skew or even fake one! That means, the interference of *tree* diagrams can already produce a signal which can be relatively large.

One interesting possibility has recently been suggested in Ref. [225]. Consider

$$e^+(p_+)\,e^-(p_-) \;\to\; \gamma^*,\,Z^* \;\to\; t(p_t)\,\bar{t}(p_{\bar{t}})\,H^0(p_{H^0}), \tag{15.23}$$

where the inital and final state momenta have been listed. The salient points of the analysis are the following:

1 In view of what was said before, the aim is to reconstruct the t and $\bar{t}$
  momenta from their decay products and to formulate a **T** odd correlation
  in them (Problem 15.5):

$$O_- \equiv \langle \vec{p}_- \cdot (\vec{p}_t \times \vec{p}_{\bar{t}}) \rangle, \tag{15.24}$$

  where

$$O_- \;\overset{\text{CP}}{\Longrightarrow}\; -O_-. \tag{15.25}$$

  We realize that the kinematics force such an observable to vanish for a
  two-body final state. We should also note that we search again for an
  asymmetry in the *production* reaction.

2 Reconstructing the t and $\bar{t}$ momenta is not an easy task. It is then tempting
  to concentrate on determining the momenta of the b and $\bar{b}$ initiated jets
  emerging from the t and $\bar{t}$ decays, respectively:

$$O_-^{(b,\bar{b})} \equiv \langle \vec{p}_- \cdot (\vec{p}_b \times \vec{p}_{\bar{b}}) \rangle. \tag{15.26}$$

Finding $O_-^{(b,\bar{b})} \neq 0$ would manifest **CP** violation; however, the strength of a
signal gets diluted in this *partial* reconstruction.

3 So far we have mainly stated that kinematics do not force such correlations
  to vanish. The important dynamical point here is that the neutral Higgs H^0
  can be emitted from the t (and $\bar{t}$) line through their Yukawa couplings as
  well as from the intermediate Z through the gauge coupling. Those three
  amplitudes can interfere and thus expose the presence of a relative weak
  phase between the gauge coupling on one hand and the t and $\bar{t}$ Yukawa
  couplings on the other. It has been stated that the resulting asymmetries
  could conceivably be of order 10% [225]!

15.3 τ **physics**

No **CP** violation has been observed yet in the leptonic sector. That is not surprising since the avenues that proved most effective in the quark sector are not open here:

- No particle–antiparticle oscillations can occur for charged leptons due to the conservation of electric charge. Neutrino oscillations on the other hand could be employed; that will be addressed in a later chapter.

- Within the SM neutrinos are degenerate in mass, i.e., massless. This additional global invariance allows us to transform the leptonic equivalent of the CKM matrix (we call it the MNS matrix for reasons explained below) into a unit matrix through appropriate re-definitions of the phases of the lepton fields. Even if some relatively modest extension of the SM breaks the degeneracy in the neutrino masses, such differences are typically tiny relative to the relevant scale set by the GIM mechanism, namely the W mass:

$$\frac{\Delta M_\nu^2}{M_W^2} \ll 1. \tag{15.27}$$

Flavour changing neutral currents are thus highly suppressed even on the quantum level.

- Muon decays proceed basically through a single channel – $\mu^- \to e^- \bar{\nu}_e \nu_\mu$ – with only two other modes having been seen. The only practical way to search for a **T** odd correlation involves the decay of polarized muons into electrons whose spin is measured:

$$O_{\mu \to e} \equiv \vec{\sigma}_e \cdot (\vec{\sigma}_\mu \times \vec{p}_e). \tag{15.28}$$

No effect has been observed yet [226]:

$$O_{\mu \to e} = 0.007 \pm 0.0022 \pm 0.007. \tag{15.29}$$

While the experimental effort behind these numbers is impressive, it is hard to see how even New Physics could have generated a signal on that numerical level.

There is, however, one area in the physics of light leptons where **T** invariance has been probed in a highly sensitive manner, namely in searches for an electric dipole moment of the electron, as described before.

The situation becomes significantly more promising for τ leptons [227, 228]:

1 The **CP** phenomenology is now multi-faceted. For there are several major decay modes; a qualitatively new feature is that hadrons can appear in the final state. The constraints imposed by **CPT** invariance then become considerably less effective; at the same time more and new types of **T** odd correlations can be constructed.

2 Due to the higher mass of the charged lepton – $m_\tau \gg m_\mu$ – and of the basic objects in the final state – $m_s \gg m_{d,u} \gg m_e$ – there is a much better chance for New Physics to create an observable impact, in particular in the context of theories with a non-minimal Higgs sector with or without SUSY.

3 At the same time, the production process $e^+e^- \rightarrow \tau^+\tau^-$ provides a clean and relatively copious source of τ leptons with two additional benefits: while the τ^+ and τ^- states are only observed through their decay products, correlations involving τ polarizations can be transformed into correlations among momenta or energies of the decay products.

Exchange of a (non-minimal) Higgs field can naturally generate a **CP** asymmetry in τ production – yet only a tiny one, since the resulting vertex correction is very roughly of order $\alpha \cdot (m_\tau/m_W)^2 \sim 10^{-6} - 10^{-5}$ only! At the same time, **T** odd correlations can be faked by the much larger electromagnetic corrections.

A similar conclusion can concretely be drawn from LEP analyses of $Z \rightarrow \tau^+\tau^-$. A **CP** asymmetry in τ production can be expressed in terms of an electric dipole moment d_τ^E, for which the following upper bound has been found [28]

$$d_\tau^E \leq 3 \times 10^{-16} \text{ e cm} \qquad (90\% \text{ CL}). \qquad (15.30)$$

This is not a significantly smaller distance scale than the 10^{-16} cm down to which leptons have exhibited pointlike behaviour.

Prospects are brighter for observing **CP** violation in τ decays. The aforementioned reaction $e^+e^- \rightarrow \tau^+\tau^-$ possesses some intriguing features in this context. The final state consists exclusively of a $\tau^+\tau^-$ pair (up to electromagnetic radiative corrections) coming from an even **CP** eigenstate. This leads to many important correlations between the decay products of the two τ leptons. Those can be used to enhance a signal and/or discriminate against background.

Let us suppose that **CP** and **T** violating τ decays exist:

$$k_l \frac{d\Gamma(\tau \rightarrow v_\tau l(\vec{\sigma},\vec{k}_l)\bar{v}_l)}{d^3 k_l} \sim C\vec{\epsilon} \cdot (\vec{\sigma}_l \times \vec{k}_l),$$

$$k_{\pi^+}k_{\pi^-} \frac{d\Gamma(\tau \rightarrow v_\tau \pi^+(\vec{k}^+)\pi^-(\vec{k}^-) + X)}{d^3 k_{\pi^+} d^3 k_{\pi^-}} \sim C'\vec{\epsilon} \cdot (\vec{k}_{\pi^+} \times k_{\pi^-}), \qquad (15.31)$$

where we have indicated the momenta and spin vectors for the corresponding particles, $\vec{\epsilon}$ is the τ polarization, and l stands for an electron or a muon.

Consider dilepton decays of a $\tau^+\tau^-$ pair

$$e^+e^- \rightarrow \quad \tau^+ \quad \tau^- \rightarrow v_\tau \mu^- \bar{v}_\mu$$
$$\hookrightarrow \quad \bar{v}_\tau e^+ v_e,$$
$$e^+e^- \rightarrow \quad \tau^+ \quad \tau^- \rightarrow \pi^+\pi^- + X$$
$$\hookrightarrow \quad \bar{v}_\tau e^+ v_e. \qquad (15.32)$$

How do the correlations given in Eq. (15.31) appear in actual production and decay of $\tau^+\tau^-$ pairs, Eq. (15.32)? It can be shown [228] that the above **T** violating correlation shows up as

$$\langle \vec{p}_+ \cdot (\vec{\sigma}_\mu \times \vec{k}_\mu)(\vec{k}_e \cdot \vec{p}_+) \rangle$$
$$\langle \vec{p}_+ \cdot (\vec{k}_{\pi^+} \times \vec{k}_{\pi^-})(\vec{k}_e \cdot \vec{p}_+) \rangle, \qquad (15.33)$$

where $\vec{p}_+$ is the positron beam momentum, and $\vec{k}_e$ is the momentum of the positron produced in the accompanying $\tau^+ \to \nu_\tau e^+ \nu_e$ decay.

In addition to the above possibility, longitudinally polarized electron or positron beams generate events which have different **T** and **CP** odd correlations.

Many options and possibilities thus exist for the **CP** phenomenology in τ decays:

1 We can search for differences in partial widths allowed under **CPT** in-variance[3], like in $\tau^- \to \nu_\tau \pi^- \pi^0$ vs $\tau^+ \to \bar{\nu}_\tau \pi^+ \pi^0$ or in $\tau^- \to \nu_\tau \pi^- \pi^- \pi^+$ vs $\tau^+ \to \bar{\nu}_\tau \pi^+ \pi^- \pi^+$ or in $\tau^- \to \nu_\tau K^- \pi^0$ vs $\tau^+ \to \bar{\nu}_\tau K^+ \pi^0$. However, conceivable asymmetries are reduced by the requirement that we must have two different amplitudes with different rescattering phases among the hadrons in the final state. That makes in particular the channel $\tau^\pm \to \nu_\tau \pi^\pm \pi^0$ fairly unpromising, since the aforementioned requirement can be satisfied through isospin violation only.

2 We can search for **T** odd observables in the $\tau \to \nu_\tau \mu \bar{\nu}_\mu, \nu_\tau \pi \pi, \nu_\tau K \pi, \nu_\tau 3\pi$ channels as stated for muon decay in Eq. (15.28).

However, before a real evaluation of the discovery potential here can be given, two tasks have to be tackled:

(a) We have to analyze how important the availabilty of polarized τ leptons really is. For the chirality of the two τ leptons produced is strictly correlated: $e^+ e^- \to \tau_L^+ \tau_R^-, \tau_R^+ \tau_L^-$. From observing one τ decay we can infer the polarization of the other τ. We then have to study in detail the relative merits of having polarized beams available versus a higher luminosity, in its impact both on statistics and systematics. For an analysis including electron beam polarization, see Ref. [228].

(b) It would be useful to consider two complementary New Physics scenarios for **CP** violation in the τ sector – like a non-minimal Higgs sector or the presence of right-handed charged currents – and undertake a systematic and concrete study of the relevant **CP** phenomenology in these two scenarios.

Neither of these tasks has been completed yet, although the necessary tools are there.

---

[3] When comparing $\tau^- \to \nu_\tau K_S \pi^-$ with $\tau^+ \to \bar{\nu}_\tau K_S \pi^+$ we have to take into account the known **CP** asymmetry in $K^0 \Rightarrow K_S$ vs $\overline{K}^0 \Rightarrow K_S$.

15.4 Résumé and call to task

Because $D^0 - \overline{D}^0$ oscillations are so slow in the SM, it is quite conceivable that New Physics will make the dominant contribution to it; it might also affect DCSD in a significant way. Once either or both of these things happen, it would be natural for those contributions to break CP invariance. This would lead to the following scenarios:

- CP asymmetries involving $D^0 - \overline{D}^0$ oscillations could be enhanced by two orders of magnitude or more and reach the *percent* level in Cabibbo allowed and once forbidden modes and even the *ten percent level* in doubly Cabibbo suppressed channels.

- An asymmetry in $D^\pm \rightarrow K_S \pi^\pm$ and in $D^\pm \rightarrow K_L \pi^\pm$ could arise that is larger (or smaller) than $2\mathrm{Re}\,\epsilon_K \simeq 3.3 \times 10^{-3}$ or that assumes a different value when the K_S is replaced by a K_L in the final state.

- We expect some singly Cabibbo suppressed channels to exhibit asymmetries around a few$\times 10^{-3}$. Unfortunately, the numerical reliability of these predictions is rather low since they depend directly on the ratio of ill-determined hadronic matrix elements; therefore at present we cannot rule out that an asymmetry of 1% or so could arise in the KM scheme. On the other hand, we can hope that future phenomenological and theoretical studies will reduce these uncertainties.

The first two scenarios represent clear evidence for the presence of New Physics; conclusions derived from the last scenario appear considerably more obscure.

Since *top* quarks decay before they can typically hadronize, they act like a very heavy lepton rather than a hadron. The CP phenomenology of K, D and B mesons is thus largely irrelevant. Qualitatively *top* and τ dynamics can be described through similar scenarios. They do provide a promising searching ground for CP violation – if New Physics intervenes; *top quarks* enjoy the advantage of large Yukawa couplings to Higgs fields. However we have not quite done our 'homework' yet for proper evaluation of the prospects and search strategies: the required experiments are certainly challenging, but what are the limitations? How important – statistically as well as systematically – is the availability of polarized $e^+ e^-$ beams for τ (or even top) studies? Which observables are most sensitive to the intervention of which class of New Physics?

Problems

15.1   Derive Eq. (15.7).

15.2   Show that $\mathrm{Im}\frac{q}{p} \sim \mathcal{O}(\lambda^4)$ and $\mathrm{Im}\overline{\rho} \sim \mathcal{O}(\lambda^4)$ in the decay given in Fig. 15.1, by drawing appropriate Feynman diagrams.

15.3 Which branching ratios do you have to measure to extract the phase shifts relevant for a non-KM difference in $D^+ \to K_S \pi^+$ vs $D^- \to K_S \pi^-$?

15.4 Derive Eq. (15.14).

15.5 Show that $O_- \Longrightarrow -O_-$ under **CP** symmetry.

15.6 Why can we consider the $\tau^+ \tau^-$ pair in the reaction shown in Eq. (15.32) to be on-shell?

15.7 Convince yourselves that direct **CP** asymmetries in once CKM suppressed decays $D \to K\overline{K}, \pi\pi$ are $\mathcal{O}(\lambda^4)$ by drawing appropriate Feynman graphs. Also discuss the presence or absence of final state interactions.

PART 3
LOOKING BEYOND THE STANDARD MODEL

16

The strong **CP** problem

16.1 The problem

It is often listed among the attractive features of QCD that it 'naturally' conserves baryon number, flavour, parity and **CP**. Actually, the last two points are not quite true[1], which had been overlooked for some time [229].

As we shall see below, the QCD Lagrangian must, in general, contain an additional term $\frac{\theta g_s^2}{32\pi^2} G_{\mu\nu} \tilde{G}^{\mu\nu}$ which violates, among many things, **CP** symmetry. The fact that this term must be present can be seen in many ways. Here we shall stick to our main story and introduce the problem in a practical fashion. Another approach is to discuss the structure of the QCD vacuum [230, 231]. This is a fascinating story, and we recommend serious student of physics to look into it [232].

16.2 QCD and quark masses

In Chapter 8, we discussed how quark masses get generated when Higgs bosons acquire a vacuum expectation value. Starting from Eq. (8.22), when a Higgs boson ϕ^0 aquires a vacuum expectation value $\langle \phi^0 \rangle = v$, the quark mass term can be written as:

$$\mathscr{L}_{\text{mass}} = v \sum_{i,j} \left((G_U)_{ij} \overline{U}_{i,L} U_{j,R} + (G_D)_{ij} \overline{D}_{i,L} D_{j,R} \right) + \text{h.c.} \tag{16.1}$$

The mass matrices $\mathscr{M}_U = v G_U$ and $\mathscr{M}_D = v G_D$ can be diagonalized as shown in Eq. (8.24), so that Eq. (16.1) can be expressed in terms of mass eigenstates:

$$\mathscr{L}_{\text{mass}} = \overline{U}_L^m \mathscr{M}_U^{\text{diag}} U_R^m + \overline{D}_L^m \mathscr{M}_D^{\text{diag}} D_R + \text{h.c.} \tag{16.2}$$

---

[1] Embedding (global) SUSY into QCD even leads to flavour changing neutral currents, see Chapter 20.

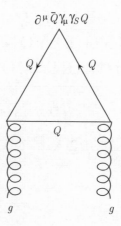

$$\partial^\mu \overline{Q} \gamma_\mu \gamma_5 Q$$

Q Q

Q

g g

Fig. 16.1.   The axial current given in Eq. (16.5) is not conserved at the quantum
level due to the anomaly diagram.

We are not finished yet! For example, just write the up quark mass term:

$$
\begin{aligned}
\mathcal{L}^U_{\text{mass}} &= \overline{U}^m_L \mathcal{M}^{\text{diag}}_U U^m_R + \overline{U}^m_R \mathcal{M}^{\text{diag}\dagger}_U U^m_L \\
&= \frac{1}{2}\overline{U}^m (\mathcal{M}^{\text{diag}}_U + \mathcal{M}^{\text{diag}\dagger}_U) U^m + \frac{1}{2}\overline{U}^m (\mathcal{M}^{\text{diag}}_U - \mathcal{M}^{\text{diag}\dagger}_U)\gamma_5 U^m. \quad (16.3)
\end{aligned}
$$

The terms proportional to $\overline{U}^m \gamma_5 U^m$ can be removed by performing the chiral
rotation

$$U^m_i \rightarrow e^{-i\frac{1}{2}\alpha_i \gamma_5} U^m_i, \qquad (16.4)$$

where we have denoted the diagonal elements of $\mathcal{M}^{\text{diag}}_U$ by $m_i e^{i\alpha_i}$. This would be
the end of the story if QCD were invariant under this chiral transformation – but
it is not! In fact, the current associated with this transformation is not conserved:

$$\partial^\mu J^{5i}_\mu = \partial^\mu \overline{U}^m_i \gamma_\mu \gamma_5 U^m_i = 2m_i i \overline{U}^m_i \gamma_5 U^m_i + \frac{g_s^2}{16\pi^2} G \cdot \tilde{G} \neq 0. \qquad (16.5)$$

So, the chiral transformation of Eq. (16.4) changes the action S:

$$S \longrightarrow S - \sum_i \int d^4x \, \partial^\mu J^{5,i}_\mu = S - i(\arg \det \mathcal{M}) \int d^4x \frac{g_s^2}{32\pi^2} G \cdot \tilde{G} \, ; \qquad (16.6)$$

where

$$\arg \det \mathcal{M} = \sum_i \alpha_i \qquad (16.7)$$

and the sum runs over terms arising from U and D quark masses. The problem
stems from the fact that the axial current, even of massless quarks, ceases to remain
conserved on the quantum level. When a *classical* symmetry is broken by *quantum*
corrections, we refer to it as an *anomaly*. The diagram which generates this chiral
anomaly is shown in Fig. 16.1 and is called the 'triangle' anomaly because it is

produced by a diagram with a triangular fermion loop, or the 'Adler–Bardeen–Bell–Jackiw' anomaly, named after its discoverers [233]. Eq. (16.6) implies that we should have started with a modified QCD Lagrangian

$$\mathscr{L}_{\text{eff}} = \mathscr{L}_{QCD} + \frac{\theta g_s^2}{32\pi^2} G_{\mu\nu}\tilde{G}^{\mu\nu}, \tag{16.8}$$

where $\mathscr{L}_{QCD}$ is given by Eq. (8.9). The QCD action remains unaffected by the simultaneous transformations

$$Q_i \to e^{-i\frac{1}{2}\alpha_i\gamma_5}Q_i, \quad m_i \to e^{-i\alpha_i}m_i, \quad \theta \to \theta - \sum_i \alpha_i = \theta - \arg\det\mathscr{M}. \tag{16.9}$$

This means that the dynamics depend on the combination

$$\bar{\theta} = \theta - \arg\det\mathscr{M}, \tag{16.10}$$

rather than θ by itself.

The anomaly is actually quite welcome, since it solves one long-standing puzzle of *strong* dynamics, the '$U(1)$ problem': in the limit of massless u and d quarks, QCD would appear to have a *global* $U(2)_L \times U(2)_R$ invariance. While the vectorial part $U(2)_{L+R}$ is a manifest symmetry, the axial part $U(2)_{L-R}$ is spontaneously broken, leading to the emergence of four Goldstone bosons. In the presence of quark masses those bosons acquire a mass as well. The pions readily play the part, but the η meson does not![2] Yet from the anomaly we infer that due to quantum corrections the axial $U(1)_{L-R}$ was never there in the first place, even for massless quarks: therefore only three Goldstone bosons are predicted for isospin, the pions!

16.2.1 *Strong* **CP** *violation*

Whatever the origin of the $G \cdot \tilde{G}$ term, its emergence in the Lagrangian causes a severe problem. For $G \cdot \tilde{G}$ – in contrast to $G \cdot G$ – violates both parity and time reversal invariance! This is best seen by expressing $G_{\mu\nu}$ and its dual through the *colour* electric and magnetic fields, $\vec{E}_a$ and $\vec{B}_a$, respectively:

$$G \cdot G \propto \sum_a |\vec{E}_a|^2 + \sum_a |\vec{B}_a|^2 \overset{\text{P,T}}{\Longrightarrow} \sum_a |\vec{E}_a|^2 + \sum_a |\vec{B}_a|^2, \tag{16.11}$$

$$G \cdot \tilde{G} \propto \sum_a \vec{E}_a \cdot \vec{B}_a \overset{\text{P,T}}{\Longrightarrow} -\sum_a \vec{E}_a \cdot \vec{B}_a, \tag{16.12}$$

since

$$\vec{E}_a \overset{\text{P}}{\Longrightarrow} -\vec{E}_a, \quad \vec{E}_a \overset{\text{T}}{\Longrightarrow} \vec{E}_a, \tag{16.13}$$

$$\vec{B}_a \overset{\text{P}}{\Longrightarrow} \vec{B}_a, \quad \vec{B}_a \overset{\text{T}}{\Longrightarrow} -\vec{B}_a; \tag{16.14}$$

---

[2] An analogous discussion can be given with s quarks included. The spontaneous breaking of the global $U(3)_{L-R}$ symmetry leads to the existence of nine Goldstone bosons, yet the η' meson is far too heavy for this role.

i.e., for $\overline{\theta} \neq 0$ neither parity nor time reversal invariance are fully conserved by QCD. This is called – somewhat sloppily – the *strong* **CP** *problem*.

Two more general comments are in order here:

- The conservation of **CP** – while not automatic – can be *imposed within the confines of QCD*, leading to $\overline{\theta} = 0$.

- On the other hand, we know the electroweak sector has to contain sources of **CP** violation, whether they are of the KM variety or not. The anomaly provides a portal for them to enter the strong sector, for the quarks acquire their masses from the Higgs mechanism driving the phase transition

$$SU(2)_L \times U(1) \rightarrow U(1)_{QED}. \tag{16.15}$$

So, once the the required chiral rotation is performed,

$$\overline{\theta} = \theta_{QCD} + \Delta\theta_{EW}, \quad \Delta\theta_{EW} = \arg \det M. \tag{16.16}$$

The second term in $\overline{\theta}$ now has no reason to vanish!

16.2.2 The neutron electric dipole moment

Since the gluonic operator $G \cdot \tilde{G}$ does not change flavour, we suspect right away that its most noticeable impact would be to generate an electric dipole moment (EDM) for neutrons. This is indeed the case, yet making this connection more concrete requires a more sophisticated argument. Let us briefly recapitulate some elementary findings on EDMs:

- They are described by an operator in the Lagrangian:

$$\mathscr{L}_{EDM} = -\frac{i}{2}d\overline{\psi}\sigma_{\mu\nu}\gamma_5\psi F_{\mu\nu}. \tag{16.17}$$

Since this operator's dimension is five, its *dimensionful* coefficient d can be calculated as a *finite* quantity.

- A non-relativistic reduction (Problem 16.1) shows that the low-energy properties of an EDM are indeed satisfied by Eq. (16.17): d describes the energy shift of a system with angular momentum $\vec{j}$ when placed in an external electric field that grows linearly with the field:

$$\Delta E = \vec{d} \cdot \vec{E} + \mathscr{O}(|\vec{E}|^2) = d\vec{j} \cdot \vec{E} + \mathscr{O}(|\vec{E}|^2). \tag{16.18}$$

- An EDM signalling **T** violation can unambiguously be differentiated against *induced* dipole moments that do not, see Chapter 3.

In the context of the strong **CP** problem we view the neutron EDM – d_N – as due to the photon coupling to a virtual proton or pion in a fluctuation of

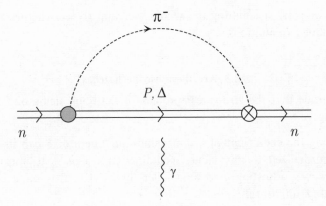

Fig. 16.2. A major contribution to the neutron EDM. The blob on the left represents a strong vertex. The blob on the right represents a **CP** or **T** violating interaction.

the neutron:

$$n \implies p^* \pi^* \implies n. \tag{16.19}$$

Of the two effective pion nucleon couplings in this one-loop process one is produced by ordinary QCD and conserves **P**, **T** and **CP**; the other one is induced by $G \cdot \tilde{G}$ with the help of Eq. (16.5). A rough guestimate can be gleaned from naive dimensional reasoning: the scale for a dipole moment (electric or magnetic) is set by e/M_N with M_N being the mass of the neutron; to be an EDM it obviously has to be proportional to $\bar{\theta}$ – and to the ratio of (current) quark masses m_q to M_N; this last point anticipates what will be discussed later, namely that the $\bar{\theta}$ dependence of observables can be rotated away for $m_q = 0$, i.e., in the chiral limit: $d_N \sim \mathcal{O}\left(\frac{e}{M_N} \frac{m_q}{M_N} \bar{\theta}\right) \sim \mathcal{O}\left(2 \cdot 10^{-15} \bar{\theta} \, \mathrm{e\,cm}\right)$. A more reliable estimate was first obtained by Baluni [234] in a nice paper using bag model computations of the transition amplitudes between the neutron and its excitations: $d_N \simeq 2.7 \cdot 10^{-16} \, \bar{\theta} \, \mathrm{e\,cm}$. In [235] chiral perturbation theory was employed instead: $d_N \simeq 5.2 \cdot 10^{-16} \, \bar{\theta} \, \mathrm{e\,cm}$. More recent estimates yield values in roughly the same range: $d_N \simeq (4 \cdot 10^{-17} \div 2 \cdot 10^{-15}) \bar{\theta} \, \mathrm{e\,cm}$ [236]. Hence

$$d_N \sim \mathcal{O}(10^{-16} \bar{\theta}) \, \mathrm{e\,cm}, \tag{16.20}$$

and we infer

$$d_N \le 1.1 \cdot 10^{-25} \, \mathrm{e\,cm}, \quad 95\% \, \mathrm{CL}, \implies \bar{\theta} < 10^{-9 \pm 1}. \tag{16.21}$$

Although θ_{QCD} is a QCD parameter, it might not necessarily be of order unity; nevertheless the truly tiny bound given in Eq. (16.21) begs for an explanation. The only kind of explanation that is usually accepted as 'natural' in our community is one based on symmetry. Such explanations have been put forward; the most widely discussed suggestions are based on a so-called Peccei–Quinn symmetry.

Yet before we start speculating too wildly, we want to see whether there are no more mundane explanations.

16.3 Are there escape hatches?

We could argue that the strong CP problem is fictitious, using one of two lines of reasoning:

- Being the coefficient of a dimension-four operator $\bar{\theta}$ can in general[3] be renormalized to any value, including zero. This is technically correct; however, adjusting θ to be smaller than $\mathcal{O}(10^{-9})$ by hand is viewed as highly 'unnatural':

  - A priori there is no reason why θ_{QCD} and $\Delta\theta_{EW}$ should practically vanish.

  - Even if $\theta_{QCD} = 0 = \Delta\theta_{EW}$ were set *by fiat*, quantum corrections to $\Delta\theta_{EW}$ are typically much larger than 10^{-9} and ultimately actually infinite. At which order this happens depends on the electroweak dynamics, though. Within the KM ansatz $\Delta\theta_{\mathrm{ren}} \neq 0$ arises first at three loops and it does not diverge before seven loops. In other models, though, the problem is more pressing. As described later, in models with right-handed currents or non-minimal Higgs dynamics $\delta\theta_{\mathrm{ren}} \neq 0$ arises at one loop already.

  - To expect that θ_{QCD} and $\Delta\theta_{EW}$ cancel so as to render $\bar{\theta}$ sufficiently tiny would require fine tuning of a kind which would have to strike even a sceptic as unnatural. For θ_{QCD} reflects dynamics of the strong sector and $\Delta\theta_{EW}$ that of the electroweak sector.

- A more respectable way out is provided by the following observation based on Eq. (16.9) and Eq. (16.10): if one of the quark masses m_i vanishes (det $M = 0$), then the unphysical phase α_i can be used to dial $\bar{\theta}$ to zero! However, most authors argue that neither the up quark nor a fortiori the down quark mass can vanish [237]:

$$m_d(1 \text{ GeV}) > m_u(1 \text{ GeV}) \simeq 5 \text{ MeV} , \qquad (16.22)$$

where the notation shows that we have to use the running mass evaluated at a scale of 1 GeV.

16.4 Peccei–Quinn symmetry

As just argued, $\bar{\theta} \leq \mathcal{O}(10^{-9})$ could hardly come about accidentally; an organizing principle had to arrange various contributions and corrections in such a way as

---

[3] Exceptions will be mentioned below.

to render the required cancellations. There is the general philosophy that such a principle has to involve some underlying symmetry. In the preceding section we have already sketched such an approach: a global chiral invariance allows us to rotate the dependence on $\bar\theta$ away; we failed however in our endeavour because this symmetry is broken by $m_q \neq 0$. Is it possible to invoke some other variant of chiral symmetry for this purpose, even if it is spontaneously broken? One particularly intriguing ansatz is to reinterpret a physical quantity that is conventionally taken to be a constant as a *dynamical degree of freedom*, that adjusts itself to a certain (desired) value in response to forces acting upon it. One early example is provided by the original Kaluza–Klein theory [238] invoking a six-dimensional 'space'–time manifold: two compactify dynamically and thus lead to the quantization of electric and magnetic charge.

Something similar has been suggested by Peccei and Quinn [239]. They augmented the SM by a global $U(1)_{PQ}$ symmetry – now referred to as the Peccei–Quinn symmetry – which is axial with the following properties:

- It is a symmetry of the *classical* theory.

- It is subject to an axial anomaly; i.e., it is broken *explicitly* by non-perturbative effects, reflecting the complexity of the QCD ground state.

- It is broken *spontaneously* as well.

The following will happen then: the spontaneous breaking of the symmetry gives rise to Goldstone bosons called *axions*. With $U(1)_{PQ}$ being axial, it exhibits a triangle anomaly leading to a coupling of the axion field to $G \cdot \tilde G$. This in turn generates a mass for the axion; more importantly it transforms the quantity $\bar\theta$ into a dynamical one, depending on the axion field. The potential due to non-perturbative dynamics induces a vacuum expectation value for the axion such that

$$\bar\theta \simeq 0 \tag{16.23}$$

emerges; i.e., $\bar\theta$ relaxes *dynamically* to a (practically) zero value.

To see that these words are more than just a nice yarn, let us consider the following Lagrangian:

$$\mathscr{L}_{PQ} = -\frac{1}{4}G \cdot G + \sum_j \left[\bar Q_j i\gamma_\mu D^\mu Q_j - (y_j \bar Q_{L,j} Q_{R,j}\phi + \text{h.c.}) \right]$$

$$+ \frac{\theta g_S^2}{32\pi^2}G \cdot \tilde G + \partial_\mu \phi^\dagger \partial^\mu \phi - V(\phi^\dagger \phi), \tag{16.24}$$

which remains invariant *classically* under

$$\phi \xrightarrow{U_{PQ}(1)} e^{i2\alpha}\phi, \quad Q_i \xrightarrow{U_{PQ}(1)} e^{-i\alpha\gamma_5}Q_i. \tag{16.25}$$

The potential $V(\phi^\dagger \phi)$ is chosen such that the axial symmetry $U_{PQ}(1)$ is broken *spontaneously* by a vacuum expectation value of ϕ:

$$\langle \phi(x) \rangle = v_{PQ} e^{i\langle a \rangle / v_{PQ}}, \tag{16.26}$$

with $\langle a \rangle$ denoting the vacuum expectation value of the axion field $a(x)$. With the quarks Q_i acquiring masses

$$m_i = y_i v_{PQ} e^{i\langle a \rangle / v_{PQ}}, \tag{16.27}$$

we obtain

$$\bar{\theta} = \theta - \sum_i \arg y_i - N_f \langle a \rangle / v_{PQ}; \tag{16.28}$$

i.e., the important new feature is that the quantity $\bar{\theta}$ – rather than being a mere parameter – depends on the dynamical field a through the latter's VEV. In the usual scenarios for the spontaneous realization of a symmetry, the phase of the scalar field remains completely undetermined, which implies the masslessness of the Goldstone bosons.

Now, a second novel feature arises: the chiral anomaly is – as just sketched – implemented through a term proportional to $aG \cdot \tilde{G}$; since it is linear in the field a, $G \cdot \tilde{G}$ acts as a non-trivial effective potential for a and the resulting dynamics determine $\langle a \rangle$. To see how this comes about, let us consider the effective Lagrangian[4]:

$$\mathscr{L}_{\text{eff}} = \mathscr{L}_{SM} + \frac{g_s^2 \bar{\theta}}{32\pi^2} G \cdot \tilde{G} + \frac{g_s^2}{32\pi^2} \frac{\xi}{v_{PQ}} aG \cdot \tilde{G} - \frac{1}{2} \partial_\mu a \partial_\mu a + \mathscr{L}_{\text{int}}(\partial_\mu a, \psi). \tag{16.29}$$

The size of the parameters v_{PQ} and ξ and the form of $\mathscr{L}_{\text{int}}(\partial_\mu a, \psi)$ describing the (purely derivative) coupling of the axion field to other fields ψ depend on how the Peccei–Quinn symmetry is specifically realized.

Because of the chiral anomaly, the expectation value for the axion field is given by (see Problem 16.3):

$$\langle a \rangle = -\frac{\bar{\theta}}{\xi} v_{PQ}. \tag{16.30}$$

The *physical* axion excitations are described by the shifted field

$$a_{\text{phys}}(x) = a(x) - \langle a \rangle, \tag{16.31}$$

in terms of which Eq. (16.29) is re-written as follows:

$$\mathscr{L}_{\text{eff}} = \mathscr{L}_{SM} - \frac{1}{2} \partial_\mu a_{\text{phys}} \partial_\mu a_{\text{phys}} + \frac{g_s^2}{32\pi^2} \frac{\xi}{v_{PQ}} a_{\text{phys}} G \cdot \tilde{G} + \mathscr{L}_{\text{int}}(\partial_\mu a_{\text{phys}}, \psi) ; \tag{16.32}$$

---

[4] Here we follow the discussion given in a review by Peccei [49].

i.e., the offending **P** and **T** violating bilinear term $\bar{\theta}G \cdot \tilde{G}$ has been traded in against a $a_{\text{phys}}G \cdot \tilde{G}$ coupling between dynamical fields:

$$\bar{\theta} = 0 \,. \tag{16.33}$$

Electroweak forces driving $K_L \to \pi\pi$ will actually move $\bar{\theta}$ away from zero – but only by an extremely tiny amount: $\bar{\theta} \sim \mathcal{O}(10^{-16})$!

16.5 The dawn of axions – and their dusk?

The story does not end here, of course, since the breaking of $U(1)_{PQ}$ gives rise to a new dynamical entity, the physical axion field [240, 241]. Its behaviour depends on two parameters: its mass and its couplings to other fields. Since its mass is controlled by the anomaly term, we guestimate on dimensional grounds (see Problem 16.3):

$$m_a^2 \sim \mathcal{O}\left(\frac{\Lambda_{QCD}^4}{v_{PQ}^2}\right) \,. \tag{16.34}$$

Since we expect on general grounds $v_{PQ} \gg \Lambda_{QCD}$, we are dealing with a very light boson. The question is how light would the axion be. As is shown below, e.g. Eq. (16.37), $1/v_{PQ}$ sets the scale also for the couplings of the physical axion, including its derivative couplings to fermionic fields ψ. So, as v_{PQ} goes up, the mass of the axion goes down – as does its coupling.

The electroweak scale $v_{EW} = \left(\sqrt{2}G_F\right)^{-\frac{1}{2}} \simeq 250$ GeV provides the discriminator for two scenarios, which will be discussed below:

- $$v_{PQ} \sim v_{EW} \quad \rightsquigarrow \quad m_a \sim \mathcal{O}(1 \text{ MeV}). \tag{16.35}$$

  In that case axions can, or even should, be seen in accelerator based experiments. Such scenarios are referred to as *visible* axions.

- $$v_{PQ} \gg v_{EW} \quad \rightsquigarrow \quad m_a \ll 1 \text{ MeV}. \tag{16.36}$$

  As explained below, such axions could not be found in accelerator based experiments, because of their minute couplings; therefore they are called *invisible* scenarios. Yet that does not mean that they necessarily escape detection! They could be of great significance for the formation of stars, whole galaxies and even the universe.

16.5.1 Visible axions

The simplest scenario involves two $SU(2)_L$ doublet Higgs fields that possess opposite hypercharge[5]. They also carry a $U(1)$ charge *in addition* to the hypercharge; this second (and global) $U(1)$ is identified with the PQ symmetry.

The couplings of the axion to fermions is purely derivative, as befits a Goldstone boson:

$$\mathscr{L}_{\text{int}} = \frac{1}{v}\partial_\mu a \left[x \sum_i \bar{u}_{R,i}\gamma_\mu u_{R,i} + \frac{1}{x}\sum_i \bar{d}_{R,i}\gamma_\mu d_{R,i} + \frac{1}{x}\sum_i \bar{l}_{R,i}\gamma_\mu l_{R,i} \right], \quad (16.37)$$

where

$$x = \frac{v_2}{v_1}, \quad v = \sqrt{v_1^2 + v_2^2}, \quad (16.38)$$

with v_i denoting the VEVs of the two Higgs doublets.

The anomaly induces non-derivative couplings to the gauge fields [242]

$$\mathscr{L}_{\text{anom}} = \frac{a}{v}\left\{ \left[N_{\text{fam}}\left(x + \frac{1}{x}\right)\frac{g_S^2}{32\pi^2}G\cdot\tilde{G} + N_{\text{fam}}\left(\frac{4}{3}x + \frac{1}{3x} + \frac{1}{x}\right)\frac{g_{EW}^2}{16\pi^2}B\cdot\tilde{B} \right] \right\}, \quad (16.39)$$

where N_{fam} denotes the number of families and $B_{\mu\nu}$ the field strength tensor of the hypercharge gauge field that couples to right-handed fermions: $B_{\mu\nu} = F_{\mu\nu}^{em} - \tan\theta_W F_{\mu\nu}^{Z^0}$. The anomaly also induces a mass for the axion [243]:

$$m_a \simeq \frac{m_\pi F_\pi}{v}N_{\text{fam}}\left(x + \frac{1}{x}\right)\frac{\sqrt{m_u m_d}}{(m_u + m_d)} \simeq 25\, N_{\text{fam}}\left(x + \frac{1}{x}\right) \text{ KeV.} \quad (16.40)$$

Such masses could conceivably reach the $\mathcal{O}(1\ \text{MeV})$ level, but not much beyond.

Its lifetime can be deduced from Eq. (16.37) and Eq. (16.39):

- If $m_a > 2m_e$, the axion decays very rapidly into electrons and positrons:

$$\tau(a \to e^+e^-) \simeq 4\cdot 10^{-9}\left(\frac{1\ \text{MeV}}{m_a}\right)\frac{x^2 \text{ or } 1/x^2}{\sqrt{1 - \frac{4m_e^2}{m_a^2}}}\text{s.} \quad (16.41)$$

- If, on the other hand, $m_a < 2m_e$, then the axion decays fairly slowly into two photons, as derived from Eq. (16.39) and Eq. (16.40):

$$\tau(a \to \gamma\gamma) \sim \mathcal{O}\left(\frac{100\ \text{KeV}}{m_a}\right)\text{s.} \quad (16.42)$$

We have presented here a very rough sketch of scenarios with visible axions since we can confidently declare that they have been ruled out experimentally. They have been looked for in beam dump experiments – without success. Yet the more telling blows have come from searches in rare decays:

---

[5] In the SM the Higgs doublet and its charge conjugate fill this role.

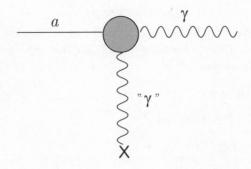

Fig. 16.3.   An 'invisible' axion may show itself while interacting with a magnetic field, converting itself into a photon.

- For long-lived axions – $m_a < 2m_e$ – we expect a dominating contribution to $K^+ \to \pi^+ +$ nothing from

$$K^+ \to \pi^+ + a, \tag{16.43}$$

with the axion decaying well outside the detector. For the two-body kinematics of Eq. (16.43) we have very tight bounds from published data [167]:

$$\mathrm{Br}(K^+ \to \pi^+ X^0) < 5.2 \cdot 10^{-10}, \quad 90\% \ \mathrm{CL} \tag{16.44}$$

for X^0 being a practically massless and non-interacting particle. Theoretically we would expect [242]:

$$\mathrm{Br}(K^+ \to \pi^+ a)\big|_{\mathrm{theor}} \sim 3 \cdot 10^{-5} \cdot (x + 1/x)^{-2}. \tag{16.45}$$

Although Eq. (16.45) does not represent a precise prediction, the discrepancy between expectation and observation is conclusive.

- We arrive at the same conclusion that *long-lived visible* axions do not exist from the absence of *quarkonia* decay into them: neither $J/\psi \to a\gamma$ nor $\Upsilon \to a\gamma$ has been seen [244].

- The analysis is a bit more involved for *short-lived* axions – $m_a > 2m_e$. Yet again their absence has been established through a combination of experiments. Unsuccessful searches for

$$\pi^+ \to a e^+ \nu, \tag{16.46}$$

figure prominently in this endeavour [245]. Likewise the *absence* of axion driven *nuclear de-excitation* has been established on a level that appears to be conclusive.

16.5.2 Invisible axions

The phenomenological conflicts just discussed can be avoided by separating the $SU(2)_L \times U(1)$ and $U(1)_{PQ}$ breaking scales. To that purpose, we introduce a complex scalar field σ that

- is an $SU(2)_L$ singlet,

- yet carries a PQ charge and

- possesses a huge VEV $v_{PQ} \gg v_{EW}$.

As can be inferred from Eq. (16.37), the couplings of such axions to gauge as well as fermion fields become truly tiny. These requirements can be realized in two distinct (sub-)scenarios:

1. Only presumably very heavy new quarks carry a PQ charge. This situation is referred to as the KSVZ axion [246]. The minimal version can do with a single $SU(2)_L$ Higgs doublet.

2. Also the known quarks and leptons carry a PQ charge. Two $SU(2)_L$ Higgs doublets are then required in addition to σ. The fermions do not couple directly to σ, yet become sensitive to PQ breaking through the Higgs potential. This is referred to as the DFSZ axion [247].

From current algebra we infer for the axion mass in either case

$$m_a \simeq 0.6 \, \text{eV} \cdot \frac{10^7 \, \text{GeV}}{v_{PQ}}. \tag{16.47}$$

The most relevant coupling of such axions is to two photons

$$\mathcal{L}(a \to \gamma\gamma) = -\tilde{g}_{a\gamma\gamma} \frac{\alpha}{\pi} \frac{a(x)}{v_{PQ}} \vec{E} \cdot \vec{B} \,, \tag{16.48}$$

where $\tilde{g}_{a\gamma\gamma}$ is a model-dependent coefficient of order unity.

Axions with such tiny masses have lifetimes easily in excess of the age of the universe. Also their couplings to other fields are so minute that they would not betray their presence – hence their name *invisible* axions – under ordinary circumstances! Yet in astrophysics and cosmology more favourable extra-ordinary conditions can arise.

Through their couplings to electrons, axions would provide a cooling mechanism to *stellar evolution*. Not surprisingly their greatest impact occurs for the lifetimes of red giants and supernovae like SN 1987a. The actual bounds depend on the model – whether it is a KSVZ or DFSZ axion – but relatively mildly only. Altogether astrophysics tells us that *if* axions exist we have

$$m_a < 3 \cdot 10^{-3} \, \text{eV}. \tag{16.49}$$

Cosmology, on the other hand, provides us with a *lower* bound through a very intriguing line of reasoning. At temperatures T above Λ_{QCD} the axion is massless and all values of $\langle a(x) \rangle$ are equally likely. For $T \sim 1$ GeV the anomaly-induced potential turns on driving $\langle a(x) \rangle$ to a value which will yield $\bar{\theta} = 0$ at the new potential minimum. The energy stored previously as *latent heat* is then released into axions oscillating around its new VEV. Precisely because the invisible axion's couplings are so immensely suppressed, the energy cannot be dissipated into other degrees of freedom. We are then dealing with a fluid of axions. Their typical momentum is the inverse of their correlation length, which in turn cannot exceed their horizon; we find

$$p_a \sim \left(10^{-6} \text{ s}\right)^{-1} \sim 10^{-9} \text{ eV} \tag{16.50}$$

at $T \simeq 1$ GeV; i.e., the axions, despite their minute mass, form a very cold fluid and actually represent a candidate for cold dark matter. Their contribution to the density of the universe relative to its critical value is [248]

$$\Omega_a = \left(\frac{0.6 \cdot 10^{-5} \text{ eV}}{m_a}\right)^{\frac{7}{6}} \cdot \left(\frac{200 \text{ MeV}}{\Lambda_{QCD}}\right)^{\frac{3}{4}} \cdot \left(\frac{75 \text{ km/s} \cdot \text{Mpc}}{H_0}\right)^2 ; \tag{16.51}$$

H_0 is the present Hubble expansion rate. For axions not to overclose the universe we thus have to require:

$$m_a \geq 10^{-6} \text{ eV} \tag{16.52}$$

or

$$v_{PQ} \leq 10^{12} \text{ GeV}. \tag{16.53}$$

This means also that we might be moving or existing in a bath of cold axions, still making up a significant fraction of the matter of the universe.

Ingenious suggestions have been made to search for such cosmic background axions. The main handle we have on them is their coupling to two photons. They can be detected by stimulating the conversion

$$\text{axion} \xrightarrow{\vec{B}} \text{photon} \tag{16.54}$$

in a strong magnetic field $\vec{B}$: the second photon (see Fig. 16.3), which is virtual in this process, effects the interaction with the inhomogeneous magnetic field in the cavity. Available microwave technology allows an impressive experimental sensitivity. No signal has been found yet, but the search continues and soon should reach a level where we have a good chance of seeing a signal [249]. A concise review can be found in Ref. [250].

16.6 Soft CP violation

Without any sightings of axions, we had better explore other solutions to the strong **CP** problem that are 'natural'. An obvious ansatz is to implement **CP** symmetry *spontaneously*, which imposes $\bar{\theta} = 0$ as the leading effect, with corrections leading to a small and *calculable* deviation from zero. This appears as a very attractive option:

- We have stated on several occasions the theoretical advantages of a *spontaneously* over a *manifestly* broken symmetry.

- Arranging for spontaneous **CP** breaking represents a quite manageable task. For once we go beyond the minimal structure of the SM, more VEVs emerge that can exhibit physical phases. Explicit examples will be given in our discussion of left–right symmetric models and non-minimal Higgs dynamics in Chapter 19.

- The resulting scenarios are intriguing in their own right.

Some challenges have to be met, though, chief among them two:

- The *cosmic domain wall problem* raises its unpleasant head (as it does for any *discrete* symmetry). By that we mean the following: as the universe cools down to a temperature below which **CP** invariance is broken spontaneously, domains of different **CP** phases emerge. Since it is a discrete symmetry, walls have to form to separate these domains. As shown in Ref. [251], the energy stored in such walls would greatly exceed the closure density for the universe. This cosmological disaster can be vitiated if the spontaneous **CP** breaking occurs *before* an inflationary period in our universe's past. For then we would live in a *single* domain! That means the breaking scale has to be very high, of the order of GUT scales.

- While $\bar{\theta}$ naturally emerges to be small in these scenarios – $\bar{\theta} \ll 1$ – we aim for truly tiny values: $\bar{\theta} \leq \mathcal{O}(10^{-9})$. Again, this favours a very high breaking scale.

It is important to note that **CP** being broken spontaneously does not suffice to enforce $\bar{\theta} = 0$ at tree level. For the emerging VEVs can still contribute to arg det M at tree level. Two strategies can be pursued to build viable models [252]:

- While still allowing for a complex quark mass matrix M, we impose a special form on M such that arg det $M = 0$ holds at tree level. The key ingredient here is the introduction of novel truly super-heavy quarks. Yet at low energies, open to experimentation, we recover effectively a CKM mechanism [253, 254].

- The quark mass matrix is actually real at tree level. *No* CKM ansatz can effectively arise at low energies then, and we need alternative sources of **CP** violation to explain $K_L \to \pi\pi$.

Alternative 'natural' solutions to the strong **CP** problem thus exist. However, the reader will be forgiven for thinking that they are more intriguing to theorists than experimentalists!

16.7 The pundits' judgement

The story of the strong **CP** problem is a particularly titillating one. It arises only if we go beyond mere renormalizability and insist that renormalization proceeds in a *natural* way, i.e., without fine tuning. Trying to resolve it has led to an impressive intellectual edifice that is based on an intriguing arsenal of theoretical reasoning, and has inspired fascinating experimental undertakings that are still going strong. The observation that neutrons have at best a tiny EDM has generated the suggestion that extremely light and weakly particles – axions – exist, that nevertheless could make up a significant fraction of the mass of galaxies or even the universe.

Neither axions nor other consequences of the strong **CP** problem have been discovered so far. Thus, like many modern novels the problem – if it is indeed one – has not found any resolution. On the other hand, it has the potential to lead the charge towards a new paradigm in high energy physics.

A cynic, however, might summarize it differently: while we certainly do not know what the solution is, we cannot be sure whether there is a problem in the first place! It thus might remind her or him of the often told story of the French officer from the period of Enlightenment who was overheard praying before a battle: 'Dear God – in case you exist – save my soul if I have got one!'

Problems

16.1 Show that the nonrelativistic reduction of Eq. (16.17) yields

$$\int d^3x \mathscr{H} = d\vec{S} \cdot \vec{E}, \quad \vec{S} = \int d^3x \psi^* \vec{\sigma} \psi. \tag{16.55}$$

where $\psi(x)$ is the wave function of a fermion at rest. Thus, it satisfies the properties of a non-relativistic EDM described in Chapter 3.

16.2 Consider the Lagrangian of Eq. (16.29). Write down the field equation for the axion field *ignoring* the anomaly term $aG \cdot \tilde{G}$. Verify that any constant value of $\langle a \rangle$ would then satisfy this equation.

Now include the anomaly term in the field equation for the axion field and derive:

$$-\partial^2 a + \partial_\mu \frac{\partial \mathscr{L}_{\text{int}}}{\partial \partial_\mu a} = \frac{\xi}{\Lambda_{PQ}} \frac{g_S^2}{32\pi^2} G \cdot \tilde{G}. \tag{16.56}$$

16.3 Show that the effective potential for a is minimized by $\langle G \cdot \tilde{G} \rangle = 0$. Using Eq. (16.29) and the fact that $\langle G \cdot \tilde{G} \rangle$ is periodic in $\bar{\theta}$, argue that $\langle G \cdot \tilde{G} \rangle$ is actually controlled by the combination $\bar{\theta} + \xi \langle a \rangle / v_{PQ}$. This leads to

$$\langle \bar{\theta} | a | \bar{\theta} \rangle = -\bar{\theta} \frac{v_{PQ}}{\xi}.$$

For example, we can derive

$$\langle G \cdot \tilde{G} \rangle \propto \sin(\overline{\theta} + \xi \langle a \rangle / v_{PQ}), \tag{16.57}$$

in the one-instanton-approximation.

16.4 Show that the mass of an axion is given by:

$$m_a^2 = -\frac{\xi}{\Lambda_{PQ}} \frac{g^2}{32\pi^2} \frac{\partial}{\partial a} \langle G \cdot \tilde{G} \rangle_{\langle a \rangle = \frac{\overline{\theta}}{\xi} \Lambda_{PQ}}$$

$$\sim \quad \mathcal{O} \left(\frac{\Lambda_{QCD}^4}{\Lambda_{PQ}^2} \right). \tag{16.58}$$

17
Setting the table for **CP** violation in the neutrino sector

The poor sleeper's impatience

A man wakes up at night,
sees it is dark outside and falls asleep again.

A short while later he awakes anew,
notices it still to be dark outside and goes back to sleep.

This sequence repeats itself a few times:
– waking up, seeing the dark outside and falling asleep –

till he cries out in despair:
'Will there never be daylight?'

A bird begins to sing.

after C.- F. v. Weizsäcker

Within the SM neutrinos are massless. As such they are mass degenerate; flavour and mass eigenstates then coincide and by definition oscillations cannot occur. Yet no reason has surfaced why neutrinos should be truly massless. In 1957, Pontecorvo indicated that if lepton charges are not conserved, then $v \leftrightarrow \bar{v}$ mixing can occur [255]. The possibility that neutrinos have non-degenerate masses and mix just like quarks had been suggested by Sakata and his collaborators [256] back in 1962[1]. They stated the potential for neutrino oscillations.

Over the years there have been many reports that neutrino oscillations of one kind or another had been seen. If true this would be quite a treat for us, since it would establish the intervention of New Physics. Very recently the strongest evidence for it has been presented, based on two different sources, namely examination of atmospheric and solar neutrinos.

To appreciate the data, we will first sketch the basic features of neutrino oscillations. A more detailed discussion will be given later.

---

[1] In the same paper they also introduced what is today known as the Cabibbo angle. Unlike Ref. [257], they did not check the validity of this scheme by examining the experimental data, though.

17.1 Basics of neutrino oscillations

Drawing on analogies to $K^0 - \overline{K}^0$, $B^0 - \overline{B}^0$, etc. oscillations we can formulate the essence of neutrino oscillations as follows:

- Weak interaction eigenstates like ν_e, ν_μ and ν_τ are non-trivial linear combinations of mass eigenstates ν_i^m:

$$|\nu_\alpha(t)\rangle = \sum_{i=1}^{3} U_{\alpha i} e^{-iE_i t} |\nu_i^m\rangle, \quad \alpha = e, \mu, \tau. \tag{17.1}$$

- The transition probabilities for $\nu_\alpha \to \nu_\beta$ are easily obtained by evaluating $|\langle \nu_\beta | \nu_\alpha(t) \rangle|^2$. It is more useful to translate time into space, i.e., to express these probabilities as function of *distance L* from the production point rather than time t:

$$
\begin{aligned}
P(\nu_\alpha \to \nu_\beta; L) \;=\; & \sum_i |U_{\alpha i}|^2 ||U_{\beta i}|^2 \\
& + \sum_{i \neq j} \left[\cos\left(\frac{m_i^2 - m_j^2}{2E} \cdot L \right) \cdot [\mathrm{Re} U_{\alpha i} U_{\alpha j}^* U_{\beta i}^* U_{\beta j}] \right. \\
& \left. + \sin\left(\frac{m_i^2 - m_j^2}{2E} \cdot L \right) \cdot [\mathrm{Im} U_{\alpha i} U_{\alpha j}^* U_{\beta i}^* U_{\beta j}] \right]. \tag{17.2}
\end{aligned}
$$

For a given value of $\Delta m_{ij}^2 \equiv m_i^2 - m_j^2$, the ratio L/E characterizes the peculiar oscillation pattern; $m_i \ll E$ has been assumed here.

- There are two classes of transitions we can probe, namely

(1) 'appearance' reactions:

$$P(\nu_\alpha \to \nu_\beta; t) > 0 \quad \text{for } \alpha \neq \beta, \tag{17.3}$$

(2) 'disappearance' reactions[2] :

$$P(\nu_\alpha \to \nu_\alpha; t) < 1. \tag{17.4}$$

17.2 Experiments

17.2.1 Solar neutrinos

There is a copious source of neutrinos in our corner of the milky way, the sun! The chain reaction producing its energy output is given by

$$
\begin{pmatrix}
pp \to {}^2\mathrm{H} + e^+ + \nu_e & [p-p], \; (0 \sim 0.4)\,\mathrm{MeV} \\
p + e^- + p \to {}^2\mathrm{H} + \nu_e & [pep], \; (1.4)\,\mathrm{MeV}
\end{pmatrix}
$$

---

[2] The disappearance of neutrinos could in principle also be caused by them decaying. This would require the intervention of New Physics as well; in any case we can distinguish the two scenarios by their dependance on t or L.

Table 17.1.   Solar neutrino experiments.

Experiments	Method	Threshold
Homestake [258]	radiochemical chlorine	0.81 MeV
SAGE [259]	radiochemical gallium	0.236 MeV
GALLEX [260]	radiochemical gallium	0.236 MeV
Kamiokande [261]	water Cerenkov	7 MeV
Superkamiokande [262]	water Cerenkov	7 MeV

$$\Rightarrow \quad {}^{2}\text{H} + p \rightarrow {}^{3}\text{He} + \gamma$$
$$\Rightarrow \quad {}^{3}\text{He} + {}^{3}\text{He} \rightarrow {}^{4}\text{He} + 2pp$$
$$\Rightarrow \quad {}^{3}\text{He} + {}^{4}\text{He} \rightarrow {}^{7}\text{Be} + \gamma$$
$$\Rightarrow \quad \left(\begin{array}{l} e^{-} + {}^{7}\text{Be} \rightarrow {}^{7}\text{Li} + \nu_{e} \quad [CNO] \,, \, (0.86, \, 0.38) \, \text{MeV} \\ p + {}^{7}\text{Be} \rightarrow {}^{8}\text{B} + \gamma \end{array} \right)$$
$$\Rightarrow \quad {}^{8}\text{B} \rightarrow {}^{8}\text{Be} + e^{+} + \nu_{e} \quad [B] \,, \, (0 \sim 15) \, \text{MeV}. \tag{17.5}$$

We have stated labels in [ ] and energy ranges in ( ) for each neutrino emitted. Experiments to detect such neutrinos have been going on for many years. The pioneering Homestake experiment was the first to observe them. There are five experiments listed in Table 17.1 counting neutrinos arriving on earth from the sun. As indicated, they employ very distinct methods and are sensitive to different parts of the neutrino spectrum. We are impressed by the experimental achievement in finding such an elusive signal. The tantalizing aspect is that while the experiments find the expected order of magnitude for the neutrino flux, all of them observe significantly fewer neutrinos than predicted.

For an interpretation of the experimental findings, we follow the argument by Bahcall [263]. The water Cerenkov detectors Kamiokande and Superkamiokande track incoming neutrinos through emission of a scattered electron. With the ensuing relatively high detection threshold of 7 MeV they can register *B* neutrinos only. From their signal we can infer how many *B* neutrinos should be observed by Homestake; it finds, however, less than half that number. So, the Homestake Chlorine experiment and the Kamiokande and Superkamiokande water Cerenkov experiments contradict each other.

The plot thickens further still. From the amount of sunlight seen, which consists mainly of photons emitted in the reaction Eq. (17.5) we can deduce the *p-p* and *pep* neutrino flux on earth. That flux is consistent with the findings of SAGE and GALLEX – yet that represents a problem! For the Gallium detectors are sensitive to the *Be*, *B* and *CNO* neutrinos as well, which surely must exist; yet there is no room for them in the data.

A detailed analysis [264, 265] of the neutrino fluxes from the different experiments comes to the same conclusions:

- We observe significantly fewer neutrinos from the sun than we expect.

- This deficit varies with the neutrino energy.

- The SM does not allow us to account for these effects.

- Ignoring any one experiment does not reconcile the remaining data with the SM. Keep in mind that these are well-developed experiments that differ vastly in their techniques and are systematically independent of each other.

17.2.2 *Atmospheric neutrinos*

When cosmic rays hit nuclei in the earth's atmosphere they produce cascades of mainly pions (and some kaons). Subsequent weak decays provide a source for neutrinos and antineutrinos:

$$\pi^- \to \quad \mu^- \quad \bar{\nu}_\mu$$
$$\hookrightarrow \quad \nu_\mu e^- \bar{\nu}_e, \tag{17.6}$$

$$\pi^+ \to \quad \mu^+ \quad \nu_\mu$$
$$\hookrightarrow \quad \bar{\nu}_\mu e^+ \nu_e, \tag{17.7}$$

with twice as many muon as electron neutrinos *produced*.

Neutrino oscillations (and/or neutrino decays) would change the ratio $(N_{\nu_\mu} + N_{\bar{\nu}_\mu})/(N_{\nu_e} + N_{\bar{\nu}_e})$ before these neutrinos enter the detector. There they are observed through their charged current reactions leading to muons and electrons. It has become customary to quote the ratio between the data and the expectations based on a Monte Carlo analysis:

$$R \equiv \frac{\left(\frac{(N_{\nu_\mu}+N_{\bar{\nu}_\mu})}{(N_{\nu_e}+N_{\bar{\nu}_e})}\right)_{\text{data}}}{\left(\frac{(N_{\nu_\mu}+N_{\bar{\nu}_\mu})}{(N_{\nu_e}+N_{\bar{\nu}_e})}\right)_{MC}} = \frac{\left(\frac{N(\mu-\text{like})}{N(e-\text{like})}\right)_{\text{data}}}{\left(\frac{N(\mu-\text{like})}{N(e-\text{like})}\right)_{MC}}. \tag{17.8}$$

Uncertainties in the neutrino flux and the cross sections drop out from R. *Without* neutrino oscillations (and/or neutrino decays) we expect $R = 1$.

Prior to June 1998, results had been put forward by five groups: Kamiokande [266], IMB-3 [267] and Soudan-2 [268] observed R to be less than one for $E_\nu < 1$ *GeV*, while Frejus [269] and NUSEX [270] observed R to be consistent with unity.

The Superkamiokande Collaboration can field better statistics and systematics. At the Neutrino 98 conference held at Takayama, Japan, in June 1998, they announced their findings [271]:

$$R = 0.66 \pm 0.06 \pm 0.08, \tag{17.9}$$

i.e., significantly below unity, which they interpret as muon deficit rather than electron excess. Thus it would fall into the 'disappearance' category.

Other elements of their data provide more specific evidence for the v_μ deficit being caused by muon neutrinos oscillating away into neutrinos they cannot observe. For the neutrinos to retain a non-negligible fraction of the primary cosmic ray energy, they have to be produced by pion decays in a thin medium, namely the atmosphere. Neutrinos reaching the detector from below thus have to travel a much longer distance L – the earth's diameter or a substantial fraction thereof – than neutrinos from above. Indeed, they see such an asymmetry:

$$\frac{N(\mu - \text{like})_{\text{up}}}{N(\mu - \text{like})_{\text{down}}} = 0.52^{+0.07}_{-0.06} \pm 0.01. \tag{17.10}$$

In addition, they have supplied information on the zenith angular dependence. The zenith angle is the polar angle between the upward direction and the neutrino direction. The length between the production point and the detector depends on the zenith angle. So, if neutrinos oscillate with an oscillation length comparable to the diameter of the earth, we expect a zenith angle dependence of R. This is indeed observed by them.

Since the Takayama conference, they have accumulated more data and the result is consistent with $v_\mu - v_\tau$ oscillation with [272]

$$\sin^2 2\theta > 0.82$$

$$5 \times 10^{-4} \text{eV}^2 < \Delta m^2 < 6 \times 10^{-3} \text{eV}^2.$$

17.2.3 Summary of experimental findings

We take these data as the most direct evidence for the SM being *incomplete*. By far the most natural explanation for the deficit is that neutrinos are not mass degenerate and that oscillations can transform the neutrinos produced into other neutrinos that are 'sterile', at least at the relevant energies. As we have seen time and again, this induces time and thus also energy dependent effects. The story as to how this can be made to fit observations is a highly intriguing one with a few more twists, that however cannot be told here in great detail (partly because the ending is not known yet).

17.3 Neutrino masses

There have been intense efforts towards determining the mass of neutrinos. Since we cannot measure their energy and momentum directly, we deduce their mass from carefully analysing the kinematics of reactions where they appear. The highest sensitivity has been achieved for the electron neutrino by studying the endpoint region in tritium beta decay where the accompanying neutrino is almost at rest, Refs. [273, 274]. The interpretation of the data is highly nontrivial, since the sample tritium is embedded into large molecules; complex molecular interactions then distort in particular the endpoint spectrum. Not surprisingly, the best fit is often obtained for negative values for the neutrino mass. For a summary of the experimental situation we refer you to a review by Fukugita and Yanagida [275]. A conservative upper bound is:

$$m_{\nu_e} \leq (10 \sim 15) \text{ eV} \qquad \text{or} \qquad \frac{m_{\nu_e}}{m_e} \leq 3.3 \cdot 10^{-4}. \qquad (17.11)$$

For the other two neutrinos we find

$$\frac{m_{\nu_\mu}}{m_\mu} \leq 1.6 \cdot 10^{-3}, \quad \frac{m_{\nu_\tau}}{m_\tau} \leq 1.35 \cdot 10^{-2}. \qquad (17.12)$$

Such tiny ratios call for an explanation. For Dirac neutrinos they are certainly unnatural, since there is no symmetry ensuring the smallness of the neutral to charged lepton mass. Even more generally, we know of no good reason why neutrinos should be mass degenerate, let alone massless. We do not see why there should be a *qualitative* difference between the quark and the lepton world in this respect. An analogue of the CKM matrix, which we shall call the Maki–Nakagawa–Sakata (MNS) matrix [276], is likely to emerge in the leptonic sector leading to neutrino oscillations (and other flavour changing neutral currents). We will come across a mechanism that naturally leads to *tiny*, though non-zero, neutrino masses. It makes use of a property peculiar to neutral fermions which we will explain now. With the mass being a scalar, Lorentz symmetry requires the bilinear term in the mass Lagrangian to combine a left and a right handed fermion field: $\mathscr{L}_m = m\overline{\psi}'_R\psi_L + \text{h.c.}$ There are two distinct ways in which this requirement can be fulfilled:

1 Both left and right handed neutrino fields exist: $m\overline{\nu}_R\nu_L + \text{h.c.}$

2 The left handed neutrino and its charge conjugate – the right handed antineutrino – are forced together: $m\overline{\nu}^C_L\nu_L$.

The merit of this representation can be seen by examining the Dirac equation. By writing:

$$\psi_R = \begin{pmatrix} \psi_+ \\ 0 \end{pmatrix}, \quad \psi_L = \begin{pmatrix} 0 \\ \psi_- \end{pmatrix}, \qquad (17.13)$$

where $\psi_\pm$ are Weyl spinors with helicity $\lambda = \pm\frac{1}{2}$, the Dirac equation becomes

$$(E \mp \vec{\sigma} \cdot \vec{p})\psi_\pm = m\psi_\mp. \qquad (17.14)$$

$\psi_\pm$ decouple from each other for $m = 0$.

Consider a general Lorentz transformations for spinors which includes the rotation and the boost. It can be written as

$$\Lambda_\pm = \exp\left[i\frac{1}{2}\vec{\sigma} \cdot (\vec{\theta} \pm i\vec{\phi})\right]. \qquad (17.15)$$

Denote two spinors, which transform as $(\frac{1}{2}, 0)$ and $(0, \frac{1}{2})$, by ψ_- and ψ_+, respectively [277]. That is,

$$\psi_\pm \to \Lambda_\pm \psi_\pm. \qquad (17.16)$$

The physical interpretation of these operators is that $\exp[i\frac{1}{2}\vec{\sigma} \cdot \vec{\theta}]$ represents the rotation operator on $\psi_\pm$ around the direction of $\vec{\theta}$ by an angle $|\vec{\theta}|$, and $\exp[\mp\frac{1}{2}\vec{\sigma}\cdot\vec{\phi}]$ represents a boost operator on $\psi_\pm$ along the direction of $\vec{\phi}$ by $|\vec{\phi}|$. Now consider another spinor $\chi_\pm$ which transforms as $\psi_\pm$. It can be shown that $\sigma_2 \chi_\pm^*$ transforms as $\chi_\mp$. Noting the identity $\Lambda_\pm^{tr}\sigma_2\Lambda_\pm = \sigma_2$, we see that

$$\mathscr{L}_m = -iM\chi_\pm^{tr}\sigma_2(\psi_\pm) + \text{h.c.} \qquad (17.17)$$

is a Lorentz scalar [278]. The choice of the phase of the mass term will be obvious below. We can convert this into a Dirac mass term by defining

$$\psi_R = \begin{pmatrix} i\sigma_2\chi_-^* \\ 0 \end{pmatrix}, \quad \psi_L = \begin{pmatrix} 0 \\ \psi_- \end{pmatrix}, \qquad (17.18)$$

and write

$$\mathscr{L}_m = -iM\chi_-^{tr}\sigma_2\psi_- + \text{h.c.} = M\overline{\psi}_R\psi_L + \text{h.c.} \qquad (17.19)$$

Similarly, writing the special case of Eq. (17.17) where $\chi_\pm = \psi_\pm$, and by defining

$$\Psi_M = \begin{pmatrix} i\sigma_2\psi_-^* \\ \psi_- \end{pmatrix}, \qquad (17.20)$$

we can write:

$$\mathscr{L}_m = -iM\psi_-^{tr}\sigma_2\psi_- + \text{h.c.} = M\overline{\Psi}_M\Psi_M. \qquad (17.21)$$

Note that the Ψ satisfies

$$\Psi_M^C = \Psi_M. \qquad (17.22)$$

This is called the Majorana condition, and Ψ_M is said to be a Majorana field. Eq. (17.21) is called a Majorana mass term.

So, we see that the most general mass matrix is given by

$$\mathscr{L} = -iM_{ij}(\chi_\pm^{tr})_i\sigma_2\psi_{\pm j}. \qquad (17.23)$$

For neutrino mass terms consider $\overline{\nu}_L\nu_R$ and $\overline{\nu}_L\nu_L^C$. These two combinations differ in one fundamental aspect: if ν carries charge q, ν^C carries charge $-q$; a term $\overline{\nu}\nu^C$ changes the charge by two units, whereas $\overline{\nu}\nu$ is neutral. A theory with left-handed neutrinos only can support neutrino masses – yet at the price of violating the corresponding lepton number. For charged fields a Majorana mass is incompatible with the conservation of electric charge.

The SM does not forbid the existence of right-handed neutrinos as $SU(2)_L$ singlets; yet with (ν_L, e^-) forming an $SU(2)_L$ doublet, a Majorana mass term requires physics beyond the SM. Consider the general mass Lagrangian for a single four-component neutrino field

$$\mathscr{L}_m = \frac{1}{2}\begin{pmatrix}\overline{v}_L^C \\ \overline{v}_R\end{pmatrix}^{\text{tr}}\begin{pmatrix} M_L & m_D^{\text{tr}} \\ m_D & M_R \end{pmatrix}\begin{pmatrix} v_L \\ v_R^C \end{pmatrix} + \text{h.c.} \tag{17.24}$$

Let us set $M_L = 0$ in Eq. (17.24), i.e., we have introduced a Majorana mass for the right handed neutrinos, but not for the left handed ones. The Dirac masses m_D are assumed to be comparable to quark masses arising from $SU(2)_L$ breaking in the SM; M_R, on the other hand, reflects the scale ≥ 1 TeV where some New Physics enters the stage; in the next chapter we will see that left–right gauge models provide a natural stage for that to happen. So, it is natural to take the mass hierarchy:

$$m_D \ll M_R . \tag{17.25}$$

The matrix in Eq. (17.24) is easily diagonalized; due to Eq. (17.25) the resulting neutrino mass eigenstates *approximately* coincide with the chiral fields

$$\begin{aligned} v_1^m &\simeq v_L \\ v_2^m &\simeq v_R, \end{aligned} \tag{17.26}$$

with masses

$$\begin{aligned} m_1 &\simeq \frac{m_D^2}{M_R} \\ m_2 &\simeq M_R. \end{aligned} \tag{17.27}$$

Thus we can conclude that v_1^m is practically left handed and very light, whereas v_2^m is almost right handed and very heavy. The phenomenon that $M_L \simeq 0 \ll m_D \ll M_R$ leads to *naturally* light neutrino states is called the *see-saw* mechanism; we know of no other way to achieve this goal [279].

   This scenario is easily generalized to more than one family. $M_{L,R}$ and m_D are then turned into matrices; see Problem (17.6).

17.4 Phases in the MNS matrix

Just as we have done for quarks, we diagonalize the neutrino mass matrix

$$v_\alpha = \sum_i U_{\alpha i} v_i^m. \tag{17.28}$$

The leptonic MNS matrix can be defined as **U** in

$$J_\mu = \overline{l}\gamma_\mu(1-\gamma_5)v = \overline{l}\gamma_\mu(1-\gamma_5)\mathbf{U}v^m. \tag{17.29}$$

where we have chosen the basis in which the charged leptons are diagonal.

   A discussion on the phase freedom for Majorana neutrinos is in order. For definiteness, let us consider the mass matrix given in Eq. (17.24). The Majorana part of the mass matrix has a structure shown in Eq. (17.21), and thus the phases of the mass eigenstate cannot be rotated freely. An alert reader may be puzzled at this point. Can very heavy particles influence low energy physics? It will be instructive to analyse what happens in the limit $M_R \to \infty$, how v_R decouples and we regain the freedom to rotate the phases of neutrino mass eigenstates. We shall see below that actually this happens!

So, since the phases of v^m have been fixed, only charged lepton phases can be changed freely in Eq. (17.29). We start out with n^2 independent parameters in U; since n phases can be removed by adjusting the charged lepton phases, we have $\frac{1}{2}n(n-1)$ rotation angles. So, we end up with

$$n^2 - n - \frac{1}{2}n(n-1) = \frac{n(n-1)}{2} \tag{17.30}$$

physical complex phases in the leptonic MNS matrix for neutrinos with a Majorana component. After we have taken advantage of redefining the phases of charged leptons, we obtain the leptonic MNS matrix, which can be written as

$$U = V^l[e^{i\phi}], \tag{17.31}$$

where $[e^{i\phi}]$ is a diagonal phase matrix, which is present as phases of v^m cannot be redefined.

17.5 CP and T violation in v oscillations

Since neutrinos can be Majorana states (or contain a Majorana component) $v_i \leftrightarrow \bar{v}_j$ as well as $v_i \leftrightarrow v_j$ transitions can occur, opening the way for unusual effects. Observing neutrino oscillations – no matter at which scale or with which pattern – would cause the transition to a new paradigm, revealing a whole new dynamical layer reflecting forces operating around a very high scale, be it a GUT scale or an intermediate scale. It is then highly appropriate to probe this new layer for manifestations of **CP** and/or **T** violation. Both searches can be performed independently of each other, thus allowing even for novel tests of **CPT** invariance. On phenomenological grounds it is conceivable that a **CPT** violation could surface there, despite previous negative searches in other systems. Of course, we would pay a heavy theoretical price for such a revolutionary discovery.

Let us denote by $P(v_\alpha \to v_\beta; t)$ the probability of finding neutrinos of type β at time t in a beam that at time $t = 0$ contained only neutrinos of type α. $P(v_\alpha \to v_\alpha; t) \neq 1$, $P(v_\alpha \to v_\beta; t) \neq 0$ for $\alpha \neq \beta$ represents manifestations of v oscillations. **CP** invariance is then probed through comparing $P(v_\alpha \to v_\beta; t)$ with $P(\bar{v}_\alpha \to \bar{v}_\beta; t)$, whereas **T** violation is searched for in $P(v_\alpha \to v_\beta; t)$ vs $P(v_\beta \to v_\alpha; t)$. Accordingly, **CPT** symmetry is studied in $P(v_\alpha \to v_\beta; t)$ vs $P(\bar{v}_\beta \to \bar{v}_\alpha; t)$. Of course, if it turns out that v oscillations can be observed only through solar neutrinos, then these tests are academic.

The appropriate formalism for going beyond these general statements is straightforward. We start from Eq. (17.1) with the weak interaction eigenstates

v_α, $\alpha = e, \mu, \tau$:

$$v_\alpha = \sum_i \mathbf{U}_{\alpha i} e^{-iE_i t} v_i^m, \tag{17.32}$$

with the implicit assumption that the parameters $\mathbf{U}_{\alpha i}$ are the result of forces operating at energy scales $\sim M$. Writing

$$E_i = \sqrt{p^2 + m_i^2} \sim p + \frac{m_i^2}{2E}, \tag{17.33}$$

we have

$$P(v_\alpha \to v_\beta; t) = \left| \sum_i e^{-i\frac{m_i^2}{2E}t} \mathbf{U}_{\alpha i} \mathbf{U}_{\beta i}^* \right|^2 = \left| \sum_i e^{-i\frac{m_i^2}{2E}t} \mathbf{V}_{\alpha i}^l \mathbf{V}_{\beta i}^{l*} \right|^2. \tag{17.34}$$

The last equality implies that the extra phases $[e^{i\phi}]$, present in **U**, when we deal with Majorana neutrinos, do not influence physical observables. *So, the transitions among light neutrinos are quite unaffected by a presence of superheavy v_R states, we have recovered the freedom in setting the phases for the light neutrino mass eigenstates; i.e., decoupling is complete.* For the antineutrinos we have the corresponding expression

$$\bar{v}_\alpha = \sum_i \overline{\mathbf{U}}_{\alpha i} \bar{v}_i^m, \tag{17.35}$$

where we have *not* assumed **CPT** invariance. Of course, as long as we are dealing with the mass terms discussed in Sec. 17.3, there is no **CPT** violation. The transition probabilities can then be expressed as follows:

$$P(v_\alpha \to v_\beta; t) = \sum_{i,j} \left[\cos\left(\frac{m_i^2 - m_j^2}{2E} \cdot t \right) \cdot [\mathrm{Re} \mathbf{V}_{\alpha i}^l \mathbf{V}_{\alpha j}^{l*} \mathbf{V}_{\beta i}^{l*} \mathbf{V}_{\beta j}^l] \right.$$

$$\left. + \sin\left(\frac{m_i^2 - m_j^2}{2E} \cdot t \right) \cdot [\mathrm{Im} \mathbf{V}_{\alpha i}^l \mathbf{V}_{\alpha j}^{l*} \mathbf{V}_{\beta i}^{l*} \mathbf{V}_{\beta j}^l] \right]. \tag{17.36}$$

Similar formulae can be written for antineutrinos:

$$P(\bar{v}_\alpha \to \bar{v}_\beta; t) = \sum_{i,j} \left[\cos\left(\frac{\overline{m}_i^2 - \overline{m}_j^2}{2E} \cdot t \right) \cdot [\mathrm{Re} \overline{\mathbf{V}}_{\alpha i}^l \overline{\mathbf{V}}_{\alpha j}^{l*} \overline{\mathbf{V}}_{\beta i}^{l*} \overline{\mathbf{V}}_{\beta j}^l] \right.$$

$$\left. + \sin\left(\frac{\overline{m}_i^2 - \overline{m}_j^2}{2E} \cdot t \right) \cdot [\mathrm{Im} \overline{\mathbf{V}}_{\alpha i}^l \overline{\mathbf{V}}_{\alpha j}^{l*} \overline{\mathbf{V}}_{\beta i}^{l*} \overline{\mathbf{V}}_{\beta j}^l] \right]. \tag{17.37}$$

Assuming **CPT** invariance to hold, we have

$$m_i = \overline{m}_i \tag{17.38}$$

$$\mathbf{V}_{\alpha i}^{l*} = \overline{\mathbf{V}}_{\alpha i}^l; \tag{17.39}$$

i.e., mass-related parameters are equated for neutrinos and antineutrinos. There are actually seven physical quantities, namely the three masses m_i and the three angles, and one intrinsic complex phase by which the mixing matrix $\mathbf{V}^l$ can be described – in analogy to the CKM matrix. Symmetries will impose constraints on these transition probabilities:

$$\begin{aligned} \mathbf{CP}: \quad & P(v_\alpha \to v_\beta; t) = P(\bar{v}_\alpha \to \bar{v}_\beta; t) \\ \mathbf{T}: \quad & P(v_\alpha \to v_\beta; t) = P(v_\beta \to v_\alpha; t). \end{aligned}$$

As expected on general grounds, these relations are violated only if the mixing matrix $\mathbf{V}^l$ contains irreducible complex phases.

For diagonal transitions, i.e. for $\alpha = \beta$, no $\mathbf{CP}$ asymmetry can arise since $\mathrm{Im}|\mathbf{V}^l_{\alpha i}|^2 |\mathbf{V}^l_{\alpha j}|^2 = 0$ holds. That is as expected, of course, since $\mathbf{CPT}$ invariance already implies

$$P(v_\alpha \to v_\alpha; t) = P(\bar{v}_\alpha \to \bar{v}_\alpha; t). \tag{17.40}$$

This discussion might very well turn out to be completely academic since either neutrino oscillations cannot be observed at all or only in 'beams' coming from astrophysical sources. However, if neutrino oscillations became observable in accelerator based experiments, we would have gained (indirect) access to a completely new dynamical regime. Probing its properties under $\mathbf{CP}$ and $\mathbf{T}$ transformations would constitute a unique opportunity and we should make every effort to exploit it!

17.6 The MSW effect

Consider solutions to the free Dirac equation,

$$i\frac{\partial}{\partial t} v_i^m = \left(p + \frac{m_i^2}{2p} \right) v_i^m, \tag{17.41}$$

where v_i^m is a mass eigenstate. In terms of weak eigenstates,

$$i\frac{\partial}{\partial t} v_\beta = p v_\beta + \sum_{i,\alpha} \mathbf{V}^l_{\beta i} \frac{m_i^2}{2p} \mathbf{V}^{l*}_{i\alpha} v_\alpha, \tag{17.42}$$

where m is a diagonal mass matrix. Let us consider the case of two families v_e and v_μ. If the Majorana mass term M_R is very large, the mixings of v_e^C and v_μ^C are negligible, and we have

$$i\frac{\partial}{\partial t} \begin{pmatrix} v_e \\ v_\mu \end{pmatrix} = \left[\left(p + \frac{\overline{m}^2}{2p} \right) \mathbf{1} + \frac{\Delta m^2}{4p} \begin{pmatrix} -\cos 2\theta & \sin 2\theta \\ \sin 2\theta & \cos 2\theta \end{pmatrix} \right] \begin{pmatrix} v_e \\ v_\mu \end{pmatrix}, \tag{17.43}$$

where $\overline{m} = \frac{1}{2}(m_1 + m_2)$, $\Delta m^2 = m_1^2 - m_2^2$.

Consider a v beam so low in energy that muons cannot be produced through scattering off nuclei and electrons[3]. Then v_e and v_μ undergo different charged

---

[3] This is the usual situation with v beams from reactors and also from the sun.

current interactions off nuclei as well as the electrons:

$$\nu_e + A \xrightarrow{NC} \nu_e + A \,,\; \nu_e + A \xrightarrow{CC} e + A' \,,\; \nu_e + e \xrightarrow{CC} e + \nu_e,$$

$$\nu_\mu + A \xrightarrow{NC} \nu_\mu + A \,,\; \nu_\mu + A \xrightarrow{CC}\!\!\!\!\!/\;\; \mu + A' \,,\; \nu_\mu + e \xrightarrow{CC}\!\!\!\!\!/\;\; \mu + \nu_e. \qquad (17.44)$$

When ν_e travels through matter, it interacts with matter through an effective interaction term

$$\mathscr{H}_{\text{eff}} = \frac{G_F}{\sqrt{2}} \bar{\nu}_e \gamma_\alpha (1 - \gamma_5) \nu_e \bar{e} \gamma^\alpha (1 - \gamma_5) e. \qquad (17.45)$$

When the electron is nearly at rest and the neutrino is relativistic, only the $\mu = 0$ component is relevant and $\bar{e}\gamma^0(1 - \gamma_5)e = n_e$; $\mathscr{H}_{\text{eff}} \simeq \sqrt{2}G_F n_e$, where n_e is the electron density. The equation of motion gets modified:

$$i\frac{\partial}{\partial t}\begin{pmatrix} \nu_e \\ \nu_\mu \end{pmatrix} = \left[\left(p + \frac{\overline{m}^2}{2p}\right)\mathbf{1} + \begin{pmatrix} -\frac{\Delta m^2}{4p}\cos 2\theta + \sqrt{2}G_F n_e & \frac{\Delta m^2}{4p}\sin 2\theta \\ \frac{\Delta m^2}{4p}\sin 2\theta & \frac{\Delta m^2}{4p}\cos 2\theta \end{pmatrix}\right]\begin{pmatrix} \nu_e \\ \nu_\mu \end{pmatrix}, \qquad (17.46)$$

where $\mathbf{1}$ is a 2×2 unit matrix. As a neutrino is emitted and travels through the sun to get out, n_e may change, as it is in general a function of r, the distance from the sun's centre. If the variation in n_e is mild, we can apply an adiabatic approximation. Suppose there is a point in the sun where

$$-\frac{\Delta m^2}{4p}\cos\frac{1}{2}\theta + \sqrt{2}G_F n_e = \frac{\Delta m^2}{4p}\cos\frac{1}{2}\theta. \qquad (17.47)$$

At this point, there will be complete mixing between ν_e and ν_μ, no matter how small $\Delta m^2 \sin\frac{1}{2}\theta$ may be.

The formalism of neutrino mixing in matter was first worked out by Wolfenstein [280] and first applied to the solar neutrino problem by Mikheyev and Smirnov [281], therefore it is referred to as the MSW effect. Note that Eq. (17.46) is very similar to Eq. (6.76). The MSW effect is thus equivalent to coherent regeneration *without* **CPT** invariance imposed! If **CPT** symmetry had not assumed such a sacrosanct position in the thinking of our community, it is quite conceivable that the MSW effect would have been found much sooner.

We shall not go into the MSW solutions. For detailed analysis of solutions to the equation of motion, Eq. (17.46), including non-adiabatic ones, see Ref. [275].

There is another intriguing aspect to the MSW effect although it might remain academic. Since the material is made up of matter rather than antimatter (at least in our cosmic neighbourhood) a difference in, say, Prob($\nu_i \to \nu_j$) vs. Prob($\bar{\nu}_i \to \bar{\nu}_j$) might be a reflection of the asymmetry in the environment instead of being a manifestation of **CP** violation. A detailed discussion of this point can be found in [282].

17.7 The Bard's song

Neutrinos have had a most remarkable career already. They were first postulated as a theoretical crutch to balance energy, momentum and angular momentum by Pauli at the VII Solvay Congress in 1933, to less than universal acclaim. At least one theorist of towering reputation – Bohr – was more inclined to limit the validity of energy conservation in the nuclear domain. It was not until 1953 that the existence of neutrinos was verified by Reines and Cowan [283]. In 1956/57 neutrinos, through their handedness, played the central role in the unfolding drama of the demise of parity as an absolute symmetry of nature. The relationship between ν_L and $\bar{\nu}_R$ gave some breathing space to **CP** remaining a true symmetry – till that notion had to be abandoned in 1964. In the 1970s neutrinos moved from their role of a ghostlike oddity – albeit with an essential impact – more into the mainstream as an experimental probe: tremendous advances in accelerator physics and technologies provided us with intense and well-collimated high energy beams of neutrinos. This led to the discovery of new fundamental forces – the neutral currents – on one hand, and on the other enabled us to probe the internal structure of nucleons with amazing accuracy.

Another story-line that had started out as a sideshow (at best) was gaining importance, namely neutrino astronomy and cosmology, first targeting the sun in 1978 by Davis and collaborators in the Homestake Gold Mine. It was realized that, according to big bang cosmology, we are bathing in a background radiation of neutrinos analogous to the microwave photon radiation; even tiny neutrino masses would have grave cosmological consequences. Huge instruments built to search for proton decay were taken over by neutrino astronomy; new and even bigger detectors were and are being assembled with it as primary purpose.

In 1987 neutrinos joined the All Universe Club when some of their brothers and sisters reached earth from supernova SN 1987A that erupted long ago in a galaxy far away, specifically 160 000 years ago in the Large Magellanic Cloud.

We have described the major success in neutrino astronomy in filtering out solar and atmospheric neutrinos. We might be at the threshold of an even bigger discovery, namely neutrino oscillations. If true, this would lead to a new paradigm of nature's fundamental design. Maybe this time we are hearing birds greeting the day they can already sense. In that case the illustrious past of neutrinos would be a mere prelude to their true glory days. A whole new world of fundamental phenomena would open up, requiring dedicated and detailed experimentation and analysis. We expect such a world to be shaped by new symmetry principles, of which we give some examples in particular in the next chapter on left–right symmetry. Those would be of great significance for a better understanding of the main subject of this book, namely **CP** violation in general and not just in neutrino oscillations.

However, the neutrino story already teaches us one important lesson of general value: *size is a decidedly poor yardstick for significance!*

Problems

17.1 Consider a model for two lepton families coupling to charged and neutral currents. Show that if the two neutrinos are mass *degenerate*, the MNS matrix is necessarily *diagonal*!

17.2 Does introducing a right handed neutrino into the SM violate some fundamental principle like gauge invariance? If not, why not?

17.3 Consider, at $t = 0$, a massive $v_L = \frac{1}{2}(1 - \gamma_5)v$ with mass m moving along the z direction. Convince yourself that this is not a solution to the Dirac equation. Write this spinor as a linear combination of the solutions to the Dirac equation. Obtain the probability for $v_L \to v_R$ occuring at time t. Note that this probability is $O\left(\left(\frac{m}{E}\right)^2\right)$ and negligible compared to transitions obtained from the mixing due to the off-diagonal term in the mass matrix.

17.4 Noting that

$$\sigma_2 \sigma_i \sigma_2 = -\sigma_i^* \tag{17.48}$$

show that

$$\sigma_2 \Lambda_\pm \sigma_2 = \Lambda_\mp^*$$
$$\sigma_2 \Lambda_\pm^{-1} \sigma_2 = \Lambda_\pm^{tr}$$
$$\text{and} \quad \Lambda_\pm^{tr} \sigma_2 \Lambda_\pm = \sigma_2, \tag{17.49}$$

where $\Lambda_\pm$ is defined in Eq. (17.15). Taking the complex conjugate of Eq. (17.16), show that

$$\sigma_2 \psi_\pm^* \to \Lambda_\mp(\sigma_2 \psi_\pm^*), \tag{17.50}$$

so we see that $\sigma_2 \psi_\pm^*$ transforms like $\psi_\mp$.

17.5 Using Eq. (17.49), show that

$$\mathscr{L}_m = -iM\chi_\pm^{tr}\sigma_2 \psi_\pm \tag{17.51}$$

is a Lorentz scalar.

17.6 Consider Majorana mass terms for three families of neutrinos, $\psi^{tr} = (v_1, v_2, v_3)$. Show that

$$\overline{\psi}^C \mathscr{M}\psi = \psi_i^\alpha \mathscr{M}_{ij} C^{\alpha\beta} \psi_j^\beta, \tag{17.52}$$

so that $\mathscr{M}$ must be symmetric.

17.7 In general we can write a Majorana field as

$$\chi = e^{i\alpha}\psi_L + e^{i\beta}(\psi_L)^C, \quad \omega = e^{i\alpha'}\psi_R + e^{i\beta'}(\psi_R^C). \tag{17.53}$$

Show that the Majorana condition becomes $\chi^c = e^{-i(\beta+\alpha)}\chi$.

18

First alternative to the KM ansatz: left–right symmetric models

18.1 Basics of left–right symmetric theories

The observed parity violation is explicitly embedded into the SM with its electroweak gauge group $SU(2)_L \otimes U(1)$. It would be quite intriguing to perceive parity as a spontaneously realized symmetry instead [284]. The simplest such scenario can be built on the gauge group [285]

$$SU(2)_L \otimes SU(2)_R \otimes U(1) \qquad (18.1)$$

with the two $SU(2)$ couplings assumed to coincide at some (high) scale [286]

$$g_L(\mu_{LR}) = g_R(\mu_{LR}) = g. \qquad (18.2)$$

At some scale below μ_{LR} the Higgs doublet field coupling to the gauge bosons of $SU(2)_R$ develops a vacuum expectation value, giving the latter a mass; at a (considerably) lower scale the process repeats itself for the gauge bosons of $SU(2)_L$. There are several highly attractive features to such a theory:

- We can entertain the hope of understanding the limitations of parity dynamically, rather than having to postulate them.

- The charge of the $U(1)$ group now finds a nice physical interpretation as $B - L$, the difference in baryon and lepton number.

- Such theories necessarily contain right-handed neutrino fields, and they allow naturally for large Majorana masses. Through the so-called 'see-saw' mechanism [279] we obtain the only known dynamical explanation for the long standing mystery of why neutrinos are nearly massless.

- Closer to our main focus: also **CP** invariance can be treated as a spontaneously realized symmetry – the limitations of which can be related to those of parity.

- The origin of **CP** violation might not reside in the family structure; this provides a counterpoint to the KM ansatz.

Since technical details can be found in excellent reviews in the literature [287,

288, 289, 290], we will merely sketch the basic structural elements of left–right symmetric models.

1 The fields of up- and down-type quarks – U and D, neutral and charged leptons – N and E – are assigned to irreducible representations of the gauge group according to the following $(SU(2)_L, SU(2)_R, U(1))$ charges:

$$U_L, D_L : \quad (\tfrac{1}{2}, 0, \tfrac{1}{3}) \qquad U_R, D_R : \quad \left(0, \tfrac{1}{2}, \tfrac{1}{3}\right)$$

$$N_L, E_L : \quad (\tfrac{1}{2}, 0, -1) \qquad N_R, E_R : \quad \left(0, \tfrac{1}{2}, -1\right), \qquad (18.3)$$

and the electric charge is given in terms of weak isospin and $B - L$ charge:

$$Q = I_{3L} + I_{3R} + \frac{B - L}{2}. \qquad (18.4)$$

2 There are three sets of gauge bosons for $SU(2)_L \otimes SU(2)_R \otimes U(1)$, namely $W_L^{\pm,0}$; $W_R^{\pm,0}$; A.

3 The Higgs sector is considerably more complex than for the SM since it has to drive a multilayered reaction:

$$SU(2)_L \otimes SU(2)_R \otimes U(1)_{B-L} \overset{A}{\Longrightarrow} SU(2)_L \otimes U(1) \overset{B}{\Longrightarrow} U(1)_{QED}. \qquad (18.5)$$

The vector bosons W_L and W_R can receive masses through the VEVs of Higgs fields Δ_L and Δ_R with quantum numbers $(1, 0, 2)$ and $(0, 1, 2)$, respectively. **P** violation is implemented spontaneously through $\langle \Delta_L \rangle \ll \langle \Delta_R \rangle$. Yet fermions cannot couple to these isovector fields. To generate their masses we introduce a third Higgs muliplet

$$\phi = \begin{pmatrix} \phi_1^0 & \phi_1^+ \\ \phi_2^- & \phi_2^0 \end{pmatrix} \qquad (18.6)$$

with quantum numbers $(\tfrac{1}{2}, \tfrac{1}{2}, 0)$ and vacuum expectation values

$$\langle \phi \rangle = \begin{pmatrix} \kappa & 0 \\ 0 & \kappa' \end{pmatrix}. \qquad (18.7)$$

With a hierarchy

$$\langle \Delta_L \rangle \ll \kappa, \kappa' \ll \langle \Delta_R \rangle \qquad (18.8)$$

that can be achieved naturally, we can realize the sequence of Eq. (18.5). Henceforth we will set $\langle \Delta_L \rangle = 0$.

4 The Higgs couplings induce $W_L - W_R$ mixing, leading to mass eigenstates

$$\begin{aligned} W_1 &= W_R \sin\zeta + W_L \cos\zeta \\ W_2 &= W_R \cos\zeta - W_L \sin\zeta, \end{aligned} \qquad (18.9)$$

with

$$\tan\zeta = \frac{\kappa\kappa'}{v_R^2}, \tag{18.10}$$

$$m_{W_1}^2 = \frac{g^2}{2}(\kappa^2 + (\kappa')^2) \ll m_{W_2}^2 = \frac{g^2}{2}(2v_R^2 + \kappa^2 + (\kappa')^2). \tag{18.11}$$

Likewise $A - W_L^0 - W_R^0$ mixing occurs, generating A_γ, Z_1 and Z_2 as fields of definite mass. Yet A, W_1 and Z_1 behave very much like the SM gauge bosons.

5 Since natural flavour conservation has been implemented, **CP** violation can enter through the charged current couplings only. The Lagrangian for quark mass eigenstates reads as follows:

$$
\begin{aligned}
\mathscr{L}_{CC} = \ & \frac{g}{\sqrt{2}} \left[\left(\cos\zeta\, \overline{U}_L^m \gamma_\mu \mathbf{V}_L D_L^m + \sin\zeta\, \overline{U}_R^m \gamma_\mu \mathbf{V}_R D_R^m \right) W_1^\mu \right. \\
+ \ & \left. \left(-\sin\zeta\, \overline{U}_L^m \gamma_\mu \mathbf{V}_L D_L^m + \cos\zeta\, \overline{U}_R^m \gamma_\mu \mathbf{V}_R D_R^m \right) W_2^\mu \right].
\end{aligned} \tag{18.12}
$$

6 There are then three sources of **CP** violation:

- The matrix $\mathbf{V}_L$ reflecting the misalignment (in flavour space) of the quark mass matrices corresponds to the usual CKM matrix.

- Its right-handed analogue $\mathbf{V}_R$, which becomes observable due to the presence of right-handed currents.

- The fact that the two VEVs κ and κ' can turn out to be complex!

These three sources are a priori independent of each other, and thus can yield a plethora of basic **CP** violating parameters. Rather than discussing the general case – hardly an enlightening enterprise – we will focus on two special cases with a particularly intriguing theoretical structure; common to both is that with parity being realized spontaneously, the Yukawa couplings form Hermitian matrices.

- *Manifest left–right symmetry*: If $\langle\phi\rangle$ remains real, the up- and down-type quark mass matrices are Hermitian; then, as discussed in Chapter 5 :

$$\mathbf{V}_L = \mathbf{V}_R = \mathbf{V}, \tag{18.13}$$

and hence the name of this variant. **CP** violation enters through the KM matrix only. We should note that the two discrete symmetries **P** and **CP** are treated on a very different footing here: for **CP** violation arises in a *hard* way, namely in the Yukawa couplings described by dimension four operators.

- *Pseudo left-right symmetry*: Alternatively, let us assume also that **CP** is realized spontaneously, which severely constrains the Yukawa couplings. **CP** violation then arises through the VEVs of ϕ acquiring a complex phase. The KM matrices for the left and right handed currents are then not identical, yet closely related:

$$\begin{aligned} \mathbf{V}_L &= T_{L,U} T_{L,D}^\dagger \\ \mathbf{V}_R &= J_U^* \mathbf{V}_L^* J_D, \end{aligned} \tag{18.14}$$

where $J_{U,D}$ are diagonal phase matrices, and T_{LQ} is defined in Eq. (8.24). While the effect enters through $W_L - W_R$ mixing, it can be parametrized in many equivalent ways. For example, we can define $W_L - W_R$ mixing as real and place the phases into $\mathbf{V}_R$. A detailed analysis shows that for the new contribution to ϵ_K from W_R, the third family can be neglected. So effectively we can set [291]

$$\mathbf{V}^R = e^{i\gamma} \begin{pmatrix} e^{-i\delta_2} \cos\theta_C & e^{-i\delta_1} \sin\theta_C \\ -e^{-i\delta_1} \sin\theta_C & e^{-i\delta_2} \cos\theta_C \end{pmatrix}. \tag{18.15}$$

A caveat should be noted here. With manifest left–right symmetry, it is difficult to generate baryon number when these theories are coupled to grand unified theories [292]. The left–right symmetry holds (if at all) only at high energy scales – and quite possibly only at scales that are immensely larger than the 1 TeV scale. Sizeable renormalization effects will then intervene and shift the balance between left and right handed couplings away from Eq. (18.2); this will obscure the underlying symmetry and make it non-manifest at the electroweak scale. Thus we should not take Eq. (18.2) and Eq. (18.14) as gospel. It should be kept in mind that a model with right handed currents *in general* contains a sizeable number of physical **CP** violating phases emanating from the mass matrices: for N families they number $[(N-1)(N-2)+N(N+1)]/2$, i.e. 1, 3, 6 for $N = 1, 2, 3$, respectively.

However, one important qualitative feature of all these models should be emphasized: *the origin of* **CP** *violation does not necessarily reside in the family structure of quark (and other fermion) fields; it can enter through W_L-W_R mixing!*.

18.2 The existing phenomenology

The predictions most germane to left–right symmetry concern the existence of (1) additional weak vector bosons $W_R^{\pm,0}$ or $W_2^\pm$ and Z_2 and (2) non-sterile right-handed neutrinos, which also opens the gateway to the realm of neutrino masses and oscillations.

In high energy collisions we routinely search for the production of new vector bosons. None have been observed so far. The numerical bound derived from the lack of a signal depends on some assumptions, in particular concerning the strength of the relevant effective gauge coupling and whether the decay $W_R^+ \to l^+ \nu_R$ can occur; as discussed later, left- and right-handed neutrinos can possess vastly different mass – which turns out to be one of the most attractive features of left–right models. Present data yield, assuming $g_L \simeq g_R$ to hold at

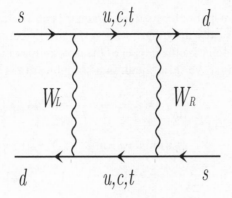

Fig. 18.1. A new box diagram which generates $\Delta S = 2$ transition.

collider scales,

$$M_{W_R} \geq 650 \text{ GeV}, \tag{18.16}$$

even for heavy right handed neutrinos. A rather model independent bound on $W_L - W_R$ mixing is obtained in deep inelastic $\bar{\nu}N \to \mu^+ X$ charged current reactions. The CCFR collaboration [293] gives:

$$|\zeta| \leq 0.04. \tag{18.17}$$

We can search for right handed currents also in $\mu^- \to e^- \bar{\nu}_e \nu_\mu$ through a detailed analysis of the generalized Michel parameters [226]. Yet that is based on the assumption that right handed neutrinos (and left handed antineutrinos) are sufficiently light for them to be produced in μ (or π, K, etc.) decays. This is, however, far from assured and actually unnatural, as explained below.

Numerically more impressive bounds on M_{W_R} can be obtained from processes that are forbidden – like $\mu \to e\gamma$ – or classically forbidden – like $K^0 - \overline{K}^0$ oscillations. Because their rates are so much suppressed in the SM, these processes are likely to exhibit a high sensitivity to the presence of New Physics. Yet, this gain in sensitivity comes at a price: the specific features of the dynamical implementation of left–right symmetry become very important. This makes the bounds unstable; i.e., they can be weakened by adopting a particular choice of the relevant parameters and/or by allowing (or engineering) sizeable cancellations among the several possible contributions. Thus they should be viewed as typical scenarios, as benchmark figures, rather than firm bounds.

18.2.1 $\Delta S = 2$ transitions

The effective Lagrangian is again obtained by integrating out the internal loop in box diagrams. Since $M_{W_L}^2 / M_{W_R}^2 \ll 1$, we consider only two types of box diagrams: one with the exchange of two W_L and the other with one W_L and one W_R, see Fig.

18.1. Furthermore, we ignore for now the top quark contributions in the $W_L - W_R$ box: they are numerically insignificant for ΔM_K in any case; as far as ϵ is concerned we want to emphasize the impact of the new source of **CP** violation coming from $W_L - W_R$ mixing. We then obtain in a straightforward manner [294, 295]:

$$\mathscr{H}_{LR}^{\Delta S=2} \simeq \frac{G_F^2}{2\pi^2} \cdot M_{W_L}^2 x_c \log x_c \frac{M_{W_L}^2}{M_{W_R}^2} (V_{cd}^{L*} V_{cs}^R V_{cd}^{R*} V_{cs}^L)$$

$$\times \left[\eta_4(\mu)\mathscr{O}_S + \eta_5(\mu) \left(\mathscr{O}_{LR} + \frac{2}{N_C}\mathscr{O}_S \right) \right], \qquad (18.18)$$

with $\eta_4 \sim 1.40$, $\eta_5 \sim 0.12$ at a renormalization scale of $\mu = 1\,\text{GeV}$,

$$\mathscr{O}_{LR} \equiv \left(\bar{d}\gamma_\mu(1 - \gamma_5)s \right) \left(\bar{d}\gamma^\mu(1 + \gamma_5)s \right)$$
$$\mathscr{O}_S \equiv \left(\bar{d}(1 - \gamma_5)s \right) \left(\bar{d}(1 + \gamma_5)s \right). \qquad (18.19)$$

The two operators $\mathscr{O}_S$ and $\mathscr{O}_{LR} + \frac{2}{N_C}\mathscr{O}_S$ possess a different chiral structure from the one encountered in the SM, which will reflect itself in their matrix elements. Factorization yields

$$\langle \overline{K}^0 | \mathscr{O}_S | K^0 \rangle \simeq 2\langle \overline{K}^0 | \bar{d}(1 - \gamma_5)s | 0 \rangle \langle 0 | \bar{d}(1 + \gamma_5)s | K^0 \rangle$$

$$= -\frac{M_K^3}{m_s^2} F_K^2,$$

$$\langle \overline{K}^0 | \mathscr{O}_{LR} + \frac{2}{N_C}\mathscr{O}_S | K^0 \rangle \sim -F_K^2 M_K - \frac{2}{N_C}\frac{M_K^3}{m_s^2}F_K^3. \qquad (18.20)$$

This simple prescription does not reveal the μ dependence of these matrix elements, which has to cancel against the μ dependence of η_4 and η_5. On the other hand, there is no dramatic change in η_4 and η_5 when varying μ in the 'reasonable' range between 0.7 and 1.2 GeV.

The following should be kept in mind: (i) Long-distance dynamics by themselves can generate about half of the observed value of ΔM_K. (ii) Short-distance physics has to be called upon to generate the operators relevant for ϵ; predictions on this quantity in terms of the basic model parameters are thus more reliable. (iii) W_L - W_R mixing can induce **CP** odd operators even without the intervention of the third family. The weight of right-handed currents will in general therefore be boosted considerably in ϵ since the d and s quarks couple so weakly to top quarks.

Accordingly, we can expect to deduce merely qualitative bounds on left–right model parameters from ΔM_K, and semiquantitative ones from ϵ. Yet even so they are instructive. For an orientation we keep just the term proportional to η_4 in Eq. (18.18) and assume manifest left-right symmetry:

$$(M_{12})_{LR} \simeq (M_{12})_{LL} \frac{6\eta_4}{B_K \eta_{cc}} e^{i(\delta_1 - \delta_2)} \cdot \log \frac{m_c^2}{M_{W_L}^2} \cdot \frac{M_{W_L}^2}{M_{W_R}^2} \cdot \frac{M_K^2}{m_s^2} ; \qquad (18.21)$$

where $(M_{12})_{LL}$ is given in Eq. (9.44). Note that the contribution from the box diagram with the exchange of a W_L and W_R contains a huge enhancement factor

$$6|\log\frac{m_c^2}{M_{W_L}^2}|\frac{M_K^2}{m_s^2} \sim 500$$

making up for its intrinsic suppression [294] by $M_{W_L}^2/M_{W_R}^2$! This is caused by the confluence of three well understood enhancements: (i) a combinatorial factor; (ii) a much softer GIM suppression – $\log(m_c/M_{W_L})$ – which actually turns out to be an enhancement since $m_c \ll M_{W_L}$; (iii) a larger matrix element for the non-chiral operator $\mathcal{O}_S$. This means that the contribution from the LR box becomes comparable to that for the LL box for $M_{W_R} \sim 1-2$ TeV. A detailed numerical computation indeed yields [291]:

- $$\frac{(\Delta M_K)_{LR}}{(\Delta M_K)_{LL}} \simeq -500\cos(\delta_2 - \delta_1)\left(\frac{M_{W_L}}{M_{W_R}}\right)^2. \tag{18.22}$$

- ϵ_m, defined in Eq. (9.50), is given by

$$|\epsilon_{LR}| = \frac{1}{2\sqrt{2}}\frac{\operatorname{Im} M_{12}}{\operatorname{Re} M_{12}} = 176\sin(\delta_2 - \delta_1)\left(\frac{M_{W_L}}{M_{W_R}}\right)^2. \tag{18.23}$$

Arguing that the long-distance contribution to ΔM_K cannot be significantly larger than the observed value and that therefore $\Delta M_K|_{\text{box}} > 0$ should hold, we infer $M_{W_R} \geq 1$ TeV from Eq. (18.22).

Eq. (18.23) shows that in ϵ the LR contribution dominates over the LL one for light W_R. A quite intriguing scenario becomes then conceivable, namely that a significant and even dominant fraction of ϵ is generated by the presence of right-handed currents – if W_R is not excessively heavy. Such an option can be realized for $M_{W_R} \leq \mathcal{O}(20)$ TeV.

To summarize, we can derive two useful bounds:

1 If we say that Eq. (18.22) is bounded by 1, then $(M_{W_L}/M_{W_R})^2 \leq 1/500$, which in turn leads to

$$\zeta \leq 6 \times 10^{-5}. \tag{18.24}$$

2 Eq. (18.23) implies

$$\frac{M_{W_L}^2}{M_{W_R}^2}\sin(\delta_2 - \delta_1) < 10^{-5}. \tag{18.25}$$

18.2.2 $\Delta S = 1$ transitions

Since **CP** odd operators emerge without the intervention of the third family, the visibility of *direct* **CP** violation in strange decays should be boosted. So why have such effects not been observed yet? No definitive answer can be given, yet certain

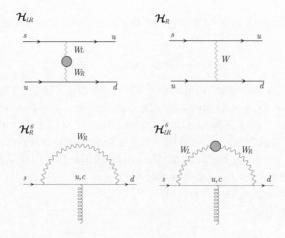

Fig. 18.2.   Feynman diagrams which give a new contribution to the $\Delta S = 1$ Hamiltonian.

tendencies can be pointed out. Here we shall concentrate on those aspects of L–R symmetry that are distinct from the SM. For a detailed numerical analysis, we refer you to Refs. [291, 296]. We limit ourselves to one-loop electroweak contributions without QCD radiative corrections. The relevant Feynman diagrams are given in Fig. 18.2:

$$\mathcal{H} = \mathcal{H}_{SM} + \mathcal{H}_R + \mathcal{H}_{LR} + \mathcal{H}_R^6 + \mathcal{H}_{LR}^6, \tag{18.26}$$

where $\mathcal{H}_{SM}$ is the SM Hamiltonian. The general expression for ϵ' is given in Eq. (7.39).

$$|\epsilon'| = \frac{1}{\sqrt{2}}\omega(\xi_0 - \xi_2) \ , \quad \xi_I = arg A_I. \tag{18.27}$$

Because of the $\Delta I = \frac{1}{2}$ enhancement, New Physics competes more favourably with the SM if it contributes to the $I = 2$ amplitude A_2.

- Since $\mathcal{H}_R^6$ and $\mathcal{H}_{LR}^6$ do not contribute to A_2, they are unlikely to cause a substantial modification to ϵ'.

-

$$
\begin{aligned}
(\xi_2)_R \ &\sim \ \left(\frac{M_L}{M_R}\right)^2 \frac{\mathrm{Im}\, \mathbf{V}_{us}^{R*}\mathbf{V}_{ud}^R}{\mathbf{V}_{us}^*\mathbf{V}_{ud}} \frac{\langle (\pi\pi)_2|(\bar{s}u)_{V+A}(\bar{u}d)_{V+A}|K\rangle}{A_2} \\
&\sim \ \sin(\delta_2 - \delta_1)\left(\frac{M_L}{M_R}\right)^2
\end{aligned}
\tag{18.28}
$$

from W_R exchange generating $\mathcal{H}_R$. Using Eq. (18.25), we obtain

$$\frac{(\epsilon')_{RR}}{\epsilon} \leq 2\cdot 10^{-4} \tag{18.29}$$

•

$$\xi_2 = \frac{\text{Im}\,\langle(\pi\pi)_2|\mathcal{H}_{LR}|K^0\rangle}{A_2}$$

$$= \sin\zeta\,[\sin(\gamma-\delta_2)-\sin(\gamma-\delta_1)]\frac{4\langle\pi\pi;I=2|\bar{s}_L\gamma_\mu u_L\bar{u}_R\gamma^\mu d_R|K^0\rangle}{\langle\pi\pi;I=2|Q_2|K^0\rangle}$$

$$(18.30)$$

from $W_L - W_R$ mixing. The matrix elements are estimated as follows:

$$\langle(\pi\pi)_2|\bar{s}_L\gamma_\mu u_L\bar{u}_R\gamma^\mu d_R|K^0\rangle \approx \frac{2}{3}\langle\pi\pi;I=2|\bar{s}_L d_R\bar{u}_R\gamma^\mu u_L|K^0\rangle \simeq$$

$$-\frac{8}{3\sqrt{6}}\langle\pi^0|\bar{s}_L d_R|K^0\rangle\langle\pi^0|\bar{u}_R d_L|0\rangle = -i\frac{1}{6\sqrt{6}}\frac{M_K^2-M_\pi^2}{m_s^0}\frac{M_\pi^2}{m_u^0}f_+F_\pi \quad (18.31)$$

$$\langle(\pi\pi)_2|Q_2|K^0\rangle = \frac{8}{3\sqrt{6}}X, \quad (18.32)$$

where $X = F_\pi(M_K^2-M_\pi^2)$. Putting these together, we have

$$\epsilon' \approx \frac{1}{5}\zeta\,[\sin(2\gamma-\delta_2)-\sin(2\gamma-\delta_1)]. \quad (18.33)$$

In a pseudo L–R scenario we have

$$\tan\zeta \simeq 2\left(\frac{M_{W_L}}{M_{W_R}}\right)^2\frac{\text{Tr}\mathbf{M}_U}{\text{Tr}\mathbf{M}_D}, \quad (18.34)$$

which, together with $M_{W_R} \geq 1.9$ TeV, leads to

$$\frac{\epsilon'}{\epsilon} < 10^{-4}. \quad (18.35)$$

With the wealth of weak phases available, no firm prediction can be made on ϵ'/ϵ [291]. However, the following comments might be helpful: (i) *Manifest L–R symmetry* forcing the left and right handed couplings into lockstep reduces predictions below present bounds. (ii) The more intriguing and novel case of having **CP** violation occur *spontaneously* only in $W_L - W_R$ mixing approximately leads to a superweak scenario with

$$\frac{\epsilon'}{\epsilon} \simeq 10^{-4}. \quad (18.36)$$

The reason for that is fairly straightforward: the concurrent enhancement factors for the W_R contribution to $\Delta S = 2$ amplitudes are basically absent in $\Delta S = 1$ amplitudes.

18.2.3 *Final state asymmetries in hyperon decays*

Here, too, there are several contributions of uncertain size. We limit ourselves to a semiquantitative sketch of where SM expectations might get modified *significantly* [296].

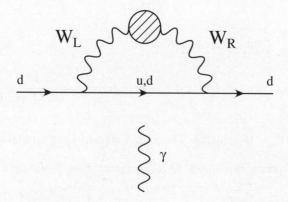

Fig. 18.3.   Feynman diagram which generates EDM in the L-R symmetric model.

$\mathscr{H}_R$, $\mathscr{H}_R^6$ and $\mathscr{H}_{LR}^6$ contributions can be ignored. The largest contribution should come from $\mathscr{H}_{LR}$:

$$\mathrm{Im}\,(\mathbf{V}_{us}^{R*}\mathbf{V}_{ud})\zeta \sim \lambda\sin(\gamma + \delta_2 - \delta_1)\zeta \leq 6 \times 10^{-5}\lambda, \tag{18.37}$$

which may have a chance to compete with the SM expectation.

18.2.4 Muon transverse polarization in $K_{\mu3}$ decays

Muon transverse polarization in $K^+ \to \mu^+\nu\pi^0$ can arise only if chirality conserving as well as violating couplings contribute with a relative weak phase, see Sec. 7.6. Yet with both W_L and W_R (or W_1 and W_2) couplings conserving chirality we need to consider box diagrams with W_L and W_R exchanges. Such a higher order effect makes $P_\perp$ unobservably tiny!

18.3 Electric dipole moments

EDM represents a highly intriguing phenomenon for left–right models – particularly so when **CP** invariance is broken spontaneously, leading to the emergence of a weak phase through $W_L - W_R$ mixing:

- With **CP** breaking occurring softly, EDMs – being coefficients of dimension five operators – become calculable finite quantities.

- The leading contribution to the EDMs is actually coming from a one-loop diagram, as shown in Fig. 18.3. This is quite different from the SM where the EDM is produced only at the three-loop level. This means that we can expect to obtain *relatively* large effects.

Evaluating the diagram of Fig. 18.3, we find for the EDM of a neutron treated as a static combination of two d and one u quarks:

$$d_n \simeq \frac{1}{3}(4d_d - d_u) = e\frac{g_L g_R}{72\pi^2 M_{W_L}^2}\sin2\zeta\sin\theta_L\sin\theta_R\left[5m_c\sin(\gamma + \delta_1) - m_s\sin(\gamma - \delta_1)\right].$$

(18.38)

For $g_L \simeq g_R$, $\sin\theta_L \simeq \sin\theta_R$ and $\zeta \leq 10^{-3}$, we arrive at an order of magnitude estimate:

$$d_n \leq 10^{-24}\gamma \text{ e cm}.$$

(18.39)

Thus

$$d_n \sim \mathcal{O}\left(10^{-27}\right) \text{ e cm}$$

(18.40)

represents a natural benchmark figure; i.e., left–right models reproducing ϵ with $\epsilon'/\epsilon \sim 10^{-4}$ can naturally yield values for the neutron EDM that is not too far below present experimental bounds, and thus appears to be within 'striking distance'!

18.4 Prospects for CP asymmetries in beauty decays

We now know from experiments that b quarks decay predominantly via left-handed currents. This was learnt from comparing the inclusive spectra of charged leptons and neutrinos in semileptonic beauty decays.

We do not anticipate any striking deviations from SM expectations in the usual $\Delta B = 1$ decay amplitudes, for example, $B \rightarrow lX, \psi K, \pi\pi$ etc. if some right-handed couplings of reduced strength were added, since the rates are not particularly sensitive to the chirality of the underlying currents. There is more sensitivity to such New Physics in $B^0 - \overline{B}^0$ oscillations and radiative B decays; yet even there the impact of right-handed couplings will not be magnified, as is the case for $K^0 - \overline{K}^0$ oscillations. For two of the enhancement factors operating for $K^0 - \overline{K}^0$ oscillations cannot be enlisted here: (i) since $M_B \simeq m_b$ in contrast to $M_K \gg m_s$ there is no significant enhancement in the matrix elements of O_{LR} and O_S operators over those of O_{LL}; (ii) $B^0 - \overline{B}^0$ oscillations are driven almost exclusively by virtual top quarks with $m_t \sim \mathcal{O}(M_W)$ and no large GIM factor like $\log m_c^2/M_W^2$ arises.

The main impact – and that could be a major one – will come from $\Delta B = 2$ Higgs couplings contributing to $\Delta M(B_d)$ and $\Delta M(B_s)$; in general those quantities can be increased or decreased relative to SM expectations, and even their ratio could be changed. Furthermore, it would be quite natural that such new contributions introduce new weak phases; this would modify **CP** asymmetries observed in B^0 decays[1].

---

[1] The relationship between the asymmetry in $B_d \rightarrow \psi K_S$ and in $B_d \rightarrow \pi\pi$ as expressed through $\phi_1 + \phi_2$ would *not* be touched, since New Physics enters through the same off-diagonal element of the B_d meson mass matrix, namely $\mathcal{M}_{12}$, rather than through a channel specific decay amplitude.

18.5 The pundits' résumé

Left–right gauge models have many attractive or at least intriguing features:

(⇑) A highly symmetric starting point gives room to the hope that parity and **CP** violation can be understood dynamically rather than being viewed as a given. This might some day allow us to calculate the basic parameters characterizing the breakings of these two symmetries and their internal relationship.

(⇑) Through the 'see-saw' mechanism it provides us with a rather natural rationale for the extreme lightness of the observed neutrinos. It relies on two main ingredients: (i) the special ability of neutrinos to acquire Majorana masses in addition to Dirac masses; (ii) the fact that parity invariance realized spontaneously can assign quite different Majorana masses to the left- and right-handed neutrinos.

(⇑) Through W_L - W_R mixing it provides us with a rather unique source for **CP** violation that quite naturally leads to $\epsilon'/\epsilon \ll \omega \simeq 0.05$ on one hand, yet at the same time and in an equally natural way suggests that electric dipole moments can emerge once present experimental bounds are pushed down by an order of magnitude or so.

A self-respecting pundit will also point out some drawbacks:

(⇓) There is no shred of evidence for direct intervention of right-handed charged currents.

(⇓) The promise that a higher degree of initial symmetry will enable us to dynamically understand its spontaneous breaking in a quantitative fashion has – well – remained that: a promise!

(⇓) When all the dust of symmetry breaking has settled down, we have to deal with a proliferation of effectively basic parameters like KM parameters for right-handed couplings, etc. Constraints due to the imposition of a discrete symmetry – like left–right symmetry – at presumably quite high energy scales get increasingly washed out at lower energies. By adjusting the plethora of such parameters we can – even without fine-tuning – easily allow the theory to take 'evasive action' when confronted with data. In that sense there are no benchmarks that make or break the theory.

So what is the final word of 'conventional wisdom'? Left–right models should be valued, at least as a gauge for how specifically certain phenomena probe the SM; they can thus act as a foil for reflections of and on the SM. To put it differently: viewing left–right models as a paradigm of New Physics is actually understating its potential value; it can act as an imagination stretcher – a commodity of which there can never be an overabundance!

Problems

18.1   Consider a model with both right- and left-handed charged quark currents

$$\mathscr{L}_{CC} = g_L W_L^\mu \overline{U}_L \gamma_\mu D_L + g_R W_R^\mu \overline{U}_R \gamma_\mu D_R, \tag{18.41}$$

where no restriction is placed on g_L/g_R. Diagonalizing the up- and down-type quark mass matrices will introduce the usual KM matrix $\mathbf{V}_L$, and a corresponding matrix $\mathbf{V}_R$ for the right-handed currents. For N quark families $\mathbf{V}_L$ and $\mathbf{V}_R$ are $N \times N$ unitary matrices, each described by $\frac{1}{2}N(N-1)$ angles and $\frac{1}{2}N(N+1)$ phases. Using the phase freedom in defining quark fields, show that $\mathbf{V}_L$ and $\mathbf{V}_R$ *together* contain $N^2 - N + 1$ irreducible and thus physical phases. Therefore for one family we have already a physical phase supporting **CP** violation.

18.2 Having parity realized spontaneously imposes constraints on the Yukawa couplings of the Higgs field ϕ with quantum numbers $(\frac{1}{2}, \frac{1}{2}, 0)$, see Eq. (18.7). Assume κ and κ', the VEVs of ϕ, to be *real*. Show that both the quark mixing angles and phases in the left and right handed sector then coincide.

18.3 Consider an $SU(2)_L \times SU(2)_R \times U(1)$ gauge theory with *spontaneous* **CP** violation realized by

$$\langle \phi \rangle = e^{i\xi} \begin{pmatrix} \kappa & 0 \\ 0 & \kappa' \end{pmatrix}, \quad \text{with } \kappa, \kappa' \text{ real}. \tag{18.42}$$

- Show that the quark mass matrices are symmetric.
- Derive the following relation between the left and right handed quark mixing matrices:

$$\mathbf{V}_R = J_U \mathbf{V}_L^* J_D^\dagger, \tag{18.43}$$

with $J_{U,D}$ being unrelated diagonal unitary matrices. (This theorem was first proved in general in Ref. [297].)

18.4 Consider an $SU(2)_L \times SU(2)_R \times U(1)$ gauge model. Show that while a Higgs field ϕ with quantum numbers $(\frac{1}{2}, \frac{1}{2}, 0)$ can produce a *Dirac* mass for neutrino fields, a Higgs field Δ_R [Δ_L] with $(0, 1, 2)$ [$(1, 0, 2)$] can generate a *Majorana* mass for right [left] handed neutrinos. The resulting neutrino mass matrix then reads

$$\mathcal{M} = \begin{pmatrix} \mu_L^M & m^D \\ m^D & M_R^M \end{pmatrix}, \tag{18.44}$$

with μ_L^M, M_R^M and m^D being controlled by the vacuum expectation values of Δ_L, Δ_R and ϕ, respectively.

Assuming the hierarchy

$$\mu_L^M \ll m^D \ll M_R^M, \tag{18.45}$$

find the eigenvectors and eigenvalues of this matrix.

18.5 Draw a Feynman diagram generating the muon transverse polarization in $K^+ \to \mu^+ \nu \pi^0$ decays.

19

CP violation from Higgs dynamics

Addressing the mysterious
with the obscure

As discussed in Chapter 8, gauge couplings must be defined to be real; yet in the presence of three families, quark mass matrices can introduce irreducible phases, which then enter the gauge interactions between quark mass eigenstates. Within the SM, the quark masses are generated by the quarks interacting with a Higgs doublet field which develops a vacuum expectation value. The seeds for these **CP** violating phases of the KM type are then in these Yukawa couplings and thus strictly speaking, driven by Higgs dynamics. Yet despite its central role, the Higgs sector with its lack of direct experimental support represents the largest 'terra incognita' in the SM, and we have considerable latitude in constructing the Higgs sector.

- No bound on the number of Higgs doublets has been derived.

- A priori quarks can couple to more than one Higgs doublet.

- The self-interaction of Higgs fields can in general be rather complex.

Several reasons have been put forward for considering an 'extended' Higgs sector, i.e., one that contains more than a single doublet. Irrespective of the original motivation for such models, they allow new sources of **CP** violation. Furthermore, **CP** breaking can then occur in a *spontaneous* as well as in a *manifest* fashion. Realistic models of **CP** violation based on Higgs dynamics have been proposed a long time ago [298, 299]. We discuss such scenarios with a dual purpose in mind. (i) We can use **CP** studies as a tool, namely as a high sensitivity probe of Higgs dynamics. (ii) Finding manifestations of Higgs driven **CP** violation would *per se* not answer the question about its fundamental origin. Yet analysing its dynamical structures will provide us with more of the missing pieces of the overall puzzle. For recent update reviews, see Refs. [300, 301, 302].

19.1 A simple example

For complex scalar fields ϕ a global phase rotation $\phi \to e^{i\beta}\phi$ leaves the theory invariant. Thus there is an arbitrary phase in the definition of **CP** transformations:

$$\mathbf{CP}\phi(t,\vec{x})\mathbf{CP}^{-1} = e^{i\delta}\phi^\dagger(t,-\vec{x}). \tag{19.1}$$

Yet with at least two complex scalar fields ϕ_1 and ϕ_2 the relative phase cannot be removed. Let us see how **CP** violation can arise then. Define

$$\mathbf{CP}\phi_j(t,\vec{x})\mathbf{CP}^{-1} = \phi_j^\dagger(t,-\vec{x}) \tag{19.2}$$

and rewrite it in terms of two Hermitian fields $\phi = s + ip$:

$$\mathbf{CP}s_j(t,\vec{x})\mathbf{CP}^{-1} = s_j(t,-\vec{x}), \quad \mathbf{CP}p_j(t,\vec{x})\mathbf{CP}^{-1} = -p_j(t,-\vec{x}) ; \tag{19.3}$$

i.e., the real fields s and p thus represent Lorentz scalar and pseudoscalar fields, respectively. Consider the following mass terms:

$$\mathcal{L}_{\text{mass}} = \mu_1^2\phi_1^\dagger\phi_1 + \mu_2^2\phi_2^\dagger\phi_2 + \mu_{12}^2\phi_1^\dagger\phi_2 + \text{h.c.},$$

where μ_{12} is assumed to be complex. The mass eigenstates obtained by diagonalizing this mass matrix are linear superpositions of the $s_{1,2}$ and $p_{1,2}$ and **CP** invariance is thus broken. In Ref. [303] an elegant, though unrealistic, model is given where such mass terms arise.

19.2 Sources of CP violation and the question of natural flavour conservation (NFC)

Consider the three-family SM with a *second* Higgs doublet field added. The two Higgs doublets contain four neutral, two positively and two negatively charged fields. One neutral and two charged fields get eaten and become longitudinal Z^0, W^+ and W^- states, leaving five *physical* Higgs fields – three neutral and two charged. The most general Higgs potential is then given by [300]

$$\begin{aligned}
\mathcal{L}_H &= \mu_1^2\phi_1^\dagger\phi_1 + \mu_2^2\phi_2^\dagger\phi_2 + (\mu_{12}^2\phi_1^\dagger\phi_2 + \text{h.c.}) \\
&\quad - \lambda_1(\phi_1^\dagger\phi_1)^2 - \lambda_2(\phi_2^\dagger\phi_2)^2 - \lambda_3(\phi_1^\dagger\phi_1\phi_2^\dagger\phi_2) - \lambda_4(\phi_1^\dagger\phi_2)(\phi_2^\dagger\phi_1) \\
&\quad - \frac{1}{2}[\lambda_5(\phi_1^\dagger\phi_2)^2 + \text{h.c.}] - (\lambda_6\phi_1^\dagger\phi_1 + \lambda_7\phi_2^\dagger\phi_2)(\phi_1^\dagger\phi_2 + \text{h.c.}). \tag{19.4}
\end{aligned}$$

The mass parameters μ_1 and μ_2 and the couplings $\lambda_1 - \lambda_4$ are necessarily real. If also μ_{12} and $\lambda_5 - \lambda_7$ happen to be real, **CP** invariance can still be broken spontaneously if $\lambda_5 > 0$ by a relative phase between the two VEVs:

$$\langle\phi_1^0\rangle = \frac{v}{\sqrt{2}}\cos\beta e^{i\delta}, \quad \langle\phi_2^0\rangle = \frac{v}{\sqrt{2}}\sin\beta. \tag{19.5}$$

If μ_{12}, λ_6 and/or λ_7 are complex, $\mathcal{L}_H$ contains explicit **CP** violation as well; it is *soft* if only μ_{12} is complex.

These sources of **CP** violation, which do *not* exist for a single Higgs doublet, are in addition to the always existing possibility that the quark–Higgs couplings are complex, which represents hard **CP** violation:

$$\mathcal{L}_Y = \sum_{i,j} \left[\overline{Q}_{iL}(\Gamma_1^D)_{ij}\phi_1 D_{jR} + \overline{Q}_{iL}(\Gamma_2^D)_{ij}\phi_2 D_{jR} \right] + [D \leftrightarrow U] + \text{h.c.} \tag{19.6}$$

The up- and down-type quarks are denoted by $U = (u,c,t)$ and $D = (d,s,b)$, respectively; the matrices $\Gamma_{1,2}^{U,D}$ contain the Yukawa couplings.

The physical interpretation becomes more transparent when the dynamics are expressed for the quark and Higgs mass eigenstates. Once the $\phi_{1,2}$ acquire VEVs we obtain quark mass matrices

$$\mathcal{M}^D = \Gamma_1^D\langle\phi_1\rangle + \Gamma_2^D\langle\phi_2\rangle, \quad \mathcal{M}^U = \Gamma_1^U\langle\phi_1\rangle + \Gamma_2^U\langle\phi_2\rangle. \tag{19.7}$$

Upon diagonalizing, $\mathcal{M}^{D,U}$ flavour-changing neutral currents (FCNC) of the type $\bar{s}d\phi$, $\bar{c}u\phi$, $\bar{b}d\phi$, etc. will in general emerge – unless the transformations $T_{L,R}^{U,D}$ diagonalize *simultaneously*,

$$\Gamma_1^{D,U}\langle\phi_1\rangle + \Gamma_2^{D,U}\langle\phi_2\rangle \text{ and } \Gamma_1^{D,U} + \Gamma_2^{D,U}. \tag{19.8}$$

In that case we have NFC. While this will not happen automatically, it can be guaranteed by imposing a discrete symmetry [304]

$$\phi_1 \rightarrow -\phi_1, \quad \phi_2 \rightarrow \phi_2, \tag{19.9}$$

$$D_R \rightarrow -D_R, \quad U_R \rightarrow U_R, \tag{19.10}$$

which prevents ϕ_1 from coupling to U_R and ϕ_2 to D_R. The same is achieved through the $U(1)$ symmetry

$$D_R \rightarrow e^{i\alpha}D_R, \quad \phi_1 \rightarrow e^{-i\alpha}\phi_1,$$

$$U_R \rightarrow U_R, \quad \phi_2 \rightarrow \phi_2. \tag{19.11}$$

The up- and down-type quark mass matrices are then controlled by a single VEV each

$$\mathcal{M}^U = \Gamma^U\langle\phi_2\rangle, \quad \mathcal{M}^D = \Gamma^D\langle\phi_1\rangle, \tag{19.12}$$

the $T_{L,R}^{U,D}$ diagonalize $\mathcal{M}$ and Γ simultaneously and no FCNC arise.

There are thus two vastly different dynamical scenarios for Higgs dynamics, which have to be analyzed *separately*, namely *with* or *without* FCNC. The difference between *spontaneous* and *manifest* **CP** violation – while fundamental – is of little *direct* importance for the phenomenological analysis; therefore we are going to refer to it only occasionally, mainly when considering the possible impact of higher loop effects.

19.2.1 Models with FCNC

In general FCNC will arise. Expressing the Yukawa couplings in terms of quark and Higgs *mass* eigenstates we obtain from Eq. (19.6) for neutral $\Delta S = 1$

transitions:

$$\mathscr{L}_Y^{\text{neut}}(\Delta S = 1) = \sum_{\alpha=2}^{4} \left[G_\alpha \bar{s} d + \overline{G}_\alpha \bar{s} i \gamma_5 d \right] H_\alpha, \tag{19.13}$$

where $H_1, \ldots, H_4$ are the neutral Higgs *mass* eigenstates with H_1 the Nambu–Goldstone boson, enabling Z^0 to become massive, and

$$G_\alpha = \frac{1}{2} \sum_\beta \left[(\tilde{\Gamma}_\beta^D + \tilde{\Gamma}_\beta^{D\dagger})_{21} R_{\beta\alpha}^{(1)} + i(\tilde{\Gamma}_\beta^D - \tilde{\Gamma}_\beta^{D\dagger})_{21} R_{\beta\alpha}^{(2)} \right]$$

$$\overline{G}_\alpha = \frac{1}{2} \sum_\beta \left[-i(\tilde{\Gamma}_\beta^D - \tilde{\Gamma}_\beta^{D\dagger})_{21} R_{\beta\alpha}^{(1)} + (\tilde{\Gamma}_\beta^D + \tilde{\Gamma}_\beta^{D\dagger})_{21} R_{\beta\alpha}^{(2)} \right], \tag{19.14}$$

with $\tilde{\Gamma}_{1,2}^D = T_L^D \Gamma_{1,2}^D (T_R^D)^\dagger$ and $R^{(1)}$ and $R^{(2)}$ denoting 2×4 matrices relating the H_i to the scalar and pseudoscalar components, respectively. Single exchange of such Higgs fields then generates

$$\mathscr{L}_{H-X}(\Delta S = 2) = \sum_\alpha \frac{-1}{M_\alpha^2} \left(G_\alpha^2 (\bar{d}s)^2 - \overline{G}_\alpha^2 (\bar{d}\gamma_5 s)^2 + 2iG_\alpha \overline{G}_\alpha (\bar{d}s)(\bar{d}\gamma_5 s) \right). \tag{19.15}$$

There are thus *tree* level transitions changing strangeness by two units – something that occurs in the SM only as a one-loop effect! Furthermore, since G_α and $\overline{G}_\alpha$ are in general complex, physical phases will arise in $\mathscr{L}_{H-X}(\Delta S = 2)$; i.e., in models with FCNC the presence of *two* Higgs doublets is sufficient to generate **CP** violation.

Invoking factorization for evaluating the relevant matrix elements should be quite adequate for our goal of obtaining order-of-magnitude estimates:

$$\langle \overline{K}^0 | \mathscr{H}_{H-X}(\Delta S = 2) | K^0 \rangle \simeq \frac{\tau^2}{M_H^2} \frac{F_K^2 M_K^3}{(m_s + m_d)^2}, \quad \frac{\tau^2}{M_H^2} \equiv \sum_\alpha \frac{\overline{G}_\alpha^2}{M_\alpha^2}, \tag{19.16}$$

where M_H is the lightest Higgs' mass. If we require that ΔM_K is not oversaturated – i.e., $2|\langle \overline{K}^0 | \mathscr{H}_{H-X}(\Delta S = 2) | K^0 \rangle| \le \Delta M_K|_{\text{exp}}$ – we find $|\tau/M_H| \ge 1$ TeV, which should be interpreted as a rough benchmark figure only.

The situation becomes somewhat more definite when we consider ϵ_K: requiring $\text{Im}\langle \overline{K}^0 | \mathscr{H}_{H-X}(\Delta S = 2) | K^0 \rangle / \Delta M_K \le \sqrt{2} |\epsilon_K|_{\text{exp}}$ we infer

$$M_H \ge (30 \cdot \text{Im}\tau) \text{ TeV}. \tag{19.17}$$

The model dependence enters through the quantity τ defined in Eq. (19.16); i.e., an a priori reasonable range for it is given by $\tau \sim 1/30 - 1$; the observed value of ϵ_K is then reproduced by the exchange of a very heavy Higgs field: $M_H \sim 1 - 30$ TeV [305]. In this type of model $\Delta S = 1$ and $\Delta S = 2$ Higgs mediated transitions are of the same order, namely tree level; with the $\Delta S = 2$ amplitude being comparable to a second order electroweak SM amplitude, *direct* **CP** violation in the $\Delta S = 1$ is then greatly suppressed and for all practical purposes we typically have a *dynamical* realization of the superweak ansatz for **CP** violation [75].

These remarks serve as general orientation rather than (semi-) quantitative predictions for these models with FCNC, since the values of τ and Imτ can conceivably vary by orders of magnitude. Furthermore, charged Higgs as 'light' as ~ 100 Gev can exist, and through complex Yukawa couplings generate **CP** asymmetries in $\Delta S = 1$ transitions as measured by ϵ'. It is, however, instructive to employ theoretical criteria to distinguish different subscenarios. They centre on the question of how NFC is violated.

1 The discrete symmetry of Eq. (19.9) is violated *softly* by the μ_{12}^2 term. **CP** violation can enter the quark sector if the symmetry transformation of Eq. (19.10) for down quark fields is modified to $\{d_{i,R}\} \rightarrow \{\eta_i d_{i,R}\}$ with η_i being (+1) for some families and (-1) for others [306]. This class of models, according to [307], is of the genuinely superweak variety: it does not generate a KM phase.

2 The invariance under Eq. (19.9) and Eq. (19.10) is violated by $\mathscr{L}_H$ as well as $\mathscr{L}_Y$ – i.e., in a hard fashion – yet by small numerical amounts. Such models contain tree level $\Delta S = 2$ transitions that violate **CP**. In addition a non-zero KM phase is induced; ϵ'/ϵ_K – while not vanishing – is expected to lie in the range 10^{-6} to 10^{-4} [308]; for all practical purposes this represents a superweak scenario as well.

3 We abandon the idea of an approximate discrete symmetry and thus allows for ϕ_1 and ϕ_2 both coupling to U_R as well as D_R, as expressed in Eq. (19.6). On the other hand, we invoke some approximate $U(1)$ family symmetries to impose a special structure on the matrices $\Gamma_{1,2}^{U,D}$: they are assumed to have small off-diagonal elements, namely $\sim \mathcal{O}(0.01)$ or $\mathcal{O}(0.1)$ of the corresponding diagonal ones. This scheme has an advantage that upon diagonalization of the quark mass matrices the known form of the CKM matrix emerges and FCNC are reduced to a phenomenologically acceptable level.

$\mathscr{L}_Y$ is then re-written in terms of complex quark and Higgs mass eigenstates. The original Higgs fields ϕ_1^0 and ϕ_2^0 are expressed through neutral fields shifted by the VEVs:

$$\frac{1}{\sqrt{2}}(v + H^0 + G^0) = \cos\beta e^{-i\delta}\phi_1^0 + \sin\beta\phi_2^0$$

$$\frac{1}{\sqrt{2}}(R + iI) = \sin\beta e^{-i\delta}\phi_1^0 - \cos\beta\phi_2^0, \qquad (19.18)$$

where G^0 denotes the Goldstone boson absorbed into the Z^0; H^0, R and I are physical neutral Higgs fields. The mass eigenstates (H_1^0, H_2^0, H_3^0) are related to (R, H^0, I) by an *orthogonal* transfomation O^H.

We divide $\mathscr{L}_Y$ into two parts. (i) If the matrices $\Gamma_{1,2}^{U,D}$ were truly diagonal no FCNC would emerge and the flavour changes mediated by $H^\pm$ exchanges would be controlled by the CKM matrix; all these terms are written into $\mathscr{L}_1$. (ii) The off-diagonal elements of $\Gamma_{1,2}^{UD}$ assumed to be small induce FCNC and additional $H^{\pm}$ couplings; they are lumped into $\mathscr{L}_2$.

Thus we write [300]:

$$\mathscr{L}_Y = (\sqrt{2}G_F)^{1/2}(\mathscr{L}_1 + \mathscr{L}_2), \tag{19.19}$$

$$\mathscr{L}_1 = \sqrt{2}\left(H^+ \sum_{i,j}^3 V_{ij}\xi_{d_j}m_{d_j}\bar{u}_L^i d_R^j - H^- \sum_{i,j}^3 V_{ji}^*\xi_{u_j}m_{u_j}\bar{d}_L^i u_R^j\right)$$

$$+ \ H^0 \sum_i^3 \left(m_{u_i}\bar{u}_L^i u_R^i + m_{d_i}\bar{d}_L^i d_R^i\right)$$

$$+ \ (R+iI)\sum_i^3 \xi_{d_j}m_{d_j}\bar{d}_L^i d_R^i + (R-iI)\sum_i^3 \xi_{u_j}m_{u_j}\bar{u}_L^i u_R^i + \text{h.c.}, \tag{19.20}$$

$$\mathscr{L}_2 = \sqrt{2}\left(H^+ \sum_{i,j,k\neq j}^3 V_{ik}\mu_{kj}^d\bar{u}_L^i d_R^j - H^- \sum_{i,j,k\neq j}^3 V_{ki}^*\mu_{kj}^u\bar{d}_L^i u_R^j\right)$$

$$+ \ (R+iI)\sum_{ij\neq i}^3 \mu_{ij}^d\bar{d}_L^i d_R^j + (R-iI)\sum_{ij\neq i}^3 \mu_{ij}^u\bar{u}_L^i u_R^j + \text{h.c.}, \tag{19.21}$$

with the factors $\xi_{d_j}m_{d_j}$, $\xi_{u_j}m_{u_j}$ and $\mu_{ij}^{d,u}$ arising mainly from the large diagonal and small off-diagonal elements, respectively, of $\Gamma_{1,2}^{U,D}$.

Since we have postulated merely the smallness – not the absence! – of various terms, we should not be surprised by the structural complexity of the weak Lagrangian and the multitude of ways in which **CP** invariance can be broken. Those can be grouped into four major categories:

- The irreducible phase in the CKM matrix induces **CP** violation into $W^{\pm}$ *and* $H^{\pm}$ exchanges.

- The factors $\xi_{d,u}$ can contain additional phases; those can generate **CP** violation also in the flavour changing R and I exchanges.

- Phases in the factor $\mu_{ij}^{u,d}$ drive **CP** violation in FCNC.

- Finally, **CP** symmetry can be broken by the Higgs mixing matrix O_H. This might seem surprising at first since O_H is orthogonal, i.e., real. Yet it comes about in the following way [302, 309]. **CP** invariance requires all couplings and VEVs to be real (or that they can be all made real by adopting an appropriate phase convention for the Higgs fields). The R and I components cannot mix then and the mass eigenstates H_i^0 are **CP** eigenstates as well. However, if the VEVs develop a relative phase, then R and I can mix, as illustrated by Eq. (19.18), and the mass eigenstates are no longer **CP** eigenstates. The most intriguing consequences arise in the Higgs gauge dynamics: while Bose statistics forbids $Z^0 \to H_i^0 H_i^0$ to proceed, it is **CP** invariance that forbids $Z^0 \to H_i^0 H_j^0$ with $i \neq j$ if both H_i and H_j are either **CP** even or **CP** odd.

Correspondingly, such models have a very rich **CP** phenomenology. In particular, direct **CP** violation can arise in the $\Delta S = 1$ transition: $H^\pm$ exchanges can generate a value for ϵ'/ϵ_K around $10^{-5} - 10^{-4}$ for $\tan \beta \sim 1$ that can rise to 10^{-3} for $\tan \beta \gg 1$! If the next round of experiments revealed a non-zero value for ϵ'/ϵ, it would be consistent with the KM ansatz of the SM as discussed before – yet it could be driven by non-minimal Higgs dynamics without NFC as well [300].

19.2.2 *Models without FCNC*

Imposing the transformations in Eq. (19.9) and Eq. (19.10) as symmetry leads to NFC; up-type and down-type quarks each couple to a different Higgs doublet only – a feature that is preserved by quantum corrections – and the absence of FCNC is achieved in a *natural* way. The first explicit model of this type had been put forward in Ref. [299]. It contains three Higgs doublets of which two couple to quarks in the following way:

$$\mathscr{L}_Y = \sum_{i,j} \left(\Gamma_{ij}^D \overline{Q}_{iL} \Phi_1 D_{jR} + \Gamma_{ij}^U \overline{Q}_{iL} \tilde{\Phi}_2 U_{jR} \right) + \text{h.c.} \tag{19.22}$$

The third doublet Φ_3 remains detached from the quarks; it may be coupled to leptons. The Yukawa couplings are assumed to be real to have **CP** invariance realized spontaneously. We postulate the Higgs potential to be such that *complex* VEVs emerge:

$$\langle \Phi_1 \rangle = e^{i\eta_1} v_1, \quad \langle \Phi_2 \rangle = e^{i\eta_2} v_2, \quad \langle \Phi_3 \rangle = v_3, \tag{19.23}$$

where without loss of generality we have defined $\langle \Phi_3 \rangle$ as real. The *neutral* Higgs couplings can be made real by a redefinition $D_{j,R} \to e^{-i\eta_1} D_{j,R}$, $U_{j,R} \to e^{-i\eta_2} U_{j,R}$ leading to

$$\mathscr{L}_{Y,\text{neut}} = \overline{D} \Gamma^D D(v_1 + \rho_1^0) + \overline{U} \Gamma^U U(v_2 + \rho_2^0), \tag{19.24}$$

where the complex fields ρ_i result from expanding Φ_i around their VEVs. Thus we have $\mathscr{M}^D = \Gamma^D v_1$, $\mathscr{M}^U = \Gamma^U v_2$ and they can be diagonalized, as is done in Eq. (8.24). Note that the same transformations diagonalize also the ρ_i^0 couplings to the quarks, thus banning the spectre of FCNC[1]. Writing the shifted fields ρ_i in

---

[1] Even so, neutral Higgs exchanges can generate **CP** or **T** violation which can, for example, give rise to electric dipole moments.

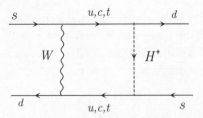

Fig. 19.1.   Diagram for the dominant contribution to $\mathrm{Im}M_{12}^K$.

terms of the Higgs mass eigenstates H_i, we find for the *charged* Higgs couplings
to the quarks

$$\mathscr{L}_{Y,\mathrm{ch}} = \frac{g}{M_W} \sum_i \left[\alpha_i \overline{D}_L^m \mathbf{V}^\dagger \mathscr{M}_U^{\mathrm{diag}} U_R^m H_i^- + \beta_i \overline{U}_L^m \mathbf{V} \mathscr{M}_D^{\mathrm{diag}} D_R^m H_i^+ \right] ; \qquad (19.25)$$

where $\mathbf{V}$ is a CKM-like mixing matrix, but it is an orthogonal matrix as all
Yukawa couplings are real – due to the condition that **CP** is broken spontaneously.
The quantities α_i and β_i contain the parameters describing the mixing among the
charged Higgs fields in addition to the phase factors and the v_i, all of which
depends on the details of the Higgs potential.

Since models with NFC contribute to $K^0 \to \overline{K}^0$ transitions only at the one-
loop level (and likewise for $D^0 \to \overline{D}^0$ and $B^0 \to \overline{B}^0$), we do not need to invoke
superheavy Higgs masses to avoid conflict with the observed value of ϵ_K. Accordingly, these models allow also for direct **CP** violation to arise on an observable
level. Their phenomenology then becomes richer: we will illustrate this through
the following observables [301, 310]: (i) ϵ_K and ϵ' in K_L decays, (ii) electric dipole
moments for neutrons and (iii) the muon transverse polarization in $K_{\mu 3}$ decays.

$$\epsilon_K$$

The expression in Eq. (7.47),

$$|\epsilon_K| \simeq \frac{1}{\sqrt{1 + (\Delta\Gamma/2\Delta M_K)^2}} \left| \frac{\mathrm{Im}M_{12}}{\Delta M_K} - \xi_0 \right|,$$

suggests that there are two components in the dynamics driving $\Delta S = 2$ transitions:

• The short distance contribution to $\mathrm{Im}M_{12}$, which is given mainly by a
  box diagram. In the case under study it contains the exchange of a W
  and a charged Higgs boson H, Fig. 19.1; assuming, for definiteness, real
  $\mathbf{V}$ and factorization for an order of magnitude estimate, we arrive at
  [310]

$$\mathrm{Im}\langle K^0 | \mathscr{H}_{\mathrm{box}} | \overline{K}^0 \rangle \simeq \frac{\mathrm{Im}(\alpha^* \beta)}{M_H^2} \frac{3G_F^2}{32\pi^2} (\sin\theta_C \cos\theta_C)^2 m_c^2 M_K^2 F_K^2$$

$$\times \frac{M_d^{\mathrm{const}}}{m_s} \langle K^0 | (\overline{d}\gamma_\mu \gamma_-s)^2 | \overline{K}^0 \rangle, \qquad (19.26)$$

where m_q and M_q^{const} are *current* and *constituent* quark masses, respectively[2]. As for the real part, we find that two W boson exchange dominates. Taking a ratio between $\text{Im}\, M_{12}$ obtained above with the two W boson exchange diagram, we obtain

$$\frac{M_{12}|_{SD}}{\Delta M} = \frac{\text{Im}\langle K^0|H|\overline{K}^0\rangle}{2\text{Re}\,\langle K^0|H|\overline{K}^0\rangle} = \frac{\text{Im}(\alpha^*\beta)}{M_H^2}\frac{3}{8}\frac{M_K^2 M_d}{m_s}. \qquad (19.27)$$

We then find

$$\left|\frac{\text{Im}\,M_{12}|_{\text{Higgs},SD}}{\Delta M_K}\right| \leq \text{few} \times 10^{-5} \text{ for } M_H \geq 80\,\text{GeV}, \ \text{Im}(\alpha^*\beta) \sim 1, \quad (19.28)$$

using the experimental value for ΔM_K; i.e., the contribution from charged Higgs to ϵ_K is quite insignificant, which holds a fortiori for ΔM_K; likewise for the effects of neutral Higgs. The situation was very different at the time such models were first put forward: charged Higgs states with a mass of a few GeV were quite conceivable then.

- If ξ_0 dominates in ϵ_K, then it leads to a sizeable value for ϵ', see Eq. (7.39):

$$\left|\frac{\epsilon'}{\epsilon_K}\right| \simeq \omega \simeq 0.05 , \qquad (19.29)$$

which is clearly in conflict with the data.

So where does ϵ_K then come from? We must turn to long distance dynamics once more and consider $K^0 \leftrightarrow \overline{K}^0$ transitions proceeding via light quark states probed off their mass shell; such a mechanism contributes to ϵ_K, yet not ϵ'. With $M_{12} = M_{12}|_{SD} + M_{12}|_{LD} \equiv M_{12}|_{SD} + DM_{12}$ we can rewrite Eq. (7.47) (see Problem 19.3) [311]:

$$\epsilon_K = \frac{1-D}{2\sqrt{2}}e^{i\pi/4}\cdot\left(\epsilon_m + 2\xi_0 + \frac{D}{1-D}\cdot\chi\right), \qquad (19.30)$$

$$\left|\frac{\epsilon'}{\epsilon_K}\right| \simeq \omega \cdot \frac{2\xi_0}{(1-D)[\epsilon_m + 2\xi_0 + D\chi/(1-D)]}. \qquad (19.31)$$

where $\epsilon_m = \arg M_{12}|_{SD}$, $\chi = \arg M_{12}|_{LD} + 2\xi_0$. The question is: can *long-distance* dynamics yield $\xi_0 \ll D\cdot\chi$ such that ϵ_K is reproduced:

$$|\epsilon_K| \simeq \frac{1}{2\sqrt{2}}D\cdot\chi \ ? \qquad (19.32)$$

Consider the *virtual* transitions

$$K^0 \rightarrow \text{`}\pi, \eta, \eta', \pi\pi\text{'} \rightarrow \overline{K}^0 . \qquad (19.33)$$

---

[2] *Constituent* quark masses – in contrast to *current* masses – are ill-defined in a quantum field theory; they are purely phenomenological constructions reflecting hadronic scales: $M_{u,d}^{\text{const}} \sim 300$ MeV.

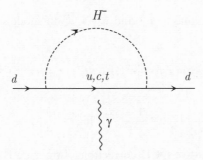

Fig. 19.2. One-loop diagram for a quark EDM.

$SU(3)$ symmetry, coupled with current algebra, implies that $\langle\pi^0|\mathcal{H}|K^0\rangle$, $\langle\eta_8|\mathcal{H}|K^0\rangle$ and $\langle(\pi\pi)_{I=0}|\mathcal{H}|K^0\rangle$ all have the same phase[3]; the lightest intermediate state contributing to χ is thus $K^0 \to$ 'η_0' $\to \overline{K}^0$; accordingly we have to study

$$|\epsilon_K| \simeq \frac{\text{Im}\langle K^0|\mathcal{H}|\eta_0\rangle\langle\eta_0|\mathcal{H}|\overline{K}^0\rangle}{2\sqrt{2}\text{Re}M_{12}}. \tag{19.34}$$

In Ref. [312] we have estimated this contribution in terms of $\text{Im}\alpha^*\beta / M_H^2$ using data on $\eta - \eta'$ mixing, the decay rates for $\eta \to \gamma\gamma$ and $\eta' \to \gamma\gamma$, $SU(3)$ symmetry and the bag model result for $\langle K^0|\mathcal{H}|\pi^0\rangle$. We found that Eq. (19.34) can be satisfied, although for $M_H \geq 40$ GeV we have to rely on more and more extreme corners of parameter space: $vv_3/v_1v_2 \geq 50$, $v = \sqrt{v_1^2 + v_2^2 + v_3^2}$. However, this leads to a direct phenomenological problem, namely that we predict too large a value for the neutron EDM.

Electric dipole moments

A non-minimal Higgs sector provides a particularly rich dynamical substratum for EDMs. There are three mechanisms for producing an EDM:

  1 Higgs exchanges generating an EDM for individual quarks or leptons;

  2 **CP** odd gluonic operators inducing a neutron EDM;

  3 analogous electroweak operators producing an EDM for electrons.

In some of these scenarios the neutron EDM results from cooperative effects among the quarks rather than individual quarks. We should also keep in mind that most of these effects can arise also for the two Higgs doublet models discussed before; the presence or absence of FCNC is not essential here.

(a) *Direct contributions:*
A quark EDM can be generated by *charged* Higgs exchange, as shown in

---

[3] η_8 [η_0] denotes the octet [singlet] component of the neutral η meson.

Fig. 19.2. With $d_N = \frac{1}{3}(4d_d - d_u)$ and $d_u \ll d_d$ in models with NFC we obtain [313]:

$$d_N = \frac{\sqrt{2}G_F e}{9\pi^2} \frac{m_c^2 m_d}{M_H^2} \mathrm{Im}(\alpha^* \beta) \cdot \left[\eta^{(c)} |V(cd)|^2 g(x_c) + \eta^{(t)} |V(td)|^2 g(x_t) \right] \tag{19.35}$$

$$g(x) = \frac{1}{(1-x)^2} \left[\frac{5}{4}x - \frac{1-3x/2}{1-x} \log x - \frac{3}{4} \right], \quad x_i = \frac{m_i^2}{M_H^2}, \tag{19.36}$$

and $\eta^{(c,t)}$ denoting radiative QCD corrections. Imposing Eq. (19.30), we can infer a *lower* bound on d_N [312]:

$$d_N \geq 5 \cdot 10^{-25} \text{ e cm}, \tag{19.37}$$

with the dependence on M_H dropping out to a large degree from this relationship of ϵ_K vs d_N. This lower bound is hardly compatible with the experimental upper bound quoted in Eq. (3.81).

(b) $G \cdot \tilde{G}$ effects

Like in the SM, we have to be concerned about the strong **CP** problem, i.e., the emergence of the **CP** odd operator $G \cdot \tilde{G}$ with coefficient $\bar{\theta}/16\pi^2$. As shown in Eq. (16.21), we infer $\bar{\theta} \leq \mathcal{O}(10^{-9})$ from the upper bound on the neutron EDM; furthermore $\bar{\theta} = \theta_{QCD} + \Delta\theta_{EW}$, where the second term denotes the electroweak renormalization of the quark mass matrix M: $\Delta\theta_{EW} \equiv \arg \det M$. Two new aspects arise relative to the SM: (i) Higgs dynamics opens up the possibility that **CP** invariance is realized *spontaneously*; θ_{QCD}, being the coefficient of a dimension four operator, has to vanish then and $\bar{\theta}$ becomes a finite and calculable quantity. (ii) Non-trivial θ renormalization occurs already on the *one*-loop level and is sizeable [314] – $\Delta\theta_{EW} \sim \mathcal{O}(10^{-3})$ – making such Higgs models appear quite unnatural.

Even if Higgs dynamics cannot be invoked to generate the lion's share of ϵ_K, it could easily provide the dominant source of observable EDMs. This can happen through the exchange of heavy charged Higgs as just described – or through a host of different mechanisms. We shall present a brief discussion of such scenarios without restricting ourselves to a specific model.

(c) *Other* **CP** *odd gluonic operators*

Whatever the origin or origins of **CP** violating electroweak forces may be, they will in general – through quantum corrections – induce a whole sequence of purely gluonic operators in increasing dimension that are *odd* under **P** and **T**:

$$\mathscr{L}_{\text{glue}}(\text{CP odd}) = \frac{\bar{\theta}}{16\pi^2} G \cdot \tilde{G} + c_{(6)} G^2 \cdot \tilde{G} + c_{(8)} G^3 \cdot \tilde{G} + \cdots \tag{19.38}$$

We have just commented on the first term; in the present discussion, we take $\bar{\theta} = 0$ and concentrate on possible other effects.

The coefficients $c_{(6)}, c_{(8)}, \ldots$ of the dimension six, eight, ... operators can be calculated as *finite* numbers. $G^2 \cdot \tilde{G}$ is likely to provide the dominant contribution

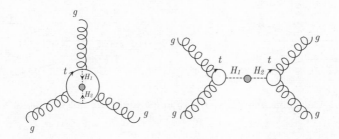

Fig. 19.3.   Diagrams which give rise to $G^2 \cdot \tilde{G}$ and $G^3 \cdot \tilde{G}$ operators.

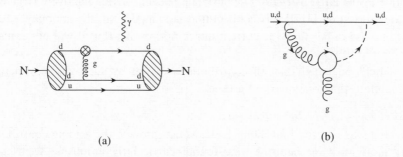

(a)   (b)

Fig. 19.4.   In (a), the neutron EDM is generated by a quark colour magnetic moment; in (b) the diagram giving rise to a quark colour dipole moment, the blob in (a) is shown.

to d_N for large values of M_H as first discussed in Ref. [315]: coming from a dimension-six operator it scales like $1/M_H^2$, whereas other contributions are suppressed by higher powers of $1/M_H$, by light quark masses and/or small mixing angles. The coefficient $c_{(6)}$ can be calculated in terms of the model parameters, namely the Higgs masses and Yukawa couplings with the latter expressed in terms of m_b, m_t and the VEVs v_i. Diagrams which give rise to these operators are shown in Fig. 19.3. No precise prediction for the resulting d_N has been extracted yet from these studies, since considerable theoretical uncertainties arise in evaluating the matrix element $\langle N|G^2 \cdot \tilde{G}|N \rangle$ upon which d_N depends. For $M_H \sim 80 - 100$ GeV we estimate [316]:

$$d_N \sim \mathcal{O}(10^{-26}) \text{ e cm}, \tag{19.39}$$

i.e., a value that is just an order of magnitude below the present experimental upper bound! This relatively large value, despite the very heavy Higgs mass, arises due to the large Yukawa couplings of the top quarks in the loop. In any case, exchanges of intermediate mass Higgs fields with $M_H \sim (1 - 2) \times M_W$ are likely to induce d_N on a level that should become observable in the foreseeable future.

At first it would appear that the contribution from *neutral* Higgs exchanges can be ignored since – in the absence of FCNC – it is proportional to the cube

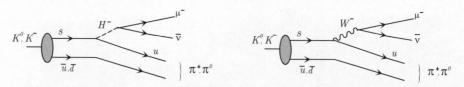

Fig. 19.5.   A diagram which gives rise to muon transverse polarization in $K \to \pi\mu\nu$.

of the u and d quark masses. Yet it induces a $G^3\tilde{G}$ term that could conceivably generate d_N as large as $\mathcal{O}(10^{-25})$ e cm! The reason for this relatively large effect is quite interesting [317]: this contribution is related – via the anomaly – to the nucleon mass rather than the current quark masses m_q; thus it will not vanish in the chiral limit $m_q \to 0$.

It should be noted that all contributions driven by such gluonic operators involve more than one quark at a time.

(d) *Colour-electric dipole moments*

Once it was realized that diquark effects can provide the leading contribution, many more such mechanisms were found. Particularly promising seems to be the so-called quark colour magnetic moment operator shown in Fig. 19.4: it represents an electroweak two-loop effect and leads to [318, 319]

$$d_N \sim \mathcal{O}(10^{-26}) \, \text{e cm}, \tag{19.40}$$

which is close to the present upper bound.

(e) *Electric dipole moment of electrons* Replacing the gluons in the colour-electric dipole operator by electroweak bosons and attaching it to a lepton, we obtain a two-loop contribution to the electron EDM, yielding [318, 319, 320]

$$d_e \sim \text{several} \times 10^{-27} \, \text{e cm}, \tag{19.41}$$

which is just below the present experimental bound $d_e = (-0.3 \pm 0.8) \cdot 10^{-26}$ e cm.

$$K \to \mu\nu\pi.$$

The transverse polarization of muons in $K_{\mu3}$ decay has been derived in Sec. 7.6.2. The two diagrams for $K^+ \to \mu^+\nu\pi^0$ giving rise to this **CP** violating effect are shown in Fig. 19.5. The amplitude for the charged Higgs exchange is given by

$$\mathcal{M} = -\frac{G_F}{\sqrt{2}} \sum_{i=1}^{2} \langle \pi^0 | m_s \frac{\alpha_i^* \beta_i}{M_{H_i}^2} \bar{s}(1-\gamma_5)u\bar{\nu}(1+\gamma_5)\mu | K^+ \rangle + \mathcal{O}(m_u). \tag{19.42}$$

Models with **CP** violation in the Higgs sector can yield a 'sizeable' $P_\perp(\mu)$ due to interference between the Higgs exchange – $F_S, F_P - -$ and the W exchange amplitude – F_{v_1}, F_A, see Eq. (7.55). 'Sizeable', however, does not mean large, since we are dealing here with *direct* **CP** violation. A rather model independent estimate on how large such an effect could be is obtained from the present bound on ϵ'/ϵ_K:

$$P_{\perp}(\mu) \leq 20 \cdot (\epsilon'/\epsilon_K) \cdot \epsilon_K \leq 10^{-4}, \tag{19.43}$$

where the factor 20 allows for the fact that ϵ'/ϵ_K is 'accidentally' suppressed by the $\Delta I = 1/2$ rule: $\omega \equiv (\text{Re}A_2/\text{Re}A_0) \simeq 1/20$. While this is much larger than anything that could be obtained in the KM ansatz, it is disappointingly small. Yet there is a relevant loophole in this generic argument: it is conceivable that the coupling of charged Higgs fields to *leptons* is strongly enhanced. As mentioned before, this happens when v_3 – the VEV of the Higgs field that couples to leptons only and gives them their mass – is greatly reduced relative to v_1, the VEV of the Higgs field coupling to the up-type quarks. In particular for $v_1/v_2 \geq \mathcal{O}(10)$ we could have [321]

$$P_{\perp}(\mu) \sim \mathcal{O}(10^{-3}), \tag{19.44}$$

without violating any constraints inferred from ϵ_K, ϵ' and d_N. What is even more striking: the forces leading to Eq. (19.44) are insignificant in $K_L \to \pi\pi$!

Since such a scenario is consistent with present phenomenology, it serves as a useful imagination stretcher – a commodity not to be belittled. For it illustrates that a detailed analysis of $K \to \mu\nu\pi$ could reveal a source of **CP** violation that escapes detection in $K \to 2\pi, 3\pi$. This level is within reach of the next round of experiments; yet ultimately it is highly desirable to reach the 10^{-4} level.

19.3 CP phenomenology with heavy fermions

19.3.1 Models with FCNC

Such models in general contain three-level operations that change the flavour quantum numbers by two units:

$$\Delta S = 2, \quad \Delta C = 2 \text{ and } \Delta B = 2. \tag{19.45}$$

$\Delta B = 2$ transitions

Rather straightforward scaling arguments suggest [322] that if Higgs induced FCNC provide a sizeable fraction of ϵ_K, they could quite naturally have a significant impact on **CP** violation in B^0 decays as well. This might sound surprising since $\Delta M(B_d)$ is a factor of 100 larger than ΔM_K, and furthermore $\text{Im}M_{12}(K) \ll \text{Re}M_{12}(K)$. However, the strength of the Higgs couplings increases with the mass of the quarks involved, as illustrated by the following generic example.

Assume tree-level FCNC to be the only source of **CP** violation in $K^0 - \overline{K}^0$ and $B^0 - \overline{B}^0$ oscillations with a more or less universal phase; the **CP** *even* parts of $M_{12}(K)$ and $M_{12}(B)$, on the other hand, receive SM contributions as well. The phases in $M_{12}(K)$ and $M_{12}(B)$ can then be related to each other [322][4]: With the notation given in Eq. (7.28),

$$\epsilon_K = \frac{1}{2}\left(1 - \frac{q}{p}\right) = \frac{i}{2}\phi\left(\frac{1}{1 - \frac{i}{2}r}\right), \tag{19.46}$$

$$\phi_K \equiv \frac{\text{Im}\langle K^0|\mathcal{H}_{\text{Higgs}}(\Delta S = 2)|\overline{K}^0\rangle}{|M_{12}(K)|} \simeq 6.5 \cdot 10^{-3} \tag{19.47}$$

from ϵ_K, and thus

$$\begin{aligned}
\phi_B &\equiv \frac{\text{Im}\langle B^0|\mathcal{H}_{\text{Higgs}}(\Delta B = 2)|\overline{B}^0\rangle}{|M_{12}(B)|} \\
&\simeq \frac{\text{Im}\langle B^0|\mathcal{H}_{\text{Higgs}}(\Delta B = 2)|\overline{B}^0\rangle}{\text{Im}\langle K^0|\mathcal{H}_{\text{Higgs}}(\Delta S = 2)|\overline{K}^0\rangle} \cdot \frac{\Delta M_K}{\Delta M_B} \cdot \sin\theta_K \\
&= \frac{B_B F_B^2}{B_K F_K^2} \cdot \left(\frac{M_B}{M_K}\right)^3 \sin\theta_K \sim 0.1 \ .
\end{aligned} \tag{19.48}$$

As seen from Eq. (11.49) and Eq. (11.50), the time dependent **CP** asymmetry is proportional to

$$As(f) \simeq \phi_B \cdot \eta_f, \tag{19.49}$$

where η_f denotes the **CP** parity of the final state f; i.e., all B^0 decays into final states that are **CP** eigenstates have to exhibit a **CP** asymmetry of equal size and a *sign* that is determined by their **CP** parity. In this superweak scenario we expect $\mathcal{O}(10\%)$ **CP** asymmetries in B decays and

$$As(\psi K_S) = -As(\pi^+\pi^-). \tag{19.50}$$

$\Delta C = 2$ transitions

Applying the scaling argument used in Eq. (19.48) to D^0 decays, we find

$$\phi_D \simeq \frac{B_D F_D^2}{B_K F_K^2} \cdot \left(\frac{M_D}{M_K}\right)^3 \cdot \sin\theta_K \sim 0.35 \cdot \frac{\Delta M_K}{\Delta M_D}. \tag{19.51}$$

$D^0 - \overline{D}^0$ oscillations can be produced by $\Delta C = 2$ tree level FCNC; the present bound on their strength is

$$\Delta M_D \leq 0.1 \cdot \Gamma_D \ . \tag{19.52}$$

Then time-dependent **CP** asymmetries, analogous to the one defined above for

---

[4] We have adopted here a phase convention such that the $\Delta S = 1$, $\Delta C = 1$ and $\Delta B = 1$ amplitudes contain *no weak* phase. This is possible in a superweak scenario.

B^0 decays, could emerge in decays like $D^0 \rightarrow K^+K^-$, $\pi^+\pi^-$ with

$$As(K^+K^-) \cdot \sin\Delta M_D t \sim 10^{-3} \cdot (t/\tau_D), \tag{19.53}$$

i.e., conceivably a few $\times 10^{-3}$ effect.

A considerably larger asymmetry would arise in the mode $D^0 \rightarrow K^+\pi^-$ since it is doubly Cabibbo suppressed *without* $D^0 - \overline{D}^0$ oscillations. The **CP** asymmetry in this channel has been written down explicitly in Eq. (15.7) and Eq. (15.8). Computing the magnitude of the asymmetry using Eq. (19.51) is left as an exercise.

19.3.2 *Models without FCNC*

Beauty decays

The prospects are not overly promising that such models could have a noticeable impact on **CP** asymmetries in B decays. A priori they could generate **CP** violation both in $\Delta B = 1$ and $\Delta B = 2$ transitions. Yet unless there are unusually enhanced Yukawa couplings of charged Higgs fields to quarks, such an effect would emerge on the percent level only for intermediate Higgs masses. Furthermore, these models are quite unlikely to provide the major source for ϵ_K; therefore whatever is behind $K_L \rightarrow \pi\pi$ could contribute to **CP** asymmetries in B decays as well, and possibly with a bigger weight, thus obscuring the effects of Higgs exchange. The most promising decay for such Higgs driven effects to emerge is in $B_s \rightarrow \psi\phi$ – within the KM ansatz we expect less than a 2% asymmetry there – and possibly rare decays like $B_d \rightarrow K\pi$, $K\rho$, etc.

Production and decay of top quarks and τ leptons

As explained in Chapter 15, production and decay of top quarks and τ leptons provide unfavourable environments for the KM mechanism to generate observable **CP** violation. There is a glimmer of hope, though, that Higgs dyamics with its enhanced Yukawa couplings to these heavy fermions could generate observable asymmetries there. In particular:

- A difference

$$\sigma(e^+e^- \rightarrow t_L\bar{t}_L) \neq \sigma(e^+e^- \rightarrow t_R\bar{t}_R) \tag{19.54}$$

could arise of order 0.1 % that can be probed by comparing the energy distributions of charged 'isolated' leptons coming from W decays:

$$\frac{d}{dE_l}\Gamma(t_L \rightarrow W^+b \rightarrow l^+X) \neq \frac{d}{dE_l}\Gamma(t_R \rightarrow W^+b \rightarrow l^+X). \tag{19.55}$$

- We can search the final state in $e^+e^- \rightarrow t\bar{t}H^0$ for **CP** odd correlations.

Details can be found in Chapter 15.

19.4 The pundits' résumé

It might appear at first sight that Higgs based models of **CP** violation represent a closed chapter of at best academic interest. (i) Models *without* FCNC built in are quite unlikely to provide a major contribution to ϵ_K. For otherwise we should have already seen non-vanishing values for ϵ' and/or d_N; furthermore charged Higgs bosons very probably should have been observed directly by now. (ii) Models *with* FCNC would seem to possess an ether-like elusiveness: the mass of neutral Higgs states with flavour-changing couplings have to exceed 1 TeV. Thus no *direct* **CP** violation would be observable in strange or beauty decays. In addition, these superheavy Higgs bosons would escape detection even at the LHC.

However, such a conclusion would miss two important points:

- The Higgs sector – its structure, its origin and even its mass scale – is the least understood part of the SM. No argument beyond that pragmatic one of convenience has so far been found in favour of limiting ourselves to the simplest possible ansatz. An elegant scheme like supersymmetry actually requires the existence of at least two different Higgs doublet fields, as discussed in the next chapter. Almost as soon as we go beyond the minimal Higgs sector we encounter additional sources of **CP** violation – whether they are 'welcome', 'needed' or not.

- Non-minimal Higgs dynamics has all the ingredients to provide the by far dominant source of d_N and d_e on a level that is observable, namely $d_N, d_e \geq 10^{-28}$ e cm; they could well lead to a transverse polarization $P_\perp$ of muons in $K \to \mu\nu\pi$ with $P_\perp \geq 10^{-4}$ and they might generate observable **CP** asymmetries in the production and decay of top *quarks*. Superweak scenarios, on the other hand, can modify significantly the large **CP** asymmetries in B decays that are expected within the KM ansatz; they also lead to **CP** asymmetries in D decays that could be as 'large' as 0.5 %. To make it quite explicit: a dedicated search for

  – electric dipole moments for neutrons, electrons and (heavy) atoms

  and detailed **CP** studies of

  – the decays of beauty and charm hadrons;
  – the muon polarization in $K \to \mu\nu\pi$ and
  – the production and decay of top quarks

  are *meaningful even – actually in particular – if Higgs exchanges are rather irrelevant in* $K_L \to \pi\pi$.

There is a dual relationship between the 'mysterious' phenomenon of **CP** violation and the 'obscure' Higgs sector:

(A) **CP** violation can serve as a most sensitive probe of the Higgs sector. As it happened with $K^0 - \overline{K}^0$ oscillations yielding the first evidence for the existence of heavy quark families, it is quite conceivable that the physical existence of Higgs fields will be first inferred from **CP** studies as outlined above. Even some of their specific properties like masses and mixing angles might be deduced from a *comprehensive* phenomenological analysis.

(B) Finding new manifestations of **CP** violation beyond ϵ_K will *by itself* not lead to a profound understanding of this fundamental property of nature. On the other hand, it is very hard – if not even outright impossible – to understand a fundamental phenomenon of which we have hardly caught a glimpse. It will be essential to prod nature into providing us with more perspectives. Higgs based models of **CP** violation at the very least can serve us well as an imagination stretcher for devising new search strategies.

Lastly, the following should be kept in mind: as described before, understanding the baryon number of the universe as a dynamically generated quantity rather than as a preset input parameter represents a fascinating intellectual challenge. As first realized by A. Sakharov, one of the essential elements of any answer to this challenge is the presence of **CP** violation. It has been suggested – as discussed in some detail in Chapter 21 – that **CP** violation that enters through Higgs dynamics is 'best' suited to generate the baryon number of today's universe – at the electroweak scale!

Problems

19.1   Consider a Higgs potential given by

$$V = -\mu^2 \sum_i |\phi_i|^2 + \lambda \sum_i |\phi_i|^4$$
$$+ c \left[(\phi_1^* \phi_2)(\phi_1^* \phi_3) + (\phi_2^* \phi_3)(\phi_2^* \phi_1) + (\phi_3^* \phi_2)(\phi_3^* \phi_1) \right]$$
$$+ \text{h.c.} \tag{19.56}$$

If the coefficients μ, λ and c are real, then V is certainly **CP** symmetric – but what about its minima? Show that for the Higgs self-coupling $c > 0$,

$$|\alpha_2 - \alpha_1| = |\alpha_3 - \alpha_2| = |\alpha_1 - \alpha_3| = \frac{2\pi}{3} ; \tag{19.57}$$

and the vacuum is not **CP** invariant.

19.2   Starting from Eq. (19.22), show that the charged Higgs interaction to quarks can be written as:

$$\mathcal{L}_{Y,ch} = \frac{g}{M_W} \left[e^{-i\lambda_1} \overline{U}_{Li}^m V_{CKM} \mathcal{M}_D^{\text{diag}} D_{Rj}^m \rho_1^+ + e^{-i\lambda_2} \overline{D}_{Li}^m V_{CKM}^\dagger \mathcal{M}_U^{\text{diag}} U_{Rj}^m \rho_2^- \right] \tag{19.58}$$

19.3   Derive Eq. (19.30) and Eq. (19.31). First show that $M_{12} = \frac{M_{12|SD}}{1-D} =$

$\frac{1}{D}M_{12}|_{LD}$. Then show that

$$\frac{\text{Im } M_{12}}{\text{Re } M_{12}} = (1 - D)\epsilon_m + D(-2\xi_0 + \chi), \qquad (19.59)$$

where $\chi = \arg M_{12}|_{LD} + 2\xi_0$. This leads to Eq. (19.30).

19.4 Using the result of Sec. 7.6.2 and Eq. (19.42), compute the transversal polarization of μ in $K \to \pi\mu\nu$.

19.5 Compute the order of magnitude of **CP** asymmetry for the doubly Cabibbo suppressed decays $D \to K^+\pi^-$ and $\overline{D} \to K^-\pi^+$ using Eq. (19.51).

20

Providing shelter for Higgs dynamics –
CP violation from supersymmetry

Our consideration of alternatives to the KM ansatz based on left–right models was driven by two complementary goals, namely to realize **CP** invariance in a spontaneous fashion and to have the dynamics subjected to a higher degree of symmetry. The motivation for analyzing non-minimal Higgs dynamics was much less profound: since no rationale more compelling than simplicity has emerged for limiting ourselves to minimal Higgs dynamics, we should be obliged to look beyond a minimalistic version – even if it served only as an imagination stretcher.

The Higgs sector is quite commonly perceived as the product of some effective, yet ultimately unsatisfactory, theoretical engineering. Two types of scenario have been suggested to provide a more appealing framework:

(A): Higgs fields are composites rather than elementary and represent an effective description of some unknown underlying dynamics. Technicolour models are one implementation of this scenario that used to be quite popular. Few definite statements can be made in such models. Yet it would be miraculous if a *minimal* Higgs sector emerged, and extra sources of **CP** violation are likely to surface following the classification given in Chapter 19.

(B): There is one very elegant theoretical scheme that provides a natural habitat for scalars – namely supersymmetry (SUSY). Since we consider it so attractive and since something definite can be said in that context, we will describe and analyse this scenario explicitly.

20.1 The virtues of SUSY

SUSY is the ultimate symmetry, it forms a bridge to gravity and it alleviates several vexing theoretical problems.

- The Coleman–Mandula theorem [323] proves, on very general grounds, that all possible symmetry groups $\mathscr{S}$ in quantum field theories can be expressed as the direct product of the Poincaré group $\mathscr{P}$ and an internal symmetry group G:

$$\mathscr{S} = \mathscr{P} \otimes G. \tag{20.1}$$

  This means that no symmetry can connect states of different spins; i.e., each irreducible representation of $\mathscr{S}$ contains only states of the same spin.

331

There is a loophole in the proof of this theorem, though: it admits only *bosonic* operators as generators of a symmetry. It was realized later that *fermionic* operators can also generate a symmetry group and that they relate fermions and bosons to each other. It thus represents the 'ultimate' symmetry of a quantum field theory and was aptly named supersymmetry.

- Once SUSY is implemented as a *local* gauge symmetry, coordinate covariance has become local as well – i.e., general relativity has to emerge. That is why local SUSY is referred to as *supergravity*. SUSY also forms an integral part of superstring theories.

- As described before, Higgs dynamics plays an essential role in theoretical engineering, since it allows us to realize (gauge) symmetries *spontaneously*. This happens through neutral Higgs fields ϕ acquiring VEVs: minimizing a potential of the type $V_H = M_\phi^2 |\phi|^2 + \lambda |\phi|^4$ with $M_\phi^2 < 0$ leads to $\langle\phi\rangle = \sqrt{-M_\phi^2/2\lambda}$. From the size of the Fermi constant G_F we infer $\langle\phi\rangle \simeq 174$ GeV for the neutral component of the $SU(2)_L$ doublet scalar; very roughly, we then expect $|M_\phi| \sim \mathcal{O}(100)$ GeV as well. However, *scalar* masses get *quadratically* renormalized: $\Delta M_\phi^2 \propto \Lambda_{UV}^2 + \cdots$ The ultraviolet cut-off Λ_{UV} represents the high sensitivity of M_ϕ to quantum corrections driven by New Physics operating at high energy scales. While $|M_\phi| \sim \mathcal{O}(100)$ GeV can be *arranged*, it requires extreme *fine tuning* since quantum corrections drive it to the high scales where New Physics – say Grand Unified Theories (GUT) or quantum gravity – enters: v_{GUT} or v_{Planck}. This is often referred to as the problem of 'unnatural' gauge hierarchies. SUSY per se does not solve this problem, but makes it much more tractable: for once a Higgs potential is chosen such that it yields the required large ratio to tree-level – i.e., classically – we can invoke SUSY to stabilize this ratio against *quantum* corrections. This is referred to as *the non-renormalization theorems of SUSY*. Light scalars then become natural by riding piggy-back on the shoulders of light fermions.

- With the *quadratic* renormalization of Higgs mass terms removed by SUSY, electroweak symmetry breaking can be induced *radiatively* if the top quark is sufficiently massive, since its Yukawa coupling g_t^Y then dominates the renormalization of the Higgs mass: $\partial M_\phi^2/\partial\log\mu = 3(g_t^Y)^2 m_t^2/8\pi^2 + \cdots$ This happens for $m_t \geq 160$ GeV or so. This feature had been noted [324] when such a mass was perceived as extravagantly high.

For SUSY to exist each state has to possess a superpartner, i.e., a state whose spin differs by half a unit[1]. For *manifest* SUSY the superpartner must be mass

---

[1] To be more precise: we do not consider *non*linear realizations of SUSY where the superpartners can be composites.

degenerate. Not a single such superpartner has been observed, i.e., SUSY breaking has to be implemented. This can be achieved by arranging for the superpartners to have masses no lighter than 100 GeV; in that case they would have escaped direct experimental detection. On the other hand, not all of their masses can exceed 1 TeV if SUSY is called upon, as sketched above, to stabilize the huge hierarchy in the GUT scales relative to the weak scales.

Intellectual honesty compels us to concede that at the moment of writing this book there is no direct evidence for SUSY being realized at all in nature: no two observed states can be considered superpartners. Even the common bon mot that half the SUSY states have been found is not quite true. For the Higgs sector in SUSY is considerably more complex than in the SM, and even the SM Higgs state has not been observed. Yet all of that can change very quickly, in particular since various indirect lines of reasoning suggest that SUSY partners enter below the TeV scale or so. Real SUSY quanta could then be produced at the FNAL and LHC colliders and high energy e^+e^- linear colliders, at present on the drawing board. Yet in the spirit of the whole book, it is reasonable to ask whether the existence of the SUSY degrees of freedom could not be 'felt' through the impact of their quantum corrections, in particular, in transitions that are highly suppressed like $K^0 - \overline{K}^0$ and $B^0 - \overline{B}^0$ oscillations. We will discuss in particular how *virtual* SUSY states will affect the CP phenomenology and rare decays.

With the increased number of layers in the dynamical structure there are many additional gateways through which CP violation can enter, in particular since SUSY has to be broken 'softly', i.e., through terms in the Lagrangian of operator dimension less than four. Three *qualitatively* new features emerge here:

- **CP** odd couplings can now arise both in chirality conserving and changing couplings.

- **CP** violation can enter flavour-diagonal couplings.

- Intriguing connections between **CP** violation and the flavour problem can be formulated.

To be able to go beyond generalities we will sketch the minimal version of SUSY still consistent with nature – the minimal supersymmetric Standard Model (MSSM) – and use it as a reference point to make generic comments on more general cases.

20.2 Low-energy SUSY

We can be brief in outlining the general structure of SUSY and the MSSM, since the details can be found in several excellent reviews like Refs. [325, 326]. We introduce R parity – a multiplicative quantum number – assigning it a value +1

Table 20.1. Superfields of MSSM with $i = 1, 2, 3$ as family index.

superfield	colour	isospin	hypercharge	lepton number	baryon number
Q_i	3	$\frac{1}{2}$	$\frac{1}{6}$	0	$\frac{1}{3}$
U_i^c	3	0	$-\frac{2}{3}$	0	$\frac{1}{3}$
D_i^c	3	0	$\frac{1}{3}$	0	$\frac{1}{3}$
L_i	0	$\frac{1}{2}$	$-\frac{1}{2}$	1	0
E_i^c	0	0	1	-1	0
H_1	0	$\frac{1}{2}$	$-\frac{1}{2}$	0	0
H_2	0	$\frac{1}{2}$	$+\frac{1}{2}$	0	0

for ordinary fields (quarks, leptons, gluons, gauge bosons, Higgs, graviton etc.) and -1 for their superpartners (squarks, sleptons, gluinos, gauginos, higgsinos, gravitinos, etc.). The theory is most concisely formulated in terms of *superfields* which (for $N = 1$ SUSY) combine fields differing by half a unit in their spin, namely $J = 0$ and $J = 1/2$ – *chiral* or *matter* superfields – or $J = 1/2$ and $J = 1$ – *gauge* or *vector* superfields; there is also a superfield combining the $J = 2$ graviton with the $J = 3/2$ gravitino. These superfields are denoted by their $R = +1$ component. *Non*-gauge interactions are introduced through a superpotential G coupling up to three superfields together. Constructing a SUSY model thus requires five steps:

- adopt a gauge group;

- choose superfields with the appropriate gauge quantum numbers;

- formulate the gauge interactions;

- construct the superpotential with them and, finally,

- implement SUSY breaking.

20.2.1 The MSSM

The superfield content of MSSM is given in Table 20.1. Each field of the SM is extended into a superfield. There is one non-trivial element: at least two *distinct* Higgs superfields H_1 and H_2 are needed, since otherwise chiral anomalies arise due to higgsino loops. The main task consists of deciding on the superpotential and on SUSY breaking.

Non-gauge interactions are introduced through the superpotential G,

$$G = \mu H_1 H_2 + Y_{ij}^u Q_i H_2 U_j^c + Y_{ij}^d Q_i H_1 D_j^c + Y_{ij}^l L_i H_1 E_j^c . \tag{20.2}$$

The dimensionless numbers $Y_{ij}^{u,d,l}$ contain – among other things – the Yukawa couplings of the ordinary fermion fields. With superfields carrying dimension one (in mass units) and G thus dimension three, we see that the mixing parameter μ has dimension one as well.

The interactions among the scalar fields can be extracted from G in a straightforward way and expressed through two types of ordinary potentials (of dimension four):

$$
\begin{aligned}
V_F &= |\mu h_1 + Y_{ij}^u \tilde{Q}_i \tilde{U}_j^c|^2 + |\mu h_2 + Y_{ij}^d \tilde{Q}_i \tilde{D}_j^c + Y_{ij}^l \tilde{L}_i \tilde{E}_j^c|^2 \\
&+ \sum_j |Y_{ij}^u \tilde{Q}_i h_2|^2 + |Y_{ij}^d \tilde{Q}_i h_1|^2 + |Y_{ij}^l \tilde{L}_i h_1|^2 \\
&+ \sum_i |Y_{ij}^u h_2 \tilde{U}_j^c|^2 + |Y_{ij}^d h_1 \tilde{D}_j^c|^2 + |Y_{ij}^l h_1 \tilde{E}_j^c|^2
\end{aligned}
$$

$$
\begin{aligned}
V_D &= \frac{g'^2}{2} \left(\frac{1}{6} \tilde{Q}_i^\dagger \tilde{Q}_i - \frac{2}{3} \tilde{U}_i^{c*} \tilde{U}_i^c + \frac{1}{3} \tilde{D}_i^{c*} \tilde{D}_i^c - \frac{1}{2} \tilde{L}_i^\dagger \tilde{L}_i + \tilde{E}_i^{c*} \tilde{E}_i^c + \frac{1}{2} h_1^* h_1 - \frac{1}{2} h_2^* h_2 \right)^2 \\
&+ \frac{g^2}{8} \left(\tilde{Q}_i^\dagger \vec{\tau} \tilde{Q}_i + \tilde{L}_i^\dagger \vec{\tau} \tilde{L}_i + h_1^\dagger \vec{\tau} h_1 + h_2^\dagger \vec{\tau} h_2 \right)^2 \\
&+ \frac{g_s^2}{8} \left(\tilde{Q}_i^\dagger \vec{t} \tilde{Q}_i - \tilde{U}_i^{c*} \vec{t} \tilde{U}_i^c - \tilde{D}_i^{c*} \vec{t} \tilde{D}_i^c \right)^2,
\end{aligned}
\tag{20.3}
$$

with $\tilde{Q}$, $\tilde{U}$, $\tilde{D}$, $\tilde{L}$ and $\tilde{E}$ now representing the squark and slepton fields and h_1 and h_2 the Higgs fields rather than the full superfields; the $\vec{\tau}$ and $\vec{t}$ denote the Pauli and Gell-Mann matrices, respectively.

SUSY obviously cannot be a manifestly realized symmetry. *If* we want to invoke SUSY to shield Higgs masses against getting *quadratically* renormalized, as mentioned before, we infer that the spectrum of SUSY partners should start below the TeV scale. A *spontaneous* breaking of SUSY could be achieved through a conventional Higgs mechanism in V_D and V_F. However, that would lead to a phenomenological conflict in the relationship between the quark and squark masses. For a sum rule given by the supertrace

$$
\text{Str } M^2 = \sum_J (-1)^{2J} M_J^2 = 0,
\tag{20.4}
$$

which holds trivially in *manifest* SUSY, is *not* modified when the spontaneous breaking of SUSY is driven by V_D or V_F (see Problem 20.1). As the sum of all fermion masses squared has to equal that of all boson masses squared, it predicts that some squarks and sleptons must be light[2]; superpartners should then have been found by now. Eq. (20.4) holds in all models in which SUSY is broken at the *tree* level.

---

[2] We might entertain the idea that gauginos, being much heavier than the corresponding gauge bosons, would allow some sfermions to be lighter than their fermions without violating Eq. (20.4). Yet it turns out that such a trade-off is not possible: Eq. (20.4) has to be satisfied sector by sector.

To avoid this phenomenological conflict we introduce explicit *soft* breaking terms into the MSSM Lagrangian, where all dimension four operators obey SUSY:

$$\mathscr{L} = \mathscr{L}_{\text{SUSY}}(d \leq 4) + \mathscr{L}_{\text{soft}}(d \leq 3). \tag{20.5}$$

This is seen merely as an 'engineering' device to generate higher masses for all SUSY partners. We have actually in mind that New Physics *beyond* MSSM dynamically generates such soft SUSY breaking. More specifically, in addition to the *visible* or low-energy sector described by MSSM, we envision a *hidden* sector characterized by very high scales where SUSY breaking originates, which then is transmitted to the MSSM through *flavour-blind* interactions. Such an ansatz effectively disassociates the scale at which the intrinsic SUSY breaking occurs – Λ_{SSB} – from the one that gives rise to the different flavours – Λ_{Flav} – with $\Lambda_{SSB} \ll \Lambda_{\text{Flav}}$. SUSY breaking masses emerge as flavour singlets; squark masses are then degenerate up to small corrections of the order of the corresponding quark masses. As described below, this severely suppresses flavour-changing neutral currents (FCNC) in the SUSY sector as well; i.e., a restrictive super-GIM mechanism arises naturally. A lively debate on how this can happen has crystalized around two basic scenarios:

1 SUSY breaking is *gravity* mediated and enters $\mathscr{L}_{\text{soft}}(d \leq 3)$ through soft mass terms, for which we find

$$M_{\text{soft}} \sim \mathcal{O}\left(\frac{\Lambda_{SSB}^2}{M_{\text{Planck}}}\right), \tag{20.6}$$

implying $\Lambda_{SSB} \sim \mathcal{O}(10^{11})$ GeV to yield $M_{\text{soft}} \sim 1$ TeV.

2 It can be *gauge* mediated instead, with the soft mass terms being generated by virtual exchanges of messenger fields from the hidden sector which possess $SU(3)_C \times SU(2)_L \times U(1)$ couplings:

$$M_{\text{soft}} \sim \mathcal{O}\left(\frac{\alpha_{\text{gauge}}}{4\pi} \frac{\Lambda_{SSB}^2}{M_{\text{mess}}}\right), \tag{20.7}$$

where α_{gauge} denotes the gauge coupling of the messenger fields and M_{mess} their typical mass scale. It turns out that $M_{\text{mess}} \sim \Lambda_{SSB} \sim \mathcal{O}(10^5)$ GeV or so is quite a natural value; i.e., the SUSY breaking occurs at much lower scales than in the gravity mediated case.

The phenomenological problem expressed by Eq. (20.4) is resolved in these schemes – at least for the time being – through the emergence of a sizeable term on the right-hand side. For example, when SUSY is implemented as a *local* symmetry, it requires a spin-3/2 fermion – called the gravitino – as the superpartner for the spin-2 graviton. One manifestation of SUSY breaking is the emergence of a finite gravitino mass leading to

$$\text{Str } M^2 = \sum_J (-1)^{2J} M_J^2 \sim m_{3/2}^2, \qquad (20.8)$$

since the graviton remains massless. Eq. (20.8) implies that superpartners of leptons and quarks are, in general, heavier than quarks and leptons characterized by $m_{3/2} \geq 100$ GeV or so.

Let us illustrate these general remarks by one example for a flavour-*blind* scenario. We make the following ansatz for soft SUSY breaking being generated at some high scale:

$$\mathcal{L}_{soft} = m_{3/2}^2 \sum_i |A_i|^2 + Am_{3/2} \left[Y_{ij}^u \tilde{Q}_i h_2^* \tilde{U}_j^c + Y_{ij}^d \tilde{Q}_i h_1 \tilde{D}_j^c + Y_{ij}^l \tilde{L}_i h_1 \tilde{E}_j^c + \text{h.c.} \right]$$

$$+ \quad Bm_{3/2}\mu h_1 h_2 + \frac{1}{2} M(\lambda_1 \lambda_1 + \lambda_2 \lambda_2 + \lambda_3 \lambda_3), \qquad (20.9)$$

where A_i denote scalar fields in general, $\tilde{Q}$, $\tilde{U}$, $\tilde{D}$, $\tilde{L}$ and $\tilde{E}$ sfermion fields, $h_{1,2}$ Higgs fields and $\lambda_{1,2,3}$ gaugino fields for the gauge groups $U(1)$, $SU(2)_L$ and $SU(3)_C$, respectively. The low energy effective Lagrangian is then obtained by running the parameters down to the electroweak scale, as described below.

In addition to the SM parameters, there are *five new* classes of dimensional parameters in MSSM: μ, describing the mixing between the two Higgs superfields, plus four more representing SUSY breaking, namely the gravitino and gaugino masses, $m_{3/2}$ and M, respectively, a higgsino mixing term, $Bm_{3/2}$, and quark–squark–higgsino couplings, $Am_{3/2}$. For simple cases $B = A - 1$ and the number of parameters is reduced to four. In addition, we must require that the electroweak symmetry breaking occurs at the right energy scale: $4M_W^2/g^2 = v_1^2 + v_2^2$, with

$$\frac{v_2}{v_1} \equiv \tan\beta \qquad (20.10)$$

being a new parameter.

To summarize this lightning review:

- There is a well defined minimal extension of the SM, the MSSM. It contains at least two Higgs doublet superfields, H_1 and H_2.

- SUSY cannot be a manifest symmetry. Realizing it spontaneously *within* MSSM leads to phenomenological disaster. Instead, we introduce *explicit soft* SUSY breaking terms. In the simplest versions they are flavour singlets; yet – as described below – they could contain a non-trivial flavour structure.

- It is understood that these explicit breaking terms put in by hand are the product of dynamics *beyond* MSSM operating at much higher scales and handed down to the low energy sector in different ways.

20.2.2 *Non-minimal SUSY models*

As just described, New Physics beyond MSSM operating in a hidden sector is invoked to generate the soft breaking terms of MSSM. In this situation minimality is not necessarily a virtue or even unambiguously defined. It is intriguing to consider scenarios with $\Lambda_{\text{Flav}} \leq \Lambda_{SSB}$, as suggested if there is a connection between SUSY breaking and the physics of flavour generation. The communication between the high energy and low energy domains no longer proceeds in a flavour blind manner and the soft breaking terms can be expected to exhibit a non-trivial flavour structure. FCNC will arise; the resulting phenomenology is less conservative and – at least potentially – much richer than in the previous scenarios. More sources of **CP** violation become relevant and there is a serious danger of violating existing bounds – unless some *approximate symmetry* is invoked to suppress the size of **CP** phases.

20.3 Gateways for CP violation

SUSY models in general introduce a host of possible sources for **CP** violation, entering through

- the superpotential[3] and/or

- the soft breaking terms.

They generate phases that at low energies can surface in charged as well as neutral current couplings and in Higgs interactions. Their actual pattern depends on the specifics of the physics driving SUSY breaking.

20.3.1 *A first glance at* **CP** *phases in MSSM*

Inspecting Eq. (20.9) as derived from a gravity induced breaking pattern allows us to arrive at the following observations:

1 Hermiticity of the potential requires $m_{3/2}^2$ to be real; $m_{3/2}$ can be made real by absorbing its remaining phase into the definition of A and B.

2 The gaugino mass term M can be chosen real by adjusting the phases of the gaugino fields λ_i.

3 The squark–quark–gluino coupling will be affected by such a choice of the gaugino phase.

4 We will also show that the phase associated with a linear combination of A and μ can be detected by experiments.

We will then analyze how these phases can affect the low-energy couplings.

---

[3] The irreducible phase contained in the Yukawa couplings of quarks that is present already in the SM is embedded here.

20.3.2 *Squark mass matrices*

The superpartners of left and right handed quarks are referred to as left and right handed squarks although the latter – being scalars – possess no chiral structure. Squark masses are greatly affected by the soft SUSY breaking. In particular the term in $\mathscr{L}_{\text{soft}}$ proportional to the trilinear scalar coupling A mixes left and right handed squarks. From Eq. (20.9) we can read off the squark mass matrices. For down-type squarks we find

$$V_{D-\text{mass}} = \begin{pmatrix} \tilde{D} \\ \tilde{D}^{c*} \end{pmatrix}^{tr} \begin{pmatrix} \tilde{\mathbf{M}}^2_{DLL} & \tilde{\mathbf{M}}^2_{DLR} \\ \tilde{\mathbf{M}}^{2\dagger}_{DLR} & \tilde{\mathbf{M}}^2_{DRR} \end{pmatrix} \begin{pmatrix} \tilde{D}^* \\ \tilde{D}^c \end{pmatrix} \tag{20.11}$$

$$\tilde{\mathbf{M}}^2_{DLL} = \left(m^2_{3/2} + (v^2_1 - v^2_2)\left(\frac{g'^2}{12} - \frac{g^2}{4}\right) \right)\mathbf{1} + \mathscr{M}_D \mathscr{M}^\dagger_D$$

$$\tilde{\mathbf{M}}^2_{DLR} = (A^* m_{3/2} + \mu^* \tan\beta)\mathscr{M}_D$$

$$\tilde{\mathbf{M}}^2_{DRR} = \left(m^2_{3/2} + (v^2_1 - v^2_2)\frac{g'^2}{6} \right)\mathbf{1} + \mathscr{M}^\dagger_D \mathscr{M}_D, \tag{20.12}$$

with $\mathscr{M}_D$ representing the mass matrix for down-type quarks. We have written the three generations of down squarks as six component vectors with $\tilde{q}^c_L = \tilde{q}^*_R$:

$$(\tilde{D}^{\text{tr}}, \tilde{D}^{c\ \text{tr}}) = (\tilde{d}_L, \tilde{s}_L, \tilde{b}_L, \tilde{d}^*_R, \tilde{s}^*_R, \tilde{b}^*_R). \tag{20.13}$$

Eq. (20.12) and its analogue for the up squark mass matrix constitute the super-GIM mechanism: the leading contribution to the squark masses is flavour independent, namely the gravitino mass, with the flavour *splittings* given by the masses for the corresponding quarks, namely the down-type quarks here. Let us say that $\mathscr{M}_D$ is diagonalized by

$$\mathscr{M}^{\text{diag}}_D = \mathbf{U}^D_L \mathscr{M}_D \mathbf{U}^{D,\dagger}_R. \tag{20.14}$$

Making a unitary transformation

$$\begin{pmatrix} \mathbf{U}^D_L e^{i\phi_A} & 0 \\ 0 & \mathbf{U}^D_R e^{-i\phi_A} \end{pmatrix} \tilde{\mathbf{M}}^2_D \begin{pmatrix} \mathbf{U}^{D\dagger}_L e^{-i\phi_A} & 0 \\ 0 & \mathbf{U}^{D\dagger}_R e^{i\phi_A} \end{pmatrix} = \begin{pmatrix} \hat{\mathbf{M}}^2_{DLL} & \hat{\mathbf{M}}^2_{DLR} \\ \hat{\mathbf{M}}^{2\dagger}_{DLR} & \hat{\mathbf{M}}^2_{DRR} \end{pmatrix}, \tag{20.15}$$

where the phase factor

$$\phi_A = \arg(A m_{3/2} + \mu \tan\beta) \tag{20.16}$$

is introduced to absorb the phase in the off-diagonal matrix elements, we obtain:

$$\hat{\mathbf{M}}^2_{DLL} = \left(m^2_{3/2} + (v^2_1 - v^2_2)\left(\frac{g'^2}{12} - \frac{g^2}{4}\right) \right)\mathbf{1} + (\mathscr{M}^{\text{diag}}_D)^2$$

$$\hat{\mathbf{M}}^2_{DRR} = \left(m^2_{3/2} + (v^2_1 - v^2_2)\frac{g'^2}{6} \right)\mathbf{1} + (\mathscr{M}^{\text{diag}}_D)^2$$

$$\hat{\mathbf{M}}^2_{DLR} = \left(|A| m_{3/2} + \mu^* \frac{v_1}{v_2} \right)\mathscr{M}^{\text{diag}}_D. \tag{20.17}$$

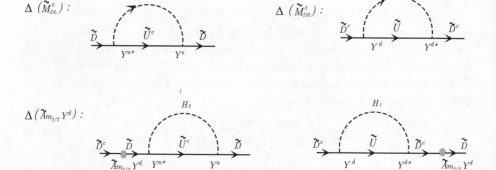

Fig. 20.1. Example of diagrams which drive the renormalization group equation for $\mathbf{M_U}$ and $\mathbf{M_D}$.

While this mass matrix mixes left and right handed squarks, it still is *diagonal* in flavour space. Now remember that Eq. (20.17) holds at some large unification scale. In the much lower energy region $\sim M_W$ and below, coupling constants acquire radiative corrections [327] shown in Fig. 20.1. They induce corrections to $\tilde{\mathbf{M}}_D^2$ proportional to $Y^U Y^{U\dagger}$, with Y^U denoting the Yukawa couplings of up-type quarks. While the precise effects depend on details of the SUSY model under study, we can draw some generic conclusions. At scale M_W the mass matrix becomes for small $\tan\beta$:

$$\tilde{\mathbf{M}}_D^2 = \begin{pmatrix} \tilde{\mathbf{M}}_{DLL}^{\mathrm{tree}} \tilde{\mathbf{M}}_{DLL}^{\mathrm{tree}\dagger} + c_1 \mathcal{M}_U \mathcal{M}_U^\dagger & A m_{3/2} \mathcal{M}_D (1 + \frac{c_2}{M_W^2} \mathcal{M}_U^\dagger \mathcal{M}_U) \\ A^* m_{3/2}(1 + \frac{c_2}{M_W^2} \mathcal{M}_U^\dagger \mathcal{M}_U) \mathcal{M}_D^\dagger & \tilde{\mathbf{M}}_{DRR}^{\mathrm{tree}\dagger} \tilde{\mathbf{M}}_{DRR}^{\mathrm{tree}} \end{pmatrix} \quad (20.18)$$

where we have kept only those contributions which give non-trivial modifications. The renormalization coefficients c_1 and c_2 can be computed from Ref. [327] within a given theory as functions of $t = \log(M_P/M_W)$ with M_P denoting the large scale where Eq. (20.9) holds; here we leave them as parameters.

Using Eq. (20.14) and Eq. (20.16) and the equivalent for up-type quarks, we perform a unitary transformation to find

$$\begin{pmatrix} U_L^{D\dagger} e^{i\phi_A} & 0 \\ 0 & U_R^{D\dagger} e^{-i\phi_A} \end{pmatrix} \tilde{\mathbf{M}}_D^2 \begin{pmatrix} U_L^D e^{-i\phi_A} & 0 \\ 0 & U_R^D e^{i\phi_A} \end{pmatrix} = \begin{pmatrix} \mathbf{W} & \mathbf{X} \\ \mathbf{Y} & \mathbf{Z} \end{pmatrix}, \quad (20.19)$$

where

$$\mathbf{W} = (\hat{\mathbf{M}}_{DLL}^{\mathrm{tree}})^2 + c_1 \mathbf{V}^\dagger (\mathcal{M}^{\mathrm{diag}})_U^2 \mathbf{V}$$

$$\mathbf{X} = |A| m_{3/2} \left(1 + \frac{c_2}{M_W^2} \mathbf{V}^\dagger (\mathcal{M}_U^{\mathrm{diag}})^2 \mathbf{V}\right) \mathcal{M}_D^{\mathrm{diag}}$$

$$\mathbf{Y} = |A| m_{3/2} \mathcal{M}_D^{\mathrm{diag}} \left(1 + \frac{c_2}{M_W^2} \mathbf{V}^\dagger (\mathcal{M}_U^{\mathrm{diag}})^2 \mathbf{V}\right)$$

$$\mathbf{Z} = (\hat{\mathbf{M}}_{DRR}^{\mathrm{tree}})^2, \quad (20.20)$$

with

$$\mathbf{V} = \mathbf{U}_L^U \mathbf{U}_L^{D\dagger} \tag{20.21}$$

being the CKM matrix.

A qualitatively new feature has emerged here due to the radiative corrections: because *in general* $\mathscr{M}_D$ *and* $\mathscr{M}_U$ *cannot* be diagonalized by the same unitary transformations, $\tilde{\mathbf{M}}_D^2$ and $\mathscr{M}_D$ cannot be diagonalized *simultaneously*. Expressing the squark–quark–gluino coupling defined for *flavour* eigenstates in terms of *mass* eigenstates will then

- reveal the emergence of *flavour changing neutral current interactions*

- in the *strong* dynamics sector, in the sense that they carry *strong* coupling strength α_S,

- the flavour structure of which is controlled by the *CKM matrix*!

To analyze the experimental consequences of the new flavour changing interactions induced by Eq. (20.18) it is convenient to work with a basis [328]

$$\begin{pmatrix} \mathbf{U}_L^D e^{i\phi_A} & 0 \\ 0 & \mathbf{U}_R^D e^{-i\phi_A} \end{pmatrix} \begin{pmatrix} \tilde{D}^* \\ \tilde{D}^c \end{pmatrix}, \tag{20.22}$$

leading to squark propagators being *not* diagonal: there is left–right squark mixing. These can be treated perturbatively using mass insertions:

$$\Delta_{LL}^2 = c_1 \mathscr{M}_U \mathscr{M}_U^\dagger$$

$$\Delta_{LR}^2 = A m_{3/2} \mathscr{M}_D \left(1 + \frac{c_2}{M_W^2} \mathscr{M}_U^\dagger \mathscr{M}_U \right)$$

$$\Delta_{RL}^2 = A^* m_{3/2} \left(1 + \frac{c_2}{M_W^2} \mathscr{M}_U^\dagger \mathscr{M}_U \right) \mathscr{M}_D^\dagger. \tag{20.23}$$

Thus we have learnt the following lesson: in MSSM we usually envision flavour-*blind* SUSY breaking to be generated at high energy scales. Evolving the effective dynamics down to low energies we find that

- FCNC emerge even in the *strong* sector described by gluino couplings;

- FCNC arise *radiatively* and thus are *reduced* considerably in strength;

- in general they do *not* conserve **CP**;

- **CP** violation occurs even in flavour *diagonal* transitions.

20.3.3 Beyond MSSM

As stated before, it is quite conceivable that the dynamics that drives SUSY breaking communicates with the low-energy softly broken SUSY model in a way that is flavour *specific* even before radiative corrections are included. Flavour changing neutral interactions will then arise that in general are not suppressed by a super-GIM mechanism and do not obey **CP** invariance. This opens the door to a very rich and multi-layered phenomenology; yet there is little we can say concretely beyond these generalities.

In Eq. (20.23) we introduced a parametrization of the FCNC and left–right mixing within MSSM in terms of an expansion in the matrices Δ_{LL}^2, Δ_{LR}^2 and Δ_{RL}^2. This procedure can be generalized for these non-minimal models by expanding the squark propagators in powers of

$$(\delta_{ij})_{LL}, \quad (\delta_{ij})_{LR}, \quad (\delta_{ij})_{RL}, \quad (\delta_{ij})_{RR}, \tag{20.24}$$

where $\delta \equiv \Delta^2/\tilde{m}^2$ with $\tilde{m}$ being an appropriate mass scale like the average squark mass. In other words: while in this treatment the quark–squark–gluino vertex conserves flavour, there is a mass insertion in the squark propagator that is *not* flavour diagonal – $\delta_{ij} \neq 0$ for $i \neq j$ – and can also mix left and right handed squarks.

On one hand, such representation allows us to express phenomenological bounds in a model independent way, while on the other, the parameters $(\delta_{ij})_{AB}$ are readily evaluated in a given model.

Within MSSM the quantities $(\delta_{ij})_{AB}$ contain an effective super-GIM filter reducing them below the 10^{-2} level or so without fine tuning. In non-minimal SUSY models, though, we have merely $|\delta| < 1$.

The story repeats itself for other quark–squark–gaugino couplings, albeit on a reduced numerical level.

20.4 Confronting experiments

The biggest *qualitative* change relative to the KM ansatz within SM is the emergence of **CP** violation in flavour-*diagonal* transitions, allowing electric dipole moments (and **CP** odd nuclear effects in general) to arise on an observable level.

20.4.1 Electric dipole moments

A tight experimental constraint can be placed on ϕ_A, the phase of the trilinear coupling A in $\mathscr{L}_{\text{soft}}$, as it generates a neutron electric dipole moment already on the one-loop level. The diagram shown in Fig. 20.2 gives (using, as before, the static approximation $d_N \simeq 2d_d + d_u$)

$$d_N = -\frac{4}{27} \frac{e\alpha_s}{\pi} \frac{M_{\tilde{g}}}{M_{\tilde{q}}^2} \operatorname{Im} [2(\delta_{11}^D)_{LR} + (\delta_{11}^U)_{LR}] F_1(x), \tag{20.25}$$

where

$$F_1(x) = \frac{1}{(1-x)^3} \left(\frac{1+5x}{2} + \frac{2+x}{1-x} \ln x \right), \quad x = \frac{M_{\tilde{g}}^2}{M_{\tilde{q}}^2}, \tag{20.26}$$

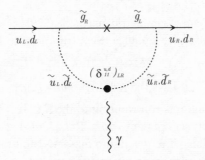

Fig. 20.2.   Feynman diagram which gives the neutron electric dipole moment.

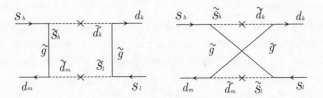

Fig. 20.3.   Example of diagrams which contribute to the $K - \overline{K}$ mass matrix.

where $M_{\tilde{g}}$ and $M_{\tilde{q}}$ denote the gluino and average squark mass, respectively. Since the dipole operator is chirality changing , a complex mixing between left and right handed squarks is needed to produce an effect: $\text{Im}\delta_{LR} \neq 0$. The experimental bound on d_N yields for $M_{\tilde{g}} \simeq M_{\tilde{q}}$

$$\text{Im}\,(\delta_{11}^{D})_{LR},\ \ \text{Im}\,(\delta_{11}^{U})_{LR} \leq \text{few} \times 10^{-6}. \qquad (20.27)$$

Within MSSM $(\delta_{11}^{D})_{LR} = (\delta_{11}^{U})_{LR} = A \cdot m_{3/2}m_d/M_{\tilde{q}}^2$; i.e., assuming $m_{3/2} \sim M_{\tilde{q}} \sim 300$ GeV as a benchmark, we infer from Eq. (20.27)

$$\arg(A) \leq \mathcal{O}(10^{-2}). \qquad (20.28)$$

While the super-GIM mechanism goes a long way towards explaining the minute size of $\text{Im}(\delta_{11})_{LR}$, the phase of A is small for $M_{\tilde{q}}$ well below 1 TeV. It is then tempting to conjecture the intervention of some symmetry to yield $\arg(A) \simeq 0$. In that case there is basically only one source for **CP** violation as in the SM, namely the irreducible KM phase (in addition to $\overline{\theta}_{QCD}$), yet it can enter through a new gateway – the renormalized squark mass matrices.

In non-minimal models *without* an effective super-GIM filter, the need for a symmetry suppressing new phases in the low energy sector appears compelling.

Of course you can also say that any improvement in experimental sensitivity for EDMs just might reveal a non-vanishing result!

Another view-point we might take is to say that the extra phases that appear in MSSM are maximal, and thus the experimental value on EDM leads to bounds on masses of supersymmetric partners [329]. For these considerations, other diagrams may also give significant contributions to EDM.

20.4.2 The $K^0 - \overline{K}^0$ mass matrix

An example of new diagrams contributing to the $K^0 - \overline{K}^0$ mass matrix is given in Fig. 20.3. For a complete set, we refer you to Ref. [330]. The resulting contributions are complicated functions of δ, the gluino mass $M_{\tilde{g}}$, and the average squark mass $\tilde{m}$. Large cancellations can occur among the various terms. Nevertheless, it is instructive to note that

$$|(\delta_{12}^D)_{LL,LR,RR}| \leq \mathcal{O}(10^{-2}), \quad \mathrm{Im}\,(\delta_{12}^D)_{LL,LR,RR} \leq \mathcal{O}(10^{-3}) \tag{20.29}$$

is inferred [331] for $M_{\tilde{g}} \sim \tilde{m} \sim 500$ GeV from ΔM_K and ϵ_K. Within MSSM with its super-GIM filter built-in, these are natural values. Again in non-minimal models most of us feel the need to envoke some kind of flavour or family symmetry to ensure the reduction from the $\mathcal{O}(1)$ level that a priori would seem natural.

20.4.3 $B^0 - \overline{B}^0$ oscillations and CP violation

With ϕ_A too small to be significant in beauty decays, see Eq. (20.28), the phases driving CP asymmetries in B transitions within MSSM still reside in the KM matrix:

$$\phi_1|_{KM} = \phi_1|_{MSSM}, \quad \phi_2|_{KM} = \phi_2|_{MSSM}, \quad \phi_3|_{KM} = \phi_3|_{MSSM},$$

$$\mathrm{Im}\frac{q}{p}\overline{\rho}_{B_s \to D_s \overline{D}_s}|_{MSSM} \simeq 0, \tag{20.30}$$

$$\left.\frac{\Delta M(B_d)}{\Delta M(B_s)}\right|_{SM} \simeq \left.\frac{\Delta M(B_d)}{\Delta M(B_s)}\right|_{MSSM}. \tag{20.31}$$

However, there can be significant (though not truly dominating) contributions to the *absolute* values of each of $\Delta M(B_d)$ and $\Delta M(B_s)$. As described in the chapter on probing the unitarity triangle, the intervention of MSSM could still be revealed in the following way: we compare $|V(td)|$ as *inferred* from the measurements of $\mathrm{Im}\frac{q}{p}\overline{\rho}_{B_d \to \psi K_S}$ – and/or $\Gamma(K^+ \to \pi^+ \nu \overline{\nu})$ and/or BR($B \to \gamma \rho/\omega$) vs BR($B \to \gamma K^*$) and/or $\Delta M(B_d)/\Delta M(B_s)$ – with what is required to reproduce the absolute value of $\Delta M(B_d)$ using the known value of m_{top}. A discrepancy would reveal the intervention of New Physics, and MSSM would be a good candidate for it.

Non-minimal SUSY on the other hand, can make large or even dominant contributions to $\Delta M(B_d)$ and $\Delta M(B_s)$ without having to satisfy Eq. (20.31); it can also yield sizeable novel weak phases, enhancing considerably CP violation in $B^0 - \overline{B}^0$ oscillations – as expressed by a_{SL}, see Eq. (11.37) – and modify the CP asymmetries in $B_d \to \psi K_S$, $B_d \to \pi^+ \pi^-$, $B_s \to D_s K$ and $B_s \to D_s \overline{D}_s$, $\psi \phi$, which

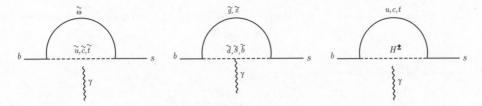

Fig. 20.4. Feynman diagram which gives the SUSY contribution to the $b \to s\gamma$ amplitude.

involve $B^0 - \overline{B}^0$ oscillations [332]. It would enter through $\Delta B = 2$ dynamics in all likelihood, and therefore:

$$\phi_1 \to \phi_1 - \phi_{NP}(B_d), \quad \phi_2 \to \phi_2 + \phi_{NP}(B_d), \quad \phi_3 \to \phi_3 + \phi_{NP}(B_s), \qquad (20.32)$$

i.e.:

$$\phi_1 + \phi_2|_{\text{meas}} = \phi_1 + \phi_2|_{KM}, \quad \phi_1 + \phi_2 + \phi_3|_{\text{meas}} = \pi + \phi_{NP}(B_s). \qquad (20.33)$$

Given the existing *theoretical* uncertainties in evaluating matrix elements of the four-fermion operators $(\bar{b}\Gamma q)(\bar{b}\Gamma q)$ – as expressed through $B_B F_B^2$ – we can infer only loose bounds from $\Delta M(B_d)$, namely $|\text{Re}(\delta_{13}^D)^2| \leq 10^{-3} - 10^{-2}$ [331].

20.4.4 $b \to s\gamma$ *(and other one-loop decays)*

Gluino exchanges mediate $b \to s\gamma$ decay through the diagrams shown in Fig. 20.4 [328][4]; they yield

$$\text{Br}(b \to s\gamma) = \frac{\alpha_s \alpha}{81\pi^2 M_{\tilde{q}}^4} m_b^3 \tau_b \left[|m_b F_3(x)(\delta_{23}^D)_{LL} + M_{\tilde{g}} F_1(x)(\delta_{23}^D)_{LR}|^2 \right.$$
$$\left. + (L \leftrightarrow R) \right]. \qquad (20.34)$$

where $F_1(x)$ is defined in Eq. (20.26) and

$$F_3(x) = \frac{1 - 8x - 17x}{12(x-1)^4} + \frac{x^2(3+x)\ln x}{2(x-1)^5}, \quad x = \frac{M_{\tilde{g}}^2}{m_{\tilde{q}}^2}. \qquad (20.35)$$

The *LR* mixing term, being enhanced relative to the *LL* term by $M_{\tilde{g}}/m_b$, which is a large number, could play a dominant role in these rare decays. In practice, however, the same *LR* squark mixing reduces the mass of the lightest stop. Requiring that the light stop is heavy enough to evade experimental detection forces us to adjust other parameters. Down-type squarks then have to be nearly degenerate, which in turn reduces the contribution of gluino–squark loops to K and B transitions. MSSM contributions to the rare decays can still be sizeable and significant, but they do not modify the SM predictions in any dramatic way:

---

[4] Similar diagrams contribute to $b \to se^+e^-$ and $b \to sg^{(*)}$.

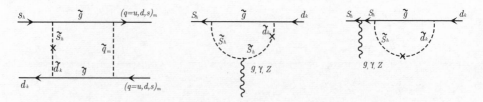

Fig. 20.5.   Example of diagrams which contribute to $\Delta S = 1$ transitions.

at most a factor of two difference in rate can arise. However, quite dramatic modifications can arise in non-minimal implementations of SUSY.

The inclusive transition $B \to \gamma X$ has been observed at CESR and LEP with [182]

$$\text{Br}(B \to \gamma X_s) = (2.62 \pm 0.60|_{\text{exp}} \pm {}^{0.37}_{0.30}|_{\text{th}}) \times 10^{-4}, \tag{20.36}$$

to be compared with the SM prediction given in Eq. (13.41). There is no clear need for a large SUSY contribution, although a sizeable one is certainly not ruled out.

For the future we should keep in mind that various **CP** asymmetries can arise naturally here.

20.4.5   Direct **CP** violation in $\Delta S = 1$ decays

There are several new diagrams for $\Delta S = 1$ decays, see Fig. 20.5. They can make significant contributions to ϵ'/ϵ_K, either enhancing or further reducing it.

None of these SUSY models – like the KM ansatz within the SM – will generate a transverse muon polarization in $K_{\mu 3}$ decays on an observable level. For that to happen we would need **CP** violation entering through a Higgs sector with more than two doublets, as described in the preceding chapter. While that can happen in a SUSY framework, it is not specific to it.

20.5 The pundits' résumé

After this crash course in SUSY technology and phenomenology, we can evaluate **CP** violation in these models in a slightly more informed and considerably more refined manner.

1 SUSY is truly a symmetry 'sans-pareille' – like no other. To realize all its attractive features we have to formulate it as a local theory – SUGRA – embedded in a GUT scheme.

2 The many new dynamical layers can support a plethora of additional sources of **CP** violation in the form of complex Yukawa couplings and other Higgs parameters, including those driving SUSY breaking.

3 Through quark–squark–gluino couplings, flavour changing neutral currents mediating even **CP** violation enter the (nominally) strong dynamics.

4 Remarkably enough, in MSSM only two observable phases emerge: one is the usual KM phase having its origin in the misalignment of the mass matrices for up- and down-type quarks; it also migrates into the squark mass matrices and controls the **CP** properties of the quark–squark–gluino couplings. The other one is ϕ_A, see Eq. (20.16), reflecting soft SUSY breaking; it is severely restricted by the experimental bound on the neutron EDM, which makes it irrelevant for $K_L \to \pi\pi$ decays.

5 Yet once we enter the vast regime of non-minimal SUSY models – even by an apparently slight modification of MSSM – the floodgates open for additional sources of **CP** violation:

- observable effects are likely or at least quite conceivable for the EDMs of neutrons and electrons.

Beyond that, completely different scenarios can occur:

(a) MSSM: the same **CP** asymmetries arise in beauty decays as with the SM implementation of the KM ansatz, although the $B^0 - \overline{B}^0$ oscillation rates are different.

(b) Non-minimal SUSY: while large effects emerge in beauty decays there are – due to $\Delta B = 2$ dynamics being modified by SUSY – sizeable deviations from the KM expectations.

6 What was just stated does not mean, though, that an interpretative chaos of 'everything goes and nobody knows' will rule. For within each scenario there are numerous non-trivial correlations among the **CP** observables, rare decay rates, the KM parameters and gross features of the sparticle spectrum.

Problems

20.1 Consider spontaneous breaking of SUSY through a conventional Higgs mechanism being implemented in V_D and V_F, as defined in Eq. (20.3). Let h_1 and h_2 – the scalar components of the superfields H_1 and H_2 – acquire VEVs v_1 and v_2, respectively. Verify that the fermion and sfermion masses are then described by the following expressions:

$$\mathscr{L}_{\text{fermion masses}} = -v_1 Y^u_{ij} u_i u^c_j - v_2 Y^d_{ij} d_i d^c_j - v_2 Y^l_{ij} l_i l^c_j, \qquad (20.37)$$

$$\begin{aligned}
\mathscr{L}_{\text{sfermion masses}} = \ & v_1^2(\tilde{u} Y^u Y^{u\dagger} \tilde{u}^* + \tilde{u}^{c*} Y^{u\dagger} Y^u \tilde{u}^c) \\
& + v_2^2(\tilde{d} Y^d Y^{d\dagger} \tilde{d}^* + \tilde{d}^{c*} Y^{d\dagger} Y^d \tilde{d}^c + \tilde{e} Y^l Y^{l\dagger} \tilde{e}^* + \tilde{e}^{c*} Y^{l\dagger} Y^l \tilde{e}^c) \\
& + \mu^* v_2 \tilde{U} Y^u \tilde{U}^c + \mu^* v_1(\tilde{d} Y^d \tilde{d}^c + \mu^* v_1 \tilde{e} Y^l \tilde{e}^c) + \text{c.c.}
\end{aligned}$$

$$+ \quad \frac{g'^2}{2}(v_1^2 - v_2^2)\left[\frac{1}{6}(\tilde{u}^*\tilde{u} + \tilde{d}^*\tilde{d}) - \frac{2}{3}\tilde{u}^{c*}\tilde{u}^c + \frac{1}{3}\tilde{d}^{c*}\tilde{d}^c\right.$$

$$- \quad \left.\frac{1}{2}\tilde{e}^*\tilde{e} - \frac{1}{2}\tilde{v}^*\tilde{v} + \tilde{e}^{c*}\tilde{e}^c\right]$$

$$+ \quad \frac{g^2}{4}(v_1^2 - v_2^2)\left[\tilde{u}^*\tilde{u} - \tilde{d}^*\tilde{d} + \frac{1}{2}\tilde{e}^*\tilde{e} - \frac{1}{2}\tilde{v}^*\tilde{v}\right] ; \qquad (20.38)$$

here we have used the following notation: u, u^c, d, d^c, l, l^c are left handed quark and charged lepton fields with their charge conjugates; $\tilde{u}$, $\tilde{u}^C$, $\tilde{d}$, $\tilde{d}^C$, $\tilde{e}$, $\tilde{e}^C$, $\tilde{v}$, $\tilde{v}^C$ denote squark, charged and neutral slepton fields, respectively, together with their charge conjugates; the $Y^{U,D,l}$ represent the Yukawa couplings.

Do Eq. (20.37) and Eq. (20.38) exhibit SUSY breaking? Evaluate the supertrace of Eq. (20.4).

20.2   Evaluate the Feynman diagram in Fig. 20.2 to reproduce Eq. (20.25).

20.3   Verify Eq. (20.33); i.e., show that if New Physics contributes to $B_d - \bar{B}_d$ oscillations such that $\left.\frac{q}{p}\right|_{B_d} = \left.\frac{q}{p}\right|_{B_d,KM} e^{-i\phi_{NP}(B_d)}$, then the new phase $\phi_{NP}(B_d)$ drops out from the sum of the two angles ϕ_1 and ϕ_2 of the KM triangle.

21

Baryogenesis in the universe

21.1 The challenge

One of the most intriguing aspects of big bang cosmology is to 'understand' nucleosynthesis, i.e., to reproduce the abundances observed for the nuclei in the universe as *dynamically* generated rather than merely dialed as input values. This challenge has been met successfully, in particular for the light nuclei, and actually so much so that it is used to obtain information on dark matter in the universe, the number of neutrinos, etc. It is natural to ask whether such a success could be repeated for an even more basic quantity, namely the baryon number density of the universe, which is defined as the difference in the abundances of baryons and antibaryons:

$$\Delta n_{\text{Bar}} \equiv n_{\text{Bar}} - n_{\overline{\text{Bar}}}. \tag{21.1}$$

Qualitatively, we can summarize the observations through two statements:

- The universe is not empty.

- The universe is almost empty.

More quantitatively, we find

$$r_{\text{Bar}} \equiv \frac{\Delta n_{\text{Bar}}}{n_\gamma} \sim \text{few} \times 10^{-10}, \tag{21.2}$$

where n_γ denotes the number density of photons in the cosmic background radiation. Actually we know more, specifically that, at least in our corner of the universe, there are practically no primary antibaryons:

$$n_{\overline{\text{Bar}}} \ll n_{\text{Bar}} \ll n_\gamma. \tag{21.3}$$

It is conceivable that in other neighbourhoods antimatter dominates, and that the universe is formed by a patchwork quilt of matter and antimatter dominated regions, with the whole being matter–antimatter symmetric. Yet it is widely held to be quite unlikely – primarily because no mechanism has been found by which a matter–antimatter symmetric universe following a big bang evolution can develop sufficiently large regions with non-vanishing baryon number. While there will be statistical fluctuations, they can be nowhere near large enough. Likewise for dynamical effects: baryon–antibaryon annihilation is by far not sufficiently

349

effective to create pockets with the observed baryon number, Eq. (21.2). For the
number density of *surviving* baryons can be estimated as [333]

$$n_{\text{Bar}} \sim \frac{n_\gamma}{\sigma_{\text{annih}} m_N M_{\text{Pl}}} \simeq 10^{-19} n_\gamma, \tag{21.4}$$

where σ_{annih} denotes the cross-section of nucleon annihilation, and m_N and M_{Pl}
the nucleon and Planck mass, respectively. Hence we conclude, for the universe
as a whole,

$$0 \neq \frac{n_{\text{Bar}}}{n_\gamma} \simeq \frac{\Delta n_{\text{Bar}}}{n_\gamma} \sim \mathcal{O}(10^{-10}), \tag{21.5}$$

which makes more explicit the meaning of the statement quoted above that the
universe has been observed to be almost empty, but not quite. Understanding this
double observation is the challenge we are going to address now.

21.2 The ingredients

The question is: under which condition can we have a situation where the baryon
number of the universe that vanishes at the initial time – which for all practical
purposes is the Planck time – develops a non-zero value later on:

$$\Delta n_{\text{Bar}}(t = t_{\text{Pl}} \simeq 0) = 0 \overset{?}{\implies} \Delta n_{\text{Bar}}(t = \text{'today'}) \neq 0. \tag{21.6}$$

We can and should actually go one step further in the task one is setting for
oneself: explaining the observed baryon number as dynamically generated *no
matter what its initial value was!*

In a seminal paper that appeared in 1967, Sakharov listed the three ingredients
that are essential for the feasibility of such a program [334, 335]:

1 Since the final and initial baryon number differ, there have to be baryon
  number violating transitions:

$$\mathscr{L}(\Delta n_{\text{Bar}} \neq 0) \neq 0. \tag{21.7}$$

2 **CP** invariance has to be broken. Otherwise for every baryon number chang-
  ing transition, $N \to f$, there is its **CP** conjugate one, $\overline{N} \to \overline{f}$, and no net
  baryon number can be generated, i.e.,

$$\Gamma(N \overset{\mathscr{L}(\Delta n_{\text{Bar}} \neq 0)}{\longrightarrow} f) \neq \Gamma(\overline{N} \overset{\mathscr{L}(\Delta n_{\text{Bar}} \neq 0)}{\longrightarrow} \overline{f}) \tag{21.8}$$

  is needed.

3 Unless we are willing to entertain thoughts of **CPT** violations, the baryon
  number and **CP** violating transitions have to proceed out of thermal equilibrium. For in thermal equilibrium time becomes irrelevant globally and
  CPT invariance reduces to **CP** symmetry, which has to be avoided, see
  above:

$$\textbf{CPT invariance} \overset{\text{thermal equilibrium}}{\implies} \textbf{CP invariance}. \tag{21.9}$$

It is important to keep in mind that these three conditions have to be satisfied *simultaneously*. The other side of the coin is, however, the following: once a baryon number has been generated through the concurrence of these three effects, it can be washed out again by these same effects.

21.3 GUT baryogenesis

Sakharov's paper was not noticed (except for [336]) for several years until the concept of Grand Unified Theories (=GUTs) emerged, starting in 1974 [337]; for those naturally provide all three necessary ingredients [338, 339]:

1 Baryon number changing reactions have to exist in GUTs. For placing quarks and leptons into common representations of the underlying gauge groups – the hallmark of GUTs – means that gauge interactions exist changing baryon and lepton numbers. Those gauge bosons are generically referred to as X bosons and have two couplings to fermions that violate baryon and/or lepton number:

$$X \leftrightarrow qq \,, \quad q\bar{l}. \tag{21.10}$$

2 Those models are sufficiently complex to allow for several potential sources of **CP** violation. Since X bosons have (at least) two decay channels open **CP** asymmetries can arise

$$\Gamma(X \to qq) = (1 + \Delta_q)\Gamma_q, \qquad \Gamma(X \to q\bar{l}) = (1 - \Delta_l)\Gamma_l,$$
$$\Gamma(\overline{X} \to \bar{q}\bar{q}) = (1 - \Delta_q)\Gamma_q, \qquad \Gamma(\overline{X} \to \bar{q}l) = (1 + \Delta_l)\Gamma_l, \tag{21.11}$$

where

$$\begin{aligned} \textbf{CPT} \quad &\Longrightarrow \quad \Delta_q\Gamma_q = \Delta_l\Gamma_l \\ \textbf{CP} \quad &\Longrightarrow \quad \Delta_q = 0 = \Delta_l \\ \textbf{C} \quad &\Longrightarrow \quad \Delta_q = 0 = \Delta_l. \end{aligned} \tag{21.12}$$

3 Grand Unification means that a phase transition takes place around an energy scale M_{GUT}. For temperatures T well above the transition point – $T \gg M_{GUT}$ – all quanta are relativistic with a number density

$$n(T) \propto T^3. \tag{21.13}$$

For temperatures around the phase transition – $T \sim M_{GUT}$ – some of the quanta, in particular those gauge bosons generically referred to as X bosons, aquire a mass $M_X \sim \mathcal{O}(M_{GUT})$ and their equilibrium number density becomes Boltzmann suppressed:

$$n_X(T) \propto (M_X T)^{\frac{3}{2}} \exp\left(-\frac{M_X}{T}\right). \tag{21.14}$$

More X bosons will decay according to Eq. (21.10) than are regenerated from qq and $q\bar{l}$ collisions, ultimately bringing the number of X bosons down to the level described by Eq. (21.14). Yet that will take some time; the expansion in the big bang cosmology leads to a cooling rate that is so rapid that thermal equilibrium cannot be maintained through the phase transition. That means that X bosons decay – and in general interactions – drop out of thermal equilibrium [335].

To the degree that the back production of X bosons in qq and $q\bar{l}$ collisions can be ignored, we find as an order-of-magnitude estimate

$$r_{\text{Bar}} \sim \frac{\frac{4}{3}\Delta_q\Gamma_q - \frac{2}{3}\Delta_l\Gamma_l}{\Gamma_{\text{tot}}}\frac{n_X}{n_0} = \frac{\frac{2}{3}\Delta_q\Gamma_q}{\Gamma_{\text{tot}}}\frac{n_X}{n_0}, \tag{21.15}$$

with n_X denoting the initial number density of X bosons and n_0 the number density of the light decay products[1]. The three essential conditions for baryogenesis are thus naturally realized around the GUT scale in big bang cosmologies, as can be read off from Eq. (21.15):

- $\Gamma_q \neq 0$ representing baryon number violation

- $\Delta_q \neq 0$ reflecting **CP** violation and

- the absence of the back reaction due to an absence of thermal equilibrium.

The fact that this problem can be formulated in GUT models and answers obtained that are very roughly in the right ballpark is a highly attractive feature of GUTs, in particular since this was *not* among the original motivations for constructing such theories.

On the other hand, it would be highly misleading to claim that baryogenesis has been understood. There are serious problems in any attempt to have baryogenesis occur at a GUT scale:

- A baryon number generated at such high temperatures is in grave danger of being washed out or diluted in the subsequent evolution of the universe.

- Very little is known about the dynamical actors operating at GUT scales and their characteristics – and that is putting it mildly. Actually even in the future we can only hope to obtain some slices of indirect information on them.

Of course it would be premature to write-off baryogenesis at GUT scales, yet it might turn out that it is best characterised as a proof of principle – that the baryon number of the universe can be understood as dynamically generated – rather than as a semi-quantitative realization.

---

[1] Due to thermalization effects we can have $n_0 \gg 2n_X$.

21.4 Electroweak baryogenesis

Baryogenesis at the electroweak scale [340] is the most actively analysed scenario at present. For it possesses several highly attractive features:

- We know that dynamical landscape fairly well.

  - In particular **CP** violation has been found to exist there.
  - A well-studied phase transition, namely the spontaneous breaking

$$SU(2)_L \times U(1) \implies U(1)_{QED} \tag{21.16}$$

  takes place.

- Future experiments will certainly probe that dynamical regime with ever increasing sensitivity, both by searching for the on-shell production of new quanta – like SUSY and/or Higgs states – and the indirect impact through quantum corrections on rare decays and **CP** violation.

However, at this point the reader might wonder: 'What about the third required ingredient, baryon number violation? At the electroweak scale?' It is often not appreciated that the electroweak forces of the SM by themselves violate baryon number, though in a very subtle way. We find here what is called an anomaly: the baryon number current is conserved on the classical, yet *not* on the quantum level:

$$\partial_\mu J_\mu^{Bar} = \partial_\mu \sum_q (\bar{q}_L \gamma_\mu q_L) = \frac{g^2}{16\pi^2} \mathrm{Tr} G_{\mu\nu} \tilde{G}_{\mu\nu} \neq 0, \tag{21.17}$$

where g denotes the $SU(2)_L$ gauge coupling, $G_{\mu\nu}$ the electroweak field strength tensor

$$G_{\mu\nu} = \tau_a \left(\partial_\mu A_\nu^a - \partial_\nu A_\mu^a + g\epsilon_{abc} A_\mu^b A_\nu^c \right) \tag{21.18}$$

(with the τ_a being the $SU(2)$ generators) and $\tilde{G}_{\mu\nu}$ its dual:

$$\tilde{G}_{\mu\nu} = \frac{1}{2}\epsilon_{\mu\nu\alpha\beta} G_{\alpha\beta}. \tag{21.19}$$

The right hand side of Eq. (21.17) can be written as the divergence of a current

$$\mathrm{Tr} G_{\mu\nu} \tilde{G}_{\mu\nu} = \partial_\mu K_\mu, \quad K_\mu = 2\epsilon_{\mu\nu\alpha\beta} \mathrm{Tr} \left(A_\nu \partial_\alpha A_\beta - \frac{2}{3} ig A_\nu A_\alpha A_\beta \right). \tag{21.20}$$

We have encountered this situation before, in our discussion of the strong **CP** problem in Chapter 16: a conservation law is vitiated through a triangle anomaly on the quantum level; although the offending term can be written as a total divergence, it still affects the physics of non-abelian gauge theories in a non-trivial way. For there is an infinity of inequivalent ground states differentiated by the value of their K charge, i.e., the space integral of K_0, the zeroth component of the current K_μ constructed from their gauge field configuration. This integral reflects differences in the gauge topology of the ground states and therefore is called the *topological charge*.

In the present context the triangle anomaly induces *baryon number violation* because of the chiral nature of the weak interactions. Eq. (21.17) and Eq. (21.20) show that the difference $J_\mu^{Bar} - K_\mu$ is conserved. The transition from one ground state to another, which represents a tunneling phenomenon, is thus accompanied by a change in baryon number. Elementary quantum mechanics tells us that this baryon number violation is described as a barrier penetration and exponentially suppressed at low temperatures or energies [341]: Prob($\Delta n_{Bar} \neq 0$) $\propto \exp(-16\pi^2/g^2) \sim \mathcal{O}(10^{-160})$ – a suppression that reflects the tiny size of the weak coupling.

There is a corresponding anomaly for the lepton number current, implying that lepton number is violated as well, with the selection rule

$$\Delta n_{Bar} - \Delta n_{lept} = 0 \,. \tag{21.21}$$

This is usually referred to by saying that $B - L$, the difference between baryon and lepton number, is still conserved.

At sufficiently high energies this huge suppression of baryon number changing transition rates will evaporate, since the transition between different ground states can be achieved classically through a motion *over* the barrier. The question then is at which energy scale this will happen and how quickly baryon number violation will become operative. Some semi-quantitative observations can be offered and answers given [342, 335].

There are special field configurations – called sphalerons – that carry the topological K charge. In the SM they induce effective multistate interactions among left handed fermions that change baryon and lepton number by three units each:

$$\Delta n_{Bar} = \Delta n_{lept} = 3. \tag{21.22}$$

At high energies where the weak bosons W and Z are massless, the height of the transition barrier between different groundstates vanishes likewise, and the change of baryon number can proceed in an unimpeded way and presumably faster than the universe expands. Thermal equilibrium is then maintained, and any baryon asymmetry existing before this era is actually washed out[2]! Rather than generating a baryon number, sphalerons act to drive the universe back to matter–antimatter symmetry at this point in its evolution.

At energies below the phase transition, i.e., in the broken phase of $SU(2)_L \times U(1)$ baryon number is conserved for all practical purposes, as pointed out above.

The value of Δn_{Bar} as observed today can thus be generated only in the transition from the unbroken high energy to the broken low energy phase. With

---

[2] To be more precise, only $B + L$ is erased within the SM, whereas $B - L$ remains unchanged.

$\Delta n_{\mathrm{Bar}} \neq 0$ processes operating there, the issue now turns to the strength of the phase transition: is it relatively smooth like a second order phase transition or violent like a first order one? Only the latter scenario can support baryogenesis.

A large amount of interesting theoretical work has been done on the thermodynamics of the SM in an expanding universe. Employing perturbation theory and lattice studies, we have arrived at the following result: for light Higgs masses up to around 70 GeV, the phase transition is first order, for larger masses it is second order [343]. Since no such light Higgs states have been observed at LEP, we infer that the phase transition is second order, thus apparently foreclosing baryogenesis occurring at the electroweak scale.

We have concentrated here on the questions of thermal equilibrium and baryon number, while taking **CP** violation for granted, since it is known to operate at the electroweak scale. Yet most authors – with the exception of some notable heretics – agree that the KM ansatz is not at all up to *this* task: it fails by several orders of magnitude. On the other hand, New Physics scenarios of **CP** violation – in particular of the Higgs variety – can reasonably be called upon to perform the task.

21.5 Leptogenesis driving baryogenesis

If the electroweak phase transition is indeed a second order one, sphaleron mediated reactions cannot drive baryogenesis, as just discussed, and they will wipe out any pre-existing $B + L$ number. Yet if at some high energy scales a lepton number is generated, the very efficiency of these sphaleron processes can *communicate* this asymmetry to the baryon sector through them maintaining conservation of $B - L$.

There are various ways in which such scenarios can be realized. The simplest one is to just add *heavy right handed Majorana* neutrinos to the SM. This is highly attractive in any case, as described in Sec. 17.3, since it enables us to implement the see-saw mechanism for explaining why the observed neutrinos are (practically) massless; it is also easily embedded into $SO(10)$ GUTs.

The basic idea is the following [344]:

- A primordial lepton asymmetry is generated at high energies well above the electroweak phase transition:

  – Since a Majorana neutrino N is its own **CPT** mirror image, its dynamics necessarily violate lepton number. It will possess at least the following classes of decay channels:

$$N \to l\overline{H}, \; \bar{l}H, \tag{21.23}$$

  with l and $\bar{l}$ denoting a light charged or neutral lepton or antilepton and H and $\overline{H}$ a Higgs or anti-Higgs field, respectively.

- A **CP** asymmetry will in general arise

$$\Gamma(N \to l\overline{H}) \neq \Gamma(\overline{N} \to \overline{l}H) \tag{21.24}$$

through a KM analogue in the neutrino mass matrix (which can be quite different from the mass matrix for charged leptons).

- These neutrino decays are sufficiently slow to occur out of thermal equilibrium around the energy scale where the Majorana masses emerge.

- The resulting lepton asymmetry is transferred into a baryon number through sphaleron mediated processes in the unbroken high energy phase of $SU(2)_L \times U(1)$:

$$
\begin{aligned}
\langle \Delta n_{\text{lept}} \rangle \quad &= \quad \frac{1}{2} \langle \Delta n_{\text{lept}} + \Delta n_{\text{Bar}} \rangle + \frac{1}{2} \langle \Delta n_{\text{lept}} - \Delta n_{\text{Bar}} \rangle \\
&\implies \quad \frac{1}{2} \langle \Delta n_{\text{lept}} - \Delta n_{\text{Bar}} \rangle .
\end{aligned}
\tag{21.25}
$$

- The baryon number thus generated survives through the subsequent evolution of the universe.

21.6 Wisdom – conventional and otherwise

We understand how nuclei were formed in the universe, given protons and neutrons. Obviously it would be even more fascinating if we could understand how these baryons were generated in the first place. We do not possess a specific and quantitative theory successfully describing baryogenesis. However, leaving it at that statement would – we believe – miss the main point. We have learnt which kinds of dynamical ingredients are neccessary for baryogenesis to occur in the universe. We have seen that these ingredients can be realized naturally:

- GUT scenarios for baryogenesis provide us with a proof of principle that such a program can be realized. In practical terms, however, they suffer from various shortcomings:

  - Since the baryon number is generated at the GUT scales, very little is and not much more might ever be known about that dynamics.

  - It appears quite likely that a baryon number produced at such high scales is subsequently washed out.

- The highly fascinating proposal of baryogenesis at the electroweak phase transition has attracted a large degree of attention – and deservedly so:

  - A baryon number emerging from this phase transition would be in no danger of being diluted substantially.

- The dynamics involved here is known to a considerable degree and will be probed even more with ever increasing sensitivity over the coming years.

However, it seems that the electroweak phase transition is of second order and thus not sufficiently violent.

- A very intriguing variant turns some of the vices of sphaleron dynamics into virtues, by attempting to understand the baryon number of the universe as a reflection of a *primary* lepton asymmetry. The required new dynamical entities – Majorana neutrinos and their decays – obviously would impact on the universe in other ways as well.

The challenge to understand baryogenesis has already inspired our imagination, prompted the development of some very intriguing scenarios and thus has initiated many fruitful studies – and in the end we might even be successful in meeting it!

PART 4
SUMMARY

22
Summary and perspectives

The discovery that the weak forces break previously unquestioned discrete symmetries – first parity **P** and charge conjugation **C**, then **CP** and **T** – had a revolutionizing impact on our perception of nature and how we analyze the elements of its grand design. We realized that symmetries should not be taken for granted; some even began questioning that sacrosanct fruit of quantum field theory, **CPT** invariance. We learnt from the violation of **CP** symmetry – *not* from that of **P** and **C** separately – that *left* and *right* or *positive* and *negative* charge are dynamically distinct rather than being mere labels based on a convention; furthermore that nature distinguishes between *past* and *future* even on the *microscopic* level. We came to understand that while **CP** violation has so far been observed only in a single system – the decays of K_L mesons – as a seemingly unobtrusive phenomenon, it represents not only a profound intellectual insight, but has also many and far-reaching concrete consequences:

- The huge predominance of matter over antimatter apparently observed in our universe requires **CP** violation if it is to be understood as *dynamically generated* rather than merely reflecting the initial conditions.

- Once the dynamics are sufficiently complex to support **CP** violation, the latter can manifest itself in numerous different ways; we can even say the floodgates open.

- The three-family SM can implement **CP** violation through the KM mechanism without requiring so-far unobserved degrees of freedom. It is highly non-trivial in that it can accommodate the data on ϵ_K and ϵ' within the uncertainties.

- Despite this phenomenological success, it is incorrect to claim the KM ansatz provides us with an understanding. Since **CP** violation enters through the quark mass matrices, its source is related to three central mysteries of the SM:

- How are fermion masses generated?[1]

- Why is there a family structure?

- Why are there three families rather than one?

- **CP** studies can be employed as a high sensitivity probe for New Physics in an indirect fashion, i.e., when new dynamical degrees of freedom enter through quantum corrections only.

CP violation is thus a fundamental as well as a mysterious phenomenon, which exists in nature, yet contains a message that has not been decoded yet. In our judgement it would be unrealistic to expect progress in answering these questions through pure thinking. We strongly believe we have to appeal to nature through experimental efforts to provide us with more pieces still missing from the puzzle. **CP** studies are essential in obtaining the full dynamical information contained in the mass matrices – and there is a wide realm open for them!

22.1 The cathedral builder's paradigm

The dynamical ingredients for numerous and multi-layered manifestations of **CP** and **T** violations do exist or are likely to exist. Accordingly, we search for them in many phenomena, namely in

- the neutron electric dipole moment probed with ultracold neutrons at ILL in Grenoble, France;

- the electric dipole moment of electrons studied through the dipole moment of atoms at Seattle, Berkeley and Amherst in the US;

- the transverse polarization of muons in $K^- \to \mu^- \bar{\nu} \pi^0$ at KEK in Japan;

- ϵ'/ϵ_K as obtained from K_L decays at FNAL and CERN and soon at DAΦNE in Italy;

- decay distributions of hyperons at FNAL;

- likewise for τ leptons at LEP, the beauty factories and BES in Beijing;

- **CP** violation in the decays of charm hadrons produced at FNAL and the beauty factories;

- **CP** asymmetries in beauty decays at DESY, at the beauty factories at Cornell, SLAC and KEK, at the FNAL collider and ultimately at the LHC.

---

[1] Or more generally: how are masses produced in general? For in alternative models **CP** violation enters through the mass matrices for gauge bosons and/or Higgs bosons.

This list makes it clear that frontline research on this topic is pursued at high energy labs all over the world – and then some; techniques from several different branches of physics – atomic, nuclear and high energy physics – are harnessed in this endeavour, together with a wide range of set-ups; lastly, experiments are performed at the lowest temperatures that can be realized on earth – ultracold neutrons – and at the highest – in collisions produced at the LHC. All of that is dedicated to one profound goal. At this point we can explain what we mean by the term 'cathedral builders' paradigm'. The building of cathedrals required interregional collaborations, front line technology (for the period) from many different fields, and commitment; it had to be based on solid foundations – and it took time. The analogy to the ways and needs of high energy physics is obvious – but it goes deeper than that. At first sight a cathedral looks like a very complicated and confusing structure with something here and something there. Yet further scrutiny reveals that a cathedral is more appropriately characterized as a complex rather than a complicated structure, one that is multi-faceted and multi-layered – with a coherent theme! We cannot (at least for first rate cathedrals) remove any of its elements without diluting (or even destroying) its architectural soundness and intellectual message. Neither can we in our efforts to come to grips with **CP** violation!

22.1.1 Present status and general expectations

- We know unequivocally that **CP** symmetry is not exact in nature since $K_L \to \pi\pi$ proceeds and presumably because we exist, i.e., because the baryon number of the universe does *not* vanish. The quantity ϵ_K has been as well measured in K_L decays as necessary. We can expect improvements in our theoretical control over the evaluation of hadronic matrix elements and more precise information on $|V_{td}|$; this will provide even more stringent tests of the KM description.

- The presence of direct **CP** violation in $K_L \to \pi\pi$ decays has been observed now, which is a discovery of the very first rank! The value of ϵ'/ϵ_K will soon be determined with an uncertainty not exceeding $\sim 2 \cdot 10^{-4}$. Yet we remain sceptical that we can theoretically interpret this signal in a precise way.

- If the KM mechanism is a significant actor in $K_L \to \pi\pi$ transitions, then there must be large **CP** asymmetries in the decays of beauty hadrons. In B^0 decays they are naturally measured in units of 10%!

- Some of these asymmetries are predicted with high *parametric* reliability.

- New theoretical technologies will allow us to translate such parametric reliability into *quantitative* accuracy.

- Any significant difference between certain KM predictions for the asymmetries and the data reveals the intervention of New Physics. There will be no 'plausible deniability'.

- We find it likely that deviations from the KM predictions will show up here. Yet to exploit this discovery potential to the fullest, we will have to harness the statistical muscle provided by beauty production at hadronic colliders.

- Searching for a muon transverse polarization in $K_{\mu 3}$ decays and for electric dipole moments represents a high sensitivity probe for New Physics.

- τ leptons, charm hadrons and top quarks provide ample fields for **CP** studies; those are far from being fully developed even theoretically.

- There is intriguing indirect evidence for the presence of New Physics coming from the patterns in the quark mass matrices, baryogenesis in the universe and neutrino physics.

- There are many models of **CP** violation based on New Physics eagerly awaiting their turn in the wings. Even in scenarios where they are quite insignificant for $K_L \rightarrow \pi\pi$, they can naturally generate observable effects for EDMs. In some of these models observable **CP** asymmetries could conceivably arise in the charm, τ decays and top sector.

22.1.2 A look back

A look back can provide us with a proper perspective. *The comprehensive study of kaon and hyperon physics has been instrumental in guiding us to the SM.*

- The $\tau - \theta$ puzzle led to the realization that parity is not conserved in nature.

- The observation that the production rate exceeded the decay rate by many orders of magnitude – this was the origin of the name 'strange particles' – was explained through postulating a new quantum number – 'strangeness' – conserved by the strong, though not the weak forces. This was the beginning of the second quark family.

- The absence of flavour-changing neutral currents was incorporated through the introduction of the quantum number 'charm', which completed the second quark family.

- **CP** violation finally led to the postulation of yet another, the third family.

All of these elements, which are now essential pillars of the SM, were New Physics at *that* time!

We take this historical precedent as clue that a detailed, comprehensive and thus neccessarily long-term program on beauty physics will lead to a new paradigm, a *new* SM!

22.2 Agenda for the future

With the grand direction set we can turn to describing the agenda for the years ahead in more concrete terms. We are helped there by the considerable progress achieved in theoretical engineering and developing a comprehensive **CP** phenomenology.

- *B* decays constitute an almost ideal, certainly optimal and unique lab, due to the dynamic interplay of all three quark families, which is greatly enhanced by the 'long' lifetimes and speedy oscillations. Powerful theoretical technologies have been developed that allow us – or will in the foreseeable future – to extract the values of $|V_{cb}|$, $|V_{ub}|$ and possibly $|V_{td}|$ with good accuracy from **CP** insensitive data.

  - They define the KM triangle and thus control the observable **CP** asymmetries. The latter are related to each other most concisely through trigonometry.

  - The CKM parameters constitute a body of fundamental quantities that in all likelihood reflect hitherto unknown dynamics, operating, presumably, at very high energy scales.

  The better the shape of the triangle is known, the more sensitive we become to the intervention of New Physics. A $\sim 10\%$ accuracy in the predictions seems achievable in the future through a combination of methods and cross checks, using the many redundancies built-in, as described before.

- It is our firm expectation that a comprehensive analysis of the weak decays of beauty hadrons will reveal the presence of New Physics and even sketch some of its salient features. Such insights will be complementary to the information acquired by a study of high $p_\perp$ physics at hadronic colliders.

- Personally we believe that, even if no deviation from the KM predictions were uncovered, we would find that the CKM parameters, in particular the angles of the KM triangle, carry special values that would give us clues about New Physics. Some very interesting theoretical work is being done about how GUT dynamics, in particular of the SUSY (or supergravity) variety, operating at very high scales, would shape the observable CKM parameters.

- A vigorous research program must be continued for *light* fermion systems, specifically in the decays of kaons and hyperons and in electric dipole moments. After all, it is conceivable of course that no **CP** asymmetries are found in B decays on a measurable level. Then we would know that the KM ansatz is *not* a significant actor in $K_L \to \pi\pi$, that New Physics drives it – but what kind of New Physics would it be? Furthermore, even if large **CP** asymmetries were found in B decays, it could happen that the signals of New Physics are obscured by the large 'KM background'. This would not be the case if electric dipole moments were found or a transverse polarization of muons in $K_{\mu3}$ decays.

- A comprehensive analysis of charm decays with special emphasis on $D^0 - \overline{D}^0$ oscillations and **CP** violation is a moral imperative! Likewise for **CP** studies with τ leptons and top quarks.

- Close feedback between experiment and theory will be essential.

22.3 Final words

In the book on what **CP** and **T** invariance and their limitations can teach us about Nature's Grand Design, we are closer to the beginning than the end:

1 There is still unfinished business in our analysis of the weak decays of strange hadrons.

2 The quest for EDMs has to be continued with renewed vigour.

3 A determined effort has to be mounted for studying the **CP** properties of τ leptons, charm hadrons and top quarks.

4 A comprehensive, detailed and high-statistics analysis of beauty decays is bound to provide us with essential information on fundamental dynamics. It represents a unique opportunity where data can be compared with predictions that are, or will be, numerically accurate by then.

Such a program of inquiry will be intriguing, exciting – but neither easy nor quick! Yet we have to keep the following in mind: insights into Nature's Grand Design which can be obtained from a comprehensive and detailed program of **CP** studies

- are of essential and fundamental importance,

- cannot be obtained any other way, and

- cannot become obsolete!

If we want to understand mass generation, one should endeavour to acquire all information concerning it, and that includes the observable content of the non-diagonal mass matrices. No direct observation of new fields – like SUSY partners – can supersede that information. On the contrary, it would be of great help by providing us with essential input for our predictions.

References

[1] R. G. Sachs, *The Physics of Time Reversal,* University of Chicago Press, Chicago, 1987.

[2] T. D. Lee, *Particle Physics and Introduction to Field Theory*, Harwood Academic Publishers GmbH, New York, N. Y., 1988.

[3] E. P. Wigner, *Nachr. Ges. Wiss. Göttingen, Math.-Physik. Kl.,* **32** 546 (1932).

[4] A. Messiah, *Quantum Mechanics Vol. II*, North-Holland Publishing Co., Amsterdam, 1965.

[5] H. A. Kramers, *Proc. Acad. Sci. Amsterdam* **33** 959 (1930); see also: F. J. Dyson, *J. Math. Phys.* **3** 140 (1962).

[6] A. Abragam and B. Bleaney, *Electron Paramagnetic Resonance of Transition Ions*, Clarendon Press, Oxford, 1970.

[7] J. J. Sakurai, *Invariance Principles and Elementary Particles*, Princeton University Press, Princeton, New Jersey, 1964.

[8] I. B. Khriplovich and S. K. Lamoreaux, *CP violation without Strangeness*, Springer, 1997.

[9] N. F. Ramsey, 'Earliest Criticisms of Assumed P and T Symmetries', M. Skalsey *et al.* (eds.), *Time Reversal – The Arthur Rich Memorial Symposium*, AIP Conf. Proc. 270, American Institute of Physics, New York, 1993.

[10] B. Heckel, *Fourty Years of Neutron Electric Dipole Moment*, M. Skalsey *et al.* (Eds.), Time Reversal – The Arthur Rich Memorial Symposium, AIP Conference Proceedings 270, American Institute of Physics, New York, 1993.

[11] I. S. Altarev *et al.*, *Phys. Lett.* **B276** 242 (1992).

[12] K. F. Smith *et. al.*, *Phys. Lett.* **B234** 191 (1990).

[13] L. I. Schiff, *Quantum Mechanics*, McGraw-Hill Inc., New York, 1968.

[14] W. Bernreuther and M. Suzuki, *Rev. Mod. Phys.* **63** 313 (1991).

[15] S. A. Murthy, D. Krause, Z. L. Li and L. R. Hunter, *Phys. Rev. Lett.* **63** 965 (1989).

[16] J. D. Bjorken and S. D. Drell, *Relativistic Quantum Mechanics,* McGraw-Hill Book Company, New York, 1964.
J. D. Bjorken and S. D. Drell, *Relativistic Quantum Fields,* McGraw-Hill Book Company, New York, 1964.

[17] G. Lüders, *Ann, Phys.* **2** 1 (57).

[18] J. D. Jackson, *Classical Electrodynamics*, John Wiley & Sons, Inc., New York, 1963.

[19] K. M. Watson, *Phys. Rev.* **D95** 228 (1954).

[20] For an exciting historical account of the progress of our field during this period, see A. Pais, 'CP violation: the first 25 years', *CP Violation in Particle and Astrophysics*, J. Tran Than Van (ed.), Edition Frontières, Gif-sur-Yvette, France, 1989.

[21] A. Pais, *Phys. Rev.* **D86** 663 (1952).

[22] M. Gell-Mann, *Phys. Rev.* **D92** 833 (1953).

[23] T. Nakano and K. Nishijima, *Prog. Theor. Phys.* **10** 580 (1953).

[24] T. D. Lee and C. N. Yang, *Phys. Rev.* **D104** 254 (1956).

[25] C. S. Wu, E. Ambler, R. W. Hayward, D. D. Hoppes and R. P. Hudson, *Phys. Rev.* **D105** 1423 (1957).

[26] M. Gell-Mann and A. Pais, *Phys. Rev.* **D97** 1387 (1955).

[27] K. Lande *et al.*, *Phys. Rev.* **D103** 1901 (1956).

[28] Particle Data Group, *Europ. Phys. J.* **C3**, 1 (1998).

[29] A. Pais and O. Piccioni, *Phys. Rev.* **D100** 1487 (1955).

[30] L. B. Okun, *Weak Interactions of Elementary Particles*, Pergamon, 1965; the Russian original had appeared in 1963, i.e., clearly *before* the discovery of CP violation.

[31] R. K. Adair, 'CP-Nonconservation – the Early Experiments', *CP Violation in Particle and Astrophysics*, J. Tran Thanh Van (ed.), Editions Frontières, 1989.

[32] F. Muller *et al.*, *Phys. Rev. Lett.* **4** 418 (1960); R. H. Good *et al.*, *Phys. Rev.* **D124** 1221 (1961).

[33] L. B. Leipuner *et al.*, *Phys. Rev.* **132** 2285 (1963).

[34] J. H. Christensen *et al.*, *Phys. Rev. Lett.* **13** 138 (1964).

[35] J. Cronin, Nishina Memorial Lecture (1997).

[36] B. Laurent and M. Roos, *Phys. Lett.* **13**, 269 (1964); *ibid.* **15**, 104.

[37] T. D. Lee and C. N. Yang, *Phys. Rev.* **98** 1501 (1955).

[38] T. D. Lee, R. Oehme and C. N. Yang, *Phys. Rev.* **106** 340 (1957). Another formulation proposed by R. G. Sachs [41] relies on the analysis of the neutral kaon propagator in which the proper self-energy diagram $\Pi^*(k^2)$ must be approximated by $\Pi^*(m_{p0}^2 + i0^+)$ near poles of the propagator.

[39] V. F. Weisskopf and E. P. Wigner, *Z. Phys.* **63** 54 (1930); *Z. Phys.* **65** 18 (1930).

[40] P. K. Kabir, Appendix A of: *The CP Puzzle*, Academic Press, 1968.

[41] For an alternative approch see: R. G. Sachs, *Ann, Phys.* **22** 239 (1963).

[42] T. D. Lee and L. Wolfenstein, *Phys. Rev.* **D138** B1490 (1965).

[43] T. D. Lee and C. S. Wu, *Ann. Rev. Nucl. Sci.* **16** 471 (1966).

[44] Y. Nir and H. R. Quinn, in: *B decays*, S. Stone (Ed.), World Scientific, Singapore, 1994.

[45] J. S. Bell and J. Steinberger, *Proc. 1965 Oxford Int. Conf. on Elementary Particles*, Rutherford High Energy Laboratory, Didcot.

[46] CPLEAR collaboration: R. Adler *et al.*, *Phys. Lett.* **B363** 243 (1995). The value for η_{+-} quoted here is a PDG average [28].

[47] NA31 collaboration: G. D. Barr, *et al.*, *Phys. Lett.* **B317** 233 (1993); E731 collaboration: L. K. Gibbons *et al.*, *Phys. Rev. Lett.* **70** 1203 (1993). The value for η_{00} quoted here is a PDG average [28].

[48] Most recent data have been taken by Geweniger *et al.*, *Phys. Lett.* **48** B483 (1974). The value for δ quoted here is a PDG average of μ and e leptonic decays [28]. For an excellent review, see K. Kleinknecht in [49].

[49] *CP Violation*, C. Jarlskog (ed.), World Scientific, Singapore, 1988.

[50] P. K. Kabir, *Phys. Rev.* **D2** 540 (1970).

[51] CPLEAR collaboration; P. Pavlopoulos, *Proc. Workshop on K Physics*, L. Fayard (ed.), Editions Frontières, Gif-sur-Yvette, France.

[52] *Proc. Workshop on Physics and Detectors for the DAΦNE*, the Frascati Φ Factory, Frascati, 1991.

[53] CPLEAR collaboration, R. LeGac, *Proc. Workshop on K Physics*, L. Fayard (ed.), Editions Frontières, Gif-sur-Yvette, France.

[54] NA31 : G. D. Barr *et al.*, *Phys. Lett.* **B317** 233 (1993).

[55] E731: L. K. Gibbons *et al.*, *Phys. Rev. Lett.* **70** 1203 (1993).

[56] http://fnpx03.fnal.gov/experiments/ktev/ktev.html

[57] S. R. Blatt *et al.*, *Phys. Rev.* **D27** 1056 (1983).

[58] N. W. Tanner and R. H. Dalitz, *Ann. Phys. (N.Y.)* **171**, 463 (1986).

[59] A. Bohm *et al.*, *Nucl. Phys.* **B9** 605 (1969).

[60] C. Geweniger *et al.*, *Phys. Lett.* **B48** 487 (1974).

[61] Recent data for for ϕ_{+-} were taken by: CPLEAR: R. Adler *et al.*, *Phys. Lett.* **B363** 243 (1995); E773: B. Schwingheuer *et al.*, *Phys. Rev. Lett.* **74** 4376 (1995); NA31: R. Carosi *et al.*, *Phys. Lett.* **B237** 303 (1990). The value quoted here is the PDG average [28] .

[62] Recent data for ϕ_{00} were taken by NA31: R. Carosi *et al.*, *Phys. Lett.* **B237** 303 (1990); E731: Karlsson *et al.*, *Phys. Rev. Lett.* **64** 2976 (1990). The value quoted here is the PDG average [28] .

[63] A. R. Zhitnitskii, *Sov. J. Nucl. Phys.* **31**, 529 (1980).

[64] L. B. Okun and I. B. Khriplovich, *Sov. J. Nucl. Phys.* **6**, 598 (1968); E. S. Ginsberg and J. Smith, *Phys. Rev.* **D8** 3887 (1973);

[65] S. W. MacDowell, *Nuov. Cim.*, **9** 258 (1958).

[66] N. Cabibbo and A. Maksymowicz, *Phys. Lett.* **9** 352 (1964); *Phys. Lett.* **11** 360 (1964); *Phys. Lett.* **14** 72 (1966).

[67] See also G. Belanger and C. Q. Geng, *Phys. Rev.* **D44** 2789 (1991).

[68] M. K. Campbell *et al.*, *Phys. Rev. Lett.* **47** 1032 (1981); S. R. Blatt *et al.*, *Phys. Rev.* **D27** 1056 (1983).

[69] J. Imazato *et al.*, *KEK-PS Research Proposal*, KEK Report 91-8.

[70] CPLEAR Collaboration, A. Angelopoulos *et al.*, *Phys. Lett.* **425B** 391 (1998.)

[71] CPLEAR Collab., A. Angelopoulos *et al.*, *Eur. Phys. J.* **C5**, 389 (1998).

[72] S. Weinberg, *Phys. Rev. Lett.* **4** 87 (1960).

[73] W. T. Ford *et al.*, *Phys. Lett.* **38B** 335 (1972).

[74] T. D. Lee and C. N. Yang, *Phys. Rev.* **D108** 645 (1957); R. Gatto, *Nucl. Phys.* **5** 183 (1958).

[75] L. Wolfenstein, *Phys. Rev. Lett.* **13** 562 (1964).

[76] M. Gell-Mann and Y. Ne'eman, *The Eightfold Way*, W. A. Benjamin Inc., New York, 1964.

[77] S. Glashow, J. Illiopolous and L. Maiani, *Phys. Rev.* **D2** 1285 (1970).

[78] K. Niu, E. Mikumo and Y. Maeda, *Prog. Theor. Phys.* **46** 1644 (1971).

[79] M. Kobayashi and T. Maskawa, *Prog. Theor. Phys.* **49** 652 (1973).

[80] N. Cabibbo, *Proc. XIIth Int. Conf. in High Energy Physics*, C. E. Mauk (ed.), University of California Press, Berkeley, 1967.

[81] For a review, see: E. S. Abers and B. W. Lee, *Phys. Rep.* **9C**, 1 (1973).

[82] Also see: T.-P. Cheng and L.-F. Lee, *Gauge Theory of Elementary Particles*, Oxford University Press, Oxford, 1982.

[83] C. Jarlskog, in: *CP Violation*, C. Jarlskog (ed.), World Scientific, Singapore, 1988.

[84] M. Gell-Mann and A. Pais, *Proc. Glasgow Conf. Nuclear and Meson Phys.*, E. H. Bellamy and R. G. Moorhouse (eds.), Pergamon Press, Oxford, 1955.

[85] G. Buchalla, A. J. Buras and M. E. Lautenbacher, *Nucl. Phys.* **B370** 69 (1992); A. J. Buras, M. Jamin and M. E. Lautenbacher, *Nucl. Phys.* **B408** 209 (1993).

[86] For an excellent discussion on these points, see A. J. Buras in Ref. [49]

[87] M. K. Gaillard and B. W. Lee, *Phys. Rev. Lett.* **33** 108 (1974).

[88] G. Altarelli and L. Maiani, *Phys. Lett.* **B52** 351 (1974).

[89] A. I. Vainshtein, V. I. Zakharov and M. A. Shifman, *Sov. Phys. JETP* **45** 670 (1977).

[90] F. J. Gilman and M. B. Wise, *Phys. Rev.* **D20** 2392 (1979).

[91] J. M. Flynn and L. Randall, *Phys. Lett.* **B224** 224 (1989).

[92] We refer you to the many text books on this subject; for example [82] or H. Georgi, *Weak interactions and modern particle theory*, The Benjamin/Cummings Publishing Co., Menlo Park, California, USA.

[93] See, for example, R. S. Chivukula, J. M. Flynn and H. Georgi, *Phys. Lett.* **B171** 453 (1986).

[94] For example, there is a strong attractive force in the $(\pi\pi)_0$ channel at around 400–800 MeV which further enhances the $\Delta I = \frac{1}{2}$ matrix elements. T. Morozumi, C. S. Lim and A. I. Sanda, *Phys. Rev. Lett.* **65** 404 (1990).

[95] P. Franzini, in: *Les Rencontres de Physique de la Vallee d'Aoste, Results and Perspectives in Particle Physics*, M. Greco (ed.), Editions Frontières, 1991.

[96] G. Buchalla, A. J. Buras and M. E. Lautenbacher, *Rev. Mod. Phys.* **68** 1125 (1996) .

[97] M. Ciuchini, *Nucl. Phys. Proc. Suppl.* **59** 149 (1997).

[98] T. Inami and C. S. Lim, *Prog. Theor. Phys.* **65** 297 (1981).

[99] For a comprehensive review on hadronic matrix elements see: A. Soni, *Nucl. Phys.* **B47** (Proc. Suppl.) 43 (1996).

[100] J. F. Donoghue, E. Golowich, B. R. Holstein and J. Trampetic, *Phys. Lett.* **179B** 361 (1986); A. J. Buras and J.-M. Gèrard, *Phys. Lett.* **192** 156 (1987); M. Lusignoli, *Nucl. Phys.* **B325** 33 (1989).

[101] We follow the arguments of [43]. While updated numbers are used, the final result hardly differs from the result obtained by Lee and Wu 30 years ago.

[102] L. Roper *et al.*, *Phys. Rev.* **D138** 190 (1965).

[103] J. F. Donoghue, X.-G. He and S. Pakvasa, *Phys. Rev.* **D34** 833 (1986).

[104] HYPERCP Collab. (E871), C. Dukes, paper presented to the 3rd Int. Conf. on Hyperons, Charm and Beauty Hadrons, Genoa, Italy, 1998.

[105] A. I. Sanda, in Proceedings of *b20: Twenty Beautiful Years of Bottom Physics*, D. Kaplan (ed.), Illinois Institute of Technology, 1997.

[106] J. H. Christenson *et al.*, *Phys. Rev. Lett.* **25** 1524 (1970).

[107] J. J. Aubert *et al.*, *Phys. Rev. Lett.* **33** 1404 (1974).

[108] J. E. Augustin, *Phys. Rev. Lett.* **33** 1406 (1974).

[109] J. W. Herb *et al.*, *Phys. Rev. Lett.* **39** 252 (1977).

[110] C. W. Darden *et al.*, *Phys. Lett.* **76B** 246 (1978).

[111] D. Besson *et al.*, *Phys. Rev. Lett.* **54** 382 (1985).

[112] D. M. J. Lovelock *et al.*, *Phys. Rev. Lett.* **54** 377 (1985).

[113] E. Fernandez *et al.*, *Phys. Rev. Lett.* **51** 1022 (1983).

[114] N. Lockyer *et al.*, *Phys. Rev. Lett.* **51** 1316 (1983).

[115] The history and up-to-date status of measurements of the lifetimes of charm and beauty hadrons is reviewed in: G. Bellini, I. Bigi and P. Dornan, *Phys. Rep.* **289** 1 (1997).

[116] H. Albrecht *et al.*, *Phys. Lett.* **B192** 245 (1987).

[117] L. B. Okun, V. I. Zakharov and M. P. Pontecorvo, *Nuov. Cim. Lett.* **13** 218 (1975).

[118] L. Wolfenstein, *Phys. Rev. Lett.* **51** 1945 (1983).

[119] L.-L. Chau and W.-Y. Keung, *Phys. Rev. Lett.* **53** 1802 (1984).

[120] C. Jarlskog, *Proc. IVth LEAR Workshop*, Villars-sur-Ollon, Switzerland, 1987.

[121] J. L. Rosner, A. I. Sanda, and M. Schmidt *Proc. Fermilab Workshop on High Sensitivity Beauty Physics at Fermilab*, A. J. Slaughter, N. Lockyer and M. Schmidt (eds.), 1987. See also C. Hamzaoui, J. L. Rosner and A. I. Sanda, same proceeding.

[122] CLEO Collaboration: M. Artuso *et al.*, *Phys. Rev. Lett.* **62** 2233 (1989).

[123] M. Voloshin *et al.*, *Sov. J. Nucl. Phys.* **46** 112 (1987); see also Ref. [115].

[124] J. S. Hagelin, *Nucl. Phys.* **B193** 123 (1981).

[125] I. I. Bigi, V. Khoze, A. Sanda and N. Uraltsev, in Ref. [49].

[126] A. Pais and S. B. Treiman, *Phys. Rev.* **D12** 2744 (1975).

[127] M. Bander, D. Silverman and A. Soni, *Phys. Rev. Lett.* **43** 242 (1979).

[128] N. Isgur and M. Wise, *Phys. Lett.* **B232** 113 (1991).

[129] I. I. Bigi, M. Shifman, and N. Uraltsev, Ann. Rev. Nucl. Part. Sci.47 591 (1997), with reference to earlier work.

[130] ARGUS Collaboration: H. Albrecht *et al.*, *Phys. Lett.* **B234** 409 (1990); *ibid.* **255** 297 (1991) .

[131] CLEO Collaboration: R. Fulton *et al.*, *Phys. Rev. Lett.* **64** 16 (1990).

[132] CLEO Collaboration: J. Bartelt *et al.*, *Phys. Rev. Lett.* **71** 41111 (1993).

[133] A. B. Carter and A. I. Sanda, *Phys. Rev.* **D23** 1567 (1981).

[134] I. I. Bigi and A. I. Sanda, *Nucl. Phys.* **B193** 85 (1981).

[135] K. Akerstaff *et al.*, *Eur. Phys. J.* **C5** 379, (1998).

[136] CDF Collaboration CDF/PUB/Bottom/CDF/4855.

[137] CDF Collaboration CDF/PUB/Bottom/CDF/4855; http://www-cdf.fnal.gov/

[138] L.-L. Chau in [49].

[139] M. Gronau and D. Wyler, *Phys. Lett.* **B265** 172 (1991); M. Gronau and D. London, *Phys. Lett.* **B253** 483 (1991).

[140] I. Dunietz, *Phys. Lett.* **B270** 75 (1991).

[141] M. Gronau, *Phys. Lett.* **B300** 163 (1993).

[142] A. I. Sanda and Z.-Z. Xing, Proceedings of 7th Int. Symp. on Heavy Flavor Physics, Santa Barbara (1997).

[143] R. Godang *et al.*, *Phys. Rev. Lett.* **80** 3456 (1998).

[144] D. Zeppenfeld, *Z. Phys.* **C8** 77 (1981).

[145] L.-L. Chau and H.-Y. Cheng, *Phys. Rev. Lett.* **53** 1037 (1984); *Phys. Lett.* **165B** 429 (1985); *Phys. Rev. Lett.* **59** 958 (1987).

[146] M. Gronau, O. F. Hernandes, D. London and J. L. Rosner, *Phys. Rev.* **D52** 6356 (1995).

[147] M. Gronau and D. London, *Phys. Rev. Lett.* **27** 3381 (1990).

[148] H. Lipkin, Y. Nir, H. R. Quinn and A. E. Snyder, *Phys. Rev.* **D44** 1454 (1991).

[149] A. E. Snyder and H. R. Quinn, *Phys. Rev.* **D48** 2139 (1993).

[150] B. Kayser, M. Kuroda, R. D. Peccei and A. I. Sanda, *Phys. Lett.* **237** 508 (1990).

[151] H. J. Lipkin, *Proc. SLAC Workshop on Physics and Detector Issues for the High-Luminosity Asymmetric B Factory*, D. Hitlin (Ed.), SLAC 373; I. Dunietz, H. R. Quinn, A. Snyder, W. Toki and H. J. Lipkin, *Phys. Rev.* **D43** 2193 (1991).

[152] N. Sinha and R. Sinha, *Phys. Rev. Lett.* **80** 3706 (1998); G. Krammer and W. F. Palmer, *Phys. Rev.* **D45** 193 (192); *ibid.* **D46** 3197 (1992) .

[153] A. S. Dighe, I. Dunietz, H. J. Lipkin and J. Rosner, *Phys. Lett.* **B369** 144 (1996); *Phys. Lett.* **165B** 429 (1985); *Phys. Rev. Lett.* **59** 958 (1987).

[154] C. P. Jessop *et al.*, *Phys. Rev. Lett.* **79** 4533 (1997).

[155] CDF/ANAL/Bottom/CDF/4672

[156] D. Atwood, I. Dunietz and A. Soni, *Phys. Rev. Lett.* **78** 3257 (1997).

[157] I. Dunietz, *Phys. Rev.* **D52** 3048 (1995).

[158] M. K. Gaillard and B. W. Lee, *Phys. Rev.* **D10** 897 (1974).

[159] F. J. Gilman and M. B. Wise, *Phys. Rev.* **D21** 3150 (1980).

[160] M. Vysotskii, *Sov. J. Nucl. Phys.* **31** 797 (1980).

[161] A. J. Buras and R. Fleischer, *Heavy Flavors II*, A. J. Buras and M. Lindner (eds.), World Scientific, Singapore, 1997.

[162] W. J. Marciano and A. I. Sanda, *Phys. Lett.* **67B** 303 (1977).

[163] For a review see A. Pich, Ref. [53].

[164] G. D. Barr *et al.*, *Phys. Lett.* **B284** 440 (1992).

[165] V. Papadimitriou *et al.*, *Phys. Rev.* **D44** 573 (1991).

[166] L. M. Sehgal, *Phys. Rev.* **D38** 808 (1988.)

[167] S. Adler *et al.*, *Phys. Rev. Lett.* **79** 2204 (1997).

[168] See [161]. In our expression for $K \to \pi \nu \bar{\nu}$ branching ratios, we ignored $SU(2)$ breaking effects. The quoted numerical result, however, contains this effect.

[169] E731 experiment at FNAL M. Weaver *et al.*, *Phys. Rev. Lett.* **72** 3758 (1994).

[170] M. McGuigan and A. I. Sanda, *Phys. Rev.* **D36** 1413 (1987).

[171] E. J. Ramberg *et.al.* (E731 Collab.), *Phys. Rev. Lett.* **70** 2525 (1993).

[172] L. M. Sehgal and M. Wanninger, *Phys. Rev.* **D46** 1035 (1992); *Phys. Rev.* **D46** 5209 (1992) (E).

[173] See also the earlier papers: A. D. Dolgov and L. A. Ponomarev, *Sov. J. Nucl. Phys.* **4** 262 (1967); D. P. Majumdar and J. Smith; *Phys. Rev.* **187** 2039 (1969).

[174] J. K. Elwood, M. B. Wise and M. J. Savage, *Phys. Rev.* **D52** 5095 (1995); **D53** 2855(E) (1996); J. K. Elwood, M. B. Wise, M. J. Savage and J. M. Walden, *Phys. Rev.* **D53** 4078 (1996).

[175] M. Arenton, Invited Talk given at HQ98, Workshop on Heavy Quarks at Fixed Target, Fermilab, Oct. 10–12, 1998, to appear in the proceedings; see also: KTeV Collab., J. Adams *et al.*, *Phys. Rev. Lett.* **80** 4123 (1998).

[176] P. Heiliger and L. M. Sehgal, *Phys. Rev.* **D48** 4146 (1993).

[177] CLEOII, *28th Int. Conf. on High Energy Physics*, Warsaw, Poland, 1996.

[178] K. G. Chetyrkin, M. Misiak and M. Munz, *Phys. Lett.* **B400** 206 (1997).

[179] A. J. Buras, M. Misiak, M. Munz and S. Pokorsky, *Nucl. Phys.* **B424** 374 (1994).

[180] C. Greub, T. Hurth and D. Wyler, *Phys. Lett.* **B380** 385 (1996); *Phys. Rev.* **D54** 3350 (1996).

[181] CLEOII Collaboration, M. S. Alam *et al.*, *Phys. Rev. Lett.* **74** 2885 (1995).

[182] For a re-analysis see: A. L. Kagan and M. Neubert, preprint hep-ph/9805303.

[183] ALEPH collaboration, 28th Int. Conf. on High Energy Physics, Warsaw, Poland, 1996.

[184] C. S. Lim, T. Morozumi and A. I. Sanda, *Phys. Lett.* **B218** 343 (1989).

[185] N. G. Deshpande, J. Trampetic and K. Panose, *Phys. Rev.* **D39** 1461 (1989); P. J. O'Donnel and H. K. K. Tung, *Phys. Rev.* **D43** R2067 (1991).

[186] A. J. Buras and M. Munz, *Phys. Rev.* **D52** 186 (1995).

[187] A. Ali, T. Mannel and T. Morozumi, *Phys. Lett.* **B273** 505 (1991).

[188] R. F. Streater and A. S. Wightmann, *PCT, Spin and Statistics, and All That*, Benjamin, New York, 1964.

[189] S. W. Hawking, *Phys. Rev.* **D14** 2460 (1975); *Commun. Math. Phys.* **87** 395 (1982).

[190] J. Ellis, J. S. Hagelin, D. V. Nanopoulos and M. Srednicki, *Nucl. Phys.* **B241** 381 (1984).

[191] J. D. Bjorken, *Ann. Phys.* **24** 174 (1963).

[192] F. A. Bais and J.-M. Frere, *Phys. Lett.* **98** 431 (1981).

[193] T. Banks and A. Zaks, *Nucl. Phys.* **B184** 303 (1981).

[194] I. I. Bigi, *Z. Phys.* **C12** 235 (1982).

[195] H. B. Nielsen, in: *Fundamentals of Quark Models*, J. M. Barbour and A. T. Davies (eds.), University of Glasgow, Glasgow, 1977, p. 528 ff.

[196] V. A. Kostelecky and R. Potting, *Nucl. Phys.* **B359** 545 (1991); *Phys. Rev.* **D51** 3923 (1995); *Phys. Lett.* **B381** 89 (1996); D. Colladay and V. A. Kostelecky, *Phys. Lett.* **B344** 259 (1995).

[197] M. Hayakawa and A. I. Sanda, *Phys. Rev.* **D48** 1150 (1993).

[198] L. Lavoura, *Ann. Phys.* **207** 428 (1991).

[199] I. Dunietz, J. Hauser and J. L. Rosner, *Phys. Rev.* **D35** 2166 (1987).

[200] CPLEAR Collaboration, A. Angelopoulos *et al.*, *Phys. Lett.* **B444** 52 (1998).

[201] CPLEAR Collaboration, A. Angelopoulos *et al.*, *Phys. Lett.* **B444** 38 (1998).

[202] CPLEAR Collaboration, A. Angelopoulos *et al.*, *Phys. Lett.* **B444** 43 (1998).

[203] CPLEAR Collaboration, R. Le Gac, Proceedings of the Workshop on K Physics, Orsay, France, 1996, L. Iconomidou-Fayard (ed.).

[204] CPLEAR Collaboration, A. Apostolakis *et al.*, CERN-EP/99-51.

[205] For a recent measurement of ϕ_{+-} see CPLEAR Collaboration, R. Adler *et al.*, *Phys. Lett.* **B369** 367 (1996).

[206] A similar result has also been obtained by E. Shabalin, *Phys. Lett.* **B396** 335 (1996).

[207] A. Einstein, B. Podolsky and N. Rosen, *Phys. Rev.* **D47** 777 (1935).

[208] M. Kobayashi and A. I. Sanda, *Phys. Rev. Lett.* **69** 3139 (1992).

[209] Plans to build the ϕ factory are also being studied at: KEK, M. Fukawa *et al.*, KEK Report No. 90-12, 1990; and UCLA, R. Adler, *et al., Proc. Workshop on Testing* **CPT** *and studying* **CP** *violation at a ϕ factory*, D. Cline (ed.) *Nucl. Phys.* **B24** (Proc. Suppl.) (1991).

[210] E791 Collaboration, E.M. Aitala *et al., Phys. Rev. Lett.* **77** 2384 (1996).

[211] I. I. Bigi and A. I. Sanda, *Phys. Lett.* **B171** 320 (1986).

[212] I. I. Bigi, in: *Proc. XIII Int. Conf. on High Energy Physics*, S. C. Loken (ed.), World Scientific, Singapore, 1986, p. 857.

[213] G. Blaylock, A. Seiden and Y. Nir, *Phys. Lett.* **B355** 555 (1995).

[214] F. Buccella, M. Lusignoli and A. Pugliese, *Phys. Lett.* **B379** 249 (1996).

[215] I. I. Bigi and H. Yamamoto, *Phys. Lett.* **B 349** 363 (1995).

[216] L. Wolfenstein, *Phys. Lett.* **B 164** 170 (1985); *Phys. Rev. Lett.* **75** 2460 (1995).

[217] For a much bolder prediction, see: G. Burdman, in: *Proc CHARM2000 Workshop*, FERMILAB-Conf-94/190.

[218] M. Fukugita, T. Hagiwara and A. I. Sanda, preprint RL-79-052, 1979.

[219] I. I. Bigi, in: *Proc Tau-Charm Factory Workshop*, SLAC, 1989, SLAC-Report-343.

[220] M. Golden and B. Grinstein, *Phys. Lett.* **B222** 501 (1989).

[221] F. Close and H. Lipkin, *Phys. Lett.* **B372** 306 (1996).

[222] Y. L. Wu and L. Wolfenstein, *Phys. Rev. Lett.* **73** 1762 (1994).

[223] I. I. Bigi, Y. Diakonov, V. Khoze, J. Kühn and P. M. Zerwas, *Phys. Lett.* **B181** 157 (1986).

[224] C. R. Schmidt and M. Peskin, *Phys. Rev. Lett.* **69** 410 (1992).

[225] S. Bar-Shalom, D. Atwood, G. Eilam and A. Soni, *Z. Phys.* **C72** 79 (1996).

[226] W. Fetscher, in: *Proc. Third Workshop on the Tau-Charm Factory*, Marbella, Spain, 1993, J. and R. Kirkby (eds.), Editions Frontières, 1994.

[227] This point has been made vigorously in: Y. S. Tsai, *Phys. Rev.* **D51** 3172 (1995).

[228] S. Y. Pi and A. I. Sanda, *Ann. Phys.* **106** 171 (1977).

[229] G. t'Hooft, *Phys. Rev. Lett.* **37** 8 (1976); *Phys. Rev.* **D14** 3432 (1976).

[230] C. G. Callan, R. Dashen and D. Gross, *Phys. Lett.* **63B** 334 (1976).

[231] R. Jackiw and C. Rebbi, *Phys. Rev. Lett.* **37** 172 (1976).

[232] This is widely discussed in text books. See, for example, [82].

[233] S. L. Adler, *Phys. Rev.* **177** 2426 (1969); J. S. Bell and R. Jackiw, *Nuov. Cim.* **60** 47 (1969); W. A. Bardeen, *Phys. Rev.* **184** 1848 (1969).

[234] V. Baluni, *Phys. Rev.* **D19** 2227 (1979).

[235] R. J. Crewther *et al., Phys. Lett.* **88B** 123 (1979); Errata, **91B** 487 (1980).

[236] See Peccei's review in Ref. [49].

[237] J. Gasser and H. Leutwyler, *Phys. Report* **87** 77 (1982).

[238] Th. Kaluza, *Sitz. Preuss. Akad. Wiss.* **Kl**, 966 (1921); O. Klein, *Z. Phys.* **37**, 895 (1926).

[239] R. Peccei and H. Quinn, *Phys. Rev. Lett.* **38** 1440 (1977); *Phys. Rev.* **D16** 1791 (1977).

[240] S. Weinberg, *Phys. Rev. Lett.* **40** 223 (1978).

[241] F. Wilczek, *Phys. Rev. Lett.* **40** 279 (1978).

[242] N. A. Bardeen, R. D. Peccei and T. Yanagita *Nucl. Phys.* **B279** 401 (1987).

[243] N. A. Bardeen and S. H. H. Tye, *Phys. Lett.* **74B** 580 (1979).

[244] S. Yamada, *Proc. 1983 Int. Symp. on Lepton and Photon Interations at High Energies*, D. G. Cassel and D. L. Kreinick (eds.), Cornell, 1983.

[245] R. Eichler *et al., Phys. Lett.* **175B** 101 (1986).

[246] For prototypes see: J. Kim, *Phys. Rev. Lett.* **43** 103 (1979); M. A. Shifman, A. I. Vainshtein and V. I. Zakharov, *Nucl. Phys.* **166** 493 (1980).

[247] Such models were first discussed in: M. Dine, W. Fischler and M. Srednicki, *Phys. Lett.* **104** 199 (1981); A. P. Zhitnitskii, *Sov. J. Nucl. Phys.* **31** 260 (1980).

[248]   For a nice update by one of the pioneers see: P. Sikivie, 'Dark Matter Axions and Caustic Rings', preprint hep-ph/9709477, 1997.

[249]   K. van Bibber *et al.*, *Int. J. Mod. Phys.* **D3** Supp. 33 (1994); S. Matsuki, I. Ogawa and K. Yamamoto, *Phys. Lett.* **B336** 573 (1994); I. Ogawa, S. Matsuki and K. Yamamoto, *Phys. Rev.* **D53** R1740 (1996); C. Hagmann *et al.*, *Phys. Rev. Lett.* **80** 2043 (1998).

[250]   G. Raffelt, preprint hep-ph/9806506, to be published in: *Proc. NEUTRINO 98, XVIII Int. Conf. on Neutrino Physics and Astrophysics*, Takayama, Japan, 4–9 June 1998, Y. Suzuki and Y. Totsuka (eds.).

[251]   Y. B. Zeldovich, I. B. Kobzarev and L. B. Okun, *Sov. Phys. JETP* **40** 1 (1975).

[252]   S. Barr and A. Zee, *Phys. Rev. Lett.* **55** 2253 (1985).

[253]   A. Nelson, *Phys. Lett.* **136B** 387 (1983).

[254]   S. Barr, *Phys. Rev. Lett.* **53** 329 (1984); *Phys. Rev.* **D30** 1805 (1984).

[255]   B. Pontecorvo, *J. Exp. Theor. Phys.* **33** 549 (1957).

[256]   Z. Maki, M. Nakagawa and S. Sakata, *Prog. Theor. Phys.* **30** 727 (1963).

[257]   N. Cabibbo, *Phys. Rev. Lett.* **10** 531 (1963).

[258]   R. Davis, Jr., *Prog. Part. Nucl. Phys.* **32**, 13 (1994).

[259]   J. N. Abdurashitov *et al.*, *Phys. Lett.* **B328** 234 (1994).

[260]   P. Anselmann *et al.*, *Phys. Lett.* **B342** 440 (1995).

[261]   Y. Suzuki, *Nucl. Phys.* (Proc. Suppl.) **38** 54 (1995).

[262]   Superkamiokande Collab., Z. Conner *et al.*, *Proc. 25th ICRC*, paper HE 4.1.22 (1997).

[263]   J. Bahcall, *XXV SLAC Summer Institute on Particle Physics, 'Physics of Leptons'*, (1997).

[264]   N. Hata, S. Bludman and P. Langacker, *Phys. Rev.* **D49** 3622 (1994).

[265]   K. H. Heeger and R. G. H. Robertson, *Phys. Rev. Lett.* **77** 3720 (1996).

[266]   Kamioka Collaboration, K. S. Hirata *et al.*, *Phys. Lett.* **B205** 416 (1988); *ibid.* **B280** 146 (1992) .

[267]   IMB Collaboration, R. Becker-Szendy *et al.*, *Phys. Rev.* **D46** 3720 (1992); D. Casper *et al.*, *Phys. Rev. Lett.* **66** 2561 (1991).

[268]   Soudan Collaboration, W. W. M. Allison *et al.*, *Phys. Lett.* **B391** 491 (1997).

[269]   Frejus Collaboration, K. Daum *et al.*, *Z. Phys.* **C66** 417 (1995).

[270]   NUSEX collaboration, M. Aglietta *et al.*, *Europhys. Lett.* **8** 611 (1989).

[271]   The Super-Kamiokande Collaboration, Y. Fukuda *et al.*, hep-ex/9805006.

[272]   Y. Fukuda, *et al.*, *Phys. Rev. Lett.* **81** 1562 (1998).

[273]   H. Backe *et al.*, in *Proc. Neutrino 92*, Granada, 1992, A. Morales (ed.), NPB (Proc. Suppl.) **31** 46 (1993).

[274]   A preliminary result was presented by C. Weinheimer, *Proc. Neutrino 98, XVIII Int. Conf. on Neutrino Physics and Astrophysics*, Takayama, Japan, 4–9 June 1998, Y. Suzuki and Y. Totsuka (eds.).

[275]   M. Fukugita and T. Yanagida, *Physics and Astrophysics of Neutrinos*, Springer-Verlag, Tokyo, 1994.

[276]   P. Ramond, *Proc. Neutrino 98, XVIII Int. Conf. on Neutrino Physics and Astrophysics*, Takayama, Japan, 4–9 June 1998, Y. Suzuki and Y. Totsuka (eds.).

[277]   L. H. Ryder, *Quantum Field Theory*, Cambridge University Press, Cambridge, 1985.

[278]   P. H. Frampton and P. Vogel, *Phys. Report* **82** 339 (1982).

[279]   M. Gell-Mann, R. Slansky and P. Ramond, in: *Supergravity*, North Holland, 1979, p. 315; T. Yanagita, in: *Proc. Workshop on Unified Theory and Baryon Number in the Universe*, KEK, Japan, 1979.

[280]   L. Wolfenstein, *Phys. Rev.* **D17** 2369 (1978).

[281]   S. P. Mikheyev and A. Yu. Smirnov, *Sov. J. Nucl. Phys.* **42** 913 (1985); *Nuov. Cim.* **C9** 17 (1986).

[282]   J. Arafune, M. Koike and J. Sato, *Phys. Rev.* **D56** 3093 (1997).

[283]   F. Reines and C. L. Cowan, Jr., *Nature* **178**, 446 (1956); C. L. Cowan *et al.*, *Science* **124**, 103 (1956).

[284]   J. C. Pati and A. Salam, *Phys. Rev.* **D10** 275 (1974).

[285]   R. N. Mohapatra and J. C. Pati, *Phys. Rev.* **D11** 566 (1975); R. N. Mohapatra, *Phys. Rev.* **D11** 2558 (1975); R. N. Mohapatra and G. Senjanovic, *Phys. Rev.* **D12** 1502 (1975).

[286]   M. Beg, *Phys. Rev. Lett.* **38** 1252 (1977); G. Senjanovic *Nucl. Phys.* **B153** 334 (1979).

[287]   R. N. Mohapatra in Ref. [49].

[288]   W. Grimus, *Fortschr. Phys. Berlin* **36** 201 (1988).

[289]   G. Ecker, W. Grimus and H. Neufeld, *Nucl. Phys.* **B247** 70 (1984).

[290]   P. Langacker and S. U. Sankar, *Phys. Rev.* **D40** 1569 (1989).

[291]   J.-M. Frere *et al.*, *Phys. Rev.* **D46** 337 (1992).

[292]   For a review see P. Langacker in Ref. [49].

[293]   S. R. Mishra *et al.*, *Phys. Rev. Lett.* **68** 3499 (1992).

[294]   G. Beall, M. Bander and A. Soni, *Phys. Rev. Lett.* **48** 848 (1982).

[295]   See also G. Barenboim, J. Bernabeu, J. Prades and M. Raidal, *Phys. Rev.* **D55** 4213 (1997).

[296]   D. Chang, X.-G. He and S. Pakvasa *Phys. Rev. Lett.* **74** 3927 (1995).

[297]   G. Branco, J. M. Frere and J. M. Gerard, *Nucl. Phys.* **B221** 317 (1983).

[298]   T. D. Lee, *Phys. Rev.* **D8** 1226 (1973); *Phys. Rep.* **96** (1979).

[299]   S. Weinberg, *Phys. Rev. Lett.* **31** 657 (1976).

[300]   A concise review is found in L. Wolfenstein and Y. L. Wu, *Phys. Rev. Lett.* **73** 1762 (1994).

[301]   For an update, see: H.-Y. Cheng, *Phys. Rev.* **D42** 2329 (2990); *Int. J. Mod. Phys.* **A7** 1059 (1992).

[302]   I. I. Bigi and A. I. Sanda in Ref. [49].

[303]   G. C. Branco, *Phys. Rev. Lett.* **44** 504 (1980); *Phys. Rev. Lett.* **73** 1762 (1994).

[304]   S. Glashow and S. Weinberg, *Phys. Rev.* **D15** 1958 (1977).

[305]   G. C. Branco and A. I. Sanda, *Phys. Rev.* **D26** 3176 (1982).

[306]   H. Georgi, *Had. J.* **1**, 155 (1978).

[307]   L. Lavoura, *Int. J. Mod. Phys.*, **A9**, 1873 (1994).

[308]   J. Liu and L. Wolfenstein, *Nucl. Phys.* **B289** 1 (1987).

[309]   A. Mendez and A. Pomarol, *Phys. Lett.* **B272** 313 (1991).

[310]   A. I. Sanda, *Phys. Rev.* **D23** 2647 (1981); N. G. Deshpande, *ibid.* **23**, 2654 (1981). J. F. Donoghue and B. R. Holstein, *Phys. Rev.* **D32** 1152 (1983).

[311]   J.-M. Frere, J. Hagelin and A. I. Sanda, *Phys. Lett.* **151B** 161 (1985).

[312]   I. I. Bigi and A. I. Sanda, *Phys. Rev. Lett.* **58** 1604 (1987).

[313]   G. Beall and N. G. Deshpande, *Phys. Lett.* **132B** 427 (1983).

[314]   R. Akhoury and I. I. Bigi, *Nucl. Phys.* **B234** 459 (1984).

[315]   S. Weinberg, *Phys. Rev. Lett.* **63** 2333 (1989).

[316]   D. Chang, W.-Y. Keung and T. C. Yuan, *Phys. Lett.* **251** 608 (1990); I. I. Bigi and N. G. Uraltsev, *Nucl. Phys.* **B353** 321 (1991).

[317]   A. Anselm *et al.*, *Phys. Lett.* **152B** 116 (1985).

[318]   M. S. Barr and A. Zee, *Phys. Rev. Lett.* **65** 21 (1990).

[319]   J. Gunion and D. Wyler, *Phys. Lett.* **248B** 170 (1990); D. W. Chang, W. Y. Keung and T. C. Yuan, *Phys. Lett.* **251B** 608 (1990).

[320]   D. Bowser-Chao, D. Chang and W. -Y. Keung, preprint hep-ph/9703435.

[321]   R. Garisto and G. Kane, *Phys. Rev.* **D44** 2038 (1991).

[322]   T. Nakada, *Phys. Lett.* **B261** 474 (1991).

[323]   S. Coleman and J. Mandula, *Phys. Rev.* **D159** 1251 (1967).

[324]   K. Inoue *et al.*, *Prog. Theor. Phys.* **C 68**, 927 (1982); *ibid.* **C 71**, 413 (1984); L. E. Ibanez and G. G. Ross, *Phys. Lett.* **B 110** 215 (1982).

References

[325] H. P. Nilles, *Phys. Rep.* **110** 1 (1982); S. P. Martin, preprint hep-ph/9709356.

[326] S. P. Martin, preprint hep-ph/9709356.

[327] S. Bertolini *et al.*, *Nucl. Phys.* **B353** 591 (1991).

[328] F. Gabbiani and A. Masiero, *Nucl. Phys.* **B322** 235 (1989).

[329] Y. Kizukuri and N. Oshimo, *Phys. Rev.* **D46** 3025 (1992).

[330] F. Gabbiani, E. Gabrielli, A. Masiero and L. Silvestrini, *Nucl. Phys.* **B477** 321 (1996).

[331] For a more detailed discussion see: A. Masiero and L. Silvestrini, 'Two Lectures on FCNC and **CP** Violation in Supersymmetry', in: *Proc. Int. School of Physics Enrico Fermi*, Course CXXXVII: 'Heavy Flavour Physics: A Probe of Nature's Grand Design', Varenna, Italy, 1997, preprint hep-ph/9711401.

[332] I. I. Bigi and F. Gabbiani, *Nucl. Phys.* **B352** 309 (1991); M. Ciuchini *et al.*, *Phys. Rev. Lett.* **79** 978 (1997).

[333] A. D. Dolgov, Ya. B. Zel'dovich, *Rev. Mod. Phys.* **53** 1 (1981).

[334] A. D. Sakharov, *JETP Lett.* **5** 24 (1967).

[335] For an updated review see: A. D. Dolgov, in: *Proc. XXVth ITEP Winterschool of Physics*, Feb. 18–27, Moscow, Russia.

[336] V. A. Kuzmin, *Pis'ma Z. Eksp. Teor. Fiz.* **12** 335 (1970).

[337] J. C. Pati and A. Salam, *Phys. Rev.* **D10** 275 (1974); H. Georgi and S. L. Glashow, *Phys. Rev. Lett.* **32** 438 (1974).

[338] M. Yoshimura, *Phys. Rev. Lett.* **41** 281 (1978).

[339] D. Touissaint, S. B. Treiman, F. Wilczek and A. Zee, *Phys. Rev.* **D19** 1036 (1979); S. Weinberg, *Phys. Rev. Lett.* **42** 850 (1979); M. Yoshimura, *Phys. Lett.* **88B** 294 (1979).

[340] V. A. Kuzmin, V. A. Rubakov and M. E. Shaposhnikov, *Phys. Lett.* **B155** 36 (1985).

[341] G. t'Hooft, *Phys. Rev. Lett.* **37** 8 (1976); *Phys. Rev.* **D14** 3432 (1976).

[342] V. A. Rubakov and M. E. Shaposhnikov, *Usp. Fiz. Nauk* **166** 493 (1996).

[343] K. Kajantie *et al.*, *Nucl. Phys.* **B466** 189 (1996).

[344] M. Fukugita and T. Yanagita, *Phys. Rev.* **D42** 1285 (1990).

Index